Environmental Burden of Disease Assessment

Environmental Science and Technology Library

VOLUME 24

For further volumes:
http://www.springer.com/series/5923

Jacqueline MacDonald Gibson • Angela S. Brammer
Christopher A. Davidson • Tiina Folley
Frederic J.P. Launay • Jens T.W. Thomsen

Environmental Burden of Disease Assessment

A case study in the United Arab Emirates

Springer

Jacqueline MacDonald Gibson
Gillings School of Global Public Health
University of North Carolina–Chapel Hill
Chapel Hill, NC, USA

Angela S. Brammer
Gillings School of Global Public Health
University of North Carolina–Chapel Hill
Chapel Hill, NC, USA

Christopher A. Davidson
Gillings School of Global Public Health
University of North Carolina–Chapel Hill
Chapel Hill, NC, USA

Tiina Folley
Gillings School of Global Public Health
University of North Carolina–Chapel Hill
Chapel Hill, NC, USA

Frederic J.P. Launay
Environment Agency–Abu Dhabi
United Arab Emirates

Jens T.W. Thomsen
Health Authority–Abu Dhabi
United Arab Emirates

ISSN 1382-3124
ISBN 978-94-007-5924-4 ISBN 978-94-007-5925-1 (eBook)
DOI 10.1007/978-94-007-5925-1
Springer Dordrecht Heidelberg New York London

Library of Congress Control Number: 2013930235

Printed on acid-free paper

Springer is part of Springer Science+Business Media (www.springer.com)

Foreword

This remarkable book is the result of the first-ever effort by a nation to commission a comprehensive model to estimate the national burden of disease from environmental pollution in air, water, soil, and food—a project especially notable for its long-term perspective in a world too often focused on short-term thinking. When we collaborated with the Environment Agency–Abu Dhabi and Health Authority–Abu Dhabi to commission the research reported here, we envisioned that the result could serve as a model for other nations wishing to understand environmental pollution impacts on public health from a holistic perspective. We believe this book fulfills the role we originally envisioned: it provides a template for other nations to use to characterize their environmental burden of disease across multiple pollutants and exposure routes.

Systematic planning of environmental investments to protect health is more important than ever as much of the world continues to recover from economic difficulties. Such planning not only ensures governments will reap the greatest health benefits from steps they take to reduce environmental pollution but also will result in substantial long-term savings in healthcare costs. As observed in an editorial in *The Lancet* advocating for countries to undertake environmental burden of disease assessments, "If policymakers don't plan for healthier environments now, many more lives will be lost unnecessarily."[1]

Using environmental monitoring data, health records, and epidemiologic information, the computer model described in this book, known as the *UAE Environmental Burden of Disease Model*, predicts how many excess deaths and visits to health-care facilities are caused by environmental contaminants each year in the United Arab Emirates. The model, created by scientists from the Gillings School of Global Public Health at the University of North Carolina–Chapel Hill, is designed to be updated and improved over time, allowing the accuracy of predictions to be continually refined as new environmental monitoring and health data become available.

[1] *The Lancet*. 2007. The environment's impact on health. *The Lancet* 369(June 23), 2052.

Other nations can build directly on the *UAE Environmental Burden of Disease Model* and the detailed information this book contains to jump start their own environmental burden of disease assessments. The model can be adapted readily by replacing UAE population, environmental monitoring, and health outcome data with comparable data for other nations. The model's modular construction, as described in this book, is designed to facilitate future use by others. The full model is encoded using the software platform *Analytica*, available from Lumina Decision Systems (http://www.lumina.com/). A stripped-down version (with UAE-specific data removed) can be requested from Dr. Jacqueline MacDonald Gibson at the University of North Carolina–Chapel Hill (jackie.macdonald@unc.edu).

As evidenced by the commissioning of the environmental health project, the UAE already had the vision and desire to become a world leader in environmental health. But with the publication of this book, the UAE now has the necessary knowledge and tools to begin to assume that leadership role. The world is shrinking, and every action taken by a nation affects its neighbors. The actions taken in response to the facts presented in this book can set an example for others to follow, both in the Middle East and throughout the world.

Centre for Environmental Health Activities
Regional Office for the Eastern Mediterranean Region
World Health Organization
Amir Johri

Department of Civil Engineering
University of Lahore
Muhammad Zulfiqar Ali Khan

Preface

This book originated from a September 2007 visit to the University of North Carolina–Chapel Hill (UNC) by H.E. Majid Al Mansouri, then Secretary General of the Environment Agency–Abu Dhabi (EAD). Accompanied by Dr. M.Z. Ali Khan and Dr. Amir Johri from the World Health Organization (WHO), H.E. Al Mansouri sought guidance on organizing his agency's fledgling environmental protection programs. At the time of his visit, the EAD was just 2 years old, having arisen from what was previously an agency focused almost exclusively on wildlife protection, the Environmental Research and Wildlife Development Agency.

H.E. Al Mansouri invited UNC's Gillings School of Global Public Health to collaborate on a study to identify the environmental pollution risks posing the largest burden of disease among the human population of the United Arab Emirates (UAE). He planned to use the results to guide the organization of his newly expanded agency. Because Abu Dhabi is the political center and geographically largest of the UAE's seven emirates, he also envisioned that the results could benefit the entire UAE. Hence, the scope of the project included not just Abu Dhabi emirate but all of the UAE.

H.E. Al Mansouri arrived at UNC with a list of 14 concerns, ranging from depletion of the stratospheric ozone layer to electromagnetic fields to carcinogens in water (see Chap. 2 for a complete list). Recognizing that his agency was understaffed to be able to manage all 14 concerns simultaneously and with equal effort, he aspired to discover ways the EAD could invest its limited resources in order to yield high benefits for public health. A key part of this process was to develop an accounting of the environmental burden of disease associated with each type of pollution on the list of concerns. This book presents the methods used to develop this accounting, along with the results for the UAE. The methods, and much of the epidemiologic evidence used to support the analysis, can be readily applied in other countries, and I hope that others will borrow freely from this book and use it as a model for their own, similar analyses.

For a U.S. scientist, the opportunity to view environmental impacts on health from a holistic perspective—rather than examining risks one at a time—was highly appealing. In the United States, environmental protection policies have arisen in piecemeal fashion, driven by acts of Congress that generally consider only one environmental medium (such as surface water or outdoor air) at a time. The result is a fragmented regulatory system that has no mechanism by which to account for the relative public health importance of pollutants in different environmental media—and that, indeed, overlooks some very important exposure routes, such as indoor air. As the eminent U.S. environmental policy scholar Richard N.L. Andrews has observed, the U.S. Congress "passed a patchwork of extraordinarily prescriptive statutes, denying EPA [the U.S. Environmental Protection Agency] authority to address some sources and to set priorities among mandates as well as denying it use of more effective and efficient policy tools."[2] Dr. Andrews (and many others) have concluded that without a new law from the U.S. Congress providing the EPA with "a coherent statutory mission and the flexibility to carry it out," adopting a multi-media perspective to U.S. environmental policymaking probably will remain an elusive goal. Hence, the opportunity to assist a visionary leader in planning a system to protect human health from environmental risks with the "big picture" in mind was highly appealing to the U.S. scientists on our team.

I hope that this book will serve as inspiration to other nations (including my own) wishing to reassess their environmental protection paradigms by viewing the environment and its impacts on human health from a more holistic perspective, considering the multiple ways in which people can be exposed to pollutants instead of addressing just one pathway at a time. There is much more work to be done to improve on the foundation established in this book, including considering the potential cumulative risks of exposure to multiple contaminants in different media. Nonetheless, I hope the model this book can serve as a starting point.

Department of Environmental Sciences
and Engineering
Gillings School of Global Public Health
University of North Carolina–Chapel Hill
Chapel Hill, NC, USA

Jacqueline MacDonald Gibson

[2] Richard, N. L. Andrews. 1995. Reform or reaction: EPA at a crossroads. *Environmental Science & Technology* 29(11):505A–510A.

Contents

List of Figures

List of Tables

Acronyms

ACS	American Cancer Society
ADFCA	Abu Dhabi Food Control Authority
ADFSC	Abu Dhabi Farmers Services Centre
ADI	acceptable daily intake
AF	attributable fraction
aPAD	acute population adjusted dose
BMI	body mass index
BPA	bisphenol A
CAREX	Carcinogen Exposure database
CCS	carbon capture and sequestration
CDC	Centers for Disease Control and Prevention (U.S.)
CMAQ	Community Multiscale Air Quality
cPAD	chronic population adjusted dose
CWM	Center of Waste Management–Abu Dhabi
DALY	disability-adjusted life year
EAD	Environment Agency–Abu Dhabi
EBD	environmental burden of disease
EC	European Commission
EFSA	European Food Safety Authority
EMF	electromagnetic field
EPA	Environmental Protection Agency (U.S.)
FDA	Food and Drug Administration (U.S.)
GBD	global burden of disease
GHG	greenhouse gases
HAAD	Health Authority–Abu Dhabi
IPCC	Intergovernmental Panel on Climate Change
LOAEL	lowest adverse effect level
MOEW	Ministry of Environment and Water (UAE)
MRLs	maximum allowable pesticide residue levels
NATO	North Atlantic Treaty Organization
NEHAP	national environmental health action plan

NOAEL	no observed adverse effect level
NOx	nitrous oxides
PAHs	polyaromatic hydrocarbons
PCBs	polychlorinated biphenyls
PDF	probability distribution function
PET	polyethylene terephthalate
ppb	parts per billion
PM	particulate matter
RfD	reference dose
RR	relative risk
SD	standard deviation
THMs	trihalomethanes
VOC	volatile organic compound
WHO	World Health Organization
WSH	water, sanitation, and hygiene

Chapter 1
Introduction

It is simple, really. Human health and the health of ecosystems are inseparable.[1]

— Gro Harlem Brundtland
Director General, World Health Organization, 1998–2003

We cherish our environment because it is an integral part of our country, our history and our heritage. On land and in the sea, our forefathers lived and survived in this environment... They were able to do so only because they recognized the need to conserve it, to take from it only what they... needed to live and to preserve it for succeeding generations.[2]

— H. H. Sheikh Zayed Bin Sultan Al Nahyan
President of the United Arab Emirates, 1971–2004
Ruler of Abu Dhabi Emirate, 1966–2004

Abstract The environmental burden of disease assessment approach described in this volume is illustrated through its application to the United Arab Emirates (UAE). The UAE occupies 83,600 km^2 along the Arabian Gulf, with an estimated 2011 population of about 7.5 million. The UAE supports a diversified modern economy and, as a result, faces environmental and public health problems similar to those of other industrial nations. The methods we illustrate build upon a series of guidelines on environmental burden of disease assessment published by the World Health Organization beginning in 2003. Although many countries have employed these guidelines to assess the burden of disease due to individual environmental risk factors, the comprehensive environmental burden of disease assessment across multiple exposure pathways and contaminants described in this book is the first of its kind.

[1] Address to the National Press Club, Canberra, Australia, October 17, 2000. Available at http://www.who.int/director-general/speeches/2000/english/20001017_camberra.html.

[2] From Sheikh Zayed in Quotes, UAE Interact, February 11, 2005. Available at http://www.uaeinteract.com/docs/Sheikh_Zayed_in_quotes/18411.htm.

J. MacDonald Gibson et al., *Environmental Burden of Disease Assessment*, Environmental Science and Technology Library 24, DOI 10.1007/978-94-007-5925-1_1,

This project was intended to serve as a model for other nations wishing to conduct similar assessments. The basic methods can be applied to any nation or subnational geographic unit (such as a state or city). Furthermore, much of the information on relationships between exposures to pollutants and the probability of becoming ill is, and will increasingly be, relevant across the globe. These relationships are specified in the *UAE Environmental Burden of Disease Model*, a multilayered computer simulation tool constructed in *Analytica* software. Other countries can adopt this model's structure, along with much of the input data, as a starting point for their own environmental burden of disease assessments. Also relevant to other nations is the process we used to prioritize risks to include in this analysis—a process that involved systematic consultations with environment and health stakeholders. Other nations can save considerable time and resources in carrying out similar assessments by using the approaches and modeling methods described in this book.

Keywords United Arab Emirates • Environmental burden of disease assessment • Environment Agency–Abu Dhabi • UAE Environmental Burden of Disease Model • Environmental risks to public health • Analytica • Multilayered environmental simulation computer modeling

The Purpose of This Report

Civilizations have long understood the inseparable connections between environmental quality and human health. Ancient Assyrian medical texts—dating from 2000 to 1000 BC and unearthed in modern-day Iraq—include numerous passages emphasizing environmental conditions as causative factors in disease. One volume (preserved on a clay tablet) advised, "He must not go into the lowlands by the river or an infectious disease will infect him." Others mention patients exposed to contaminated river water prior to the onset of medical symptoms (Scurlock and Anderson 2005).

Although the ancients understood environmental risks to health in a qualitative sense, the tools to quantify these risks are new relative to the time span of human history (see Box 1.1). Late-twentieth-century research in environmental and health sciences, along with modern computing tools, make it possible to estimate the number of deaths and illnesses attributable to environmental pollution in a population exposed to such pollution. In this book, we show how to combine environmental and public health data, modern literature on environmental causes of disease, and sophisticated computing tools to determine the number of premature deaths and illnesses that could be prevented if exposure to modern pollutants were reduced. The results provide an indication of the pollutants and routes of exposure to these pollutants that induce the greatest health burden in the population under study. Hence, these results can be used to identify high-impact opportunities for improving public health through environmental interventions.

The environmental burden of disease assessment approach described in this volume is illustrated through its application to the United Arab Emirates (UAE).

Box 1.1 Historical Development of Tools for Environmental Burden of Disease Assessment

Quantifying the number of deaths and illnesses attributable to environmental pollution as is carried out in this book would not be possible without (1) an understanding of probability theory, (2) instruments to measure pollutant concentrations in the environment, (3) mathematical relationships (derived from epidemiologic and toxicologic studies) linking these measured concentrations to specific illnesses, (4) baseline public health data, and (5) computing tools that allow simulation of population exposures to pollutants under different scenarios, such as under different levels of pollutant emissions to the environment. None of these tools existed prior to the Renaissance. And, while Renaissance-era discoveries laid the foundation the techniques employed in this book, much of the knowledge underlying the analysis presented here was unavailable before the end of the twentieth century.

Probability theory—which provides the foundation for modern, quantitative risk assessment—is a product of Renaissance-era mathematicians (Bernstein 1996). Prior to the Renaissance, decision-makers faced with risky choices and uncertainties consulted with priests or other religious authorities, rather than employing the tools of probability to quantify risks. In his authoritative history of risk assessment, *Against the Gods: The Remarkable Story of Risk*, Bernstein writes,

> The revolutionary idea that defines the boundary between modern times and the past is the mastery of risk: the notion that the future is more than a whim of the gods. … Until human beings discovered a way across that boundary, the future was a mirror of the past or the murky domain of oracles and soothsayers who held a monopoly over knowledge of anticipated events (Bernstein 1996).

In addition to probability theory, quantitative environmental health risk assessment requires instruments to measure pollutant levels, mathematical relationships linking exposure to measured contaminants to specific health effects (derived from toxicologic and epidemiologic studies), and baseline public health statistics. Before the nineteenth century, a lack of instrumentation for measuring pollutant levels and a lack of systematic public health record-keeping posed a major obstacles to making progress in understanding the relationships between exposure to environmental hazards and disease (Covello and Mumpower 1985). Although the science of toxicology dates to the sixteenth century, when the physician Paracelsus established the fundamental principle of toxicology (the dose makes the poison), and although historians have identified early epidemiologic investigations dating from the sixteenth century, research to quantify the relationship between the dose of a contaminant received and the probability of becoming ill is quite recent (Graham 1995). Covello and Mumpower note, "It is surprisingly easy to forget that it was not until the work of Pasteur in the late 19th century that scientists

(continued)

Box 1.1 (continued)

first began to comprehend adequately the concept of infection or the causal relationship between the environment and biological agents of infectious disease." For chemicals, dose-response assessment emerged in the last quarter of the twentieth century with the "convergence of interest in the carcinogenic effects of radiation and chemicals" (Graham 1995).

A final ingredient for the type of analysis presented in this book is computing tools for simulating human exposure to pollutants at different geographic scales and under different pollution scenarios. Such tools became widely available only with the advent of personal computers toward the end of the twentieth century.

The UAE occupies 83,600 km^2 along the Arabian Gulf (see Fig. 1.1), with an estimated 2011 population of about 7.5 million. The UAE supports a diversified modern economy and, as a result, faces environmental and public health problems similar to those of other industrial nations.

The methods we illustrate build upon a series of guidelines on environmental burden of disease assessment published by the World Health Organization (WHO) beginning in 2003 (Prüss-Ustün et al. 2003). Although many countries have employed these guidelines to assess the burden of disease due to individual environmental risk factors, the comprehensive environmental burden of disease assessment across multiple exposure pathways and contaminants described in this book is the first of its kind.

The UAE launched its environmental burden of disease assessment project with the encouragement of the WHO Centre for Environmental Health Activities in Amman, Jordan, in June 2008. The project was funded through the Environment Agency–Abu Dhabi (EAD) and involved collaboration of many other government agencies in the UAE. In addition to helping the UAE's leaders identify the leading environmental risks to UAE public health as part of a process of updating environmental policies, this project was intended to serve as a model for other nations wishing to conduct similar assessments. Hence, this book fully documents the burden of disease assessment process as implemented in the UAE. The quantitative assessments were conducted by an international team of environment and health scientists commissioned by EAD. We hope this documentation will enable other nations to conduct similar assessments, building on the methods and data we illustrate.

The basic methods in this book apply to any nation or subnational geographic unit (such as a state or city). Furthermore, much of the information on relationships between exposures to pollutants and the probability of becoming ill is, and will increasingly be, relevant across the globe. These relationships are specified in the *UAE Environmental Burden of Disease Model*, a multilayered computer simulation tool constructed in *Analytica* software (version 4.1, Lumina Decision Systems). Appendix B explains how to use this model, and Appendix C documents the model

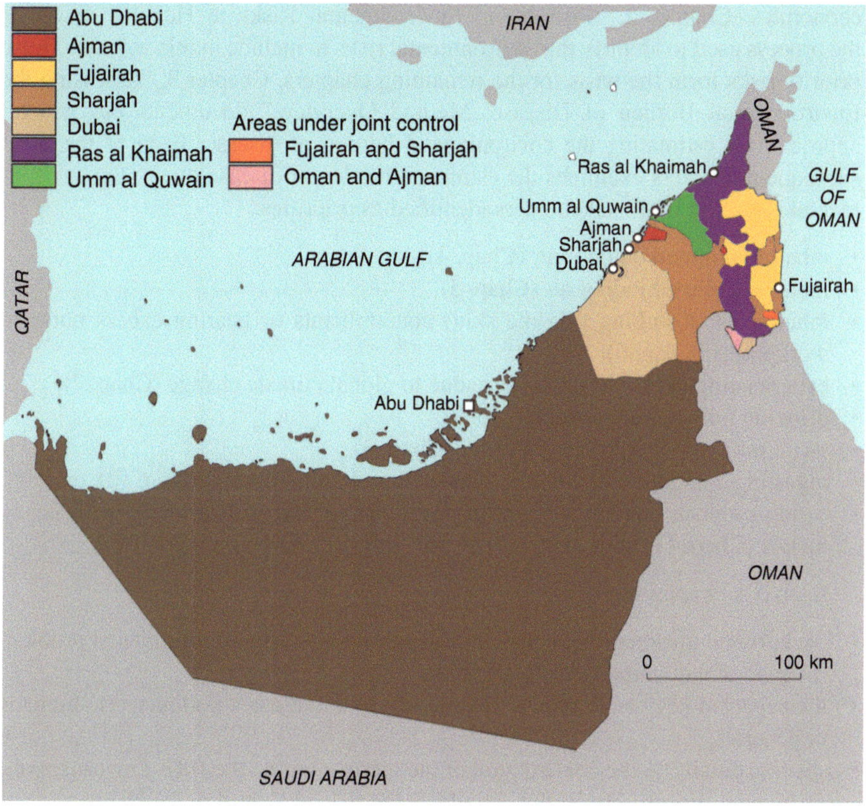

Fig. 1.1 The UAE comprises seven emirates and borders Saudi Arabia and Oman on the Arabian Gulf. From bottom left, the emirates are Abu Dhabi, Dubai, Sharjah, Ajman, Umm al Quwain, Ras al Khaimah, and Fujairah

input parameters. Other countries can adopt this model's structure, along with much of the input data, as a starting point for their own environmental burden of disease assessments. Also relevant to other nations is the process we used to prioritize risks to include in this analysis—a process that involved systematic consultations with environment and health stakeholders. Other nations can save considerable time and resources in carrying out similar assessments by using the approaches and modeling methods described in this book.

How This Book Is Organized

This chapter briefly reviews the rationale for conducting national-level environmental burden of disease assessments. It then provides background information on the UAE and why it serves as a particularly useful case study nation for environmental burden of disease assessment for modern, industrial nations and emerging economies. Chapter 2, "Prioritizing Environmental Risks to Health," describes the process used to identify the environmental risks to include in this volume. Those priority risks form the basis for the remaining chapters. Chapter 3, "Assessing the Environmental Burden of Disease: Method Overview," documents our general approach for estimating the environmental burden of disease. Each of the eight subsequent chapters examines the estimated burden of disease due to the environmental risks that UAE stakeholders identified as priorities:

- inhaling polluted outdoor air (Chap. 4),
- inhaling polluted indoor air (Chap. 5),
- inhaling or absorbing (via the skin) contaminants or hearing excess noise in workplaces (Chap. 6),
- experiencing excess heat exposure due to global climate change (Chap. 7),
- drinking polluted tap water (Chap. 8),
- swimming in polluted coastal water (Chap. 9),
- ingesting contaminants from polluted soil and groundwater (Chap. 10), and
- eating produce contaminated with pesticides or fish contaminated with heavy metals (Chap. 11).

Each risk chapter includes

- background information on the nature and sources of the environmental problem (in general and in the UAE in particular),
- the expected key health effects of exposure to the relevant pollutants (to humans in general),
- specific details on the construction of the module within the *UAE Environmental Burden of Disease Model* that simulates the risk addressed in the chapter,
- the estimated burden of disease due to that risk factor (including sensitivity and uncertainty analysis),
- information on future data needed to improve the disease burden estimate, and
- overall conclusions.

Chapter 12 ("Applying Environmental Burden of Disease Models to Strengthen Public Policy") summarizes the overall estimates of the environmental burden of disease from all these risks, identifies the leading causes of environmentally related diseases, recommends priorities for collecting data to improve future burden of disease estimates, and discusses important steps for increasing the capacity of UAE institutions to collect additional environmental data and update the burden of disease estimates in the future.

Why Assess a Nation's Environmental Disease Burden?

The motivation for conducting national-level environmental burden of disease assessments emerged from a WHO project to compile global statistics on the incidence, prevalence, survival, and long-term impacts on quality of life of all of the predominant diseases affecting human populations. The World Bank commissioned this global assessment in the 1980s as a source of impartial information to inform cost-effectiveness analyses of its public health programs (Stein et al. 2007). In 1990, the WHO published its results, Global Burden of Disease 1990. This study provided the first-ever comprehensive global overview of morbidity and mortality due to 130 diseases and conditions. In a retrospective analysis of the project's impacts, a prominent demographer who tracks global trends in population health compared the study's influence to the invention of the microscope:

> Like the microscope, the Global Burden of Disease 1990 [report] brought diseases into much sharper focus. … As a sophisticated measuring device, it could not be ignored by any serious student of epidemiology or development (Preston 2006).

In a brief history of the global burden of disease assessment process, Stein notes that the project's results enabled policymakers—previously frustrated "by fragmented, incomplete, incomparable, and often advocacy-driven health information"—to "directly compare the burden of different diseases, set priorities, and evaluate the cost-effectiveness of their interventions."

Having quantified the global disease burden, the WHO recognized that another layer of information—beyond an understanding of global disease patterns—was needed to inform decisions about strategies to improve global public health. In addition to understanding how disease patterns vary by country and region, policymakers also needed to understand the causes of these diseases, in order to determine the interventions with the greatest potential for improving the human condition. Murray and co-authors, investigators in the global burden of disease project, have written, "Data on disease or injury outcomes alone, such as death or hospitalization, tend to focus on the need for palliative or curative services. Reliable and comparable analysis of risks to health, on the other hand, is key for preventing disease and injury" (Murray et al. 2003).

Subsequently, WHO employed crude population and environmental measures collected at low resolution to develop the first approximation of the global environmental burden of disease. This analysis concluded that environmental risk factors contribute to 24% of the global disease burden (Prüss-Ustün and Corvalán 2007). Subsequently, the WHO identified the need for higher-resolution environmental burden of disease studies (e.g., at the national or subnational level)—ones that would accurately reflect local conditions and recognize the global variation in environmental conditions and baseline disease patterns. As a result, the WHO developed a series of guidebooks with methods for computing the burden of disease for several different environmental risk factors, including outdoor air pollution; indoor smoke from solid fuels; occupational carcinogens, airborne particulates, and noise; solar ultraviolet radiation; climate change; water, sanitation, and hygiene; lead; and mercury (see Chap. 3). The project we document in this book is the first comprehensive national-scale implementation of these guidelines. For some exposure pathways and contaminants, detailed WHO guidelines were not available, and in those cases we adapted the WHO's methods to the particular risk.

A comprehensive environmental burden of disease assessment such as that documented in this book can serve many purposes (Prüss-Ustün et al. 2003):

- ***Prioritizing environmental risks to health***: Governments rarely, if ever, have sufficient financial and other resources to tackle all health risk factors at once or to reduce levels of all contaminants in all media to concentrations known not to cause harm. An environmental burden of disease assessment can identify combinations of contaminants and exposure pathways with the most substantial public health impacts, and this information can be considered in planning which problems to address first. Of course, priority-setting requires not only quantitative information about disease burden but also value judgments because values-based trade-offs are inevitable when deciding to fund some programs but not others, or to emphasize some risks over others. Chapter 2 describes a method for combining the quantitative output of environmental burden of disease assessments with a stakeholder deliberation process in order to set priorities.
- ***Assessing trends in environmental impacts on health over time***: Environmental burden of disease assessments, if performed regularly, can serve as indicators of a nation's progress toward reducing the preventable environmental burden of disease.
- ***Predicting the health benefits of environmental interventions***: Environmental burden of disease assessments also can be used to predict the public health benefits of interventions (e.g., increasing wastewater treatment capacity, promoting fuel-efficient vehicles), to improve environmental quality, and/or reduce human exposure to environmental contaminants. This information, in turn, can be used to compare the costs and benefits of alternative environmental interventions a country may be considering.
- ***Identifying high-risk populations***: If performed at sufficient geographic and population resolution, environmental burden of disease assessments can highlight locations and population subgroups that may be at elevated risk due to environmental exposures, in comparison to the national average. Chapter 4 illustrates this potential by highlighting regions of the UAE at highest risk for health effects brought on by air pollution.
- ***Assessing health under future environmental scenarios***: Building on international research to develop future climate change scenarios, environmental burden of disease assessments can predict how a nation's disease burden may shift in response to global warming; Chap. 7 illustrates such an analysis. These assessments also can be used to illustrate how population behavior changes may affect health risks due to environmental exposures. For example, Chap. 9 illustrates how changing recreational swimming behaviors affects the predicted burden of disease due to exposure to fecal pathogens in contaminated marine waters, and Chap. 11 illustrates how food consumption behaviors affect risks of methylmercury exposure.
- ***Setting environment and public health research priorities***: The process of preparing a comprehensive national environmental burden of disease assessment inevitably reveals gaps in environmental and public health data, as well as in scientific knowledge. Thus, the results can be used to guide decisions about which data collection initiatives and research topics to fund. For example,

Chap. 8 illustrates the need for data on concentrations of microbiological and chemical in the UAE's potable water distribution system, in order to assess more accurately the burden of disease due to drinking water contamination.

- ***Informing environmental and public health policy decisions***: Ultimately, environmental burden of disease assessments provide a tool that governments can use to help chart a path toward sustainable development.

  Despite these benefits, few nations have undertaken environmental burden of disease assessments across multiple exposure pathways, and, as Chap. 2 describes, none has undertaken an analysis as comprehensive as the study presented in this book.

Overview of the UAE

The UAE consists of seven emirates (Fig. 1.1) united under a federal constitution but retaining substantial governance autonomy. Table 1.1 shows population data for the emirates from the end of 2007. The emirate-level population data were obtained from the UAE Ministry of the Economy at the launch of the project described in this book (in June 2008). These estimates were based on the 2005, the most recent confirmed census data in the UAE. Recent (2011) population projections from the UAE National Bureau of Statistics suggest that the net rate of migration into the country may have exceeded previous projections. Accordingly, to estimate the 2011 population, the bureau adopted an exponential growth model, which resulted in the estimated 2011 population of 7.5 million, an increase of 66% over the 2007 estimate. According to the National Bureau of Statistics, the current distribution of this population by emirate is not known because of uncertainty over where expatriates reside:

> The resident data … are classified by place of visa issuance or cancellation, therefore the net migration by emirate does not reflect the actual geographical distribution of the population and efforts are underway to find an alternative methodology depending on other sources to find the estimate of the population by emirate (UAE National Bureau of Statistics 2011).

The analyses presented in this book use the 2007 estimates, which are based on the 2005 census and the population projections available when this project began (in June 2008). Notably, total population estimates for the UAE vary by source. For example, the CIA World Factbook estimates the 2011 UAE population as 5.1 million. These discrepancies reflect the difficulties of conducting an accurate census in a nation with a large transient population. In the UAE, this transient population consists of expatriate workers who fill the demand for the variety of jobs, from construction laborer through financial manager, created by the UAE's economic boom. As is notable in Table 1.1, the majority (81%) of the UAE population consists of expatriate workers.

The UAE formed in 1971 from the Trucial States, a protectorate established by England early in the nineteenth century to protect British trade routes to India (Rizvi 1993). England held power until 1971, when it withdrew from the Arabian Peninsula of its own volition. Until the discovery of oil in Abu Dhabi in 1958, the Trucial States were among the Arab world's poorest inhabitants (Rizvi 1993), with "no

Table 1.1 UAE population, end of 2007

Emirate	Citizens	Expatriates	Total
Abu Dhabi	372,000	1,121,000	1,493,000
Dubai	141,000	1,337,000	1,478,000
Sharjah	143,000	739,000	882,000
Ajman	41,000	183,000	224,000
Umm Al Quwain	16,000	36,000	52,000
Ras Al Khaimah	91,000	131,000	222,000
Fujairah	60,000	77,000	137,000
Total	864,000	3,624,000	4,488,000

Fig. 1.2 Maqta Crossing in Abu Dhabi, 1965, before the building of the first bridge at the site. The sandbar in the distance is the site of downtown Abu Dhabi today (photo courtesy of George Bell, http://maribelecosystems.com/OldAbuDhabiandDubaiPhotos.html)

electrical grid, indoor plumbing, telephone system, public hospital, or modern school" (Walters et al. 2006). Abu Dhabi City, the capital, had no paved roads or even a paved runway for its airport (see the historic image in Fig. 1.2).

In spite of—or perhaps especially because of—its late start toward development, the UAE is an ideal demonstration nation for comprehensive environmental burden of disease assessment. The UAE transitioned from one of the world's poorest to one of the most highly developed nations in less than a generation—a time frame difficult to comprehend even for Emiratis who have lived through this transformation (Fig. 1.3). As a result, the UAE is a microcosm for studying the effects of industrial development on environmental health. In addition, its modern governmental institutions are still evolving—a condition that affords these institutions more flexibility to address environmental problems in creative ways than

Fig. 1.3 Downtown Abu Dhabi, 2011 (photo by Wouter Kingma)

long-established bureaucracies in nations that followed a slower, more traditional development trajectory. As well, care for the environment is a deep-rooted cultural tradition, as illustrated in the quote by the nation's late founder, H.H. Sheikh Zayed Bin Sultan Al Nahyan, on the first page of this chapter. Moreover, the leadership has a history of addressing problems of all sorts—including environmental problems—in innovative ways.

The UAE's Development: A Brief History

When the British withdrew, Sheikh Zayed bin Sultan Al Nahyan united several tribes of the Trucial States to form the UAE. Since then, it has developed at an unprecedented rate, fueled by its oil resources and guided by Sheikh Zayed's ambitious vision. All told, the UAE has about 8% of known global oil reserves and 3.4% of natural gas reserves (Davidson 2009). From the beginning, Sheikh Zayed decided that this vast oil wealth should enhance the living standard of all the new nation's citizens, rather than benefiting only the ruling elite. First priorities for his development vision included establishing modern transportation networks, building homes for all citizens, and educating citizens, who were mostly illiterate when the country was founded (Walters et al. 2006).

As a result of Sheikh Zayed's visionary early investments, only 40 years after its founding the UAE ranks among the top echelon of the world's nations, in the category "very high human development" on the Human Development Index, a United Nations

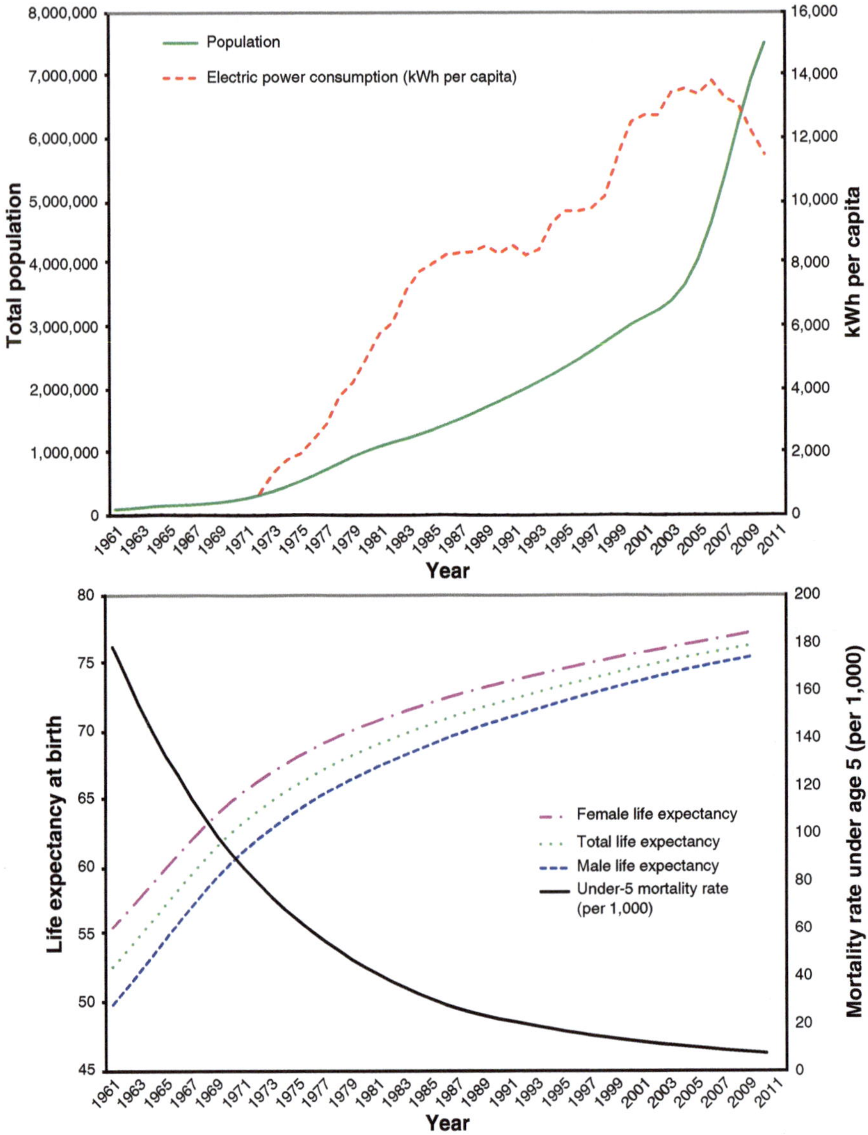

Fig. 1.4 Development in the UAE has led to dramatic population increases (*top*), decreases in infant mortality (*bottom*), and increases in life expectancy (*bottom*). Data adapted from the World Bank Databank (http://data.worldbank.org/country/united-arab-emirates)

measure of prosperity that considers population longevity, educational attainment, and income (Walters et al. 2006). Life expectancy has increased from less than 45 in the 1950s to 75.9 in 2010, and infant mortality rates have plummeted (Fig. 1.4). The UAE's gross national income per capita in 2010 was $59,993—the third highest in the world behind Lichtenstein and Qatar (United Nations Development Program 2011).

Were its wealth concentrated solely in the oil industry, the UAE might not be a useful case study for other nations to emulate in environmental burden of disease assessment, since it would lack industrial diversity and could proceed with simpler solutions to decrease pressure on the environment. However, also due to Sheikh Zayed's vision, the nation's leaders decided early on to broadly diversify the economy, recognizing that oil would not last forever and that relying on oil as the sole basis for development would leave the nation highly vulnerable to oil price shocks (Davidson 2009). Hence, the government invested substantial funds in launching state-owned, export-oriented manufacturing companies in a variety of industries and also developed industrial zones to lure foreign investment. The UAE is home to major industrial facilities that produce metals, plastics, fertilizers, petrochemicals, cement, microelectronics, and many other products. While crude oil revenue accounted for 68% of the UAE's economy in 1975, the oil sector's share of GDP dropped to less than 22% by 1998 (Walters et al. 2006).

The UAE also has invested in amenities to attract tourists and consumers. Abu Dhabi is constructing Middle East versions of the Guggenheim and Louvre, both designed by world-renowned architects. Dubai has constructed deluxe five-star hotels on artificial islands, water parks, indoor snow ski parks, and many other amenities designed to attract tourists. As another marker of the flourishing consumer and tourist culture, the UAE has the highest per-capita square footage of shopping mall space in the Middle East and the second highest in the world (Walters et al. 2006).

Beyond diversifying its industrial base and establishing tourism and consumer infrastructure, the UAE also has invested heavily overseas, as another cushion against oil price shocks. A handful of sovereign wealth funds managed by Abu Dhabi control an estimated $1 trillion in overseas investments, making Abu Dhabi the world's largest holder of sovereign wealth (Davidson 2009). Its sovereign wealth funds are the largest shareholder in Citigroup; own a 75% stake in New York's historic Chrysler building; own 5% of the Italian car manufacturer Ferrari; hold an 8.1% stake in Advanced Micro Devices, the second-largest microprocessor manufacturer in the world; and possess major shares of many other long-established businesses and real estate ventures around the world (Davidson 2009).

The UAE's development has brought improvements in the standard of living of its population more dramatic than anyone likely could have conceived a generation ago. At the same time, however, this transformation has stressed the natural environment and, in so doing, created new risks for public health, as described in detail in Chaps. 4, 5, 6, 7, 8, 9, 10 and 11. The government has recognized these risks, and the study presented in this book represents but one of many initiatives the UAE has adopted in response. Another prominent example is Abu Dhabi's development of Masdar City, a sustainable, zero-carbon-emission city designed to demonstrate renewable and clean energy technologies.

The UAE's Institutional Flexibility

In the 1990s the UAE government began enacting laws and forming institutions to address its newly emerging environmental threats. The Federal Environment Agency was established in 1993, and the first environmental law, Federal Law 24 for the Protection and Development of the Environment, was enacted 6 years later. The law stated as one of its principal goals "control of all forms of pollution and avoidance of any immediate or long-term harmful effects resulting from economic, agricultural, industrial, development or other programmes aiming at improving life standards." The law contained broad provisions for environmental protection, covering air, water, soil, and natural reserves (UAE Federal Government 1999).

Authority to regulate environmental quality was further strengthened with the establishment in 2005 of EAD, expanding the mission of what had formerly been a wildlife protection agency to include a broad range of environmental concerns. In the UAE, unlike in other federations, the wealthier emirates, Abu Dhabi and Dubai, hold considerably more power than the federal government. In fact, Abu Dhabi's revenues heavily subsidize the federal government as well as government functions of the five smaller emirates. As such, EAD, in many respects, is more influential than its federal-level counterpart, and its emergence from what had been a natural resources protection agency signified the growing priority the UAE government was placing on environmental protection, writ large.

Because the UAE's environmental laws and institutions are little more than a decade old, they are still in flux. Agency reorganizations occur regularly. For example, in 2009, the government abolished the Federal Environment Agency and transferred its responsibilities to the new Ministry of Environment and Water. Because they are young, the UAE's government agencies are more nimble than those in many other developed nations and can react rapidly to address problems. By comparison, scholars of environmental policy have cited institutional inflexibility as a major barrier to implementing a more efficient, evidence-based approach to reducing environmental risks to health in the United States. For example, in a review of two unsuccessful U.S. government initiatives intended to streamline environmental regulatory processes, Hoffmann and others conclude that institutional inertia impedes attempts to improve the efficiency of U.S. environmental regulatory programs (Hoffman et al. 2002). Addressing environmental problems in new and more efficient ways, Hoffmann and others write, requires

> rethinking what has been ingrained within the government bureaucracy over the past 30 years. Some may resist this … process as contrary to their conception of the underlying purpose of the agency or as a threat to their own political interests, competencies, skills, or personal security (Hoffman et al. 2002).

Hoffman and others continue, "Believing that [the U.S.] is rapidly approaching the point of diminishing returns on … environmental regulation, many see the existing policy regime as possibly the greatest obstacle to continued environmental improvement." In the UAE, unlike in the United States and other long-developed nations, it may be possible to develop highly efficient systems for protecting the environment and public health, without encountering seemingly intractable institutional inertia.

The UAE's Culture of Environmentalism

The UAE has a culture of concern for environmental stewardship, stemming both from its Islamic foundation and its Bedouin roots. "The Holy Qur'an makes frequent reference to animals and plants and instructs all Muslims to study and appreciate living and nonliving things around them," writes Aspinall in a history of environmental protection in the UAE (Aspinall 2001). Further, environmental conservation was essential for Bedouin survival in the desert. As a result, "Despite the irony of Abu Dhabi's being one of the largest hydrocarbon exporters, support for nature and improving the environment have historically been key legitimacy resources for the emirate's rulers," according to Davidson (2009).

One indicator of the cultural roots of environmentalism is the global recognition that H.H. Sheikh Zayed received for his conservation activities. He was the first sitting head of state (of any country) to receive the World Wildlife Foundation's Gold Panda Award. Posthumously, he received the Champion of the Earth Award from the United Nations Environment Programme. Commenting on the former award, Sheikh Zayed reflected a commonly held local view (Aspinall 2001):

> With God's will, … we shall continue to work to protect our environment and our wildlife, as did our forefathers before us. It is a duty—and, if we fail, our children, rightly, will reproach us for squandering an essential part of their inheritance and of our heritage.

Already, the UAE has demonstrated an ability to be creative in redressing environmental damages brought on by its rapid development. Perhaps no example better illustrates this creativity than the program to restore the endangered houbara bustard (Muller 2011). This migratory bird historically provided an important food source for Bedouins, who hunted with the assistance of pet falcons. While no longer essential for survival, hunting houbara remains popular in the UAE. In the 1980s, Sheikh Zayed noticed declining populations of houbara on his hunting expeditions, and he established the National Avian Research Center (NARC) to study the causes. Habitat fragmentation and increasing human populations, along with excessive hunting and illegal trapping, were among the key factors of the bird's disappearance, the new NARC found. Subsequently, under Sheikh Zayed's leadership, the NARC developed an innovative captive breeding program that now releases about 1,000 birds a year with a survival rate exceeding 50%. By contrast, other breeding programs have struggled to raise more than 20–100 at a time in captivity (Van Heezik and Seddon 2001). The NARC has not only replenished houbara populations in the UAE but also is working to regenerate the species across its habitat, which spans multiple continents (Launay et al. 1997).

The success of the houbara restoration program illustrates the UAE's propensity to innovate, rather than relying solely on conventional approaches for environmental problem solving. The environmental burden of disease assessment presented in this book is but another example of such an innovation. Rather than adopting a piecemeal approach to reducing environmental threats to human health, the UAE has sought to develop an evidence-based process that quantifies environmental pollution risks across all exposure media and that incorporates the multiple values of stakeholders in prioritizing those risks. This approach, as documented in the following chapters, offers the potential to achieve the kind of efficiency in environmental protection long sought but not yet achieved in other nations.

References

Aspinall, Simon. 2001. Environmental development and protection in the UAE. In *United Arab Emirates: A new perspective*, ed. I. Al Abed and P. Hellyer, 277–304. London: Trident Press.

Bernstein, Peter L. 1996. *Against the Gods: The remarkable story of risk.* New York: Wiley.

Covello, V.T., and J. Mumpower. 1985. Risk analysis and risk management: an historical perspective. *Risk Analysis* 5(2): 103–120.

Davidson, C. 2009. Abu Dhabi's new economy: Oil, investment, and domestic development. *Middle East Policy* XVI(2): 59–79.

Graham, J.D. 1995. Historical perspective on risk assessment in the federal government. *Toxicology* 102(1–2): 29–52.

Hoffman, A.J., H.C. Riley, J.G. Troast, and M.H. Bazerman. 2002. Cognitive and institutional barriers to new forms of cooperation on environmental protection: Insights from XL and habitat conservation plans 2002. *American Behavioral Scientist* 45(5): 830–845.

Launay, F., D. Roshier, R. Loughland, and S.J. Aspinall. 1997. Habitat use by houbara bustard (*Chlamydotis undulate macqueenii*) in arid shrubland in the United Arab Emirates. *Journal of Arid Environments* 35: 111–121.

Muller, J. 2011. Saving the houbara. *Abu Dhabi Week*, July 27.

Murray, C.J.L., M. Ezzati, A.D. Lopez, A. Rodgers, and S. Vander Hoorn. 2003. Comparative quantification of health risks: Conceptual framework and methodological issues. *Population Health Metrics* 1(1): 1.

Preston, S.H. 2006. Foreword to Global Burden of Disease and Risk Factors, ed. A.D. Lopez, C.D. Mathers, M. Ezzati, D.T. Jamison, and C.J.L. Murray. Washington, D.C. : The International Bank for Reconstruction and Development.

Prüss-Ustün, A., and C. Corvalán. 2007. How much disease burden can be prevented by environmental interventions? *Epidemiology* 18(1): 167–178.

Prüss-Ustün, A., C. Mathers, C. Corvalán, and A. Woodward. 2003. *Introduction and methods: Assessing the environmental burden of disease at national and local levels. Protection of the human environmental.* Geneva: World Health Organization.

Rizvi, S.N.A. 1993. From tents to high rise: Economic development of the United Arab Emirates. *Middle Eastern Studies* 29(4): 664–678.

Scurlock, J.A., and B.R. Andersen. 2005. *Diagnoses in Assyrian and Babylonian medicine: Ancient sources, translations, and modern medical analyses.* Champaign: University of Illinois Press.

Stein, C., T. Kuchenmüller, S. Hendrickx, A. Prüss-Ustün, L. Wolfson, D. Engels, and J. Schlundt. 2007. The global burden of disease assessments—WHO is responsible? *PLoS Neglected Tropical Diseases* 1(3): e161.

UAE Federal Government. 1999. Protection and development of the environment. Federal Law 24.

UAE National Bureau of Statistics. 2011. Methodology of estimating the population in UAE. http://www.uaestatistics.gov.ae/ReportPDF/Population%20Estimates%202006%20-%20 2010.pdf.

United Nations Development Program. 2011. Human Development Report 2011. Sustainability and equity: A better future for all. New York, NY: Palgrave MacMillan.

Van Heezik, Y., and P. Seddon. 2001. Born to be tame. *Natural History* 110.5 (June 1): 58–63.

Walters, T.N., A. Kadragic, and L.M. Walters. 2006. Miracle or mirage: Is development sustainable in the United Arab Emirates? *Middle East Review of International Affairs* 10(3): 77–91.

Chapter 2
Prioritizing Environmental Risks to Health

Abstract This chapter discusses in detail the process we used to engage stakeholders in further refining the scope of issues to consider in this environmental burden of disease assessment. First, we provide background on innate human cognitive biases that affect our perceptions of risk and how these biases pose challenges to rational priority setting. Then, we describe previous international experiences in prioritizing environmental risks to health for policymaking. Next, we describe the systematic approach used here to prioritize environmental risk factors—an approach that compensates for cognitive biases, incorporates scientific information, systematically involves multiple stakeholders, and builds on international experiences. Finally, we describe how we implemented this ranking process and how the results led to the eight environmental risk factor categories that are the subjects of the remaining chapters of this book: outdoor air pollution, indoor air pollution, occupational exposures, climate change, drinking water contamination, coastal water pollution, soil and groundwater contamination, and produce and seafood contamination.

Keywords Comparative risk assessment • Risk summary sheets • Categorizing environmental health risks • Environmental and public health stakeholders • Deliberative approach for ranking risks • Risk analysis research • Setting priorities for environmental health risk management • Cognitive biases in risk perception

Environmental Factors Included in This Report

According to the broadest medical definition, environmental risks to health encompass every type of nongenetic illness and also nongenetic illnesses triggered by environmental exposures (Smith et al. 1999). According to this definition, diseases and deaths from accidents, crime, stress, wars, earthquakes, tsunamis, and a multitude of other causes are "environmental." However, this all-encompassing view is not practical from an environmental policymaking perspective, because environmental policymaking bodies have no control over these kinds of risks. For this

J. MacDonald Gibson et al., *Environmental Burden of Disease Assessment*,
Environmental Science and Technology Library 24, DOI 10.1007/978-94-007-5925-1_2,

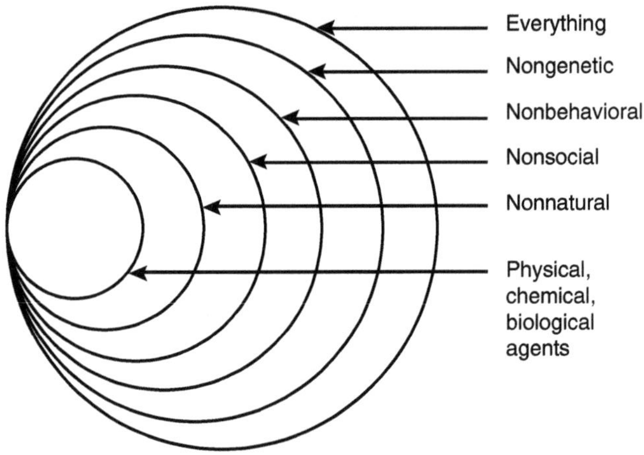

Fig. 2.1 "Environmental risk" may be defined very broadly as all illnesses (since even genetic illnesses may have an environmental trigger if one considers a sufficiently long time horizon) or according to various narrower definitions. The environmental factors considered in this book are those in the inner circle of the diagram: physical, chemical, and biological pollutants released by human activities (Smith et al. 1999)

project, we adopt the narrow definition of "environmental factor" proposed by Smith et al. (1999): we focus on physical, chemical, and biological pollutants (see Fig. 2.1) released to the environment as a result of human activities because these are the risk factors most amenable to control by an environment agency.

Defining environmental factors as pollutants released by human activities still results in a potentially very large number of risks to consider. For example, the database of synthetic chemicals maintained by the U.S. Environmental Protection Agency under the Toxic Substances Control Act lists more than 83,000 chemicals. Therefore, the first step of the UAE environmental burden of disease assessment project was to identify which specific pollutants to consider. Our starting point was a list of 14 categories of concerns provided by the Environment Agency—Abu Dhabi (EAD) when it commissioned this study:

1. Air (indoor and outdoor air pollution)
2. Water resources and water quality
3. Food safety and security
4. Land pollution
5. Waste management (solid, hazardous, and health-care)
6. Noise pollution
7. Chemical safety
8. Vector-borne diseases (such as those carried by rodents)
9. Climate change
10. Electromagnetic fields
11. Nonionizing radiation

12. Depletion of the stratospheric ozone layer
13. The built environment
14. Hazards in the workplace

We reframed these categories approximately according to the routes by which people are exposed to the risk. From an environmental policy perspective, focusing on exposure routes provides a useful organizing framework. This framing is consistent with commonly accepted principles for risk assessment, first articulated by the National Academy of Sciences (1983) and now accepted as the international best practice in risk assessment. Humans can be exposed to contaminants in the environment by (1) breathing contaminated air; (2) drinking, inhaling, or absorbing (via dermal exposure) contaminants in water; (3) ingesting contaminated soil; (4) eating contaminated food; (5) absorbing chemicals at full strength through the skin; and (6) coming into contact with harmful levels of sound waves, electromagnetic radiation, ultraviolet radiation, or heat in the ambient environment. Table 2.1 organizes the 14 risk categories identified by the EAD according to the major exposure routes (air, water, soil, food, occupational environments, ambient environment).

Table 2.1 excludes two of the 14 risks: vector-borne disease and nonionizing radiation. An analysis of infectious disease information from the UAE revealed no evidence of diseases that would be expected if animal vectors were a sanitation concern. Similarly, no existing evidence warranted including nonionizing radiation in indoor air: background radiation levels across the UAE currently are lower than global averages. While risks of exposure to nonionizing radiation may develop in the future once the UAE constructs its recently commissioned nuclear power plants, this exposure route was not considered because it is not relevant at present due to the absence of nuclear power in the UAE. Also worth noting is that while the EAD's list of concerns identified the built environment and waste management as separate risk factors, our analysis considers these factors via their effects on pollutant concentrations in air, water, and soil.

While the framework in Table 2.1 narrowed the environmental burden of disease analysis to a more tractable scope, we wished to further refine the scope and prioritize issues for assessment based on input from UAE residents with multiple viewpoints, representing multiple organizations with stakes in environment and health. Hence, before delving into the detailed burden of disease analysis, we conducted preliminary risk assessments of the factors shown in the middle column of Table 2.1 and then engaged stakeholders in a systematic, deliberative process to prioritize those risks. Appendix A contains the preliminary risk assessments (called "risk summary sheets").

The remainder of this chapter discusses in detail the process we used to engage stakeholders in further refining the scope of issues to consider in this environmental burden of disease assessment. First, we provide background on innate human cognitive biases that affect our perceptions of risk and how these biases pose challenges to rational priority setting. Then, we describe previous international experiences in prioritizing environmental risks to health for policymaking. Next, we describe the systematic approach used here to prioritize environmental risk factors—an approach that compensates for cognitive biases, incorporates scientific information, systematically

Table 2.1 Framework for categorizing environmental health risks

Exposure route	Environmental factors considered in this report	Related issues identified by EAD
Air (breathing)	Outdoor air pollution	Outdoor air pollution, built environment
	Indoor air pollution in residential environments	Indoor air pollution, built environment
Water (drinking, bathing, inhaling droplets)	Drinking water contamination	Water, waste management, built environment
	Coastal water pollution	Water, built environment
Soil (dermal contact followed by ingestion)	Soil and associated groundwater pollution due to solid and hazardous waste disposal	Waste management, land pollution, built environment
Food (eating)	Seafood contamination	Food
	Produce (fruits and vegetables) contamination	Food
Occupational (inhaling, absorbing, hearing)	Pollutants in industrial, construction, and agricultural work environments	Chemical safety
Ambient environment	Ambient noise above healthful levels	Noise pollution
	Ultraviolet radiation above natural levels as a result of stratospheric ozone depletion	Stratospheric ozone depletion
	Electromagnetic fields from power lines	Electromagnetic fields
	Excess heat due to global climate change	Global climate change

involves multiple stakeholders, and builds on international experiences. Finally, we describe how we implemented this ranking process and how the results led to the eight environmental risk factor categories that are the subjects of the remaining chapters of this book: outdoor air pollution, indoor air pollution, occupational exposures, climate change, drinking water contamination, coastal water pollution, soil and groundwater contamination, and produce and seafood contamination.

Challenges of Ranking Environmental Risks

Human cognitive limitations, along with differences in individual values, pose major challenges to efforts to set national risk priorities. As has been well documented in scientific research on risk perception, humans tend to underestimate high risks and overstate low risks. Figure 2.2 shows the results of a historic experiment that documented this tendency (Lichtenstein et al. 1978). In this study, 39 community leaders in Oregon (United States) were asked to estimate the number of deaths per year in the United States due to a variety of risk factors. The dots on Fig. 2.1 show

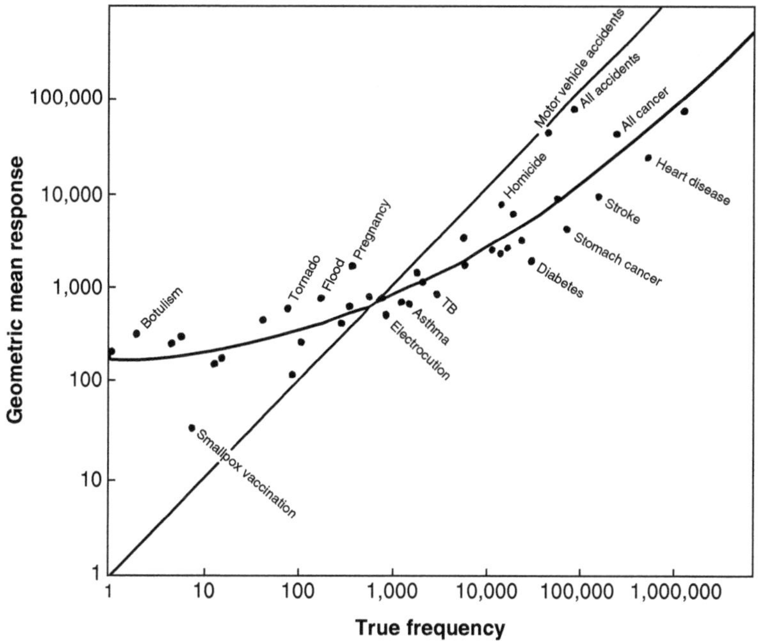

Fig. 2.2 Relationship between judged frequency and the actual number of deaths per year for 41 causes of death. Respondents were told that about 50,000 people per year die from motor vehicle accidents in the United States. They were then asked to estimate the frequency of the other causes of death shown on the graph. If people were perfect judges of risk, the experimental data points would have fit the straight line shown in the graph. Instead, as shown, respondents overestimated the frequency of low-probability causes of death (for example, death due to tornado) and underestimated the frequency of common causes of death (such as heart disease) (Lichtenstein et al. 1978)

the geometric mean estimates across all the participants. If, on average, these well-educated participants could judge risks accurately, then the dots would fit the straight line from the origin. Instead, as shown, the points trace a curve that is above the line for low-frequency events (botulism, tornadoes, floods) and below it for high-frequency events (e.g., heart disease, stroke, cancer). This illustrates a systematic bias toward underestimating high risks and overestimating low risks. The authors of the study, along with subsequent authors who repeated these results with a variety of population samples, attribute this bias in part to the familiarity of high-risk events due to their frequent occurrence and the novelty of low-risk events due to their infrequent occurrence (Lichtenstein et al. 1978; Hastie and Dawes 2001). High-frequency occurrences, such as death due to cardiovascular disease, seldom make the news, while low-frequency events dominate headlines due to their rarity.

In general, due to human cognitive limitations and the evolutionary development of the human brain, our intuitive perceptions of risk are biased, and these biases interfere with accurate assessment of risks. Similar research also has demonstrated that even scientific experts are subject to this type of bias when assessing risks unaided by systematic analysis. As Fischhoff et al. (1982) suggest in a review of research on risk perception, "There is no particular reason to believe that the thought processes of experts are appreciably different from those of laypeople. ... When forced to go beyond the limits of available data or to convert their incomplete knowledge into judgments usable by risk assessors, they may fall back on intuitive processes just like everyone else."

Because of innate biases in perceiving the magnitude of risks, a systematic process is needed to separate fact from emotion when identifying environmental priorities, while also accounting for the values of the society at risk. Yet, deciding on such a process is not a simple matter. The difficulty arises from the multiattribute nature of risk. That is, when describing and thinking about risks, people tend to be worried about multiple attributes, or characteristics, of the risk. For example, they may be concerned about the number of deaths per year caused by the risk, their own ability to control the risk, how much is known about the risk, the number of illnesses per year, and many other factors described in more detail later in this chapter.

If risks were described only in terms of one attribute, such as fatalities, setting priorities would be a simpler task. The estimates of fatalities for each risk could be assessed, risks could be ranked from high to low on this measure, and greater emphasis could be placed on finding and implementing solutions to those risks suspected of causing the greatest number of fatalities. But research has shown that people disagree about which characteristics are most important. For example, some people may be most concerned about fatalities because of the severity of the outcome, while others may be more concerned about injuries or illnesses because of the burden they impose on society. Similarly, some people may be more concerned about risks that cause fatalities due to controllable exposures while others may be more concerned about risks that cause fatalities due to uncontrollable exposures. Thus, while a sound understanding of science is necessary for setting environmental priorities, it is not sufficient. Priority setting also requires understanding individual preferences for which types of risk to manage. In other words, risk management should be based on a sound understanding of what science can say about environmental health risks *and* informed judgments about those risks that reflect the values of people who will be affected by the risk management strategy. These scientific assessments and informed judgments must be considered in a systematic way to develop a justifiable ranking of risks.

International Experiences in Prioritizing Environmental Risks

Over the past two decades, many countries around the world have undertaken projects to prioritize environmental risks to public health. In order to build on these experiences, our science team systematically reviewed the processes used for environmental health priority setting in the United States, European nations, and Australia. With a few notable exceptions, the processes used to set priorities in these other nations were ad hoc and struggled both to address the cognitive biases described in the previous section and to incorporate the perspectives of multiple stakeholders.

Setting Environmental Priorities in the United States

The United States undertook its first project to set national environmental priorities in 1986. Prior to that, the nation's spending on environmental problems followed a piecemeal approach, reacting to problems as they arose rather than following a strategic, long-range view of environmental protection. As the U.S. Environmental Protection Agency (EPA) Science Advisory Board notes in a 1990 report,

> For the past 20 years, EPA has been basically a 'reactive' agency. As environmental problems were identified, the public conveyed its concern to Congress, and Congress passed laws to try to solve the problems. ... EPA then implemented the laws using the resources ... allocated by Congress. This reactive mode, although understandable when seen in its historical context, has limited the efficiency and effectiveness of EPA's environmental protection efforts.

The Science Advisory Board suggests that the reactive approach interfered with efforts to anticipate environmental problems and take preemptive actions to prevent serious problems from occurring. Further, the board notes, "At EPA there has been little correlation between the relative resources dedicated to different environmental problems and the relative risks posed by those problems" (U.S. EPA Science Advisory Board 1990).

In 1986, Lee Thomas, then the administrator of the EPA, initiated an effort to undertake a holistic assessment of the environmental problems in the United States and identify the most important unresolved issues—what he called the nation's "unfinished business" in environmental protection. The resulting report, *Unfinished Business: A Comparative Assessment of Environmental Problems*, was released in 1987. In the preface, Thomas observes, "In a world of limited resources, it may be wise to give priority attention to those pollutants and problems that pose the greatest risks to our society. That is the measure this study begins to apply. It represents, in my view, the first few sketchy lines of what might become the future picture of environmental protection in America" (U.S. EPA 1987).

To carry out the study, Thomas appointed a group of 75 senior EPA staff members, both scientists and managers. The group first developed a list of 31 environmental problems to consider, ranging from global warming to releases from chemical storage tanks. Then, the group divided into four teams, each focused on a different risk endpoint: (1) cancer effects, (2) noncancer health effects, (3) ecological effects, and (4) welfare effects. Each group developed its own ranking of the 31 problems.

Lacking any formal guidance or a scientific basis for setting priorities, each of the four EPA work groups devised its own method for ranking the 31 risks. For example, the cancer work group based its ranking on quantitative estimates of the population cancer risk and maximum individual risk for each of the 31 problems. The group reviewed this information in a 1-day meeting and, based on that review, rank-ordered the 31 problems from 1 through 26 (with some ties, resulting in a total of 26 rankings rather than 31). The noncancer group, on the other hand, identified a few select chemicals with noncancer health endpoints for each problem area. Then, the group developed a qualitative scheme for ranking the problem areas as high, medium, or low based on consideration of the number of people exposed, severity of the health endpoint, and exposure level for the selected chemicals.

The result of EPA's *Unfinished Business* project was four sets of risk rankings, each presented in a different format, with no attempt to combine the rankings across the four area of effects. While the Science Advisory Board and other reviewers lauded the EPA's effort to set priorities, the EPA's process had a number of limitations. First, EPA did not seek participation from experts or stakeholders outside the agency. The Science Advisory Board, National Academy of Sciences, and other organizations subsequently recommended that the public should participate in future risk-ranking efforts (U.S. EPA Science Advisory Board 1990; National Research Council 1996), pointing out that risk ranking must incorporate both science and value judgments and that the values of the affected public therefore should be considered in any ranking.

An additional limitation of EPA's approach was that the 31 categories of risks were not consistently defined. Some were based on sources of pollution (such as nonpoint source discharges to surface water), while others were based on specific pollutants (such as hazardous/toxic air pollutants). This led to double counting of some kinds of risks. For example, contaminants in drinking water might arise from the category "hazardous waste sites" but also would be included in the category "drinking water as it arrives at the tap." Another very important limitation was the lack of a systematic, consistent process for combining quantitative assessment of risk information with the judgments of the involved EPA staff in order to arrive at the risk rankings. As noted above, the work groups (even the two focused on human health) used different metrics for comparing risks and different systems for reporting the resulting priorities.

The ad hoc nature of the process used to generate the priorities in *Unfinished Business* reflects the state of knowledge available at the time about how to perform comparative risk assessment. No guidance was available in the scientific literature on how best to address questions such as how to combine quantitative information with the value judgments inherent in any risk decision or how to overcome biases in risk perception when providing participants in the ranking exercise with a basis for performing the rankings. New research from the fields of risk assessment, decision analysis, and the psychology of risk perception was available to inform a priority-setting process, but scientists had not yet assembled the results of this recent research into a process for setting priorities.

Setting Environmental Priorities in Europe and Central Asia

The 1994 World Health Organization (WHO)/Euro Second Ministerial Environment and Health Conference, held in Helsinki, provided the impetus for European nations to prioritize environmental risks to health and develop strategic plans to reduce those risks. More than 50 European and Central Asian nations participated and signed the Helsinki Declaration on Action for Environment and Health. Under this declaration, each nation committed to developing a national environmental health strategy and action plan. By 2002, 43 nations had developed such plans.

We were able to obtain documentation on 17 of these plans. Table 2.2 summarizes information about these plans, including information about the method for setting priorities, whether the process used quantitative risk information, and whether stakeholders outside of the lead government environmental agencies were involved.

Table 2.2 shows that European and Central Asian nations used vastly different approaches for prioritizing risks. There was wide variation in the methods and metrics used to set priorities, the involvement of stakeholders, and the use of quantitative risk information. Like the U.S. EPA, environmental and public health agencies in Europe and Central Asia struggled with the lack of a formal, validated method for setting priorities.

The vast differences in the methods used by these nations are exemplified in the different approaches used by Switzerland and Sweden. Switzerland involved a range of stakeholders in the ranking process. Representatives from government agencies at all levels (local and federal), nongovernment organizations, and the science sector participated in a working group to assess priorities. Staff from the Federal Office of Public Health and the Swiss Agency for the Environment, Forests, and Landscape first identified 17 environmental topics to consider, from promotion of sustainable agricultural production to reduction of air pollution. Each member of the working group then provided qualitative judgments of each of the 17 topics according to impacts on ecosystems, impacts on public health, scientific evidence supporting a causal association between the risk and negative effects, economic burden, political feasibility of addressing the risk, perception in society, and relationships to other European environmental programs. These qualitative assessments formed the basis for a subsequent ranking of topics for action.

Sweden relied to a much greater extent than Switzerland on quantitative risk information. But the group of stakeholders involved was narrower than Switzerland's, including only scientists and government agency officials, with no representation from nongovernment organizations or the industrial sector. The Swedish process grouped risks into categories roughly corresponding to exposure route: outdoor air, indoor air, water, food, soil, household chemicals, noise, ionizing radiation, nonionizing radiation, and injuries. Risk analysts estimated the number of cases attributed to different kinds of hazards (e.g., microbes in drinking water or radon in indoor air) to which people might be exposed through these different exposure routes. A work group of 30 scientists and government officials then reviewed this risk information and developed the following algorithm for categorizing each hazard as low, medium, or high priority:

- **Low**. Risks causing severe effects such as death, cancer, or long-term suffering at a rate of less than one case per year in Sweden. Also in this category were less severe illnesses and annoyances occurring at rates of less than 100 per year.
- **Medium**. Risks causing 1–100 severe cases per year and 100–10,000 less severe cases per year.
- **High**. Risks causing more than 100 severe cases per year or more than 10,000 less severe cases per year.

Table 2.2 Overview of 17 environmental health plans from Europe and Central Asia

Country	Document title (year)	Strategic planning approach	Use of quantitative risk information	Stakeholder involvement
Albania	National Environmental Health Action Plan (1998)	Working group of experts from government agencies and research institutes prepared qualitative assessments of environmental risks to health and developed recommended actions to address these risks.	No	No
Austria	Austrian National Environmental Health Action Plan (1999)	Representatives from three federal ministries assessed available information on environmental risks to health and developed possible solutions.	No	No
Czech Republic	National Environmental Health Action Plan (1998)	Working group of experts representing seven government organizations assessed existing information on environmental quality and health outcomes and based priorities on their judgments of this information.	No	No
Estonia	The National Environmental Health Action Plan of Estonia (1999)	A group of 37 experts from government agencies and research institutes conducted quantitative risk assessments. Recommended actions were prioritized based on whether risks in Estonia are higher than in other European nations.	Yes	Limited—stakeholders represented government agencies and research centers
Finland	Finnish Environmental Health Action Plan (1999)	Stakeholder committee quantified deaths and illnesses attributable to specific environmental risks. Met 18 times to prioritize risks and develop objectives and actions.	Yes	Yes
France	National Environment and Health Action Plan 2004–2008 (2004)	Expert group prepared report that analyzed exposure to environmental risks and made recommendations for action priorities.	Yes	Limited—stakeholders were invited to review and comment on proposed actions

(continued)

Table 2.2 (continued)

Country	Document title (year)	Strategic planning approach	Use of quantitative risk information	Stakeholder involvement
Lithuania	National Environmental Health Action Plan Lithuania (2001)	Thirteen working groups of experts developed recommended actions based on qualitative assessment of environmental health risks.	No	Limited—involvement appears to have been limited to experts
Malta	National Environmental Health Action Plan (2006)	The Ministry of Health, the Elderly, and Community Care developed recommended actions based on a qualitative assessment of environmental conditions.	No	No
Netherlands, The	Environmental Health Action Plan: Implementing More Powerful Policy (2002)	Stakeholder working groups assessed bottlenecks and gaps in existing environmental health policies and broad actions to address the identified needs.	No	Yes
Poland	Long-Term Programme: Environment and Health (2001)	Broad outline of steps needed to quantify the environmental burden of disease, formulate environmental policies, and establish medical prevention programs was developed by the Ministry of Health and Ministry of Environment.	No	No
Romania	The Romanian National Environmental Health Action Plan (1997)	Prepared by a steering committee composed of representatives from seven government agencies, academic institutions, and nongovernment organizations. Proposed actions were solicited from a wide variety of stakeholders through use of formal "action proposal forms".	No	Yes
Slovak Republic	National Environmental Health Action Plan for the Slovak Republic II (2000)	Experts in public health developed the plan based on an analysis of the state of implementation of a 1997 action plan. The group considered data on trends in environmentally related health effects and on pollutant levels in the environment but did not carry out formal risk assessments.	No	No

(continued)

Table 2.2 (continued)

Country	Document title (year)	Strategic planning approach	Use of quantitative risk information	Stakeholder involvement
Sweden	Environment for Sustainable Health: An Action Plan for Sweden (1996)	Work group of 30 scientists and government officials quantified risks and then ranked them as high, medium, or low priority based on estimated number of health outcomes per year due to the risk.	Yes	Limited—involved scientists and government agency representatives but not industry groups or nongovernment organizations
Switzerland	Sustainable Development: Action Plan Environment and Health (1997)	Interagency working group identified and ranked 17 areas where additional measures to protect environmental health are needed. Group then identified potential interventions and goals (targets).	No	Yes
Ukraine	National Environmental Health Action Plan of Ukraine (1999)	Expert team identified broad categories of environmental risks to health and measures to reduce each risk.	No	No
United Kingdom	National Environmental Health Action Plan (1996)	The Department of Environment, Food and Rural Affairs Actions recommended actions based on a qualitative assessment of environmental conditions. This action plan was overtaken by the development of the 1999 U.K. Sustainable Development Strategy.	No	Limited—stakeholders were invited to submit comments
Uzbekistan	National Environmental Health Action Plan of the Republic of Uzbekistan (1999)	A stakeholder working group recommended actions based on a qualitative assessment of the current state of environmental health and of measures already in place to reduce environmental health risks.	No	Yes

Based on this ranking, the work group identified the leading health outcomes that might be substantially attributed to environmental exposures and that should form the basis for identification of interventions to reduce risk.

Based on a review of available European and Central Asian national environmental health strategy and action plans, it is clear that the European nations faced the same limitation the EPA did in attempting to establish environmental priorities: the lack of a scientifically credible and publicly acceptable process to guide the priority-setting exercise—one that would help address innate human cognitive limitations in judging risks and that would reflect both quantitative risk information and the values of society. Of course, a one-size-fits-all approach to environmental health planning would not be suitable. Rather, a flexible method is needed to reflect differences in the political and consultative processes of the respective countries.

Setting Environmental Priorities in Australia

Following Europe, Central Asia, and the United States, Australia undertook its first environmental health strategic planning exercise in 1999, producing the *National Environmental Health Strategy* (Commonwealth of Australia 1999). The primary goal of Australia's strategic planning exercise was to unite local, state, and federal government agencies spanning the environment and health sectors; priority setting was a secondary goal. The strategy notes,

> Australia's ability to predict and reduce environmental threats to health has been impeded by the fragmentation of management across government and key organizations. Different jurisdictions have differing operational approaches to environmental health, resulting in reduced awareness of existing activities, lack of coordinated actions and duplication of effort.

Australia's strategy considered seven areas at the interface between environmental quality and public health: (1) water quality, (2) air quality, (3) food, (4) contaminated land, (5) waste, (6) vector-borne diseases, and (7) the built environment. It is worth noting that some of these categories (water, air, food) represent exposure routes, while others (waste) represent sources of pollutants. Within each area, the strategy document describes current problems in qualitative terms without defining specific priorities for action. The extent to which stakeholders were involved in developing the document is unclear from the text.

Australia's 1999 strategy appears to be based entirely on qualitative information. Except for a few case studies that collected limited data on the health outcomes of environmental risks at some locations, no formal risk assessment was conducted to support the strategy. An updated version, issued in 2007, identified as a top priority the use of a risk assessment and management approach in future environmental and health strategic planning (although the 2007 strategy, like the 1999 version, relied on a qualitative process) (Commonwealth of Australia 2007).

The Deliberative Approach for Ranking Risks

Nations have struggled to conceive of a systematic, scientifically sound process for ranking environmental health risks. Previous environmental health priority-setting projects have included important features, including in some cases the use of quantitative risk assessments and the involvement of various stakeholders from the public, private, and nonprofit sectors. However, none of the previous environmental health strategic planning projects provides a model that is clearly superior to the others and that meets the criteria of scientific defensibility and broad participation. In fact, participants at a workshop convened by the North Atlantic Treaty Organization (NATO) and WHO in 1997 concluded, "At present there are large differences in the application of risk assessment methods across Europe as part of the NEHAP [national environmental health action plan] process, both in terms of qualitative and quantitative methods" (Briggs et al. 1999). Workshop participants recommended "NATO and other responsible organizations should provide financial support for a systematic and detailed review of differences in the risk assessment methods used for NEHAPs." Of course, each nation needs some flexibility to tailor its environmental burden of disease assessment and planning process to reflect local conditions. Nonetheless, ideally, all these assessments would incorporate both scientific input and stakeholder deliberations in a systematic process.

In the wake of projects around the world to set environmental risk priorities, the U.S. Office of Science and Technology Policy, part of the Executive Office of the President, asked Resources for the Future to organize a workshop of researchers involved in comparative risk assessment. The goal was to develop a systematic process for prioritizing risks across government agencies (Davies 1996). In preparation for this workshop, researchers at Carnegie Mellon University developed a proposal for a risk-ranking method that included both quantitative and qualitative metrics and a systematic process for stakeholder participation (Fischhoff 1995; Morgan et al. 1996). This proposal led to a substantial body of research that further developed the method, pilot tested it, and refined it according to the findings of the pilot tests (see Morgan et al. 1999, 2000, 2001; Long and Fischhoff 2000; DeKay et al. 2001; Florig et al. 2001; Willis et al. 2004). The resulting publications referred to the method as the "deliberative method for ranking risks."

The researchers who developed the deliberative method sought to design a ranking approach that would

- use existing scientific knowledge (including recent findings from research in decision theory, risk analysis, and the psychology or risk communication);
- help those doing the ranking "to systematically consider all relevant information," including available quantitative risk information;
- help participants construct rankings consistent with their own values;
- ensure that participants understand and are satisfied with the procedure; and
- determine the level of agreement and sources of disagreement among participants in the risk-ranking process.

The method built on a body of research in decision analysis, risk analysis, and risk perception that had begun in the mid-1970s. Risk analysis research had resulted in improved methods for characterizing the quantitative aspects of risk (such as, for example, the number of deaths and illnesses that might result) using improved understanding of pollutant transport in the environment, health effects of pollutant exposures, and increasingly powerful computer simulation tools (see, for example, Ramaswami et al. 2005). Risk perception research had led to improved understanding of the kinds of attributes that people consider to be important when making risky decisions. For example, Slovic, Fischhoff, and Lichtenstein showed that while a large number of factors may influence people's thinking about risks, many of these factors are highly correlated. Due to these correlations, factors that influence risk perception can be categorized into three groups, described as "number," "knowledge," and "dread" (1980). "Number" refers to the number of people killed or injured due to the risk (which may be characterized in different ways). "Knowledge" refers to aspects of lay and expert knowledge about the risk, as well as properties of the risk such as whether it is imminent or delayed. "Dread" is a term researchers have used to characterize factors such as individual controllability, catastrophic potential, and the proportion of the population affected. The deliberative method builds on this research by seeking to characterize risks according to several metrics within each of these groups, while relying on quantitative risk assessment methods to estimate attributes such as the number of deaths and illnesses that can be expressed in quantitative terms. It also builds on research to develop approaches to support decision-making for complex problems in which multiple attributes may be important—in particular, on "multiattribute utility theory" (Keeney and Raiffa 1993).

The deliberative method for ranking risks was extensively tested in a set of experiments involving 218 professional risk managers (Florig et al. 2001). In the pilot tests, participants were divided into small groups of 3–7 participants each and asked to rank subsets of 22 risks to students at a fictitious middle school. Analysis of the results showed high correlations among the rankings produced by each group, as well as high levels of satisfaction with the process among participants.

Recently, a variety of national and international entities have employed the deliberative method in risk-ranking exercises. For example, in Canada, the Consumer and Market Demand Agricultural Policy Research Network used the method to rank food safety risks (Webster et al. 2008). The U.S. National Research Council currently is using the method to rank products regulated by the U.S. Food and Drug Administration in terms of health risks.[1] The U.S. Army Corps of Engineers (2008) is using the methods to rank hurricane mitigation opportunities on the Louisiana Gulf Coast. The UAE is the first country in the world to employ the deliberative method for ranking risks as part of the development of a national environmental health strategy—even though the method initially was developed with that purpose in mind. As such, it provides a valuable model for other nations to study and consider implementing.

[1] For information about this project, see http://www8.nationalacademies.org/cp/projectview.aspx?key=BEST-K-08-03-B.

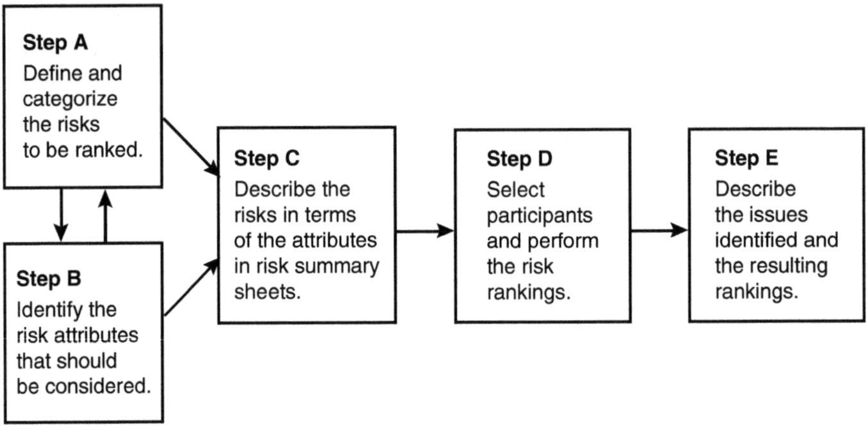

Fig. 2.3 Steps in the deliberative method for ranking risks

The deliberative method for ranking risks involves five steps, shown in Fig. 2.3 and explained in further detail below.

Step A: Define and Categorize Risks

The first step in the ranking process involves sorting risks into a limited number of categories that can be meaningfully compared. Morgan et al. (1996) recommend as one option the categorization of risks based on the route by which people are exposed. This approach was adopted for the UAE, as explained below.

Step B: Identify the Risk Attributes

A large number of attributes may influence people's perceptions of risky events (Florig et al. 2001). Examples of risk attributes include number of deaths or illnesses, age of those most affected, latency of illness, nature of the illness, and ability to control exposure to the risk. The second step of the deliberative method involves identifying the risk attributes that should be considered when comparing the risks. As noted above, empirical studies on risk perception have demonstrated that while a wide variety of attributes are important in people's perceptions of risk, the attributes can be sorted into three categories: number, knowledge, and dread. Within each category, the attributes are highly correlated, so that if a risk-ranking exercise considers just a few attributes from each category, the ranking results will not depend significantly on the specific attributes that are assessed (Morgan et al. 1996). Table 2.3 shows examples of attributes in each of these three categories. The second

Table 2.3 Examples of attributes that can be used to compare risks

Category	Example attributes	Units
Number	Number of deaths per year*	Deaths/year
	Number of more serious long-term illnesses per year*	Cases/year
	Number of less serious long-term illnesses per year*	Cases/year
	Number of more serious short-term illnesses per year*	Cases/year
	Number of less serious short-term illnesses per year*	Cases/year
	Expected number of annual person-years lost due to death	Person-years
	Expected number of annual person-years lost due to nonfatal illnesses	Person-years
	Total expected number of annual person-years lost	Person-years
	Chance in a million of death per year for the average individual*	Probability
	Chance in a million of death per year for the individual at highest risk*	Probability
Knowledge	Time between exposure and health effects (degree to which impacts are delayed)*	Constructed scale (e.g., <1 year, 1–10 years, >10 years)
	Degree to which risk is observable	Constructed scale (e.g., unobservable, with difficulty, with ease)
	Uncertainty in number of deaths, illnesses (degree to which risk is known)*	Constructed scale (e.g., low, medium, high)
	Degree to which risk is reversible	Constructed scale (e.g., yes, frequently, no)
	Degree of scientific understanding of risk*	Constructed scale (e.g., low, medium, high)
Dread	Individual controllability*	Constructed scale (e.g., low, medium, high)
	Catastrophic potential (e.g., greatest number of deaths in a single event)*	Number (or other appropriate measure)
	Outcome equity (number who receive benefits from risk divided by half the sum of number who receive benefits and number at risk)	Constructed scale (e.g., high = 1–3, medium = 4–10, low > 10)
	Intergenerational risk	Constructed scale (e.g., negligible, modest, large)

Adapted from Morgan et al. (1996)

*Attributes previously identified as the most suitable indicators in each category by Florig et al. (2001)

step in the deliberative method is to select two or more attributes within each category to use as metrics for comparing the different risks. Previous research has identified attributes that are the most suitable indicators in each category (Florig et al. 2001). For the UAE risk-ranking exercise, we relied on these previously established attributes, shown with asterisks in Table 2.3.

Step C: Describe the Risks in Terms of the Attributes

The third step in the priority-setting process is to summarize, based on available information, how each risk measures along each of the selected attributes. This information is presented in a standard format that was extensively researched by the Carnegie Mellon scientists who developed the deliberative method. The format is based on concepts from modern risk communication (Florig et al. 2001). The summary sheets are designed to contain sufficient information so that readers without special expertise can comprehend the information, while also providing additional details for more knowledgeable experts. The risk summary sheets are each four pages in length. The first page contains a brief summary of the risk, followed by a table showing how the risk measures along each selected attribute. The use of consistent attributes and units for each risk category facilitates comparisons. The interior text (pages 2–4) contains additional information about the risk in general and in the specific context considered in the ranking exercise. Figure 2.4 shows an example of a risk summary sheet taken from one of the Carnegie Mellon pilot tests of the method. The summary sheets are designed so they can easily be spread out on a table when comparing risks (Florig et al. 2001).

Step D: Perform the Risk Rankings

The fourth step in the deliberative risk-ranking method involves holding workshops or focus groups with small groups of stakeholders (generally 8–12 per group) to review the risk summary sheets and establish priorities both individually and in groups. At the beginning of the workshop, all participants attend a brief (less than an hour) lecture on the psychology of risk perception and the multiattribute nature of risk (Morgan et al. 2001). The purpose of the session is to make participants cognizant of factors that may bias their risk perceptions and cause them to overlook quantitative information about risks. Individuals are then provided with an opportunity to rank risks. They rank risks in two ways: holistically, based on a review of the risk summary sheets, and using a multiattribute utility process, in which they rank the importance of the attributes and the workshop facilitators then compute what the individual's ranking should have been, in order to reflect these preferences for attributes. The groups then meet and develop group rankings. Finally, individuals have a chance to reconsider their rankings based on the group discussions and multiattribute utility analysis of their preferences.

School Bus Accidents

Summary:

Most school bus-related deaths occur among students who are outside the bus either getting on or getting off. Half of school bus injuries occur among students on the bus. At Centerville Middle School half of the 430 students ride the school bus, almost identical to the national average. Accidents involving more than one death are very rare. Because CMS buses use the Alvarez Expressway and cross the C&LL rail line, the risk of a catastrophic bus accident in Centerville is estimated to be between four and six times higher than the national average.

School bus accident risk for Centerville Middle School*

Student deaths	*Low estim.*	**Best estimate**	*High estim.*
Number of deaths per year	.0001	**.0002**	.0004
Chance in a million of death per year for the average student	.25	**0.5**	1
Chance in a million of death per year for the student at highest risk	0.5	**1**	2
Greatest number of deaths in a single episode		**20-50**	
Student illness or injury			
More serious long-term cases per year	.0002	**.0006**	.002
Less serious long-term cases per year	.0004	**.0015**	.004
More serious short-term cases per year	.001	**.002**	.006
Less serious short-term cases per year	.002	**.005**	.015
Other factors			
Time between exposure and health effects		**immediate**	
Quality of scientific understanding		**high**	
Combined uncertainty in death, illness, injury		**1.6 (low)**	
Ability of student/parent to control exposure		**moderate**	

*See "Notes on the Numbers" for definitions and explanations of assumptions.

Fig. 2.4 Layout of the front page of a four-page risk summary sheet from the Carnegie Mellon pilot test of the method, showing the risk name, a summary paragraph, and a table of key risk attributes. Additional pages include a short narrative that describes the risk in national and local contexts and a description of actions offi cials have taken to address the risk (Florig et al. 2001)

Step E: Describe the Resulting Rankings

Following the workshop, the results are analyzed to assess areas of agreement and disagreement. Risk problems that are consistently ranked high or low are noted. Additionally, risk problems for which rankings are highly divergent (with a combination of high and low rankings and thus low interindividual or intergroup correlations) are identified. Averages of the individual and group rankings, along with confidence intervals around these averages, are computed.

Implementation of the Deliberative Method in the UAE

The science team employed the five-step risk-ranking process to prioritize environmental health risks with stakeholders in the UAE. This section describes the details of implementing the method in the UAE.

Step A: Define and Categorize Risks in the UAE

For the reasons explained in the introduction to this chapter, this study categorized risks as shown in the middle column of Table 2.1.

Step B: Identify the Risk Attributes for the UAE

The science team selected a set of risk attributes based on the results of previous research on the risk-ranking method. Table 2.3 shows the attributes.

Step C: Describe the Risks in the UAE in Terms of the Attributes

To assess each risk according to the attributes in Table 2.2, the science team reviewed more than 400 government reports and scientific publications that described exposures and health consequences to environmental hazards. Where possible, the literature review focused on studies specific to the UAE. When UAE-specific literature did not exist, the team next turned to studies of environmental health risks in other Gulf countries based on an assumption that similarities in economic conditions and environmental policy institutions would result in comparable outcomes. Finally, in some cases, the review had to rely on information about exposures and health effects in Europe or the United States if a case could be made that the exposures could be expected to be comparable between the two regions.

The results of this literature review and the resulting risk summary sheets are included in Appendix A along with referenced descriptions of the calculations that were conducted to develop the risk estimates provided in the summary sheets. The risk estimates describe what was known about environmental health risks in the UAE at the time that the priority-setting exercise was conducted. They represent the first iteration of science team's effort to assess the environmental burden of disease. Subsequently, after using this information to identify priorities, the science team gathered additional information needed to construct the detailed environmental burden of disease models described in Chaps. 4, 5, 6, 7, 8, 9, 10, and 11 of this report.

The risk estimates in Appendix A are not identical to the estimates provided in the remainder of this report. The differences stem from the fact that the burden-of-disease modeling conducted by the science team utilized more sophisticated modeling techniques and incorporated new data. Despite these differences, the uncertainty ranges are largely consistent between the risk summary sheets and burden of disease estimates. A future environmental health strategic planning effort in the UAE could update the risk summary sheets in Appendix A with the results described later in this report and then repeat the ranking exercise.

Step D: Perform the Risk Rankings for the UAE

In January 2009, the science team hosted workshops in Abu Dhabi and Dubai with stakeholders from throughout the UAE to elicit their concerns about environmental health risks in the UAE. Invitations to participate in these workshops were sent to organizations selected by the Environment Agency—Abu Dhabi and the project steering committee because of the role they would ultimately need to have in helping to develop and implement interventions to reduce environmental effects on health. Expatriates and UAE citizens were included among the invited participants, because expatriates hold important positions in some of the key stakeholder organizations. Seventy-three people from the federal government, emirate governments, private-sector organizations (including key industries), and nongovernmental organizations participated in the workshops, representing five of the seven emirates in the UAE and a diverse range of expertise and perspectives on managing environmental risks. Table 2.4 summarizes the characteristics of participants, obtained though surveys completed at the end of the workshops. It is important to keep in mind that although the participants represented diverse viewpoints, they were not selected at random. The ranking of risks that would result from a random sample of the UAE's population may, therefore, be different. However, the results from this project provide an indication of the priorities of the key institutional stakeholders that would be involved in interventions to reduce environmental impacts on health.

The workshops were conducted in groups of 8–20 people. While the goal was for each group to include about ten participants, the group sizes had to be adjusted to accommodate the availability of participants. In an attempt to ensure that diverse views were represented in each group, we separated participants from the same organization into different groups when possible.

Table 2.4 Characteristics of participants in the priority-setting workshops

Sector	Emirate	Expertise
Emirate government (39)	Abu Dhabi (18)	Environment (18)
Federal government (11)	Ajman (1)	Health (12)
Private sector (3)	Dubai (11)	Food (2)
Nongovernmental organizations (2)	Fujairah (1)	Education (1)
Other (2)	Sharjah (2)	Petroleum (1)
		Other industry (2)

Note: The number of responses in this table does not sum to the total number of participants, 73. Some participants chose not to answer these questions on the exit survey, and others left the workshops before the forms were distributed

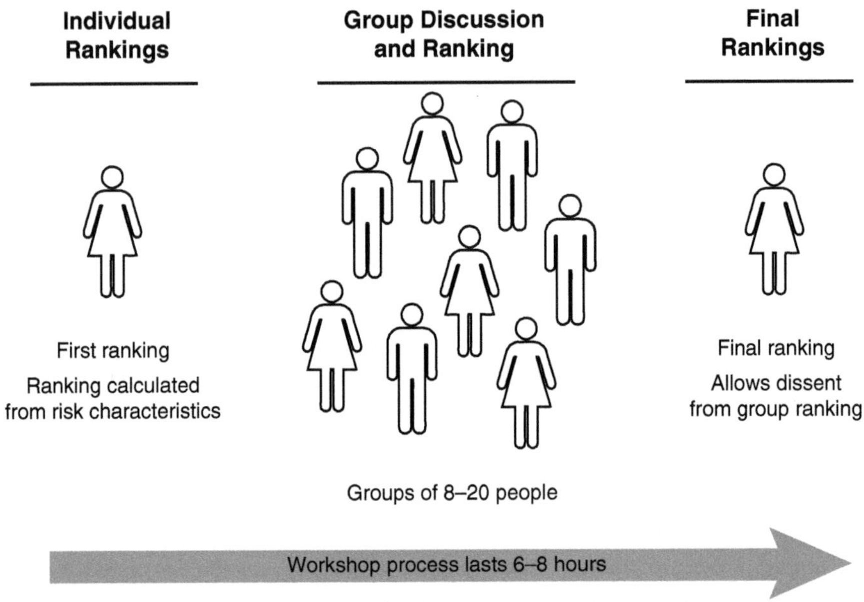

Fig. 2.5 Overview of the process used during the risk-ranking workshops

The agenda used throughout the workshops, depicted in Fig. 2.5, was designed to elicit informed and reliable measures of participants' concerns. Throughout each six- to eight-hour workshop, participants were asked to provide several rankings of the 14 risks that reflected how concerned they were about the risks. The workshops provided opportunities for people to familiarize themselves with scientific information about the risks by reading the risk summary sheets before providing a first ranking. Then, they were assisted in developing rankings based on their levels of concern about the risk attributes. Later, participants had the opportunity to learn from each other about the risks through group discussions and group ranking of the risks.

To avoid problems associated with group process leading to forced consensus, the workshops concluded by allowing participants to provide a final individual ranking. This allowed participants to adjust their first ranking to reflect new insights

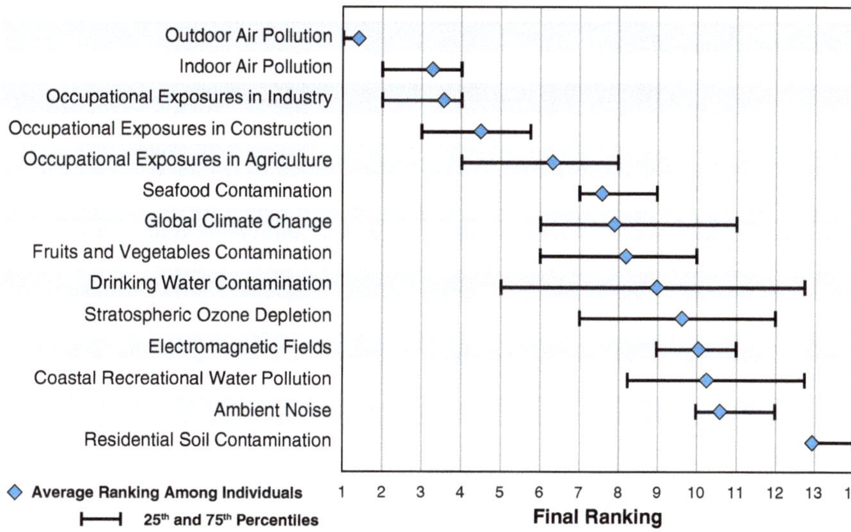

Fig. 2.6 Final rankings of 14 environmental health hazards by 56 stakeholders in environmental health in the UAE

they gained about the risks or their preferences through the structured ranking or group discussion processes. Finally, to aid in interpretation of results, participants were asked to answer several questions about their views on the process and the rankings that resulted.

Step E: Describe the Resulting Rankings in the UAE

The rankings obtained from the workshops provide insights into which environmental health risks residents of the UAE are most concerned about and why. Figure 2.6 summarizes these results by displaying the average results of the 56 participants[2] who submitted final rankings.

The workshop results demonstrate substantial agreement about which environmental health risks are of most concern in the UAE. Risks from outdoor air pollution, indoor air pollution, and occupational exposures were consistently ranked as being of greater concern than other risks.

Producing consensus was not the goal of the risk-ranking workshops. The different values and interests of participants can lead to valid disagreements about which risks are of greatest concern. However, disagreements could also be the result of confusion or misunderstandings about definitions of environmental health risks or their consequences. Thus, effective risk communication can lead to greater agreement

[2]Because some individuals were unable to participate in the entire eight-hour workshop and the recorded rankings of others erroneously omitted some risks, this analysis could not be conducted on rankings from 17 of the original 73 participants.

Table 2.5 Agreement among individuals' first and final rankings as measured through mean pairwise correlations of results

	Mean pairwise correlation among individuals' rankings						
	Current study					Previous studies	
Ranking step	Group A	Group B	Group C	Group D	Group E	Willis et al. (2004)	Morgan et al. (2001)
Agreement among first rankings	0.45	0.52	0.35	0.45	0.17	0.39	0.59
Agreement among final rankings	0.70	0.87	0.45	0.93	0.79	0.87	0.86
p-values for test that mean correlation among final rankings is larger than among first	<0.0001	<0.0001	0.22	<0.0001	<0.0001	0.0102	<0.0001

by reducing confusion and misunderstanding. One goal of the workshops was to increase participants' knowledge of risks and eliminate misunderstanding or misconceptions. The average correlations among participants' rankings within a group provide a measure of agreement among participants. These data suggest that agreement increased through the ranking workshops for each of the groups (see Table 2.5).

Comments made during the group discussion phase of the workshops revealed reasons for agreement about the level of concern about these risks. In discussions, participants consistently said that outdoor and indoor air pollution were high-priority risks because all residents in the UAE are exposed, individuals can do little to avoid exposure (especially for outdoor air), and estimates of these risks placed them among the leading contributors to the environmental burden of disease. Some participants noted that indoor air pollution risks can be avoided by not smoking tobacco, while others mentioned that the uncertainty about indoor air pollution risks is high because of the lack of UAE-specific data on indoor air quality. (UNC and UAE University recently completed a study to measure indoor air quality in 700 Emirati homes that will improve understanding of this risk.)

Discussions of occupational risks suggested that participants' concerns were motivated by a different set of characteristics. In these cases, concerns were less motivated by the estimates of overall burden of disease and more motivated by the higher levels of risk to which each segment of the workforce is exposed. Compounding this issue, participants noted that while mechanisms exist for workers to protect themselves from occupational exposures, workers in the UAE frequently lack awareness of the risks and proper use of personal protective technologies and work practices for avoiding hazardous exposures.

The risk-ranking results reveal similar levels of agreement about risks that were consistently viewed as low priority. These risks included stratospheric ozone depletion, electromagnetic fields, coastal water pollution, ambient noise, and residential soil contamination. In each of these cases, participants noted through discussions about some of the risks that the scientific evidence about the existence of the risks is weak (e.g., risks from electromagnetic fields), that exposures to the risks are easily avoidable (e.g., using sunscreen to avoid UV exposures or avoiding swimming in the ocean at certain times), that there is little evidence of exposure (e.g., residential soil contamination), or that the consequences are relatively minor and treatable (e.g., exposures to coastal water pollution or ambient noise).

Finally, discussions also provided insight into reasons for agreement and disagreement about the relative ranking of risks from environmental exposure. Participants disagreed most about the relative risks associated with drinking water. Individuals who viewed drinking-water risks as being of great concern frequently noted the importance of clean water in the UAE, where water resources are scarce. They also pointed out that many UAE residents drink bottled water rather than treated water from the tap, in part because of a lack of awareness about the quality of treated drinking water and belief that reports of the high quality of the drinking water may not be consistently reliable. Others viewed risks from drinking water to be of low concern, frequently citing both the high level of treatment that drinking water receives at desalination plants and the relatively low incidence in the UAE of diseases that are frequently associated with poor drinking water quality, such as cholera and dysentery.

Assessing the Quality and Level of Support for the Ranking Results

When considering whether and how to use the results of the risk-ranking workshops to shape strategic plans for managing environmental health risks, it is important to ask whether the workshops achieved their objective of eliciting informed, reliable judgments of participants' concerns. At the same time, it is important to ask whether participants viewed the workshops as achieving this objective and would support using the results of the workshops in further planning efforts. Answers to these questions can be found in responses that participants provided to evaluation surveys and inferences drawn from the ranking results.

Assessing Whether Rankings Represent Informed Judgments

The workshop process described in this chapter was designed to supplement participants' knowledge about environmental health risks in several ways. The participants were provided with concise summaries about the risks, guided through a structured ranking process to help them better understand the characteristics of the risks, and provided an opportunity to discuss the risks with other participants. Responses to the evaluation survey confirmed that each of these learning opportunities contributed to the participants' knowledge of the risks (see Fig. 2.7).

The workshops incorporated opportunities for participants to express their personal judgments about and learn about the risks by participating in a group exercise.

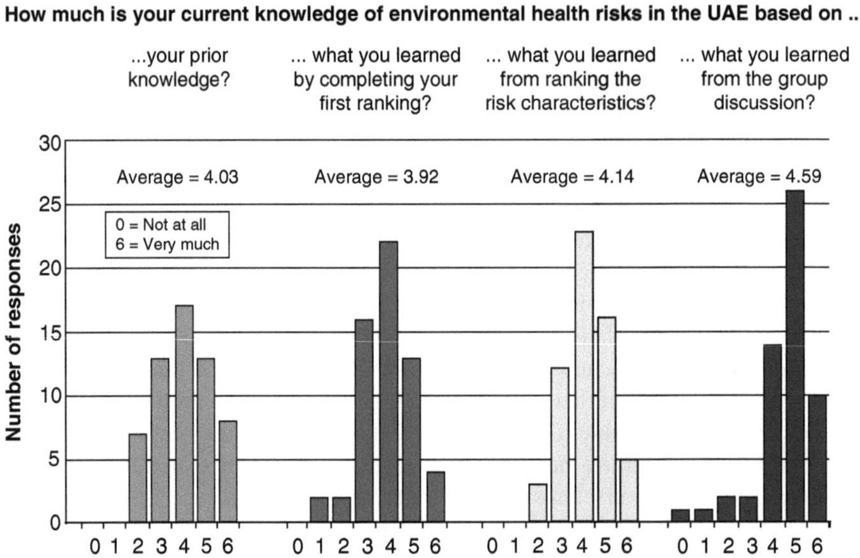

Fig. 2.7 Participants' responses to questions of how information sources contribute to their current knowledge of environmental health risks in the UAE

To the extent that participants viewed the group process as instructive, one would expect that each person's final rankings were influenced by the group rankings. This is a desirable outcome when the influence indicates participants came to a common understanding of the scope and the expected consequences of the risks. It is an undesirable outcome if the influence represents a forced consensus generated by the group discussion process.

Responses to evaluation questions suggest that participants' final individual risk rankings were influenced both by the participants' first ranking and the group ranking to which they contributed (see Fig. 2.8). By trying to predict individuals' final rankings using each participant's first and group ranking, it is possible to assess whether the rankings themselves confirm the survey responses presented in Fig. 2.8. A regression model that describes this relationship while controlling for the effect of group membership confirms that participants' individual and group rankings both were found to be predictive of their final individual ranking.[3] As further evidence that the participants learned from group discussions but did not feel that they were forced to adopt one view or another, they generally judged the group ranking portion of the workshop to be open and engaging (see Fig. 2.9).

[3]The Wald Chi-squared statistic for the model regressing first and group rankings on the final individual rankings and including group membership as a random effect variable was 101 ($n=44$; p-value 2.2×10^{-6}). Regression coefficients for the first ranking and group ranking were 0.39 and 0.65, respectively with associated standard errors of 0.08 and p-values of less than 0.01.

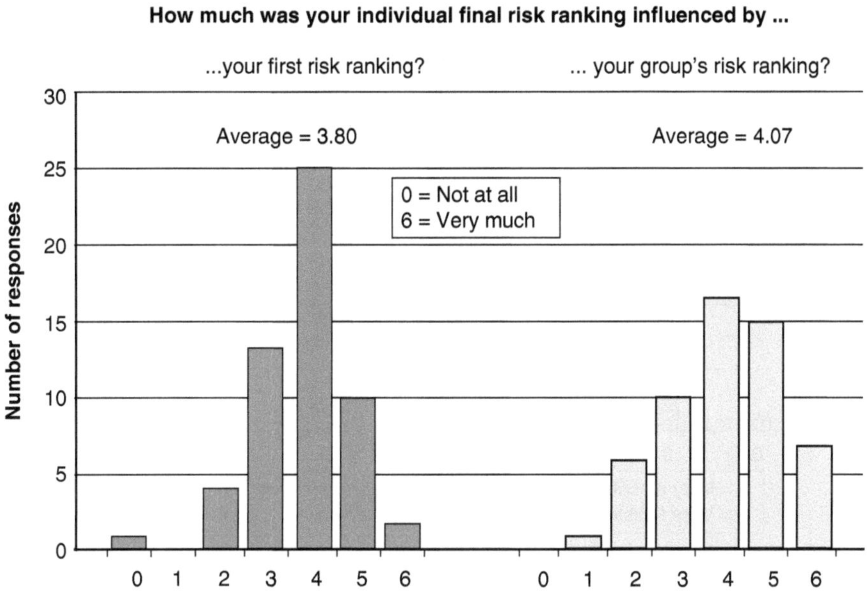

Fig. 2.8 Participants' responses to questions about whether their final individual rankings were influenced by their first ranking and their group's ranking

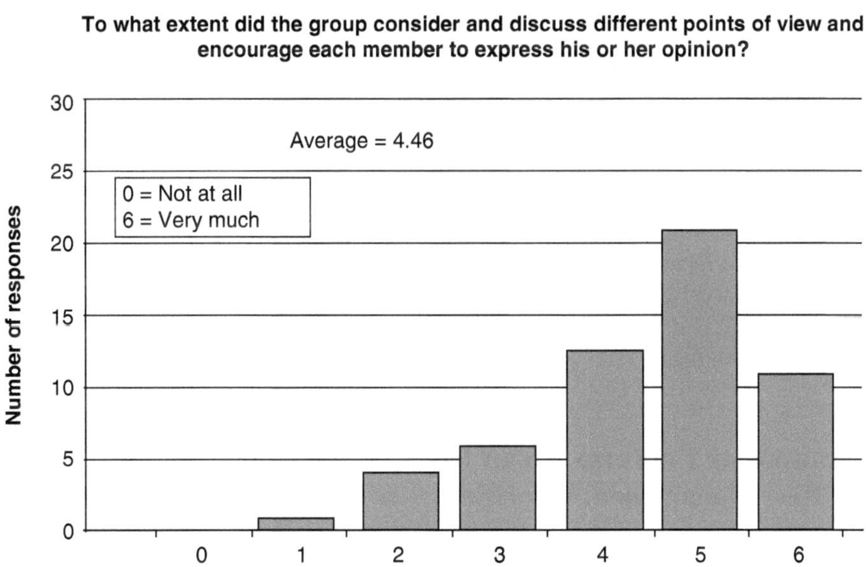

Fig. 2.9 Participants' responses to questions about whether their group considered and openly discussed different points of view during the group ranking process

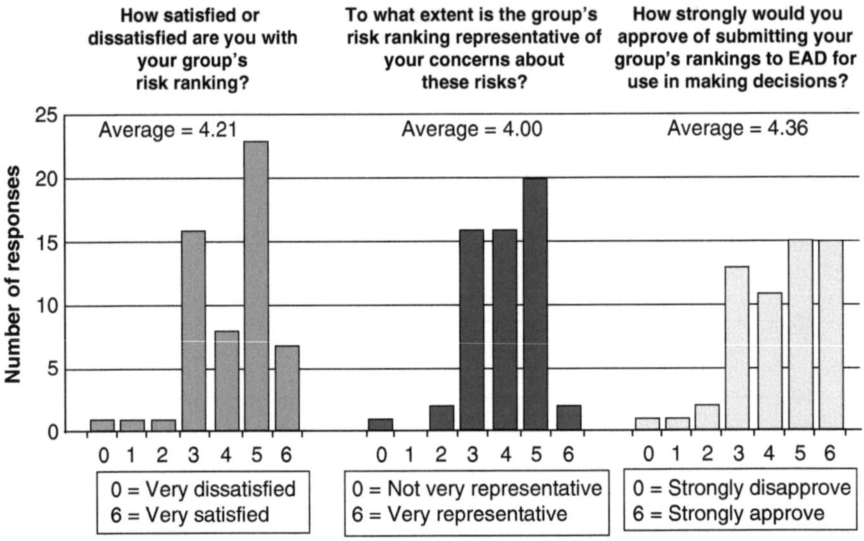

Fig. 2.10 Participants' responses to questions about their views of the results of the risk-ranking workshop

Assessing Participants' Satisfaction with the Results

Possibly the most important measure of whether the risk-ranking workshops captured informed and reliable judgments of participants' concerns about environmental health risks is the extent to which they support using the results of the workshop to inform further risk management planning and policy. To answer this question, participants were asked how satisfied they were with the group rankings, whether the group rankings were representative of their concerns, and whether they would support submitting the results from the ranking workshops to the Environment Agency–Abu Dhabi (EAD) for use in decision-making. Responses indicated that participants were generally satisfied with the group rankings and that they strongly supported EAD using the results to guide future decisions about how to manage environmental health risks (see Fig. 2.10).

Translating Concerns About Risks to Risk Management Priorities

The combined evidence about how participants viewed the stages of the ranking workshop and the results suggests that the workshops achieved the goal of capturing informed judgments about environmental health risks in the UAE. It also suggests that the results provide a sound basis upon which to set priorities for risk management strategies.

However, it is not simple to translate these assessments of concern regarding risks into risk management efforts. Simply because people agree that a risk is of low concern is not reason enough to forego efforts to manage the risks. Low-cost efforts to greatly reduce or better understand these risks may be prudent. Similarly, because people agree that a risk is of great concern is not justification of all efforts to manage that risk. If effective or affordable alternatives to manage the risks do not exist, it may be more prudent to invest in research to generate better alternatives in the future than to direct resources to costly, ineffective solutions.

For these reasons, the concerns about environmental health risks that are captured in the results presented in this chapter are one perspective on how environmental health policy priorities should be set in the UAE. In subsequent phases of this study, these assessments were integrated with views from across the UAE about the feasibility, effectiveness, and attractiveness of strategies for managing environmental health risks. As the process of implementing these strategies continues, it may be useful to revisit the concerns of UAE residents about environmental health risks. Doing so would provide an opportunity to judge how risk management efforts are being perceived, identify risks that are emerging as new concerns of residents, and broaden the outreach process to engage all of the emirates and more residents in the process of managing risks from exposures to hazards in the UAE environment.

The ranking exercise was intended to identify not only priorities for risk management but also priorities for additional, detailed risk analysis. The workshops revealed strong consensus that the three highest priority environmental health risks in the UAE are outdoor air pollution; indoor air pollution; and occupational exposures in industry, construction, and agriculture. These risks were retained for further analysis.

Upon discussion of the results from the workshops with EAD and WHO, the science team also further investigated five additional environmental hazards: climate change, drinking water contamination, coastal water pollution, soil and groundwater contamination (from solid and hazardous wastes), and contamination of food (fruits, vegetables, and seafood). EAD, the WHO, and the science team decided that these environmental health areas—although not the highest environmental health priorities for the UAE—could become increasingly important if development in the UAE continues at its current, rapid pace. Thus, the result of the ranking exercise plus the follow-up discussions with EAD and WHO was a list of eight environmental risk areas that became the focus of subsequent analysis:

1. Outdoor air pollution
2. Indoor air pollution
3. Occupational exposures in industry, construction, and agriculture
4. Global climate change
5. Drinking water contamination
6. Coastal water pollution
7. Soil and groundwater contamination due to solid and hazardous waste
8. Contamination of produce and seafood with environmental pollutants

Subsequent chapters in this report describe the science team's research to characterize the environmental burden of disease in these areas in greater detail than was possible prior to the risk-ranking exercise. Chapter 3 describes the overall method used to assess the environmental burden of disease. As the chapter explains, the method is based on a substantial body of research and resulting guidelines developed by WHO. Each of the subsequent eight chapters addresses one of the eight risk categories.

References

Briggs, D.J., R. Stern, and T.L. Tinker. 1999. *Environmental health for all: Risk assessment and risk communication for national environmental health action plans*. Dordrecht: Kluwer.

Commonwealth of Australia. 1999. National Environmental Health Strategy. Publication number 2592. Melbourne: enHealth. http://enhealth.nphp.gov.au/strategy/nehs/index.htm.

Commonwealth of Australia. 2007. National Environmental Health Strategy 2007–2012. Publication number P3-2333. Melbourne: enHealth. http://www.health.gov.au/internet/main/publishing.nsf/content/ohp-environ-envstrat.htm.

Davies, J.C. 1996. Comparative risk analysis in the 1990s: The state of the art. In *Comparing environmental risks: Tools for setting government priorities*, ed. J.C. Davies. Washington, D.C.: Resources for the Future.

DeKay, M.L., H.K. Florig, P.S. Fischbeck, M.G. Morgan, K.M. Morgan, B. Fischhoff, and K.E. Jenni. 2001. The use of public risk ranking in regulatory development. In *Improving regulation: Cases in environment, health, and safety*, ed. P.S. Fischbeck and R.S. Farrow, 208–230. Washington, D.C.: Resources for the Future.

Fischhoff, B. 1995. Ranking risks. *Risk – Health Safety and Environment* 6: 189–200.

Fischhoff, B., P. Slovic, and S. Lichtenstein. 1982. Lay foibles and expert fables in judgments about risk. *The American Statistician* 36(3): 240–255.

Florig, H.K., M.G. Morgan, K.M. Morgan, K.E. Jenni, B. Fischhoff, P.S. Fischbeck, and M.L. DeKay. 2001. A deliberative method for ranking risks (1): Overview and test bed development. *Risk Analysis* 21(5): 913–921.

Hastie, R., and R. Dawes. 2001. *Rational choice in an uncertain world: The psychology of judgment and decision making*. Thousand Oaks, Calif.: Sage Publications.

Keeney, R.L., and H. Raiffa. 1993. *Decisions with multiple objectives: Preferences and value tradeoffs*. Cambridge, U.K.: Cambridge University Press.

Lichtenstein, S., P. Slovic, B. Fischhoff, M. Layman, and B. Combs. 1978. Judged frequency of lethal events. *Journal of Experimental Psychology: Human Learning and Memory* 4(6): 551–578.

Long, J., and B. Fischhoff. 2000. Setting risk priorities: A formal model. *Risk Analysis* 20: 339–351.

Morgan, M.G., B. Fischhoff, L. Lave, and P. Fischbeck. 1996. A proposal for ranking risk within federal agencies. Chapter 6. In *Comparing environmental risks: Tools for setting government priorities*, ed. J.C. Davies. Washington, D.C.: Resources for the Future.

Morgan, K.M., M.L. DeKay, and P.S. Fischbeck. 1999. A multi-attribute approach to risk prioritization. *Risk Policy Report* 6(6): 38–40.

Morgan, M.G., H.K. Florig, M.L. DeKay, and P.S. Fischbeck. 2000. Categorizing risks for risk ranking. *Risk Analysis* 20: 49–58.

Morgan, K.M., M.L. DeKay, P.S. Fischbeck, M.G. Morgan, B. Fischhoff, and H.K. Florig. 2001. A deliberative method for ranking risks (II): Evaluation of validity and agreement among risk managers. *Risk Analysis* 21(5): 923–937.

National Academy of Sciences. 1983. *Risk assessment in the federal government*. Washington, D.C.: National Academies Press.

National Research Council. 1996. *Understanding risk: Informing decisions in a democratic society*. Washington, D.C.: National Academies Press.

Ramaswami, A., J.B. Milford, and M.J. Small. 2005. *Integrated environmental modeling: Pollutant fate, transport, and risk in the environment*. Hoboken: Wiley.

Slovic, P., B. Fischhoff, and S. Lichtenstein. 1980. Facts and fears: Understanding perceived risk. In *Societal risk assessment: How safe is safe enough*, ed. R. Schwing and W. Albers Jr.. New York: Plenum Press.

Smith, K.R., C.F. Corvalán, and T. Kjellstrom. 1999. How much global ill health is attributable to environmental factors? *Epidemiology* 10(5): 573–584.

U.S. Army Corps of Engineers. 2008. Risk-informed decision framework appendix, draft. Louisiana coastal protection and restoration technical report. U.S. Army Corps of Engineers, New Orleans District, Mississippi Valley Division. http://lacpr.usace.army.mil/\Report\Draft Appendices\Risk Informed Decision Framework Appendix.pdf. February.

U.S. Environmental Protection Agency (EPA). 1987. Unfinished business: A comparative assessment of environmental problems. EPA Number: 230287025a. NTIS PB88-127048. Alexandria, Va.: National Technical Information Service.

U.S. Environmental Protection Agency (EPA) Science Advisory Board. 1990. Reducing risk: setting priorities and strategies for environmental protection. Relative Risk Reduction Strategies Committee. Report SAB-EC-90-021. Washington, D.C.: U.S. EPA.

Webster, K.D., C.G. Jardine, L. McMullen, and SB. Cash. 2008. Risk ranking: Investigation expert and public differences in evaluating food safety risks. Project Report. Research Project Number CMD-08-02. Edmonton: University of Alberta, Department of Rural Economy.

Willis, H.H., M.L. DeKay, M.G. Morgan, H.K. Florig, and P.S. Fischbeck. 2004. Ecological risk ranking: Development and evaluation of a method for improving public participation in environmental decision making. *Risk Analysis* 24: 363–378.

Chapter 3
Assessing the Environmental Burden of Disease: Method Overview

Abstract The purpose of environmental burden of disease (EBD) studies is to assess what fraction of the global, national, or regional burden of disease is attributable to selected environmental risks, using an explicit, widely recognized methodology. The method used to estimate the EBD in the United Arab Emirates is based on a method developed in the 1990s by the World Health Organization in the first global burden of disease study. The approach is based on determining the attributable fraction—the proportion of death or disability attributable to a specific risk (e.g., air pollution) or health condition (e.g., high blood pressure). To estimate the environmental burden of disease in the UAE, the research team that conducted this study constructed an innovative computer model, the *UAE Environmental Burden of Disease Model,* coded in *Analytica* software. The model, the first of its kind, is designed to facilitate comparing the importance of different risks and testing the effects of various environmental interventions on the UAE's overall disease burden. The model is divided into subcomponents, each corresponding to one of the eight environmental risk areas retained for analysis as a result of the priority-setting exercise described in Chap. 2. This chapter describes the principles underlying the model, based on steps including exposure assessment, determination of the exposure-response relationship, estimation of mortality and morbidity, calculation of the attributable fraction, determination of the disease burden attributable to the risk, and uncertainty and sensitivity analysis. Estimation of the burden of disease in the UAE with an easy-to-understand computer model is a state-of-the-art method for analyzing the fragmentary data that were available on the disease distribution in the UAE and for communicating the results effectively. This innovative model allows comparison of the relative importance of various sources of ill health and examination of the effects of alternative interventions on the disease burden. The model also makes it very easy to update future burden of disease estimates when new data become available, and it allows UAE officials to test the effect of various intervention options. Because all assumptions, decisions about input variables, and specific methods are clearly stated in each step, changes to the model structure can be made easily should future research and new data prove it necessary.

J. MacDonald Gibson et al., *Environmental Burden of Disease Assessment*,
Environmental Science and Technology Library 24, DOI 10.1007/978-94-007-5925-1_3,

Because resources are always limited, the model can facilitate identification of the most important risks and prioritize competing actions to recognize the ones with the greatest potential to reduce the burden of disease.

Keywords Attributable fraction • UAE Environmental Burden of Disease Model • Global environmental burden of disease framework • World Health Organization • Exposure assessment • Exposure-response relationship • Estimation of mortality and morbidity • Uncertainty and sensitivity analysis • Disability-adjusted life years (DALYs)

The Purpose of Environmental Burden of Disease Studies

Accurate and systematically analyzed information on the distribution of death and disability in the population and the potential causes and risks for various health conditions is crucial for health policy and planning in any country. Unfortunately, this information is often fragmented and incomplete and does not easily lend itself to making consistent comparisons regarding the relative importance of different diseases and risks. In addition, the lack of usable data related to certain health conditions and risks may lead those conditions and risks to be inadvertently disregarded in decision-making, even though they may be responsible for a significant portion of the overall burden of disease. The purpose of environmental burden of disease (EBD) studies is to assess what fraction of the global, national, or regional burden of disease is attributable to selected environmental risks, using an explicit, widely recognized methodology.

The method used to estimate the EBD in the United Arab Emirates is based on a method developed in the 1990s by the World Health Organization (WHO) in the first global burden of disease (GBD) study. The goal of the study was to introduce a universal approach for analyzing the often incomplete and inconsistent information on the distribution of death and disability in populations. The approach is based on determining the attributable fraction (AF)—the proportion of death or disability attributable to a specific risk (e.g., air pollution) or health condition (e.g., high blood pressure). The formal EBD approach is based on a well-established framework derived from the principles of modern epidemiology (Prüss-Üstün et al. 2003). This framework allows comparisons of the burden of disease arising from different risks among different regions and populations. Since its introduction, the GBD framework has been widely accepted as the preferred method for assessing national burdens of disease. Several practical guides are available to assist in the correct use of the method (Mathers et al. 2001; WHO 2003–2007).

To estimate the environmental burden of disease in the UAE, the research team that conducted this study constructed an innovative computer model, the *UAE Environmental Burden of Disease Model,* coded in *Analytica* software (version 4.1, Lumina Decision Systems). The model, the first of its kind, is designed to facilitate comparing the importance of different risks and testing the effects of various environmental interventions on the UAE's overall disease burden. The model is divided into subcomponents, each corresponding to one of the eight environmental risk areas retained for analysis as a result of the priority-setting exercise described in Chap. 2. This chapter describes the principles underlying the model. Subsequent chapters present details for each subcomponent. Appendix B contains instructions for using the model.

History of the Global Burden of Disease Framework

Initial 1990 Global Burden of Disease Study

The need for an explicit framework to comparatively estimate the relative importance of diseases, injuries, and risks that cause premature death and disability was addressed in 1992 when the initial global burden of disease study was commissioned by the World Bank (Lopez et al. 2006). This study, carried out in collaboration with WHO and the Harvard School of Public Health, assessed the GBD for ten risks, using data from 1990. It was the first study to produce comprehensive estimates of mortality and morbidity by age and gender, both globally and for eight regions of the world. The main improvements over earlier studies that had attempted to estimate the global burden of disease (Hakulinen et al. 1986; Lopez 1993) were that the 1990 GBD study included 100 specific causes of death, as opposed to the broad categories used previously, and that it addressed nonfatal health outcomes in addition to mortality (Lopez et al. 2006). The 1990 GBD estimate also included low and middle income countries for which limited data were available, as well as diseases for which data involved substantial uncertainty. The study also incorporated methods to assess the reliability of input data (Mathers et al. 2006a).

One of the main features of the original GBD study was the introduction of a new metric, the disability-adjusted life year (DALY). The DALY is a time-based summary measure of population health that combines the effect of premature death and disability into one metric that can be used to compare the burden of disease across different regions and populations. Quantification of nonfatal health outcomes in the 1990 study was important because the study revealed that certain conditions, such as neuropsychiatric conditions and injuries, were major contributors to the disease burden when measured in DALYs but were grossly underestimated when only mortality was taken into account (Mathers et al. 2006a). The methods and results of the GBD 1990 study have been discussed in several journal articles (Murray and Lopez 1996a, 1997a, b, c, d) and other publications (Murray and Lopez 1996b, c, 1998).

Improvements to the Approach

The 1990 GBD study has been updated since its publication. Subsequent GBD studies featured major improvements in methodology and covered more risks and health outcomes. The next version, the 2000 GBD study (WHO 2002), incorporated comparative risk assessment, a systematic evaluation of the changes in population health that would result from modifying the exposure distribution for a particular risk or group of risks, relative to other risks using the attributable fraction approach (Lopez et al. 2006). The study assessed the burden of disease attributable to five environmental and five occupational risks among a total of 26 environmental, occupational, social and behavioral risks, and quantified the disease burden for 135 major causes or groups of causes. Inclusion of as many causes of disease as possible is important because the exclusion of disease causes due to data limitations easily translates into "no burden" and results in a misleading picture of population health (Mathers et al. 2002). To provide more definition, individual countries were used as a unit of measurement. Approaches for nations in different stages of health development, based on categorization by data availability, were standardized, which improved the comparability of estimates across populations (Lopez et al. 2006). Uncertainty and sensitivity analyses were incorporated into the method in an attempt to systematically quantify some of the uncertainty associated with both national and global estimates of the disease burden (Lopez et al. 2006; Mathers et al. 2006b). The GBD studies, most recently updated for the year 2004 (WHO 2004), have produced information that can be used in national burden of disease studies when local data are unavailable or incomplete. The studies project alternative scenarios of mortality and morbidity over the next 30 years by cause, age, gender, and region (Mathers et al. 2002). The WHO and the Institute for Health Metrics and Evaluation at the University of Washington, together with other academic partners, are currently working on a new GBD study in an effort to update previous global disease burden estimates (WHO 2010; Institute for Health Metrics and Evaluation 2007). The goal of this study is to produce more accurate disease burden estimates and to assess trends since 1990 by using improved methodology and new health data, particularly from developing countries.

National Burden of Disease Studies Using the Global Burden of Disease Framework

National burden of disease studies are becoming increasingly common as countries attempt to prioritize health interventions. Financial and human resources limitations often preclude combatting every disease to the fullest possible extent. Hence, information produced by a burden of disease study can help prioritize health conditions for which interventions may yield the greatest gains in health.

The first national burden of disease studies using the GBD approach were carried out in the 1990s and early 2000s, after the initial 1990 GBD study. The first countries to estimate the national or regional burden of disease included Australia (Mathers et al. 1999); Mauritius (Vos et al. 1995); Mexico (Lozano et al. 1995); the state of Andhra Pradesh, India (Mahapatra 2002); Thailand (Bundhamcharoen et al. 2002); South Africa (Bradshaw et al. 2003); Turkey (Baskent University 2005); and the United States (Michaud et al. 1996). Burden of disease studies have been conducted in several European countries as well, including The Netherlands (Melse et al. 2000), Serbia (Jankovic et al. 2007), and France (Lapostolle et al. 2008), as well as other countries around the world such as Chile (Concha-Barrientos et al. 1996), New Zealand (Pakari and Roa 1999), Egypt (Egypt Ministry of Health and Population 2004) and Canada (Public Health Agency of Canada 2006). Some countries, including Australia (Begg et al. 2007) and Mexico, have already repeated their studies and updated their original disease burden estimates. In addition to the national-level studies, some countries, including Australia (Begg et al. 2007), Mexico (Stevens et al. 2008), and Iran (Naghavi et al. 2009), have estimated the disease burden on a subnational level. This is particularly useful when environmental and health policy decisions are made on the state or provincial level rather than on the government level. In addition, different regions of a country may have very different patterns of disease, particularly in middle-income countries (Stevens et al. 2008).

National burden of disease studies have been organized more commonly by disease, not by risk. These studies attempt to estimate the relative impact of various health conditions to the total disease burden by including most of the diseases or disease groups within the country, often ranging from approximately 100 to 200 or more health conditions. Fewer studies have examined the burden of disease by risk. The current study in the UAE is the first in the world for which a comprehensive computer model was developed to quantitatively estimate the disease burden resulting from pollutants in multiple environmental media. It also is the first to apply, comprehensively, WHO guidance documents for estimating the environmental burden of disease for a broad range of environmental risk factors.

Estimating the Environmental Burden of Disease in the UAE Step by Step

The General Method for Assessing Environmental Burden of Disease

The method for estimating the environmental burden of disease in the UAE closely follows the attributable fraction approach used by the WHO in its GBD estimates. To assist countries in carrying out national and local EBD studies using the formal framework, the WHO has published a series of practical documents that provide guidance in estimating the disease burden related to several environmental risks, including outdoor air pollution, climate change, and occupational exposures (WHO 2003–2007). The first guide in the series introduces the GBD concept and the general methodology for estimating the EBD (Prüss-Üstün et al. 2003). The other guides focus on specific environmental risks, providing information on data needed, a step-by step method for each risk, and numerical examples. Box 3.1 lists the risks the series covers (Prüss-Üstün et al. 2003; WHO 2003–2007).

Box 3.1 Risk Factors Covered by WHO's Environmental Burden of Disease Series

- Outdoor air pollution
- Indoor smoke from solid fuels
- Lead and mercury
- Water, sanitation, and hygiene
- Climate change
- Occupational factors
- Airborne particulate matter
- Carcinogens
- Ergonomic stressors
- Injuries
- Noise (including community noise)
- Sharps injuries in health-care workers
- Malnutrition and poverty
- Solar ultraviolet radiation
- Recreational water quality
- Fluoride, arsenic, and nitrates in drinking water
- Selected risk factors and injuries in European children and adolescents

The approach (Prüss-Üstün et al. 2003) for assessing the EBD associated with each of the selected risks in the UAE is based on the following steps, discussed in more detail below:

1. **Exposure assessment** (estimating exposure to the environmental risk within the UAE population)
2. **Determination of the exposure-response relationship** (for the particular risk)
3. **Estimation of mortality and morbidity** (collecting estimates of total mortality and morbidity in the UAE for the selected health conditions)
4. **Calculation of the attributable fraction** (AF)
5. **Determination of disease burden attributable to the risk** (calculated by multiplying the total disease burden by the AF)
6. **Uncertainty and sensitivity analysis**

Exposure Assessment

Exposure assessment starts with the selection of specific exposure indicators for each risk. For example, for the disease burden related to outdoor air pollution in the UAE, the indicators considered in this study include the concentrations of particulate

Table 3.1 Exposure indicators and adverse health conditions for each risk considered in this study

Risk	Exposure indicators	Adverse health conditions
Outdoor air pollution	PM_{10}, daily average (μg/m^3)	All-cause mortality (all ages) and respiratory mortality (<5 years)
		Respiratory and cardiovascular morbidity (all ages)
	$PM_{2.5}$, annual average (μg/m^3)	All-cause, cardiopulmonary, and lung cancer mortality (>30 years)
	Ground-level ozone, daily (24-h) average (ppb)	Total nonaccidental, cardiovascular, and respiratory mortality (all ages)
		Respiratory morbidity (all ages)
	Ground-level ozone, annual average of daily maximum concentration (ppb)	Respiratory mortality (>30 years)
Indoor air pollution	PM_{10}, $PM_{2.5}$	Asthma (<5 years)
	Benzene, formaldehyde	Asthma (<3 years)
	Radon	Lung cancer
	Environmental tobacco smoke (ETS)	Lung cancer and lung cancer mortality, leukemia, cardiovascular disease and cardiovascular disease mortality, asthma (<18 years), lower respiratory tract infection (<6 years)
	Bio-aerosols (mold)	Childhood (6–12 years) and adult asthma
	Incense use	Respiratory tract cancer and respiratory tract cancer mortality
Occupational exposures	Carcinogens[a] and leukemogens[b]	Lung cancer, leukemia, malignant mesothelioma
	Particulate matter	Asthma, chronic obstructive pulmonary disease, asbestosis, silicosis
	Noise	Noise-induced hearing loss
Climate change	Heat exposure	Cardiovascular disease
Drinking water contamination	Disinfection by-products	Bladder, rectal, and colon cancer
	Microbial contamination	Gastroenteritis
Coastal water contamination	Microbial contamination	Gastroenteritis
Food contamination	Methylmercury in seafood	Neurological disorders
	Pesticides in fruit and vegetables	Pesticide poisoning

[a]Arsenic, asbestos, beryllium, cadmium, chromium, diesel exhaust, nickel, silica
[b]Benzene, ethylene oxide

matter (daily average for PM_{10} and annual average for $PM_{2.5}$) and ground-level ozone (daily average and annual average of the daily maximum). Exposure indicators for risks covered in this study and the related health conditions are listed in Table 3.1.

The next part involves estimating the distribution of exposure in the population. For example, exposure to low and high levels of carcinogens in occupational settings in the UAE is estimated in reference to the relevant Permissible Exposure Limit (PEL). If it is not feasible to establish a numerical relationship between the exposure and the related health outcome, the population can be divided into defined exposure scenarios (Prüss-Üstün et al. 2003). One example is the assessment of occupational exposure to particulate matter, which divides the population into occupational groups with corresponding health risks. The exposure distribution can be either continuous (such as the probability of being exposed to a specific concentration of contaminants) or discrete (such as the probability of high, medium, and low exposure levels).

In an ideal case, the exposure data should be derived from local measurements. If reliable and representative local data are not available, data from studies performed in countries where exposure scenarios are likely to be similar can be used. For the UAE, limited exposure information was available for many of the selected risks. For other risks, data had to be derived from international studies and extrapolated to the UAE, using appropriate assumptions about the representativeness of the data for the UAE population. Different exposure variables, the related probability distributions, and the sources of data used to derive the estimates (whether international or local) are listed in Appendix C and described in more detail in subsequent chapters.

Determination of the Exposure-Response Relationship

The second step includes selecting health outcomes to be included in the analysis and determining the relative risk for the exposure. Relative risk, i.e., the ratio of the probability of a given health outcome occurring in an exposed population versus a nonexposed population, can be determined from a systematic review of the epidemiologic literature. Usually results from other populations can be applied to the local population, but if there is a strong reason to believe that the information from the literature does not apply, then local data should be used (Prüss-Üstün et al. 2003). In practice, all relative-risk information used in the UAE study was derived from the international epidemiologic literature because local environmental epidemiologic studies have not been completed. Regardless of the source of the data, exposure indicators and exposure limits used in the epidemiologic study should correspond with the exposure measures used in the exposure assessment step. Values and probability functions related to the relative-risk estimates used in the study, including the sources of data, are presented in Appendix C.

Estimation of Mortality and Morbidity

Mortality and morbidity in the general population need to be estimated in order to estimate the burden of disease related to each risk. Data on mortality and morbidity can be gathered from national records such as vital registration systems, censuses, health examination surveys, and disease registries. If these are unavailable or incomplete, data can be derived from studies conducted in other countries, international epidemiologic studies, or regional or country-specific estimates from the WHO.

The International Classification of Diseases (ICD) is the most widely used standard to classify diseases and health-related conditions in the word. The ICD coding system, published by WHO, was designed to facilitate international comparability of mortality and morbidity statistics. The system is used by countries to compile and report basic health statistics, and it has many applications in clinical medicine, epidemiologic research, and health management. The most recent revision, ICD-10, has been in use since 1994 (WHO 2009). The previous edition, ICD-9, was published in 1977.

In this study, mortality and morbidity data related to each risk were requested from the UAE according to ICD-9 and ICD-10 coding. This research used health data obtained from the Health Authority–Abu Dhabi (HAAD). HAAD is the main agency for promoting public health in Abu Dhabi. HAAD provided four categories of previously collected health outcome data for the purposes of this study: mortality, morbidity, birth, and cancer. Except for data regarding cancer and certain birth outcomes, the data were provided as lists of individual patient encounters with a few descriptive parameters. Out of concern for medical privacy and confidentiality of the patients, patient identifiers were removed and the data securely stored.

Disease-specific morbidity data were available for only 73% of the Abu Dhabi population and had to be extrapolated from this data set for the other emirates (HAAD 2009). Diseases included in this study and corresponding ICD codes and baseline rates are listed in Tables 3.2 and 3.3.

To facilitate comparison of the disease burden across various diseases, risks, and populations, several summary measures of population health that combine the impact of death and disability into one metric have been proposed. Health expectancy metrics, which estimate the average time in years that a person could expect to live in a certain state of health, include, for example, health-adjusted life expectancy, disability-free life expectancy, and disability-adjusted life expectancy. The other category of population summary measures is health gap metrics, which measures the difference between the actual observed population health and some ideal or reference status. These include, for example, quality-adjusted life years, health-adjusted life years, healthy life years, and disability-adjusted life years (DALYs) (Mathers et al. 2001). The best known health gap measure is the DALY, proposed by Murray and Acharya (1997) and used in the formal EBD framework since the first GBD study, calculated as shown (Prüss-Üstün et al. 2003):

$$DALY = YLL + YLD \tag{3.1}$$

Where:

YLL = years of life lost due to premature death
YLD = years lived with disability

Table 3.2 Diseases included in the study, the corresponding ICD-10 codes, and number of annual deaths at baseline in the data received for Abu Dhabi emirate

Environmental risk	Cause of death	ICD-10 code(s)	Number of deaths, Abu Dhabi emirate[a]	Estimated number of deaths, UAE[b]
Outdoor air pollution[c]	All-cause mortality	N/A	2,949	8,865
	Cardiopulmonary disease (adults >30)	J44	10	30
	Respiratory disease (adults >30)	J00–99	68	203
	Respiratory disease (children <5)	J00–99	9	27
	Lung cancer (adults >30)	C34	38	113
	Cardiovascular and respiratory mortality	I00–79, J00–99	848	2,550
Indoor air pollution	Cardiovascular disease	I00–79	769	2,310
	Lung cancer	C33–4	40	120
	Respiratory tract cancer	C33–4[d]	40	120
Occupational exposures	Asthma	J45	3	10
	Asbestosis	501	0	0
	Chronic obstructive pulmonary disease	J44	12	37
	Leukemia	C91–5	43	130
	Lung cancer	C33–4	40	120
	Malignant mesothelioma	C45	2	7
	Silicosis	502	0	0
Climate change	Cardiovascular disease	I00–79	769	2,310
Drinking water contamination	Bladder cancer	C67, C68	8	23
	Colon cancer	C18	27	80
	Gastroenteritis	A00–9	2	7
	Rectal cancer	C19–21	10	30

[a] Number of deaths in 2008 (HAAD 2009)
[b] Extrapolated to the entire UAE population from Abu Dhabi emirate mortality data
[c] Outdoor air figures based on UAE Ministry of Health (2008) report, not HAAD data
[d] Lung cancer data was used for respiratory tract cancer calculations due to the lack of an ICD code specific to respiratory tract cancer

The YLL component is defined as the number of deaths at each age multiplied by the standard life expectancy at the age of death, whereas YLD is calculated by multiplying the number of incident cases in the population by disability weight and average duration of disability. YLL and YLD can be calculated as (Prüss-Üstün et al. 2003):

$$YLL = N(L) \tag{3.2}$$

Table 3.3 Diseases included in the study, the corresponding ICD-9 codes, and number of annual health-care facility visits at baseline in the data received for Abu Dhabi emirate

Environmental risk	Cause of visit	ICD-9 code(s)	Number of health-care facility visits, Abu Dhabi data	Estimated number of health-care facility visits, UAE[a]
Outdoor air pollution[b]	Cardiovascular disease	390–448	132,021	307,667
	Respiratory diseases	480–6; 490–7; 507	92,271	176,048
Indoor air pollution	Asthma (≥18 year old)	493	10,774	32,388
	Asthma (6–12 year old)	493	2,117	6,363
	Asthma (≤6 year old)	493	4,617	13,879
	Cardiovascular disease	390–448	135,021	307,667
	Leukemia	204–208.9	464	1,520
	Lower respiratory tract infection (≤6 year old)	480–92	4,656	13,996
	Lung cancer	162	133	444
	Respiratory tract cancer	162[c]	133	444
Occupational exposures	Asthma (all ages)	493	24,052	72,301
	Asbestosis	501	1	3
	Chronic obstructive pulmonary disease	490–2, 494, 496	8,812	27,213
	Leukemia	204–208.9	464	1,520
	Lung cancer	162	133	444
	Malignant mesothelioma	163	8	28
	Noise-induced hearing loss	388.12	0	29,788[d]
	Silicosis	502	3	8
Climate change	Cardiovascular disease	390–448	135,021	307,668
Drinking water contamination	Bladder cancer	188	275	930
	Colon cancer	153	671	2,191
	Gastroenteritis	008–9, 558.9	28,230	81,100
	Rectal cancer	154	196	639
Coastal water pollution	Gastroenteritis[e]	008–9, 558.9	28,230	81,100

[a] Based on partial Abu Dhabi health-care facility visit data

[b] Outdoor air figures based on UAE Ministry of Health (2008) report, not HAAD data

[c] Lung cancer data was used for respiratory tract cancer calculations due to the lack of an ICD code specific to respiratory tract cancer

[d] Noise-induced hearing loss estimates use prevalence rates from Mathers et al. (2000), not HAAD data

[e] The underreporting of gastroenteritis in health records is common, as those afflicted often do not seek medical attention (Palmer et al. 1997). Thus, the ***Coastal Water*** module models one scenario with the baseline health data from HAAD (above) and another scenario with increased gastroenteritis incidences

Where:

N = number of deaths

L = standard life expectancy at age of death, in years

$$YLD = I(DW)(L) \tag{3.3}$$

Where:

I = number of incident cases
DW = disability weight
L = average duration of illness

The equations for *YLL* (3.4) and *YLD* (3.5) below expand the basic formulas by taking into account various social preferences such as age weighting, time discounting, and disability weighting, as discussed below. Values of r = 0.03, K = 1, and β = 0.04 were used in the initial 1990 GBD study (Murray and Lopez 1996b, adapted from Fox-Rushby and Hanson 2001).

$$YLLs[r,K,\beta] = \frac{KCe^{ra}}{(r+\beta)^2}\left\{e^{-(r+\beta)(L+a)}[-(r+\beta)(L+a)-1] - e^{-(r+\beta)a}[-(r+\beta)a-1]\right\} + \frac{1-K}{r}(1-e^{-rL}) \tag{3.4}$$

The formula for *YLD*[*r*; *K*, *β*] differs only in the addition of *D* (the disability weight):

$$YLDs[r,K,\beta] = D\left\{\frac{KCe^{ra}}{(r+\beta)^2}\left\{e^{-(r+\beta)(L+a)}[-(r+\beta)(L+a)-1] - e^{-(r+\beta)a}[-(r+\beta)a-1]\right\} + \frac{1-K}{r}(1-e^{-rL})\right\} \tag{3.5}$$

Where:

K = age weighting modulation factor
C = constant
r = discount rate
a = age of death
β = parameter from the age weighting function
L = standard expectation of life at age a
D = disability weight

Estimating YLL only requires population-level information on the number of deaths and the age at death, but estimating YLD is much more complex. The data required to assess the disability component include disease incidence, duration of the disease, age at onset, and distribution of disease by severity class. All of this information is needed by age and gender. Estimates of incidence, remission, and case-fatality rates or relative risks are also required by age and gender (Mathers et al. 2001).

Numerous questions have been raised about combining the disease burden arising from death and disability into a summary measure, starting with what is the best metric to use. Expressing disease burden in DALYs has been controversial because it requires several decisions about social values. The social choices regarding age weighting and time discounting have garnered the most criticism (Mathers et al. 2001; Lopez et al. 2006; Anand and Hanson 1997). Age weighting means that a year of healthy life lived is valued differently at different ages. In the GBD framework, a year of life lived is weighted lower at young and old ages compared with other ages (Mathers et al. 2001). Time discounting, on the other hand, rates the value of healthy life gained now versus in the future. For example, if a 3% time discount rate used in the GBD study is applied, then a year of healthy life gained 10 years from now is worth 24% less than one healthy year gained now (Mathers et al. 2006a). Other social value choices include how severity scores for disabilities are assigned, how long healthy people should be expected to live, and whether the amount of healthy life lost at death at various ages should be considered to be the same for all populations despite the different life expectancies in different population groups (Mathers et al. 1997). The DALY method has been critically discussed by Anand and Hanson (1997), Murray and Acharya (1997), Williams (1999), and Murray and Lopez (2000).

Even though DALYs can be useful in comparing results between studies, the amount of specific data on disability in the population such as disease incidence, duration of the disease, and age at onset makes calculating DALYs a challenge. Registries exist for mortality in many countries, but information is not systematically collected for nonfatal health conditions, which hampers the estimation of the YLD component. This is also the case in the UAE. Information for nonfatal health conditions was available only as the number of visits to health-care facilities, and from this data set it was not possible to determine the incidence of disease, which is the basis for calculating YLD. The lack of data that would have been necessary for DALY calculations combined with the controversy surrounding the social value choices discussed above led to the decision to express the burden of disease as the number of deaths and the number of health-care facility visits, instead of as total DALYs, for this project. Attempting to calculate DALYs with too little information was considered an additional source of uncertainty. A recent national burden of disease study in France pointed out that the source of input data and the choice of social values may have a considerable effect on the disease burden estimates (Lapostolle et al. 2008).

Calculation of the Attributable Fraction

The proportion of death or disability attributable to the risks of interest can be expressed as the attributable fraction (AF). In some rare cases, the health outcome is directly related to a single environmental risk. For example, in the case of the fibrotic respiratory disease called silicosis, which is solely caused by exposure to silica, the AF is considered to be 100%. But most diseases have several potential causes, and the relative impact of these causes must be determined. The concept of population attributable fraction was first introduced by Levin in 1953. By definition, the attributable fraction is the proportion of disease attributable to a given exposure.

The AF is also described as the fraction of disease in a population that might be prevented if exposure to a causative agent were eliminated (Coughlin et al. 1994). The attributable fraction for health effects resulting from a specific environmental risk can be calculated thus:

$$AF = \frac{\sum_i P_i RR_i - 1}{\sum_i P_i RR_i} \quad (3.6)$$

Where:

AF = Attributable fraction
P_i = Proportion of the population at exposure category "i", including the unexposed population
RR_i = Relative risk at the exposure category "i", compared to the reference level

Since in most cases the disease burden results from exposure to a diverse mix of environmental, social, and behavioral risks, the AFs for these risks often add up to more than 1. If the AF is seen as a measure of how much ill health might be avoided if exposure to a risk did not occur, this may sound counterintuitive. However, AFs are interdependent and cannot simply be added together except under special circumstances, such as when risks do not overlap (Rowe et al. 2004). Each AF describes the change that is theoretically possible to achieve if that particular exposure were eliminated, assuming that each risk is the first to be eliminated and that the other exposures remain unchanged. However, in practice, changing one risk may cause changes in the other risks (Prüss-Üstün et al. 2003). Rowe et al. (2004) examine a hypothetical situation in which risks for a disease were eliminated in different sequences. Their study shows that for individuals who have many risk factors, it is possible to prevent disease in more than one way and these prevented cases may be counted more than one time. Smith et al. (1999) give an example of how AFs could logically add up to more than 100%, involving methods to prevent 1,000 annual deaths from car accidents. For example, assume that 20% of the deaths could be prevented if the use of headlights were required during the day, 40% if stricter speed limits were applied, 50% if more stop-lights were installed, and 90% if more speed bumps were installed. The total of these attributable fractions, 200%, indicates that some of these prevention methods overlap. Thus, if 200 lives could be saved by requiring daytime use of headlights and 400 by applying stricter speed limits, could 600 lives be saved through implementing both strategies? The answer is no because once one intervention is applied, the overall situation is altered and the potential of the other interventions to reduce the burden of disease is reduced. After the speed-limit campaign is implemented, the benefit from the headlight campaign might be reduced from 200 lives saved to perhaps only 100 because many of the people whose lives were saved by the first intervention may also have been saved by the second. Therefore, AFs adding up to more than 1 should not be incorrectly interpreted as a more-than-100% potential reduction in disease burden.

As discussed by Rockhill et al. (1998), it is important to correctly interpret and communicate the meanings of AFs when conducting a burden of disease study. Misunderstandings regarding the interpretation of the AF have occurred. For example, the AF has been confused with the proportion of cases associated with any risks or with the proportion of people exposed to a given risk. Understanding how selection of the exposure levels affects the AF is important. The limit between exposed and unexposed populations should be set to a level at which it is possible for the exposed group to be moved to the unexposed group. Otherwise, the AF does not have much value in practice when different prevention options are considered (Rockhill et al. 1998). On the other hand, if the exposure limit is set to a level at which almost the entire population is considered exposed, shifting all of this population to the background exposure category to reduce the disease burden would be unrealistic. It is also misleading to imply that the AF directly means the proportion of disease that is explained or caused by the risk. The AF is influenced by what is considered a risk in the analysis and by the selected level of exposure that differentiates exposed and unexposed people. As an extreme example, discussed by Rockhill et al. (1998), if age greater than 15 years is considered a risk for developing breast cancer, essentially all cases of breast cancer could be "explained" when comparing the rate of breast cancer in people under 15 years of age with that in people who are 15 years or older. In practice, however, most people 15 years or older do not develop breast cancer.

Despite the potential for misunderstanding regarding the meaning of AFs, the AF method is still a valid means for estimating the burden of disease. Calculating AFs makes it possible to compare the relative importance of various risk factors to the disease burden, thus facilitating the identification of risk factors and adverse health conditions responsible for the greatest burden of disease in the population. Determining AFs is also the basis for estimating what the effect on the disease burden would be if exposures to environmental hazards were lowered by implementing control measures. When using AFs to prioritize activities to reduce the disease burden, it should be kept in mind that AFs typically do not add up to 1 and that the AF assumes that each risk is the first to be eliminated, after which the potential of reducing the disease burden by eliminating the remaining risks will change (Rowe et al. 2004).

Determination of Disease Burden Attributable to the Risk

To estimate the burden of disease attributable to each risk, the observed mortality and morbidity in the general population are multiplied with the corresponding attributable fraction:

$$D_{attrib} = AF(D_{total}) \tag{3.7}$$

Where:

D_{attrib} = Attributable disease burden

AF = Attributable fraction
D_{total} = total disease burden in the population

Uncertainty and Sensitivity Analyses

Estimating the environmental burden of disease requires collection and analysis of a substantial amount of data. Because several potential sources of uncertainty exist, arising both from the input data itself and the method used for the calculations, uncertainty and sensitivity analyses are an important part of a complete burden of disease study. The purpose of uncertainty analysis is to quantitatively measure the uncertainty within the disease burden estimates, whereas the purpose of sensitivity analysis is to help identify variables that potentially have the largest impact on the results.

General sources of uncertainty in burden of disease studies include measurement error, systematic biases, and uncertainty related to modeling and extrapolation of data (Prüss-Üstün et al. 2003; Mathers et al. 2006b). In the exposure-assessment step, the sample size may be small or it may not represent the study population adequately. The choice of exposure indicators and inaccuracy in exposure measurements may also introduce error. In the exposure-response step, uncertainty is introduced when extrapolating relative-risk information from the epidemiologic literature to the study population. Additional uncertainty arises from matching the measures used in the exposure assessment with the measures applied in the relative-risk estimates. Mortality and morbidity data at baseline often must be collected from various sources with varying degrees of reliability and may have to be extrapolated from a small subgroup to the whole population. For example, in the UAE, health data were available only for part of the population of Abu Dhabi emirate. Data to represent the rest of the country had to be extrapolated from that data set. The fact that traditional burden of disease studies address risks individually, even though in real life people are often exposed to multiple risks simultaneously, is also a source of inaccuracy since the potential synergistic effects of multiple hazards or health conditions cannot be quantitatively measured. Because disease burden is not expressed as DALYs in this study, uncertainty and sensitivity analyses for the variables related to the DALY calculations are not covered here, but they are discussed in detail by Mathers et al. (2006b).

Quantifying uncertainty related to many factors, such as the extrapolation of data to populations for which limited information is available, is difficult (Prüss-Üstün et al. 2003). The *UAE Environmental Burden of Disease Model* is designed to assist with analyzing uncertainty and also assessing the sensitivity of the disease burden estimates to the model's underlying assumptions. This is the first time that such a model has been constructed to facilitate analysis of the complex data and presentation of the results in a national burden of disease study. Use of the model allowed the insertion of a range of likely values for each uncertain input variable and the inclusion of confidence intervals for input variables when they were available, such as for the relative risks or exposure estimates. Monte Carlo analysis was then applied to estimate the distribution of the disease burden estimates, and 2.5% and 97.5% fractiles were determined from the graphs to provide the lower and upper bounds of 95% confidence intervals.

Sensitivity analysis is performed by varying the input variables systematically within certain limits, one at a time. It is important for decision-making and for future analysis efforts because it helps identify which input variables have the largest effect on the output variables (i.e., potential reduction in the burden of disease). Sensitivity analysis also helps identify areas in which the model would most benefit from more specific local data, such as new exposure data gathered locally in the UAE.

Modeling the Burden of Disease in the UAE

As mentioned at the beginning of this chapter, the *UAE Environmental Burden of Disease Model*, contains a separate module for each risk.[1] The risks include outdoor air pollution, indoor air pollution, occupational exposures, climate change, drinking water contamination, coastal water contamination, and food contamination. Risks from groundwater and soil contamination could not be estimated quantitatively due to a lack of data on concentrations of pollutants in groundwater and soil in the UAE and a lack of information necessary for estimating which populations, if any, might be exposed.

Each risk module includes a link to the input variables used in the analysis so that data can be inserted and the impact of different variations tested as new information becomes available. Each risk module is also linked to the relevant baseline health data and includes a separate link for details describing the step-by-step flow of each model. The detailed exposure models contain descriptions of the sources of input data, all assumptions, and the steps of the calculations, which facilitate understanding of the structure of the model and the ability to make improvements. Figure 3.1 presents a top-level abstraction of the overall model.

Using this innovative model to illustrate the burden of disease in the UAE can support environmental and health policy decision-making in many ways. The model can help identify which environmental risks cause the greatest burden of disease in the UAE and demonstrate what the effect on disease burden would be if concentrations of or exposures to specific pollutants were reduced by implementing various control methods. The model can also help identify what data should be collected to better characterize the health effects caused by environmental pollutants in the UAE.

[1] Appendix B provides instructions on how to use the model.

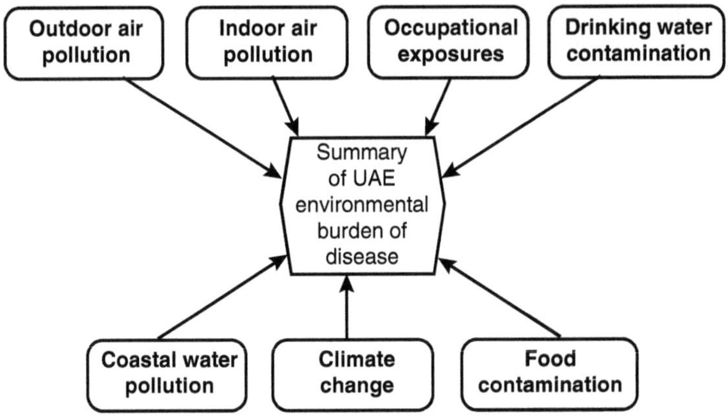

Fig. 3.1 Individual modules jointly contribute to overall estimates of environmental burden of disease in the UAE

Using Environmental Burden of Disease Studies in Decision-Making

Even though EBD studies can be beneficial in environmental and health decision-making, they have limitations that need to be taken into consideration, as discussed in detail by Prüss-Üstün et al. (2003). First, the current EBD methodology does not adequately address the complexity of the environment when multiple exposures and diseases are considered. Even though people are exposed to several environmental hazards simultaneously, exposures are addressed individually. This procedure does not take into account the potential synergistic effect of multiple hazards. Similarly, the disease burden is typically assessed for individual diseases, even though disability resulting from diseases occurring simultaneously may be greater than the sum of the impacts from the individual diseases. In addition, EBD studies only measure potential impacts on human health. They do not cover other factors such as economic gains or benefits to the environment resulting from actions taken to reduce the disease burden (Prüss-Üstün et al. 2003). Moreover, not all aspects of risk and the population's perception of risks are covered by EBD studies. Many factors, including equity between different populations; perceived degree of control over exposure; uncertainty related to a given risk; and social, cultural, and economic factors, can shape people's perspectives and influence priorities (Prüss-Üstün et al. 2003; WHO 2002). The method used to prioritize risks for the UAE and identify which risks to include in the study did account for these other factors, as explained in Chap. 2.

For EBD studies to be useful in decision-making, it is important that uncertainties in the model are expressed clearly because EBD studies are only as accurate as the input data used in the calculations. Sensitivity analysis is essential in order to demonstrate how different assumptions in the model impact the output values and may change the resulting intervention priorities. It should also be kept in mind that exposures and health conditions for which information is readily available and which can be rapidly measured could be favored over those that are more difficult to measure (Prüss-Üstün et al. 2003). The absence of data should not lead to the perception that the parameter is insignificant or that it should be ignored in decision-making. When comparing the disease burden attributable to different risks, a larger AF does not automatically mean that the risk is more important and should be given priority when selecting preventive measures. For example, a large AF may reflect a broad definition of exposure. In addition to the AF, other factors must also be taken into account when prioritizing actions, such as the relationship among exposures (i.e., will other risks change when one is changed?), the feasibility and cost-effectiveness of potential interventions, and the potential risks and benefits related to the interventions (Levine 2007).

Despite the potential limitations, EBD studies can still be very useful in decision-making. Estimation of the burden of disease in the UAE with an easy-to-understand computer model is a state-of-the-art method for analyzing the fragmentary data that was available on the disease distribution in the UAE and for communicating the results effectively. This innovative model allows comparison of the relative importance of various sources of ill health and examination of the effects of alternative interventions on the disease burden. The model also makes it very easy to update future burden of disease estimates when new data become available, and it allows UAE officials to test the effect of various intervention options. Because all assumptions, decisions about input variables, and specific methods are clearly stated in each step, changes to the model structure can be made easily should future research and new data prove it necessary. Because resources are always limited, the model can facilitate identification of the most important risks and prioritize competing actions to recognize the ones with the greatest potential to reduce the burden of disease in the UAE.

References

Anand, S., and K. Hanson. 1997. Disability-adjusted life years: A critical review. *Journal of Health Economics* 16(6): 685–702.

Baskent University. 2005. Burden of disease final report. Baskent University and the School of Public Health, Refik Saydam Hygiene Center. Ankara, Turkey: Ministry of Health.

Begg, S., T. Vos, B. Barker, C. Stevenson, L. Stanley, and A.D. Lopez. 2007. *The burden of disease and injury in Australia 2003*. Canberra: The Australian Institute of Health and Welfare.

Bradshaw, D., P. Groenewald, R. Laubscher, N. Nannan, B. Nojilana, R. Norman, D. Pieterse, et al. 2003. Initial burden of disease estimates for South Africa, 2000. *South African Medical Journal* 93(9): 682–688.

Bundhamcharoen, K., Y. Teerawattananon, T. Vos, and S. Begg. 2002. *Burden of disease and injuries in Thailand: Priority setting for policy*. Bangkok: Ministry of Public Health.

Concha-Barrientos, M., X.P.A. Sanhueza, and J.S. Vergara. 1996. *La carga de enfermedad en Chile*. Chile: Ministry of Health.

Coughlin, S.S., J. Benichou, and D.L. Weed. 1994. Attributable risk estimation in case–control studies. *Epidemiological Reviews* 16(1): 51–64.

Egypt Ministry of Health and Population. 2004. Key findings–burden of disease and injury (DALYs). Egyptian Doctor's Guide. http://www.drguide.mohp.gov.eg/NewSite/E-Learning/BurdenOfDisease/key3.asp

Fox-Rushby, J.A., and K. Hanson. 2001. Calculating and presenting disability adjusted life years (DALYs) in cost-effectiveness analysis. *Health Policy and Planning* 16(3): 326–331.

Hakulinen, T., H. Hansluwka, A.D. Lopez, and T. Nakada. 1986. Global and regional mortality patterns by cause of death in 1980. *International Journal of Epidemiology* 15(2): 226–233.

Health Authority–Abu Dhabi (HAAD). 2009. 2008 Health data for Abu Dhabi emirate.

Institute for Health Metrics and Evaluation. 2007. Global burden of disease study. http://globalburden.org/

Jankovic, S., H. Vlajinac, V. Bjegovic, J. Marinkovic, S. Sipetic-Grujicic, L. Markovic-Denic, N. Kocev, et al. 2007. The burden of disease and injury in Serbia. *European Journal of Public Health* 17(1): 80–85.

Lapostolle, A., A. Lefranc, I. Gremy, and A. Spira. 2008. Sensitivity analysis in summary measure of population health in France. *European Journal of Public Health* 18(2): 195–200.

Levin, M.L. 1953. The occurrence of lung cancer in man. *Acta - Unio Internationalis Contra Cancrum* 9(3): 531–541.

Levine, B. 2007. What does population attributable fraction mean? *Preventing Chronic Disease* 4(1): 1–5.

Lopez, A.D. 1993. Causes of death in the industrialized and developing countries: Estimates for 1985–1990. In *Disease control priorities in developing countries*, ed. D. Jamison, W.H. Mosely, A.R. Measham, and J.L. Bobadilla, 15–30. New York: Oxford University Press.

Lopez, A.D., C.D. Mathers, M. Ezzati, D.T. Jamison, and C.J.L. Murray. 2006. Measuring the global burden of disease and risk factors, 1990–2001. In *Global burden of disease and risk factors*, ed. A. Lopez, C. Mathers, M. Ezzati, D.T. Jamison, and C.J.L. Murray, 1–13. New York: Oxford University Press.

Lozano, R., C.J.L. Murray, J. Frenk, and J. Bobadilla. 1995. Burden of disease assessment and health system reform: Results of a study in Mexico. *Journal of International Development* 7(3): 555–564.

Mahapatra, P. 2002. *Estimating national burden of disease: The burden of disease in Andhra Pradesh, 1990s*. Hyderabad: Institute of Health Systems.

Mathers, C., T. Vos, and C. Stevenson. 1999. *The burden of disease and injury in Australia*. Canberra: The Australian Institute of Health and Welfare.

Mathers, C., A. Smith, and M. Concha. 2000. Global burden of hearing loss in the year 2000. *Global Burden of Disease 2000*. World Health Organization. http://www.who.int/healthinfo/statistics/bod_hearingloss.pdf

Mathers, C.D., T. Vos, A.D. Lopez, J. Salomon, and M. Ezzati. 2001. National burden of disease studies: A practical guide. In *Global program on evidence for health policy*. Ed. 2.0, Geneva: World Health Organization.

Mathers, C.D., C. Stain, D. Ma Fat, C. Rao, M. Inoue, N. Tomijima, C. Bernard, A.D. Lopez, and C.J.L. Murray. 2002. Global burden of disease 2000: Version 2 methods and results. In *Global programme on evidence for health policy discussion paper No. 50*, 108. World Health Organization.

Mathers, C.D., A.D. Lopez, and C.J.L. Murray. 2006a. The burden of disease and mortality by condition: Data, methods, and results for 2001. In *Global burden of disease and risk factors*, ed. A. Lopez, C. Mathers, M. Ezzati, D.T. Jamison, and C.J.L. Murray, 45–93. New York: Oxford University Press.

Mathers, C.D., J.A. Salomon, M. Ezzati, S. Begg, S. Vander Hoorn, and A.D. Lopez. 2006b. Sensitivity and uncertainty analyses for burden of disease and risk factor estimates. In *Global burden of disease and risk factors*, ed. A. Lopez, C. Mathers, M. Ezzati, D.T. Jamison, and C.J.L. Murray, 399–424. New York: Oxford University Press.

Melse, J.M., M.L. Essink-Bot, P.G. Kramers, and N. Hoeymans. 2000. A national burden of disease calculation: Dutch disability-adjusted life-years. Dutch Burden of Disease Group. *American Journal of Public Health* 90(8): 1241–1247.

Michaud, C.M., M.T. McKenna, S. Begg, N. Tomijima, M. Majmudar, M.T. Bulzachelli, S. Ebrahim, et al. 1996. The burden of disease and injury in the United States. *Population Health Metrics* 4: 11.

Murray, C.J.L., and A.K. Acharya. 1997. Understanding DALYs (disability-adjusted life years). *Journal of Health Economics* 16(6): 703–730.

Murray, C.J.L., and A.D. Lopez. 1996a. Evidence-based health policy—Lessons from the global burden of disease study. *Science* 274(5288): 740–743.

Murray, C.J.L., and A.D. Lopez. 1996b. *The global burden of disease: A comprehensive assessment of mortality and disability from diseases, injuries and risk factors in 1990 and projected to 2020*, The global burden of disease and injury series. Cambridge, MA: Harvard University Press.

Murray, C.J.L., and A.D. Lopez. 1996c. *Global health statistics: A compendium of incidence, prevalence and mortality estimates for over 200 conditions*, The global burden of disease and injury series. Cambridge, MA: Harvard University Press.

Murray, C.J.L., and A.D. Lopez. 1997a. Alternative projections of mortality and disability by cause 1990–2020: Global burden of disease study. *The Lancet* 349(9064): 1498–1504.

Murray, C.J.L., and A.D. Lopez. 1997b. Global mortality, disability, and the contribution of risk factors: Global burden of disease study. *The Lancet* 349(9063): 1436–1442.

Murray, C.J.L., and A.D. Lopez. 1997c. Mortality by cause for eight regions of the world: Global burden of disease study. *The Lancet* 349(9061): 1269–1276.

Murray, C.J.L., and A.D. Lopez. 1997d. Regional patterns of disability-free life expectancy and disability-adjusted life expectancy: Global burden of disease study. *The Lancet* 349(9062): 1347–1352.

Murray, C.J.L., and A.D. Lopez. 1998. *Health dimensions of sex and reproduction: The global burden of sexually transmitted diseases, HIV, maternal conditions, perinatal disorders, and congenital anomalies*, The global burden of disease and injury series. Cambridge, MA: Harvard University Press.

Murray, C.J.L., and A.D. Lopez. 2000. Progress and directions in refining the global burden of disease approach: A response to Williams. *Health Economics* 9(1): 69–82.

Naghavi, M., F. Abolhassani, F. Pourmalek, M.M. Lakeh, N. Jafari, S. Vaseghi, N.M. Hezaveh, and H. Kazemeini. 2009. The burden of disease and injury in Iran 2003. *Population Health Metrics* 7: 9.

Pakari, H., and K. Roa. 1999. *Our health, our future: The health of New Zealanders 1999.* Wellington: Ministry of Health.

Palmer, S., H. Houston, B. Lervy, D. Riberio, and P. Thomas. 1997. Problems in the diagnosis of foodborne infection in general practice. *Epidemiology and Infection* 117: 479–484.

Prüss-Üstün, A., C. Mathers, C. Corvalán, and A. Woodward. 2003. *Introduction and methods: Assessing the environmental burden of disease at national and local levels*, WHO environmental burden of disease series, No. 1. Geneva: World Health Organization.

Public Health Agency of Canada. 2006. Population health impact of disease in Canada. http://www.phac-aspc.gc.ca/phi-isp/index-eng.php

Rockhill, B., B. Newman, and C. Weinberg. 1998. Use and misuse of population attributable fraction. *American Journal of Public Health* 88(1): 15–19.

Rowe, A.K., K.E. Powell, and W.D. Flanders. 2004. Why population attributable fractions can sum to more than one. *American Journal of Preventive Medicine* 26(3): 243–249.

Smith, K.R., C.F. Corvalán, and T. Kjellstrom. 1999. How much global ill health is attributable to environmental factors? *Epidemiology* 10(5): 573–584.

Stevens, G., R.H. Dias, K.J.A. Thomas, J.A. Rivera, N. Carvalho, S. Barquera, K. Hill, and M. Ezzati. 2008. Characterizing the epidemiological transition in Mexico: National and subnational burden of diseases, injuries, and risk factors. *PLoS Medicine* 5(6): e125.

UAE Ministry of Health. 2008. *Summary statistics 2007*. Abu Dhabi: Preventive Medicine Sector.

Vos, T., M. Tobias, H. Gareboo, F. Rousetty, S. Huttley, and C.J.L. Murray. 1995. *Mauritius health sector reform, national burden of disease study. Final report of consultancy.* Port Luis: Ministry of Health and Ministry of Economic Planning and Development.

Williams, A. 1999. Calculating the global burden of disease: Time for a strategic reappraisal? *Health Economics* 8(1): 1–8.

World Health Organization (WHO). 2002. Reducing risks: Promoting healthy lives. World Health Report 2002. Geneva: WHO.

World Health Organization (WHO). 2003–2007. Practical guidance for assessment of disease burden at national and local levels. Environmental burden of disease series, Nos. 1–16. Geneva: World Health Organization. http://www.who.int/quantifying_ehimpacts/national/en/

World Health Organization (WHO). 2004. The global burden of disease: 2004 update. http://www.who.int/healthinfo/global_burden_disease/2004_report_update/en/index.html

World Health Organization (WHO). 2009. International Classification of Diseases (ICD). http://www.who.int/classifications/icd/en/

World Health Organization (WHO). 2010. Health statistics and health information systems. *Global Burden of Disease (GBD) 2005 study*. http://www.who.int/healthinfo/global_burden_disease/GBD_2005_study/en/index.html

Chapter 4
Burden of Disease from Outdoor Air Pollution

Abstract Anthropogenic outdoor air pollution caused a substantial number of premature deaths in the United Arab Emirates in 2008, and this mortality number is estimated to be the greatest among the eight priority environmental risk areas in this book. In this chapter we quantify the burden of disease, including premature deaths and health-care facility visits, associated with outdoor air pollution, specifically ambient particulate matter (PM) and ozone from anthropogenic sources, and we discuss the uncertainties associated with the estimates. The negative impacts of PM and ozone on public health have been well documented, particularly the mortality effect of PM. For morbidity, scientific studies have linked exposure to PM and ozone to a variety of health problems, particularly respiratory and cardiovascular diseases. Two different approaches were used to estimate outdoor PM and ozone concentrations across the UAE: the measurement-based approach and the air-quality-model-based approach. The measurement-based approach relies on data from 10 fixed monitoring stations in Abu Dhabi emirate. The model-based approach uses Community Multiscale Air Quality modeling software to predict air quality based on estimates of air pollutant emissions and meteorological conditions. Using the measurement-based approach, this research estimates that in 2008 the total number of premature deaths in the UAE caused by exposure to ambient particulate matter was approximately 650. These account for about 7% of the total deaths occurring in the UAE in 2008. About 77 deaths were attributable to ground-level ozone in 2008. With respect to excess illness, in 2008 PM_{10} exposure caused a mean estimate of 15,000 health-care facility visits for respiratory and cardiac illnesses, accounting for about 3% of total medical visits. Ground-level ozone caused a mean estimate of 9,800 respiratory health-care facility visits in 2008, accounting for about 6% of total respiratory health-care facility visits in that year. Thus, in total, PM appears to cause a larger disease burden in the UAE than ozone. Using the CMAQ model-based approach, the estimated death numbers attributable to PM were smaller than the measurement-based estimates, whereas the estimated death

J. MacDonald Gibson et al., *Environmental Burden of Disease Assessment*, Environmental Science and Technology Library 24, DOI 10.1007/978-94-007-5925-1_4,

numbers attributable to ozone are greater than those using the measurement-based approach. Also, the estimated health-care facility visits attributable to PM are smaller than the measurement-based estimates.

Keywords Outdoor air pollution • Ambient particulate matter • Ground-level ozone • Premature deaths and health-care facility visits • Respiratory and cardiovascular diseases • Attributable fraction • Relative risk • Concentration-response coefficients • Community Multiscale Air Quality (CMAQ) modeling software • Environmental burden of disease • United Arab Emirates

Overview: Nature and Sources of Outdoor Air Pollution

Current scientific evidence, derived largely from studies in western industrial countries but increasingly in the developing world, demonstrates that exposure to outdoor air pollution causes a wide spectrum of adverse health outcomes, from acute respiratory symptoms to premature death.

The air pollutant considered to have the greatest health impact is particulate matter (PM), a term comprising microscopic particles (or aerosols) suspended in the air that vary in size and chemical composition, may be solid or liquid, and derive from a variety of sources. PM is commonly measured as the mass concentration of all particles smaller than 10 μm in diameter (PM_{10}) or 2.5 μm in diameter ($PM_{2.5}$). Both have consistently been associated with premature death and cardiopulmonary disease.

Another important air pollutant is tropospheric ozone (O_3), which has been linked to similar adverse health effects. Ozone is a secondary pollutant formed by a complex set of reactions involving nitrogen oxides (NO_x) and volatile organic compounds (VOCs) in the presence of sunlight. Emissions of these gas-phase precursors come from a variety of sources, including industry, motor vehicles, and some natural sources. Although associations also have been reported for other common air pollutants such as sulfur dioxide, carbon monoxide, nitrogen dioxide, and toxic compounds, given the relative wealth of epidemiologic evidence, this assessment of the national burden of outdoor air pollution on human health in the United Arab Emirates (UAE) focuses on PM and ozone.

Poor air quality is apparent in the UAE. Some evidence includes degraded visibility and ambient measurements of air pollutants. For example, in Abu Dhabi emirate in 2007, PM_{10} was observed to exceed the emirate standard on about a third of the days of the year at each of ten monitoring stations maintained by the Environment Agency–Abu Dhabi (EAD) (Whitford 2008). Sources of air pollution include large industries, frequent dust storms that can transport pollutants from other continents, and the rapidly growing fleet of motor vehicles, particularly in the urban areas of Abu Dhabi and Dubai. The highest observed PM_{10} concentrations have exceeded the standard severalfold and are clearly associated with periodic dust storm events. While most of the coarse PM (that between 2.5 and 10 μm in diameter) is the result of natural and primary emissions (dust and sea salt), most $PM_{2.5}$ is typically anthropogenic and related to combustion (both primary and secondary). Preliminary measurements of $PM_{2.5}$ concentrations by the EAD suggest that 40% of PM_{10} may be $PM_{2.5}$ in urban areas and 20% in rural and suburban areas (Sivertsen 2010). With regard to ozone, the climate in the UAE (intense sunlight) and rapidly growing emissions of NO_x and VOCs from motor vehicles and industry result in an environment conducive to high ozone production. Monitoring data from the EAD network did not record a single 1-h average ozone concentration in 2007 exceeding the EAD standard (Whitford 2008), but it is unclear whether the measurements accurately reflect the exposure of the population in the UAE and, if ozone formation is being suppressed, whether it could increase rapidly in the future as emissions change. Therefore, possible health impairments induced by ozone are still a matter of concern.

In this chapter, we quantify the burden of disease, including premature deaths and health-care facility visits, associated with ambient PM and ozone from anthropogenic sources and discuss the uncertainties associated with the estimates.[1] The second section of this chapter presents a review of the current epidemiology literature on the health effects of PM and ozone. The methods used to quantify the burden of disease in the UAE are described in the third section. The estimated mortality and morbidity resulting from exposure to PM and ozone, as well as uncertainty and sensitivity analyses, are presented in the fourth section. Data needed to improve future burden of disease estimates are covered in the fifth section, and the major conclusions of this study are summarized in the sixth section.

Key Health Effects of Outdoor Air Pollution

Epidemiologic Evidence of Health Effects Resulting from Exposure to Particulate Matter

Time-series studies of the short-term effects of air pollutants examine the relationship between daily changes in air pollution (24-h average concentrations in most cases) and daily occurrence of mortality or morbidity in an area. The key advantage of the time-series method is that it potentially reduces the confounding effects of many factors that otherwise might be difficult for researchers to control. Specifically, several important confounding factors (e.g., smoking habits, health-care status, activity patterns, socioeconomic status, and working and living environment) do not vary considerably over time. Time-series studies have consistently reported significant associations between daily mortality and daily exposures to both PM_{10}

[1]The estimates presented here are updated from those published in an earlier analysis we conducted and reported in Li et al. (2010). The Li et al. estimates employed 2007 baseline health data, whereas in this chapter the estimates are derived from 2008 health data.

and $PM_{2.5}$ and thus provide compelling evidence that PM increases mortality rates (Ostro 2004). Recent multicity studies and meta-analyses in the United States indicate that the increase in daily all-cause mortality falls into a range of approximately 0.2–0.8% per 10 μg/m³ increase in daily PM_{10} (Dominici et al. 2005; Levy et al. 2000; Samet et al. 2000; Schwartz et al. 2002; Zeka et al. 2005). A meta-analysis of 33 European studies suggests a mean increase in the risk of premature mortality of 0.6% per 10 μg/m³ PM_{10} (Anderson et al. 2004). Consistent associations also have been reported by studies conducted in cities outside of Western industrialized nations and in developing countries, but the effects tend to be slightly greater than those reported in the United States and Europe. For example, the following all-cause mortality effect estimates have been reported for total populations and a 10 μg/m³ change in PM_{10} (with 95% confidence intervals): 1.7% (1.1%, 2.3%) – Bangkok, Thailand; 1.83% (0.9%, 2.7%) – Mexico City, Mexico; 1.1% (0.9%, 1.4%) – Santiago, Chile; 0.8% (0.2%, 1.6%) – Incheon, South Korea; 1.6% (0.5%, 2.6%) – Brisbane, Australia; and 0.95% (0.32%, 1.6%) – Sydney, Australia (Ostro 2004).

Cohort studies follow a group of initially healthy people for a long period (e.g., 10–20 years) to observe how they develop diseases or die. These studies are valuable in investigating the possible long-term chronic effects of exposure to air pollution. In the United States, two large-scale cohort studies—the Harvard Six Cities Study and the American Cancer Society (ACS) Study—have been conducted in the past two decades. Both studies observed increased mortality associated with an increase in time-average $PM_{2.5}$ levels, but not PM_{10}, suggesting that long-term adverse health effects are influenced by the fine portion of PM (Dockery et al. 1993; Krewski et al. 2000; Laden et al. 2006; Pope et al. 1995, 2002). Compared with the acute effects observed by time-series studies, chronic effects reported by cohort studies are generally larger. Künzli et al. (2001) reported that the estimates of health effects attributable to air pollution based on cohort studies are generally 5–10 times larger than those based on time-series studies. In the United States, the ACS study is most commonly used as a basis for assessing health impacts of ambient PM because it has the largest study size. In an extended analysis of the ACS study, Pope et al. (2002) reported results using 16 years (1982–1998) of follow-up data for approximately 500,000 adults (age 30 and older) throughout the United States. It was found that each 10 μg/m³ elevation in $PM_{2.5}$ is associated with increases of 4, 6, and 8% in all-cause, cardiopulmonary, and lung-cancer mortality, respectively. A new extended follow-up and spatial analysis of the ACS study, conducted to clarify outstanding scientific issues arising from earlier ACS analyses, reported results consistent with those from other studies, further supporting the hypothesis that long-term exposure to ambient $PM_{2.5}$ increases mortality in the general population (Krewski et al. 2009).

The Arabian Gulf region frequently experiences episodes of high PM concentrations that are dominated by windblown desert dust. In these episodes, coarse particles are likely to comprise a greater proportion of the total PM_{10}. For instance, a study in the Coachella Valley, an arid region in southern California, reported a $PM_{2.5}$ to PM_{10} ratio of 0.35 (Ostro et al. 2000), compared with ratios of 0.50–0.65 that are regularly found in many urban areas (Ostro 2004). Fine particles ($PM_{2.5}$) have generally been considered to be more toxic and to pose a greater health risk than coarse particles. It is uncertain whether a unit of PM concentration (e.g., 1 or 10 μg/m³) in areas significantly impacted by desert dust causes similar health effects as it does in other places. A few time-series studies of the acute health effects of exposure to PM have been conducted in the United States in arid, desert areas with obvious windblown-dust episodes similar to those in the Gulf region. The following increases in daily mortality associated with a 10 μg/m³ increase in PM_{10} (with 95% confidence intervals) were reported: 0.41% (−0.42%, 0.81%) – Coachella Valley, California (Ostro et al. 2000); 0.8% (0.3%, 1.3%) – Salt Lake City, Utah (Pope et al. 1999); and no significant association – Spokane, Washington (Slaughter et al. 2005). Although the third study did not link mortality with PM in windblown-dust episodes, findings from the first two studies are consistent with those from most studies of PM_{10}-related mortality in the United States. Consequently, while the health effects of desert dust relative to other components of PM remain uncertain and will continue to be an important topic for future research, the available literature does not justify treating desert dust differently from other components of PM when assessing health effects.

Epidemiologic Evidence of Health Effects Resulting from Exposure to Ozone

Scientific evidence increasingly indicates that ground-level ozone, even at low levels, can damage health, as indicated by increased rates of hospital admissions, exacerbation of respiratory illness, and premature mortality. Meta-analyses of city-specific studies have consistently reported associations between ozone and daily mortality (Bell et al. 2005; Ito et al. 2005; Levy et al. 2005). A recent large time-series study examined the link between ambient ozone and short-term mortality for 95 large U.S. urban communities from 1987 to 2000 and reported that a 10 ppb increase in daily average ozone was associated with a 0.52% increase in daily non-accidental mortality (Bell et al. 2004). A meta-analysis of studies in Europe found a 0.3% increase in all-cause mortality per 10 μg/m³ (5 ppb) increase in 8-h ozone (Anderson et al. 2004). While cohort studies had not clearly identified a relationship between ozone and mortality in the past, a recent follow-up of the ACS cohort reported that long-term ozone exposure was significantly associated with an increase in deaths from respiratory causes, with an increase in relative risk of 0.04 per 10 ppb increment in ozone concentration (Jerrett et al. 2009). This evidence collectively supports an association between ground-level ozone and mortality, with respect to both short-term and long-term exposure, over a wide range of concentrations.

The negative impacts of PM and ozone on public health have been well documented, particularly the mortality effect of PM. For morbidity, scientific studies have linked exposure to PM and ozone to a variety of health problems, particularly

Table 4.1 Outdoor air exposure indicators and related adverse health effects considered in this study

Exposure indicators	Adverse health effects
PM_{10}, daily average (μg/m^3)	All-cause mortality (all ages) and respiratory mortality (<5 years old); respiratory and cardiovascular morbidity (all ages)
$PM_{2.5}$, annual average (μg/m^3)	All-cause, cardiopulmonary, and lung cancer mortality (>30 years old)
Ground-level ozone, daily (24-h) average (ppb)	Total nonaccidental, cardiovascular, and respiratory mortality (all ages); respiratory morbidity (all ages)
Ground-level ozone, annual average of daily maximum concentration (ppb)	Respiratory mortality (>30 years old)

respiratory and cardiovascular diseases. PM exposure has been linked to increased respiratory symptoms, decreased lung function, aggravation of asthma, development of chronic bronchitis, irregular heartbeat, and nonfatal heart attacks. Ground-level ozone exposure has been linked to respiratory symptoms such as airway irritation, coughing, pain when taking a deep breath, and wheezing and breathing difficulties during exercise or outdoor activities; aggravation of asthma; increased susceptibility to respiratory illnesses like pneumonia and bronchitis; and permanent lung damage with repeated exposures. Key uncertainties at present include the magnitude and variability of risk estimates, possible thresholds or discontinuities in the concentration-response function, the extent to which findings in one location can be generalized to other locations, and a lack of clear understanding of the underlying biological mechanisms. For PM, scientists are not certain whether responsibility for health effects lies with the mass of PM or particular chemical components within the PM. For ozone, questions remain about whether ozone causes health effects directly or is an indicator of other constituents produced by atmospheric photochemistry that may influence health (Bell et al. 2006). Table 4.1 presents the adverse health effects resulting from exposure to PM_{10}, $PM_{2.5}$ and ground-level ozone that were considered in this study.

Method for Estimating the Burden of Disease from Outdoor Air Pollution

Overview of Methods to Quantify the Disease Burden Attributable to Outdoor Air Pollution

Quantitative assessment of the health impacts of outdoor air pollution in the UAE is based on the following components: exposure assessment based on measurement at monitoring stations or on model-based estimates of concentrations; determinations of the size of the population groups exposed, the type of health effects of interest, and the baseline incidence of those health effects; and concentration-response functions abstracted from epidemiologic literature. These factors were combined in an integrated procedure to estimate the attributable burden of disease due to outdoor air pollution. These methods are similar to those used previously by the World Health Organization (WHO) and others (Ostro 2004; Cohen et al. 2004) to assess the global burden of disease and the U.S. Environmental Protection Agency's (EPA) health impact assessment approach to quantifying the benefits of proposed actions to improve air quality (U.S. EPA 1999). For ozone, our methods follow the recommendations of the U.S. National Research Council (2008).

The health impact function to quantify the disease burden attributable to outdoor air pollution is:

$$\Delta y = y_0 (1 - e^{-\beta \Delta x}) = I_0 P (1 - e^{-\beta \Delta x}). \tag{4.1}$$

Where:

Δy = Attributable mortality or morbidity (deaths or health-care facility visits per year)
y_0 = Baseline incidence (current prevalence of cause-specific deaths or health-care facility visits per year), equal to the baseline incidence rate (I_0) multiplied by the potentially affected population (P)
β = Concentration-response coefficient from epidemiologic studies (% increase in cause-specific mortality or morbidity per 1 μg/m^3 PM or per 1 ppb O_3)
Δx = Change in concentration of the pollutant of interest (PM_{10}, $PM_{2.5}$, or O_3; μg/m^3 for PM and ppb for O_3)

Equation 4.1 can be derived from the attributable fraction (AF), defined as the fraction of the disease burden attributable to a risk, and relative risk (RR). WHO has proposed calculating AF due to exposure to outdoor air pollution using the following equation (Cohen et al. 2004; Ostro 2004):

$$AF = \frac{RR - 1}{RR} \tag{4.2}$$

Where: RR = Relative risk

Estimating the burden of disease using attributable fraction methodology is discussed in more detail in Chap. 3. In epidemiologic studies on the health effects of outdoor air pollution, RR is commonly calculated by:

$$RR = e^{\beta \Delta x}. \tag{4.3}$$

Hence, Eq. 4.4 shows the health impact function by deriving our original equation above:

$$\Delta y = y_0 AF = y_0 \frac{RR - 1}{RR} = y_0 \frac{e^{\beta \Delta x} - 1}{e^{\beta \Delta x}} = y_0 (1 - e^{-\beta \Delta x}). \tag{4.4}$$

The next section discusses the health endpoints associated with PM or ozone considered in this study and the corresponding concentration-response coefficients (β), followed by discussions of exposure assessment, natural background levels of pollutants, exposed population, and baseline mortality.

Health Endpoints and Concentration-Response Coefficients

For mortality, based on WHO's recommendation, we considered the short-term health effects of exposure to PM_{10}, including all-cause mortality in the general population and respiratory mortality in children younger than 5, using daily average PM_{10} concentrations as the exposure indicator. We estimated the effects of $PM_{2.5}$ on long-term mortality using estimates from Pope et al. (2002) that include all-cause, cardiopulmonary, and lung-cancer mortality for adults over 30. Annual average concentration was used to estimate the long-term effects of $PM_{2.5}$. In aggregating the total mortality attributable to PM, only all-cause mortality attributable to long-term exposure to $PM_{2.5}$ in adults over 30 and premature mortality from respiratory disease attributable to the short-term exposure to PM_{10} in children younger than 5 were summed, in order to avoid double-counting, whereas the remaining health endpoints were assessed to provide additional information for decision makers. For ozone, we estimated all-cause nonaccidental mortality due to short-term exposure (using the daily average concentration as the exposure indicator) and respiratory mortality due to long-term exposure (using the annual average of daily maximum concentration as the exposure indicator), based on the best available literature.

Based on the epidemiologic literature and available baseline health-care facility visit data, we selected respiratory and cardiovascular health-care facility visits to assess the morbidity attributable to short-term PM_{10} exposure and respiratory health-care facility visits for ozone. We used concentration-response coefficients from two representative studies conducted in the United States. Specifically, for PM_{10}, we used Ostro and Chestnut's (1998) report, which assessed the total health benefits (including respiratory and cardiovascular hospital admissions) of reducing PM air pollution in the United States. For ozone, we used the 2001 report by Levy et al., which assessed the public health benefits of reduced ozone concentrations in Houston, Texas.

Although local epidemiologic studies may better reflect the influence of pollutant characteristics and baseline health status on the associations between air pollution and health, studies conducted at various locations across the United States and Europe that involve a wide range of underlying conditions, and studies conducted in cities outside the developed world, all report generally consistent effect estimates. Given this, WHO argues that it is reasonable to extrapolate existing estimates to areas where studies have not been undertaken (Ostro 2004), which is the case for the UAE. Based on WHO's suggestion, we selected representative multicity studies and meta-analyses conducted in the United States as the bases for assessing the health impact of outdoor air pollution in the UAE. Also, uncertainty may be introduced in transferring hospital admission concentration-response coefficients from U.S. epidemiologic studies to the UAE, due to possible differences in the cultural meaning of hospital admissions. The UAE health-care facility visit data we used (see Chap. 3) included visits to both full-sized hospitals and smaller clinics, denoting each time an insurance company was notified of a patient encounter with a health-care facility—regardless of whether the visit involved a large operation or a small checkup. We concluded that the UAE data have very similar meaning to (and can be reasonably compared to) common definitions of hospital/clinic admission used outside the UAE (e.g., in the United States). One possible uncertainty is if the large immigrant worker population in UAE is unlikely to go to health-care facilities, then the same health effects would not be seen in the baseline health rates compared with the United States, which may result in an underestimate of the disease burden attributable to outdoor air pollution.

Tables 4.2 and 4.3 summarize mortality and morbidity outcomes considered in this study, their concentration-response coefficients (β), and sources in the literature. The concentration-response coefficients were assumed to be normally distributed in the model based on the general epidemiologic literature. Since it remains controversial whether the mortality or morbidity effects of PM and ozone are independent or considerably correlated, mortality and health-care facility visits attributable to the two pollutants are reported separately to avoid possible double-counting. We likewise do not add short-term and long-term health effects among the same study population, as it is generally expected that short-term effects will be a subset of long-term effects.

Exposure Assessment

Two different approaches were used to estimate outdoor PM and ozone concentrations across the UAE: the measurement-based approach and the air-quality-model-based approach.

Measurement-Based Approach

The measurement-based approach relies on data from 10 fixed monitoring stations in Abu Dhabi emirate. At the time this study was conducted, these were the only quality-assured air quality data available to us. Since then, the Abu Dhabi network has expanded, and efforts have begun to link monitoring networks across the emirates. Monitor data were available for 2 years: 2007 and 2008. Here we describe the methods used for space-time interpolation of these measurements to produce estimates of exposure to outdoor air pollutants throughout the UAE.

The first step of the measurement-based estimation is based on a classical (in the sense that it minimizes the mean-square error) geostatistical procedure called kriging. Given a finite set of n observations $Y=(y(s_1,t),\ldots,y(s_n,t))'$ from a random field

Table 4.2 Cause-specific mortality and concentration-response coefficients selected to estimate the disease burden attributable to outdoor air pollution in the UAE

Exposure indicator (unit)	Exposure type	Mortality health outcome	Reference (study type)	Concentration-response coefficients β[a] Mean (95% CI[b])	Standard deviation[c]	Age group
PM_{10}						
Daily average (μg/m³)	Short term	All-cause	Ostro (2004) (Meta-analysis and expert judgment)	0.08 (0.06, 0.1)	0.01	All
		Respiratory[d]	Ostro (2004) (Meta-analysis)	0.166 (0.034, 0.3)	0.07	<5
$PM_{2.5}$						
Annual average (μg/m³)	Long term	All-cause	Pope et al. (2002) (Multicity cohort study)	0.6 (0.2, 1.1)	0.2	>30
		Cardiopulmonary[e]		0.9 (0.3, 1.6)	0.3	>30
		Lung cancer[f]		1.4 (0.4, 2.3)	0.5	>30
OZONE						
Daily (24-h) average (ppb)	Short term	Total nonaccidental[g]	Bell et al. (2004) (Multicity time-series study)	0.052 (0.027, 0.077)	0.0128	All
Daily (24-h) average (ppb)	Short term	Cardiovascular and respiratory[h]	Bell et al. (2004) (Multicity time-series study)	0.064 (0.031, 0.098)	0.17	All
Annual average of daily 1-h maximum concentration (ppb)	Long term	Respiratory	Jerrett et al. (2009) (Multicity cohort study)	0.4 (0.1, 0.67)	0.1	>30

[a]% increase in incidence per 10 μg/m³ PM or 100 ppb ozone
[b]Confidence interval
[c]Derived using 95% confidence intervals provided by the original references.
[d]International Classification of Diseases (ICD-9) codes 470–8, 480–8, and 490–3 (WHO 2009b)
[e]ICD-9 codes 401–40 and 460–519
[f]ICD-9 code 162
[g]The study excluded deaths from injuries or other external causes.
[h]ICD-9 codes 390–448, 480–6, 490–7, and 507

Table 4.3 Morbidity (health-care facility visits) and concentration-response coefficients selected to estimate the disease burden attributable to outdoor air pollution in the UAE

Exposure indicator (Unit)	Exposure type	Morbidity health outcome	Reference	Concentration-response coefficients β[a] Mean (95% CI)	Standard deviation	Age group
PM_{10}						
Daily average (μg/m^3)	Short term	Respiratory health-care facility visits[b]	Ostro and Chestnut (1998)	0.084 (0.047, 0.12)	0.019	All
		Cardiovascular health-care facility visits[c]		0.03 (0.024, 0.035)	0.0031	
OZONE						
Daily (24-h) average (ppb)	Short term	Respiratory health-care facility visits[b]	Levy et al. (2001)	0.34 (0.22, 0.46)	0.06	All

[a]% increase in incidence per 10 μg/m^3 PM or per 100 ppb ozone
[b]ICD-9 codes 480–6, 490–7, and 507
[c]ICD-9 codes 390–448

we would like to estimate the variable of interest (i.e., PM_{10} concentration) at a location that has not been observed, say $y(s_0,t)$. In the context of Gaussian processes, kriging can be described using the following model for the data:

$$Y = \mu + \varepsilon, \varepsilon \sim N(0, \Sigma) \tag{4.5}$$

That is, the observed data can be expressed as the sum of a deterministic component (μ) and a stochastic error (ε) coming from a normal distribution with covariance matrix Σ. When μ is constant, the procedure is referred to as ordinary kriging. When covariates are included (typically spatial covariates such as latitude and longitude), the procedure is known as universal kriging. The covariance matrix Σ is generally obtained from a space-time covariance function $C_X(r,\tau)$ that roughly measures the degree of association between observations that are close in space and in time.

For the UAE, we chose to use a linear combination of two separable exponential covariance functions, that is:

$$C_X(r,\tau) = \sigma_X^2 \left[\alpha \exp\left(-\frac{3r}{a_{r1}}\right) \exp\left(-\frac{3\tau}{a_{\tau 1}}\right) + (1-\alpha) \exp\left(-\frac{3r}{a_{r2}}\right) \exp\left(-\frac{3\tau}{a_{\tau 2}}\right) \right] \tag{4.6}$$

Estimating the covariance function can provide insight about the distribution of pollutant concentrations and answer the question of how much concentration can vary in a small area or a small time. The advantage of using two different exponential functions is that it allows separate modeling of the variability due to

Table 4.4 Rate parameters for the PM_{10} and ozone concentration space-time covariance functions

Pollutant	a_{r1} (degree)	$a_{\tau1}$ (days)	a_{r2} (degree)	$a_{\tau2}$ (days)
PM_{10}	0.9	120	1	15
Ozone	0.1	200	0.5	3

human intervention (which typically has a short spatial range a_{r1} and long term persistence $a_{\tau1}$) and the variability explained by atmospheric conditions (typically having a long spatial range a_{r2} and short time span $a_{\tau2}$). Table 4.4 reports the estimates of the parameters in Eq. 4.6 for PM_{10} and ozone. The parameters were estimated using a nonlinear least-squares procedure.

In the kriging process, the covariance function determines the level of spatial association among air quality estimates in different locations. In order to visualize the estimates we created a grid of cells (30 × 30 arc sec, approximately 0.865 km^2) and drew maps using the contour of the UAE at the proper spatial coordinates. The two maps in Fig. 4.1 report estimates for daily PM_{10} concentration on two different days. The color scale on the right side of each plot indicates that darker colors are associated with higher concentrations. Figure 4.2 reports analogous plots for ozone concentration on June 21, 2007, and May 6, 2008. As in Fig. 4.1, darker colors are associated with higher concentrations.

We conducted space-time analysis to estimate concentrations in each created grid cell. When entering ambient concentration estimates into the *UAE Environmental Burden of Disease Model* (model details discussed later), the original 30 × 30 arc sec grid cells are combined in groups of 64, resulting in an approximately 55 km^2 grid resolution. This grid resolution reduces the size of input data without suffering a significant decrease in the desired precision of the model forecast. We kept the 30 × 30 arc sec resolution, however, when mapping mortality due to outdoor air pollution. Attributable deaths were calculated for each grid cell and then aggregated across each emirate and over the entire UAE. This approach captures the temporal (at a daily level for short-term health effects) and spatial (as fine as 55 km^2 in the model and 0.865 km^2 in the maps) variations of input variables within the health impact function, including pollutant concentration and population.

At the time we carried out this study, no data were available to calculate of the fraction of PM_{10} that is comprised of $PM_{2.5}$. Thus, for the measurement-based approach, $PM_{2.5}$ concentrations were estimated based on an empirical $PM_{2.5}/PM_{10}$ ratio of 0.35, as suggested by WHO for arid, desert regions. A recent study on the characterization of PM for three sites in Kuwait reported that $PM_{2.5}$ comprised 47% of PM_{10} at two sites and 41% at the third (Brown et al. 2008). Although there are not sufficient data to make any definite conclusions on exposures, preliminary measurements from $PM_{2.5}$ monitors purchased by EAD in 2009 suggest that 40% of PM_{10} may be $PM_{2.5}$ in urban areas and 20% in rural and suburban areas (Sivertsen 2010). These facts suggest that using WHO's 0.35 scaling factor is reasonable for the UAE.

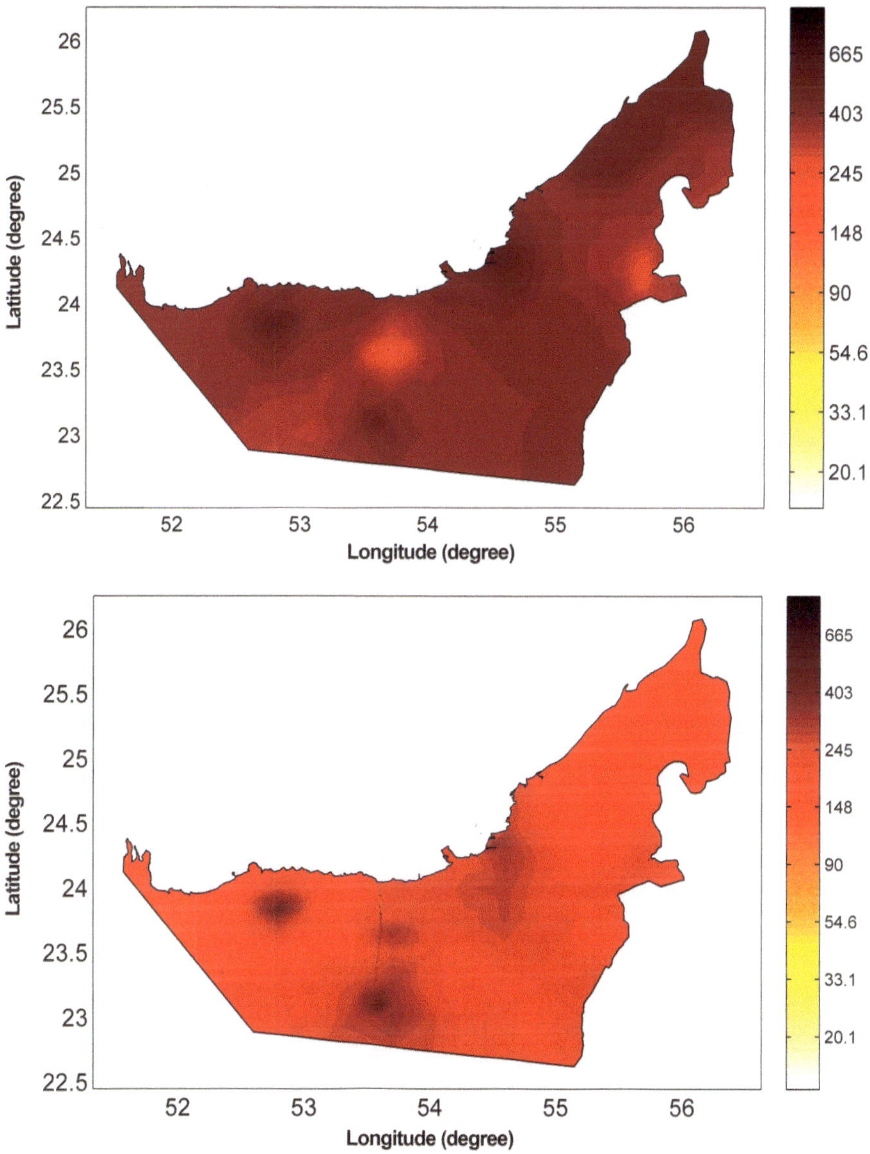

Fig. 4.1 PM_{10} concentration maps for the UAE on May 22, 2007, and April 5, 2008 (maps created by Prahlad Jat, University of North Carolina, Chapel Hill)

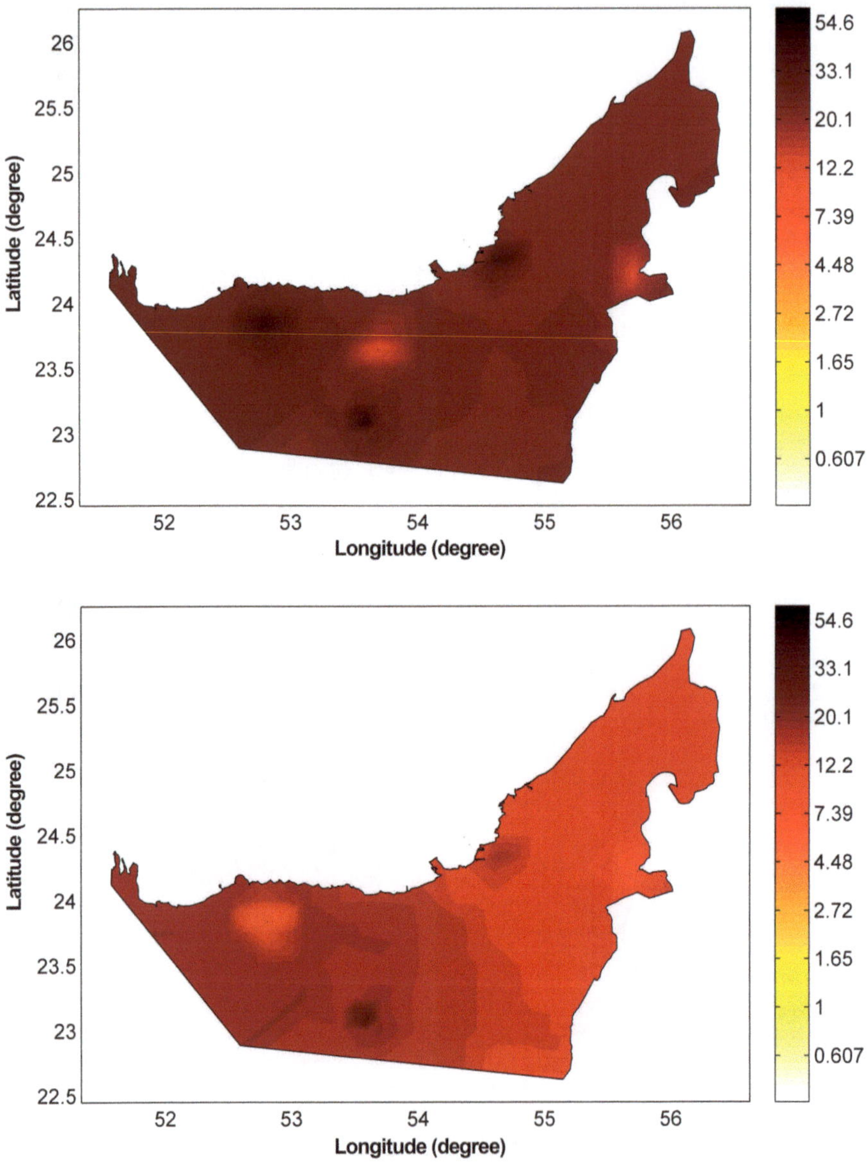

Fig. 4.2 Ozone concentration maps for the UAE on June 21, 2007, and May 6, 2008 (maps created by Prahlad Jat, University of North Carolina, Chapel Hill)

Model-Based Approach

As an alternative to the measurement-based approach, we also estimated air pollution exposures by constructing a mechanistic air quality model. This model predicts air quality based on estimates of air pollutant emissions and meteorological conditions. It uses Community Multiscale Air Quality (CMAQ) modeling software. CMAQ has been approved by the U.S. EPA for regulatory applications and is now one of the leading air quality models internationally, having been applied in several continents. CMAQ produces hourly estimates of the concentrations of many pollutants in three dimensions over large regions. Ground-level concentrations of daily average PM_{10} are used to estimate short-term mortality, and concentrations of annual average $PM_{2.5}$ are used to estimate long-term mortality. For ground-level ozone, the annual average of daily 1-h-maximum ozone concentrations predicted by CMAQ was used as the indicator of long-term exposure in this study, based on the literature. CMAQ also generates daily-average ozone concentrations. Using these daily-average predictions to assess short-term ozone exposure, however, is computationally intensive and beyond the scope of this project. Therefore, deaths due to short-term exposure to ozone were not estimated using the model-based approach. While the CMAQ model uses 36 km^2 grid resolution, we have translated the grid estimates output from CMAQ to match both the finer and coarser grids used in the measurement-based approach in our model (approximately 0.8645 km^2 for generating maps and 55 km^2 for running simulations to estimate the uncertainty in our burden of disease estimates). CMAQ results are only available for 7 months (summer: May 1–August 31, 2007, and winter: January 1–March 31, 2008). Due to data limitations, for estimates of short-term effects (PM_{10}), daily mortality was calculated for the 7 months for which data were available and then scaled to the entire year; for estimates of long-term effects ($PM_{2.5}$ and ozone), the 7-month average concentration was used to represent the corresponding annual average concentration in each grid cell.

Ambient concentrations of pollutants in each grid cell were assumed to be log-normally distributed in our model. As explained, the means and standard deviations in each grid cell were separately estimated using both the measurement-based approach and the model-based approach.

Natural Background Levels of PM and Ozone

For each pollutant, a natural background level that reflects the nonanthropogenic concentration is needed to determine the disease burden attributable to manmade air pollution. Here, "background level" does not mean a threshold concentration below which there are no health effects. Current scientific evidence does not clearly indicate a threshold for either PM or ozone for estimating health impacts. Instead, some studies report that the association between PM or ozone and mortality persists at levels close to zero (Schwartz et al. 2002; Jerrett et al. 2009). This study assumed no threshold below which pollutants do not cause harm. The natural background level was subtracted from the ambient pollutant concentration to estimate the disease

Table 4.5 Numerical assumptions for the natural background of PM and ground-level ozone

	Measurement-based approach		CMAQ model-based approach	
Pollutant (unit)	Lower bound	Upper bound	Lower bound	Upper bound
PM_{10} (μg/m³)	10	90	10	50
$PM_{2.5}$ (μg/m³)	5	35	5	15
Ground-level ozone (ppb)	0	25	0	25

burden due to anthropogenic pollution rather than pollution from natural sources, which is not easily controlled by humans and thus of less interest.

WHO recommends natural background levels of 10 and 5 μg/m³ for PM_{10} and $PM_{2.5}$, respectively, based on observations in typical urban areas in the United States (Ostro 2004). These values might be too low for the UAE, where dust storms can significantly increase the natural background PM levels. Given this, the PM levels recommended by WHO were employed as the lower bounds, and the upper bounds were chosen based on previous air quality studies of desert regions in the United States. For ozone, 0 ppb was assumed to be the lower bound and 25 ppb was applied as the upper bound (Anenberg et al. 2010). The natural background of a pollutant was assumed to be uniformly distributed in our model. Table 4.5 lists the assumptions for the natural background variable.

Size of the Exposed Population

The UAE population was mapped on the same grid created for estimating pollutant concentrations, and the population within each grid cell (approximately 55 km² for the model and 0.865 km² for mapping) was assumed to be exposed to the same pollutant concentration. The population within each grid cell was obtained from the LandScan Global Population Dataset (Oak Ridge National Laboratory 2007). The total population in each emirate, and in the whole UAE, obtained from this database compares well to the population estimates by the UAE Ministry of Health (Table 4.6, with a UAE total population during the study year of nearly 4.5 million).

Baseline Number of Illnesses

As Chap. 3 describes, mortality data for 2008 were obtained from the Health Authority–Abu Dhabi and were extrapolated to the UAE population as a whole, adjusting for differing mortality rates among different demographic groups (ethnicities, gender). Corresponding to Table 4.2, cause-specific deaths in specific age groups were obtained, and mortality rates were estimated by dividing deaths by the age-specific population. Table 4.6 summarizes the estimated 2008 UAE population of three age groups: all ages, adults over 30, and children under 5. Given that age-specific

Table 4.6 Approximate population estimates by emirate and age group, end of 2007

Population	Abu Dhabi	Dubai	Sharjah	Ajman	Umm Al Quwain	Ras Al Khaimah	Fujairah	UAE total
Total	1,493,000	1,478,000	882,000	224,000	52,000	222,000	137,000	4,488,000
Over 30	708,000	716,000	427,000	108,000	25,000	107,000	66,000	2,173,000
Under 5	77,000	100,000	60,000	15,000	4,000	15,000	9,000	305,000

UAE Ministry of Health (2008)

population estimates were only available for Abu Dhabi emirate (Statistics Center–Abu Dhabi 2008) and the entire UAE (Ministry of Health 2008), the UAE-wide age distribution was applied to the population in each of the six remaining emirates to derive age-specific population.

Table 4.7 summarizes the total number of deaths and medical visits for the health end points considered in this chapter. In Figure 4.3, baseline mortality rates are illustrated in a map to demonstrate how they vary across the UAE.

Finally, using the ambient concentration in a grid cell (approximately 55 km^2), the background level, concentration-response coefficient, baseline incidence rate and population, the model calculates the annual number of deaths and health-care facility visits in each grid cell attributable to an air pollutant of interest. The attributable deaths or health-care facility visits in each grid cell are then aggregated for the total disease burden due to a pollutant in an emirate as well as in the entire UAE. The next section describes the details of the model itself.

Structure of Model

The outdoor air pollution portion of the *UAE Environmental Burden of Disease Model* is constructed based on the equations presented earlier. Figure 4.4 shows the top-level diagram of the module. The model is divided by the two types of air pollutants analyzed in this study: PM and ground-level ozone. The PM and ground-level ozone dashboards contain all the input variables in the model, and users can easily change the values of input parameters and observe outputs. The two ***Details*** nodes contain the model details. ***Outdoor Air Pollution Globals*** contains indexes and assumptions that are globally used in the model, as shown in Fig. 4.5 and Tables 4.6 and 4.8.

The ***Particulate Matter Details*** module contains two submodules for the two indicators of PM: PM_{10} and $PM_{2.5}$. The ***Ground-Level Ozone Details*** module also contains two submodules for the two types of measurements of ozone: ***Daily Average*** (indicator of short-term exposure) and ***Annual Average*** (indicator of long-term exposure). The structures of all these submodules are quite similar since the same mathematical formulations are used. The details of the PM_{10} submodule are discussed next as an example. Figure 4.6 shows the influence diagram of the PM_{10} submodule, followed by Table 4.9, which contains the details of each node.

Table 4.7 Estimated total number of selected cause-specific mortality and health-care facility visits in the UAE, 2008

	All-cause mortality[a]	All-cause mortality in adults over age 30[b]	Cardiopulmonary mortality in adults over age 30[c, d]	Respiratory mortality in adults over age 30[c, d]	Lung/trachea/bronchus cancer mortality in adults over age 30[c, d]	Respiratory mortality in children under age 5[d, e]	Respiratory health-care facility visits[f]	Cardiopulmonary health-care facility visits[f]
Abu Dhabi	2,949	2,075	412	68	38	9	97,271	135,021
Dubai	2,919	2,053	415	67	37	9	39,864	92,227
Sharjah	1,742	1,225	244	40	22	5	23,322	50,190
Ajman	442	311	62	10	6	1	5,802	12,247
Umm Al Quwain	103	72	14	2	1	0	1,252	2,446
Ras Al Khaimah	439	308	65	10	6	1	5,289	9,666
Fujairah	271	190	41	6	3	1	3,249	5,869
UAE	8,865	6,235	1,254	203	113	27	176,048	307,667

[a] Fraction of total deaths among the total population in each emirate
[b] Fraction of total deaths occurring in adults over age 30 among the estimated total population of adults over age 30
[c] Calculated similarly to all-cause mortality in adults over age 30; see note (b)
[d] See Table 4.2 for ICD-9 codes
[e] Calculated similarly to all-cause mortality in adults over age 30, except using mortality and population data for the 0–4 age group
[f] See Table 4.3 for ICD-9 codes

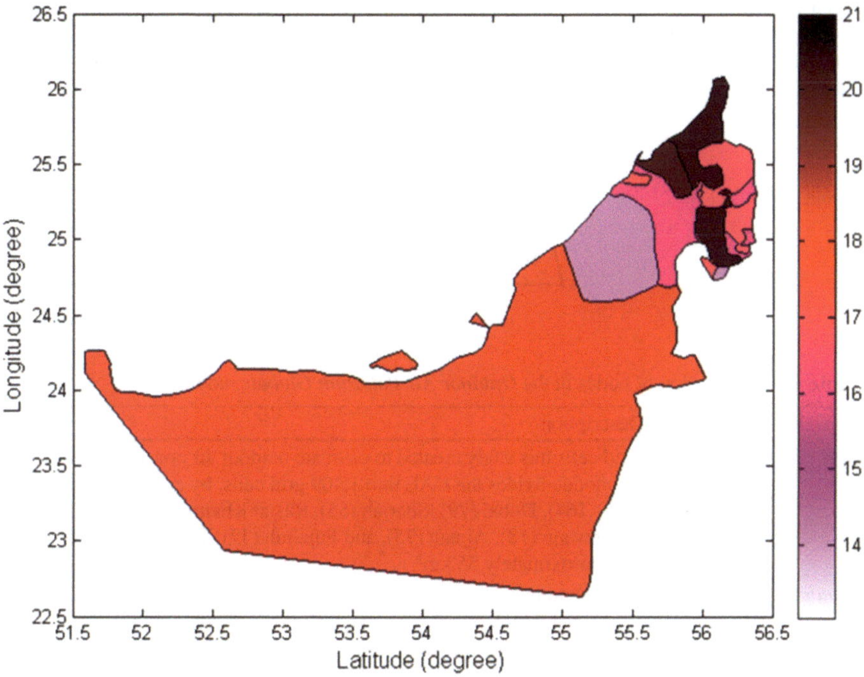

Fig. 4.3 2008 all-cause baseline mortality rates in the UAE (mortality/10,000 people) (map created by Prahlad Jat, University of North Carolina, Chapel Hill)

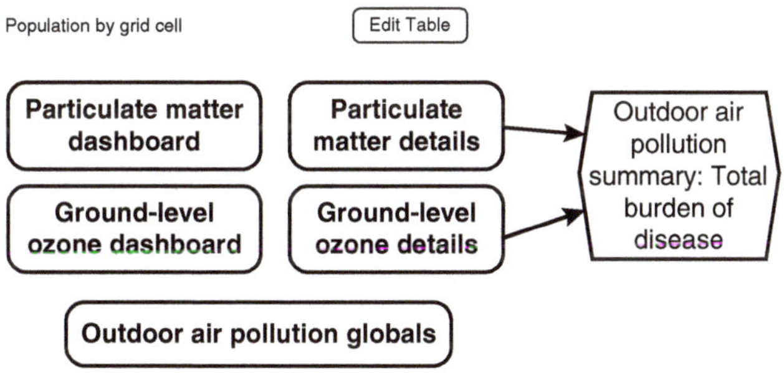

Fig. 4.4 Top-level diagram of the ***Outdoor Air Pollution*** module

The structure and variables are the same in the other three submodules (***$PM_{2.5}$***, ***Ozone – Daily Average***, and ***Ozone – Annual Average***) as in the ***PM_{10}*** submodule. The differences only lie in the numerical assumptions and health endpoints considered.

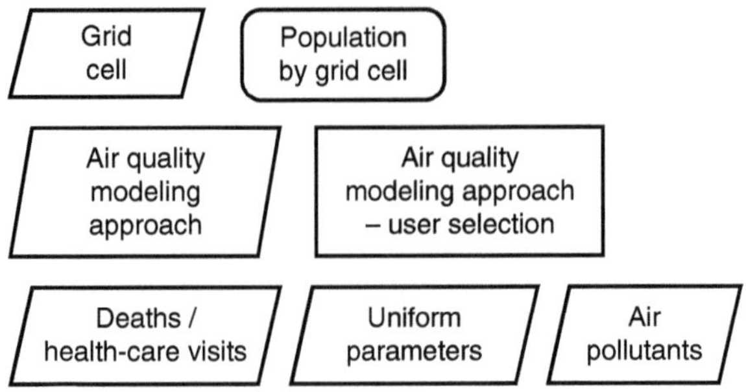

Fig. 4.5 Details of the ***Outdoor Air Pollution Globals*** module

Table 4.8 Description of nodes in the ***Outdoor Air Pollution Globals*** module

Name of node	Description
Grid cell	Grid cells this study created to estimate outdoor air concentrations. The model divides the UAE into 1,409 grid cells, by emirate: Abu Dhabi (1,164), Dubai (79), Sharjah (55), Ras al Khaimah (51), Umm al Quwain (18), Ajman (75), and Fujairah (37). The resolution is approximately 55 km^2.
Population by grid cell	Population in each of the 1,409 grid cells, estimated using Landscan Global Population Dataset (Oak Ridge National Laboratory 2007).
Air quality modeling approach	The two approaches used to estimate outdoor air quality in the UAE: measurement-based and CMAQ model-based.
Air quality modeling approach – user selection	A decision node for users to select the air quality modeling approach (measurement-based or model-based) desired in a run. The model is set to run one modeling approach at a time in order to reduce the burden on computer memory.
Deaths/health-care facility visits	The two types of health endpoints included in this model: deaths and health-care facility visits.
Uniform parameters	Key parameters of a uniform distribution, including low and high values. The natural background levels of all the air pollutants included in this study are assumed to be uniformly distributed.
Air pollutants	An index for the two air pollutants included in the burden of disease assessment: PM and ground-level ozone.

Estimated Burden of Disease

Mean Estimates by the Measurement-Based Approach

Particulate Matter (PM)

Table 4.10 summarizes the mean estimates of cause-specific deaths and health-care facility visits resulting from exposure to ambient PM by emirate and in the entire UAE, based on our statistical analysis of measurements of air pollutant concentrations. All estimates are reported to two significant figures.

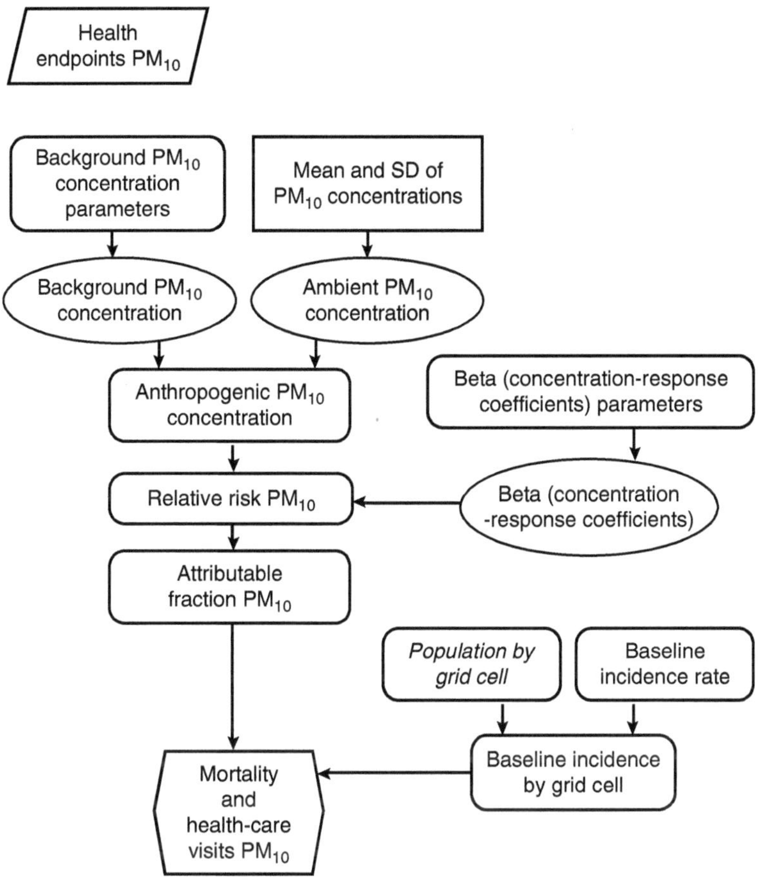

Fig. 4.6 Influence diagram of the $\boldsymbol{PM_{10}}$ submodule

Table 4.10 shows that short-term PM_{10} exposure is estimated to cause 420 all-cause premature deaths annually in the general population of the UAE and two respiratory deaths in children under 5 years. Although the β value for respiratory deaths in children under 5 is significantly greater than that for all-cause mortality, the latter estimate is rather small due to the low baseline incidence (in 2008, there were a total of 27 respiratory deaths in children under 5 in the UAE).

Long-term $PM_{2.5}$ exposure is estimated to cause 650 all-cause deaths in adults over 30 in the UAE in 2008. We also estimate 180 cardiopulmonary deaths (roughly 40% of all-cause $PM_{2.5}$ deaths) and 23 lung cancer deaths (roughly 4% of all-cause $PM_{2.5}$ deaths). Although the β value, and consequently the attributable fraction, for lung cancer mortality is greater than those for all-cause and cardiopulmonary deaths, lung cancer only accounts for a rather small fraction of all-cause $PM_{2.5}$ mortality due to the low baseline incidence (113 lung cancer deaths occurred in the UAE in adults over 30 in 2008). Due to the greater β value, all-cause deaths attributable to long-term

Table 4.9 Description of nodes in the ***PM_{10}*** submodule

Name of node	Description	Equation
Health endpoints PM_{10}	Number of deaths associated with short-term exposure to ambient PM_{10}: all-cause mortality and respiratory mortality in children under 5 (Table 4.7).	
Background PM_{10} concentration parameters	This study assumes a wide range of natural background that is uniformly distributed. In the measurement-based approach, the minimum and maximum values are assumed to be 10 and 90 μg/m³; in the CMAQ-based approach, the minimum and maximum values are assumed to be 10 and 50 μg/m³.	
Background PM_{10} concentration	This variable specifies the shape of the probability density function of the background PM_{10} concentration as uniform and calls the parameters entered in the parent node ***Background PM_{10} concentration parameters***.	*Uniform (low, high)*
Mean and SD of PM_{10} concentrations	Means and standard deviations (SD) of PM_{10} concentrations in each grid cell, estimated by the UAE outdoor air quality modeling group (using the measurement-based or the model-based approach).	
Ambient PM_{10} concentration	Annual ambient PM_{10} concentrations in 2007–2008 in each grid cell, defined as lognormally distributed and characterized by mean and standard deviation.	*Lognormal (mean, SD)*
Anthropogenic PM_{10} concentration	Ambient PM_{10} concentration as a result of anthropogenic pollution, estimated by subtracting the background concentration from the ambient concentration.	*Ambient concentration – Natural background*
Beta (concentration-response coefficients) parameters	Concentration-response coefficient from literature, defined as normally distributed and characterized by mean and standard deviation.	
Beta (Concentration-response coefficients)	This variable specifies the shape of the probability density function of the concentration-response coefficients as normal and calls the parameters entered in the parent node ***Beta (concentration-response coefficients) parameters***.	*Normal (mean, SD)*
Relative risk PM_{10}	In epidemiologic studies, relative risk (RR) is defined as the exponential function e to the product of anthropogenic concentration and concentration-response coefficient (β, see Eq. 4.3).	$RR = e^{\beta\Delta x}$

(continued)

Table 4.9 (continued)

Name of node	Description	Equation
Attributable fraction PM$_{10}$	By WHO's definition (Ostro 2004), attributable fraction (AF), meaning the fraction of a certain health outcome that is attributable to exposure to the environmental health risk of interest (see Eq. 4.2).	$AF = \frac{RR-1}{RR}$
Population by grid cell	An alias of the node ***Population by grid cell*** in the ***Outdoor Air Pollution Globals*** Module.	
Baseline incidence rate	Calculated by dividing the total incidence (in this case, the total number of deaths) in an emirate by the total population in that emirate.	$\frac{Health\ endpoints\ PM_{10}}{Population}$
Baseline mortality	Baseline mortality among the population studied for 2008, equal to baseline mortality rate multiplied by the study population.	*Baseline incidence rate × Population*
Mortality PM$_{10}$	Annual number of deaths attributable to exposure to ambient PM_{10} in 2008, equal to baseline mortality multiplied by AF. Deaths in each grid cell are aggregated to obtain the total mortality due to PM_{10} by emirate as well as in the whole UAE.	*Attributable fraction × Baseline mortality*

Table 4.10 Measurement-based estimates of annual cause-specific deaths and health-care facility visits attributable to ambient PM in the UAE (2008)

Health endpoint	Abu Dhabi	Dubai	Sharjah	Ajman	Umm Al Quwain	Ras Al Khaimah	Fujairah	UAE total
PM_{10}								
Short-term, all-cause mortality	150	140	78	21	5	21	11	420
Short-term, respiratory mortality in children <5	1	1	0	0	0	0	0	2
Respiratory health-care facility visits	5,100	1,900	1,100	280	61	260	140	8,900
Cardiovascular health-care facility visits	2,600	1,700	8,708	220	44	170	94	5,700
$PM_{2.5}$								
Long-term, all-cause mortality in adults >30	230	210	120	31	7	31	17	650
Long-term, cardiopulmonary mortality in adults >30	63	61	35	9	2	9	5	180
Long-term, lung cancer mortality in adults >30	8	8	4	1	0	1	1	23

exposure to $PM_{2.5}$ are roughly 1.5 times those attributable to short-term exposure to PM_{10}, even if the former only represent deaths among adults over 30.

WHO suggests aggregating cardiopulmonary and lung cancer mortality related to long-term exposure and respiratory mortality in infants and children related to short-term exposure when reporting the total burden of disease estimates for outdoor pollution (Ostro 2004), mainly because these cause-specific deaths are not affected by regional differences in baseline mortality rates when extrapolating U.S. concentration-response coefficients to another population. However, we believe that cardiopulmonary and lung cancer mortality may not be able to capture all the premature deaths caused by PM exposure. Given this, we combined all-cause deaths in adults related to $PM_{2.5}$ and respiratory deaths in children related to PM_{10} in aggregating the total PM-related premature deaths. Therefore, anthropogenic PM is estimated to cause about 650 premature deaths annually, which accounts for about 7% of the total deaths in the UAE (a total of 8,865 deaths occurred in the UAE in 2008). This estimate is still conservative because people ages 5–30 are not included. If only the cause-specific deaths due to long-term exposure to $PM_{2.5}$ (cardiopulmonary and lung cancer deaths) and respiratory deaths in children due to PM_{10} short-term exposure are included, the PM-mortality estimate is 210 deaths per year, accounting for approximately 2% of the total deaths in 2008. This estimate agrees with WHO's estimate of 200 PM-related deaths in 2008, based on assumptions including a mean urban PM_{10} concentration of 109 $\mu g/m^3$, a background PM_{10} level of 10 $\mu g/m^3$, a total population of 4.5 million and the same cause-specific deaths (cardiopulmonary and lung cancer deaths in adults and respiratory deaths in children) (Ostro 2004; WHO 2009a).

Regarding health-care facility visits, it is estimated that PM_{10} exposure caused approximately 8,900 respiratory health-care facility visits and approximately 5,700 cardiovascular health-care facility visits in 2008, which accounted for about 5% of total annual respiratory and 2% of total annual cardiovascular health-care facility visits in the UAE. Combining the estimated health-care facility visits for cardiovascular or respiratory illnesses produces a total of 15,000 estimated visits due to anthropogenic PM annually in the UAE.

We next illustrate the results by mapping the all-cause deaths due to long-term exposure to $PM_{2.5}$. As mentioned before, we used the finer grid cells (30×30 arc sec resolution) in mapping the deaths due to outdoor air pollution. The attributable fraction (AF) in Eq. 4.2 for $PM_{2.5}$ exposure can be visualized by mapping AF over the entire UAE for the study year 2008 as shown in Fig. 4.7. The figure indicates that the Abu Dhabi City area has the highest AF value in the UAE, owing to the higher levels of PM pollution in this area.

Figure 4.8 shows all-cause deaths per 100,000 population per year (mortality rate) caused by $PM_{2.5}$ exposure. It shows that the area around Abu Dhabi Island/Middle Region has the highest rates due to its higher level of $PM_{2.5}$ (resulting in high attributable fractions as shown in Fig. 4.7) and the relatively high baseline all-cause mortality rate in Abu Dhabi emirate as well (see Table 4.7). Contrastingly, Dubai emirate has the lowest $PM_{2.5}$-related mortality rates due to its lowest baseline mortality rate among all the emirates and a lack of air quality measurements in Dubai City area, where peak PM concentrations would be expected. If urban concentrations

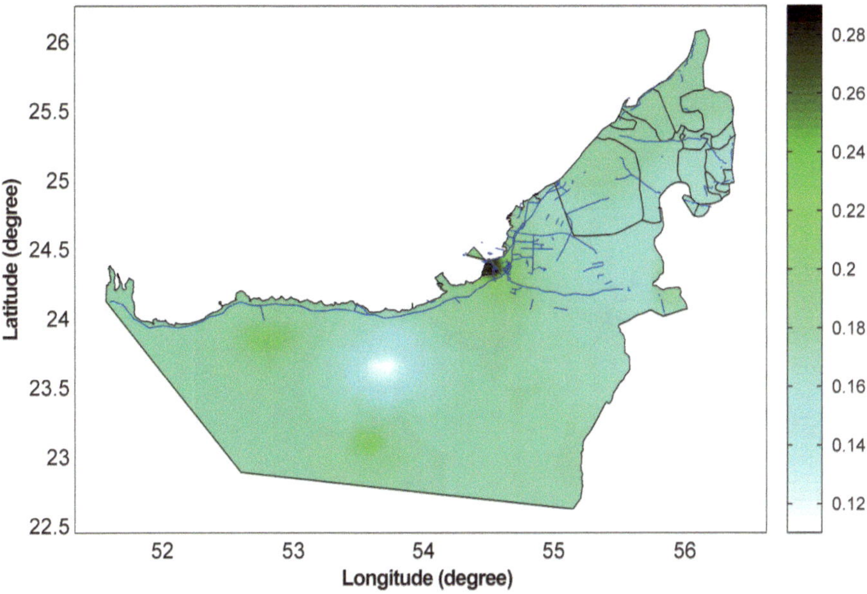

Fig. 4.7 2008 attributable fraction of all-cause mortality in adults over 30 years due to long-term exposure to $PM_{2.5}$ (μg/m^3). *Blue lines* show state and international highways with four lanes or more, and *black lines* show the emirate boundaries within the UAE (map created by Prahlad Jat, University of North Carolina, Chapel Hill)

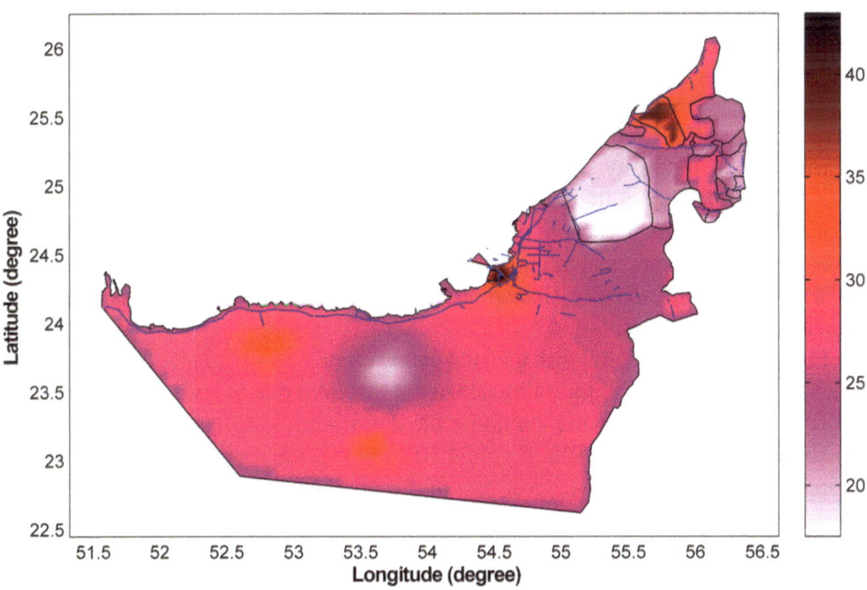

Fig. 4.8 2008 attributable all-cause mortality rate (mortality/100,000 people) due to long-term exposure to $PM_{2.5}$ (map created by Prahlad Jat, University of North Carolina, Chapel Hill)

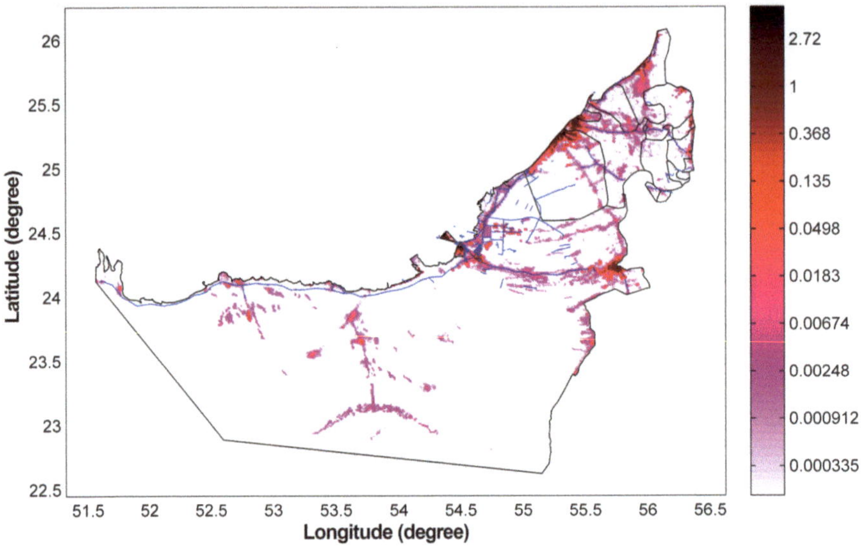

Fig. 4.9 2008 attributable all-cause mortality density (mortality/30×30 arc sec^2), in log scale, due to long-term exposure to $PM_{2.5}$ (μg/m^3) (map created by Prahlad Jat, University of North Carolina, Chapel Hill)

in Dubai are similar to those in Abu Dhabi, this measurement-based method would estimate many more premature deaths in the Dubai City area.

Mortality density, defined as the total annual deaths attributable to a pollutant ($PM_{2.5}$ in this case) in 2008 within each grid cell (30×30 arc sec resolution), is visualized in Fig. 4.9. This figure shows the combined effect of pollution and population distribution as mortality density (i.e., deaths per grid cell). It shows that coastal areas of both Dubai emirate and the city of Abu Dhabi have the highest mortality density, compared with the rest of the UAE. Although Dubai emirate has the lowest baseline mortality rate, its high population density along the coastal area contributes to the high mortality density, since mortality density integrates the effects of population and pollution distribution.

Figures 4.7, 4.8, and 4.9 give a visual overview of the attributable fraction, attributable mortality rate, and mortality density in the UAE due to $PM_{2.5}$. However these plots do not provide information about the day-to-day variations of deaths caused by short-term exposure to PM. Figure 4.10 shows the temporal variations of attributable deaths caused by PM_{10} over part of the study period in the entire UAE.

Ozone

The mean estimates of cause-specific deaths resulting from exposure to ambient ozone by emirate and in the entire UAE, based on our statistical analysis of measurements

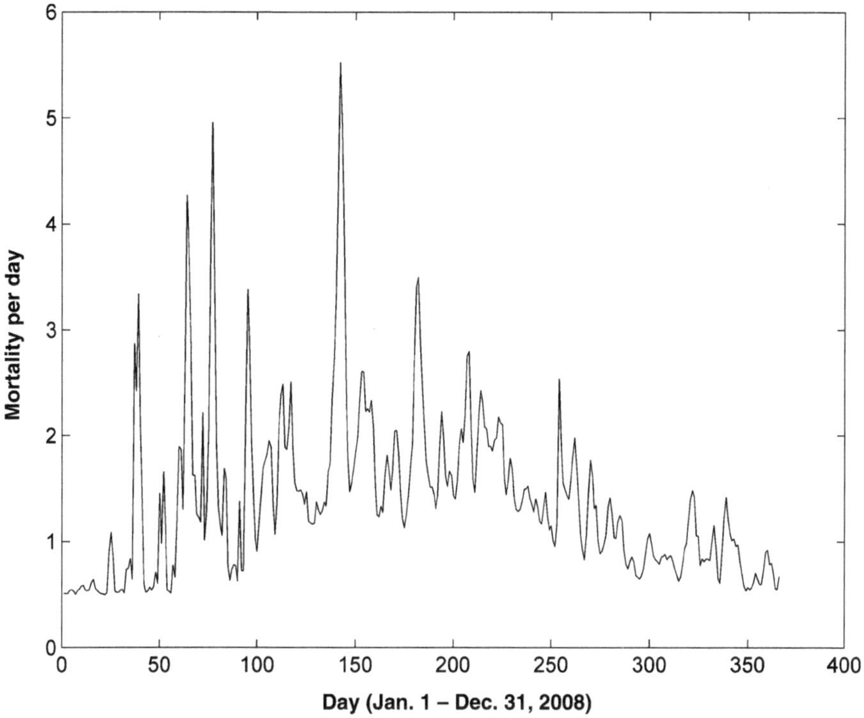

Fig. 4.10 Daily deaths due to short-term exposure to ambient PM_{10} in the UAE (January 1 – December 31, 2008)

of air pollutant concentrations, are summarized in Table 4.11. As shown in Table 4.11, ground-level ozone was estimated to cause 77 all-cause deaths and 27 cardiovascular and respiratory deaths in 2008 due to short-term exposure. These account for approximately 0.8% of total all-cause deaths in the UAE and 1% of cardiovascular and respiratory deaths. Long-term exposure is estimated to cause 27 respiratory deaths in adults over 30. The long-term mortality estimate is smaller than the short-term all-cause mortality estimate as the former is based on respiratory deaths in adults.

We selected all-cause deaths caused by short-term ozone exposure to represent the total ozone premature deaths, given that only respiratory mortality in adults over 30 is included in estimating the effects of long-term exposure. Therefore, anthropogenic ozone was estimated to cause 77 premature deaths in 2008. This result demonstrates that premature deaths caused by ground-level ozone are less frequent than those caused by PM.

Ground-level ozone was estimated to cause approximately 9,800 respiratory health-care facility visits in 2008, which accounts for about 6% of total respiratory health-care facility visits that year.

Table 4.11 Measurement-based estimates of annual cause-specific deaths and health-care facility visits attributable to ground-level ozone in the UAE (2008)

Health endpoint	Abu Dhabi	Dubai	Sharjah	Ajman	Umm Al Quwain	Ras Al Khaimah	Fujairah	UAE total
All-cause mortality caused by short-term exposure	27	25	15	4	1	4	2	77
Cardiovascular and respiratory mortality caused by short-term exposure	10	9	5	1	0	1	1	27
Respiratory mortality in adults over 30 caused by long-term exposure	8	9	6	1	0	1	1	27
Respiratory health-care facility visits caused by short-term exposure	5,600	2,200	1,300	310	68	290	180	9,800

Mean Estimates by the CMAQ Model-Based Approach

Tables 4.12 and 4.13 summarize the mean estimates (rounded to two significant figures) of cause-specific mortality attributable to ambient PM and ozone, by emirate and in the entire UAE, using ambient concentration estimates from CMAQ modeling. Table 4.12 shows that PM mortality estimates based on CMAQ modeling results are about 25% of the estimates based on measurements shown in Table 4.10, due to the fact that PM_{10} from CMAQ modeling is generally significantly lower than the concentrations found at monitoring sites, as is the modeled $PM_{2.5}$ compared with observed $PM_{2.5}$, calculated as a fraction of the measured PM_{10}. Also, using CMAQ modeling results, PM_{10} was estimated to cause roughly 1,800 respiratory health-care facility visits and roughly 1,100 cardiovascular health-care facility visits. Similar to the mortality estimates, the CMAQ-based estimates of health-care facility visits attributable to PM_{10} exposure are smaller than the measurement-based estimates because of the smaller PM_{10} estimates from the CMAQ results. These results emphasize the importance of further work to calibrate the CMAQ model to ambient measurements. Such a model is necessary to estimate how decreasing pollutant emissions from specific sources, such as cars or power plants, will affect PM levels in the ambient air. Such predictions, in turn, are needed to estimate the health benefits of reducing ambient PM concentrations. The current CMAQ model, which was developed in pilot version for this research project and has not been fully calibrated, currently would underestimate the health benefits of policies to decrease PM concentrations.

Table 4.13 shows ozone mortality estimates based on CMAQ results. Only long-term effects are quantified, using the annual average of daily 1-h-maximum

Table 4.12 CMAQ model-based estimates of annual cause-specific deaths and health-care facility visits attributable to ambient PM in the UAE (2008)

Health endpoint	Abu Dhabi	Dubai	Sharjah	Ajman	Umm Al Quwain	Ras Al Khaimah	Fujairah	UAE total
PM_{10}								
Short-term, all-cause mortality	29	26	17	5	1	6	2	88
Short-term, respiratory mortality in children under 5	0	0	0	0	0	0	0	1
Respiratory health-care facility visits	1,000	380	230	75	17	79	28	1,800
Cardiovascular health-care facility visits	510	320	180	58	12	53	18	1,100
$PM_{2.5}$								
Long-term, all-cause mortality in adults over 30	53	48	29	10	2	11	4	160
Long-term, cardiopulmonary mortality in adults over 30	16	14	9	3	1	4	1	47
Long-term, lung cancer mortality in adults over 30	2	2	1	0	0	0	0	6

Table 4.13 CMAQ model-based estimates of annual respiratory mortality in adults over 30 attributable to long-term exposure to ground-level ozone in the UAE (2008)

Health endpoint	Abu Dhabi	Dubai	Sharjah	Ajman	Umm Al Quwain	Ras Al Khaimah	Fujairah	UAE total
Respiratory mortality in adults over 30[a]	12	14	8	2	0	2	1	40

[a]Natural background assumption: 0 ppb

concentrations. Ozone mortality estimates based on CMAQ results are higher (approximately 1.5 times) than the estimates based on measurements in Table 4.11, since CMAQ-modeled ozone concentrations are generally greater than those found at monitoring sites. Again, this results illustrates the need to calibrate the CMAQ model.

In summary, the CMAQ-based estimates of mortality are smaller than the measurement-based estimates for PM, but the CMAQ-based estimates are greater for ozone. For $PM_{2.5}$, the CMAQ-based approach models $PM_{2.5}$ concentrations that are not measured directly, and therefore the CMAQ estimates may be more credible in this case. CMAQ modeling may also better represent anthropogenic pollution that is central to air quality regulation. Nonetheless, in order to do so, the model needs to be better calibrated against observational data.

Table 4.14 Uncertainty analysis results of total mortality and health-care facility visits attributable to PM or ground-level ozone in the UAE (2008)

	Measurement-based approach		CMAQ-based approach	
Health endpoint	Mean	95% CI	Mean	95% CI
PM_{10}				
Short-term, all-cause mortality	420	(150,780)	88	(14, 210)
Short-term, respiratory mortality in children under 5	2	(0, 6)	1	(0, 2)
Respiratory health-care facility visits	8,900	(2,900; 18,000)	1,800	(280; 4,700)
Cardiovascular health-care facility visits	5,700	(2,100; 10,000)	1,100	(180; 2,700)
$PM_{2.5}$				
Long-term, all-cause mortality in adults over 30	650	(150; 1,400)	160	(27, 410)
Long-term, cardiopulmonary mortality in adults over 30	206	(25, 519)	50	(8, 120)
Long-term, lung cancer mortality in adults over 30	19	(3, 42)	6	(1, 16)
Ozone				
All-cause mortality caused by short-term exposure	77	(21,160)	N/A	
Cardiovascular and respiratory mortality caused by short-term exposure	27	(7,57)	N/A	
Respiratory mortality in adults over 30 caused by long-term exposure	27	(13, 44)	40	(21, 60)
Respiratory health-care facility visits caused by short-term exposure	9,800	(3,100; 19,000)	N/A	

Uncertainty Analysis Results

We considered uncertainties involved in three input variables in the model: natural background of a pollutant, mean concentrations of a pollutant in each grid cell, and concentration-response coefficients (β). Table 4.14 summarizes the uncertainty analysis results of attributed deaths or health-care facility visits in the entire UAE, including the mean estimates (as shown in Tables 4.10, 4.11, 4.12 and 4.13) and their 95% confidence intervals.

All-cause mortality due to $PM_{2.5}$ and respiratory mortality in children due to PM_{10} are included in the aggregate total of PM-related premature deaths. To represent the number of premature deaths due to ozone, we used all-cause deaths caused by short-term ozone exposure. For the model-based approach, we relied on respiratory mortality caused by long-term exposure since daily average estimates were not available. Therefore, the uncertainty analysis does not show short-term health effects estimates for ozone as calculated with the CMAQ-based approach.

Table 4.15 Summary of total mortality and health-care facility visits attributable to PM or ground-level ozone in the UAE, 2008

	Measurement-based approach		CMAQ model-based approach	
Health outcome	Mean	95% CI	Mean	95% CI
PM				
Deaths	650	(150; 790)	160	(27; 410)
Health-care facility visits	15,000	(5,400; 27,000)	3,000	(500; 7,100)
Ozone				
Deaths	77	(21; 160)	40	(21; 60)
Health-care facility visits	9,800	(3,100; 19,000)	N/A	

Table 4.16 Sensitivity analysis: effects of changing key model input variables on burden of disease estimates

		Ambient pollutant concentration		Natural background concentration		Concentration-response coefficient	
Health outcome	Original estimate	−25%	25%	−25%	25%	−25%	25%
PM							
Deaths	650	420	920	780	530	510	800
Health-care facility visits	15,000	9,800	21,000	17,000	13,000	12,000	17,000
Ozone							
Deaths	77	51	110	88	66	57	94
Health-care facility visits	9,800	6,400	13,000	11,000	8,500	7,400	12,000

Sensitivity Analysis Results

The wide confidence intervals around the estimated burden of disease due to outdoor air pollution (as summarized in Table 4.15) indicate that these estimates are subject to considerable uncertainty. As noted above, we estimated how uncertainty in three of the key inputs—the measured ambient concentrations, the natural background levels of pollutants, and the concentration-response coefficients—propagates through our risk model and affects the estimated burden of disease. Next we estimate the sensitivity of our predictions to changes in the key input variables by adjusting each key input one at a time by ±25%, while holding all the remaining variables at their baseline values. The sensitivity analysis was conducted only in the measurement-based approach. Table 4.16 presents the results of the sensitivity analysis. These results illustrate that the estimates presented here are most sensitive to assumptions about the pollutant concentrations in each grid cell. A 25% increase or decrease in estimated concentration causes the estimated burden of disease to change by about +40% and – 35%, respectively. These results illustrate that ambient air quality measurement errors would result in considerable under- or overestimation of the burden of disease.

The estimates are less sensitive to the assumed background concentration of pollutants than to the total ambient pollution levels. As Table 4.16 shows, a 25% change in the estimated natural background level of pollutants could change the estimates of mortality or health-care by about 15–20%, depending on the health outcome. As discussed previously in this chapter, higher natural background levels of PM might exist in the UAE due to the significant dust storms in the region. For PM, when higher background levels are applied, the total mortality and health-care facility visits attributed to anthropogenic air pollution both decrease.

The burden of disease estimates change approximately in direct proportion with the assumed concentration-response coefficient. Adjusting β upwards or downwards by 25% results in a change of ±25% in the burden of disease estimate. In the future, conducting epidemiologic studies to determine concentration-response coefficients specific to the region will be important to enable more accurate estimates of the burden of disease attributable to outdoor air pollution. Previous research has shown that concentration-response coefficients can vary by region (Aunan and Pan 2004), but since no UAE-specific studies were available, this study relied on international estimates of these coefficients. As Table 4.16 shows, however, altering the assumptions about these coefficients could have an important effect on the estimated burden of disease due to outdoor air pollution, though this effect is not as large as that of the assumed total pollutant concentrations in the ambient air.

Information Needed to Improve Future Burden of Disease Predictions

Presently, there is significant uncertainty associated with ambient concentration estimates. This can be attributed to uncertainties in ambient concentration estimates from both measurement uncertainties and modeling results. These uncertainties can be reduced with ambient monitoring data from other emirates, as well as detailed estimates of air pollutant emissions from all the major pollutant sources throughout the UAE. With these data, CMAQ predictions would be greatly improved.

Given that frequent dust storms occur in the area, natural background levels of PM could be considerably higher than those found in other places. This natural background may show a strong daily variability associated with windblown dust loadings. Sensitivity analyses suggest that increases in natural background assumptions result in significant decreases in attributable mortality estimates for both pollutants. Separating the natural and anthropogenic contributions to PM, however, may be difficult given that human activities may increase the contribution of dust to PM. Future research should consider using the air quality model to simulate atmospheric concentrations in the absence of anthropogenic emissions, and define anthropogenic contributions to air pollution accordingly. In addition, there is significant uncertainty in published estimates of concentration-response coefficients. This study relies on concentration-response coefficients from representative epidemiologic studies in the United States. The bias caused by extrapolation of findings from one location to another might be reduced in the future as local epidemiologic studies become available and are incorporated into health impact assessments.

Finally, although this study estimates respiratory and cardiovascular health-care facility visits attributable to outdoor air pollution, scientific studies have linked exposure to other minor morbidity outcomes that may not lead to health-care facility visits, such as acute respiratory symptom days and minor respiratory symptom days (Ostro and Chestnut 1998). These health effects can also result in significant economic loss due to loss of productivity from lost work days. Future research could include these morbidity estimates as the relevant baseline health statistics become available so that a more comprehensive assessment of the disease burden due to outdoor air pollution in the UAE could be made.

Conclusions

Air pollution has been a major public health concern throughout the world, particularly in urban areas where population density and pollutant concentrations are greatest. Current scientific evidence generally considers PM and ground-level ozone as the two air pollutants with the greatest health impact on humans. This study quantitatively assesses the national burden of anthropogenic PM and ozone air pollution on public health in the UAE, focusing on premature mortality and health-care facility visits. An integrated approach was used, similar to methods used previously to estimate the disease burden due to outdoor air pollution elsewhere. Grid cells with fine resolution (approximately 55 km^2 for the model and 0.865 km^2 for the maps) were created within the UAE to reflect both the spatial and temporal variations of pollution levels. Mean pollutant concentrations in each grid cell were estimated using either a measurement-based approach (using only data from EAD and no data from the other three emirates known to have air-monitoring programs) or a CMAQ model-based approach.

Using the measurement-based approach, this research estimates that in 2008 the total number of premature deaths in the UAE caused by exposure to ambient particulate matter was approximately 650 (95% confidence interval 150–790). These account for about 7% (95% confidence interval 2–9%) of the total deaths occurring in the UAE in 2008. About 77 deaths (95% confidence interval 21–160) were attributable to ground-level ozone exposure (about 0.9%, with a range of 0.2–2%, of the total deaths) in 2008. With respect to excess illness, we estimate that in 2008 PM_{10} exposure caused 5,400–27,000 health-care facility visits (mean estimate: 15,000) for respiratory and cardiac illnesses. The mean estimates account for about 3% of total medical visits for these conditions. Ground-level ozone was estimated to cause 3,100–19,000 respiratory health-care facility visits (mean estimate: 9,800) in 2008. The mean estimate accounts for about 6% of total respiratory health-care facility visits in that year. Thus, in total, PM appears to cause a larger disease burden in the UAE than ozone. This study reports the mortality as well as health-care facility visit estimates due to PM and ozone separately, since current scientific evidence is unclear as to whether health effects of the two pollutants are independent. If we include only PM-related deaths, our best estimate of total deaths caused by outdoor air pollution is 650, which is possibly conservative if some of the mortality related to PM is independent of that related to ozone.

Using the CMAQ model-based approach, the estimated death numbers attributable to PM were smaller than the measurement-based estimates, whereas the estimated death numbers attributable to ozone are greater than those using the measurement-based approach. Also, the estimated health-care facility visits attributable to PM are smaller than the measurement-based estimates.

Regardless of the significant uncertainty associated with the estimated outdoor air pollutants, some general conclusions can be drawn from the overall disease burden estimates:

- Anthropogenic outdoor air pollution caused a substantial number of premature deaths in the UAE in 2008, and this mortality number is estimated to be the greatest among the eight priority environmental risk areas in the UAE.
- Particulate matter causes a significantly greater number of premature deaths than does ground-level ozone.
- Both particulate matter and ground-level ozone cause substantial health-care facility visits annually due to respiratory and cardiovascular diseases.

References

Anderson, H.R., R.W. Atkinson, J.L. Peacock, L. Marston, and K. Konstantinou. 2004. *Meta-analysis of time-series studies and panel studies of particulate matter (PM) and ozone (O3): Report of a WHO task group*. Copenhagen: World Health Organization.

Anenberg, S.C., L.W. Horowitz, D.Q. Tong, and J.J. West. 2010. An estimate of the global burden of anthropogenic ozone and fine particulate matter on premature human mortality using atmospheric modeling. *Environmental Health Perspectives* 118(9): 1189–1195.

Aunan, K., and X.C. Pan. 2004. Exposure-response functions for health effects of ambient air pollution applicable for China—A meta-analysis. *The Science of the Total Environment* 329(1–3): 3–16.

Bell, M.L., A. McDermott, S.L. Zeger, J.M. Samet, and F. Dominici. 2004. Ozone and short-term mortality in 95 U.S. urban communities, 1987–2000. *Journal of the American Medical Association* 292(19): 2372–2378.

Bell, M.L., F. Dominici, and J.M. Samet. 2005. A meta-analysis of time-series studies of ozone and mortality with comparison to the National Morbidity, Mortality, and Air Pollution Study. *Epidemiology* 16(4): 436–445.

Bell, M.L., R.D. Peng, and F. Dominici. 2006. The exposure-response curve for ozone and risk of mortality and the adequacy of current ozone regulations. *Environmental Health Perspectives* 114(4): 532–536.

Brown, K.W., W. Bouhamra, D.P. Lamoureux, J.S. Evans, and P. Koutrakis. 2008. Characterization of particulate matter for three sites in Kuwait. *Journal of Air and Waste Management* 58: 994–1003.

Cohen, A.J., H.R. Anderson, B. Ostro, K.D. Pandey, M. Krzyzanowski, N. Kunzli, K. Gutschmidt, et al. 2004. Mortality impacts of urban air pollution. In *Comparative quantification of health risks: Global and regional burden of disease attributable to selected major risk factor*, 2nd ed, ed. M. Ezzati, A.D. Lopez, A. Rodgers, and C.J.L. Murray. Geneva: World Health Organization.

Dockery, D.W., C.A. Pope III, X. Xu, J.D. Spengler, J.H. Ware, M.E. Fay, B.G. Ferris, and F.E. Speizer. 1993. An association between air pollution and mortality in six U.S. cities. *The New England Journal of Medicine* 329(24): 1753–1759.

Dominici, F., A. McDermott, M. Daniels, S.L. Zeger, and J.M. Samet. 2005. Revised analyses of the National Morbidity, Mortality, and Air Pollution Study: Mortality among residents of 90 cities. *Journal of Toxicology and Environmental Health. Part A* 68(13–14): 1071–1092.

Health Authority–Abu Dhabi. 2008. *Health Statistics 2007.*

Ito, K., S.F. De Leon, and M. Lippmann. 2005. Associations between ozone and daily mortality: Analysis and meta-analysis. *Epidemiology* 16(4): 446–457.

Jerrett, M., R.T. Burnett, C.A. Pope III, K. Ito, G. Thurston, D. Krewski, Y. Shi, E. Calle, and M. Thun. 2009. Long-term ozone exposure and mortality. *The New England Journal of Medicine* 360(11): 1085–1095.

Krewski, D., R.T. Burnett, M.S. Goldberg, K. Hoover, J. Siemiatycki, M. Jerrett, M. Abrahamowicz, and W.H. White. 2000. *Reanalysis of the Harvard Six Cities Study and the American Cancer Society Study of particulate air pollution and mortality: A special report of the Health Effects Institute's particle epidemiology reanalysis project.* Cambridge, MA: Health Effects Institute.

Krewski, D., M. Jerrett, R.T. Burnett, R. Ma, E. Hughes, Y. Shi, M.C. Turner, et al. 2009. Extended follow-up and spatial analysis of the American Cancer Society study linking particulate air pollution and mortality. *Research Report. Health Effects Institute* 140: 5–114.

Künzli, N., S. Medina, R. Kaiser, P. Quénel, F. Horak Jr., and M. Studnicka. 2001. Assessment of deaths attributable to air pollution: Should we use risk estimates based on time series or on cohort studies? *American Journal of Epidemiology* 153(11): 1050–1055.

Laden, F., J. Schwartz, F.E. Speizer, and D.W. Dockery. 2006. Reduction in fine particulate air pollution and mortality: Extended follow-up of the Harvard Six Cities Study. *American Journal of Respiratory and Critical Care Medicine* 173: 667–672.

Levy, J.I., J.K. Hammitt, and J.D. Spengler. 2000. Estimating the mortality impacts of particulate matter: What can be learned from between-study variability? *Environmental Health Perspectives* 108(2): 109–117.

Levy, J.I., T.J. Carrothers, J.T. Tuomisto, J.K. Hammitt, and J.S. Evans. 2001. Assessing the public health benefits of reduced ozone concentrations. *Environmental Health Perspectives* 109(21): 9–20.

Levy, J.I., S.M. Chemerynski, and J.A. Sarnat. 2005. Ozone exposure and mortality: An empiric Bayes metaregression analysis. *Epidemiology* 16(4): 458–468.

Li, Y., J. MacDonald Gibson, P. Jat, G. Puggioni, M. Hasan, J. West, W. Vizuete, K. Sexton, and M. Serre. 2010. Burden of disease attributed to anthropogenic air pollution in the United Arab Emirates: Estimates based on observed air quality data. *The Science of the Total Environment* 408(23): 5784–5793. doi:10.1016/j.scitotenv.2010.08.017.

National Research Council. 2008. *Estimating mortality risk reduction and economic benefits from controlling ozone air pollution.* Washington, D.C.: National Academies Press.

Oak Ridge National Laboratory. 2007. LandScan global population dataset. Oak Ridge, Tenn. http://www.ornl.gov.landscan

Ostro, B. 2004. Outdoor air pollution: Assessing the environmental burden of disease at national and local levels. Environmental Burden of Disease series, No. 5. Geneva: WHO.

Ostro, B.D., and L. Chestnut. 1998. Assessing the health benefits of reducing particulate matter air pollution in the United States. *Environmental Research (Section A)* 76: 94–106.

Ostro, B.D., R. Broadwin, and M.J. Lipsett. 2000. Coarse and fine particles and daily mortality in the Coachella Valley, California: A follow-up study. *Journal of Exposure Analysis and Environmental Epidemiology* 10(5): 412–419.

Pope III, C.A., M.J. Thun, M.M. Namboodiri, D.W. Dockery, J.S. Evans, F.E. Speizer, and J.C.W. Heath. 1995. Particulate air pollution as a predictor of mortality in a prospective study of U.S. adults. *American Journal of Respiratory and Critical Care Medicine* 151: 669–674.

Pope III, C.A., R.W. Hill, and G.M. Villegas. 1999. Particulate air pollution and daily mortality on Utah's Wasatch Front. *Environmental Health Perspectives* 107(7): 567–573.

Pope III, C.A., R.T. Burnett, M.J. Thun, E.E. Calle, D. Krewski, K. Ito, and G.D. Thurston. 2002. Lung cancer, cardiopulmonary mortality, and long-term exposure to fine particulate air pollution. *Journal of the American Medical Association* 287(9): 1132–1141.

Samet, J.M., S.L. Zeger, F. Dominici, F. Curriero, I. Coursac, D.W. Dockery, J. Schwartz, and A. Zanobetti. 2000. The national morbidity, mortality, and air pollution study, Part II: Morbidity and mortality from air pollution in the United States. *Research Report. Health Effects Institute* 94(II): 5–70.

Schwartz, J., F. Laden, and A. Zanobetti. 2002. The concentration-response relation between $PM_{2.5}$ and daily deaths. *Environmental Health Perspectives* 110(10): 1025–1029.

Sivertsen, B. 2010. Air quality monitoring and management in Abu Dhabi. Norwegian Institute for Air Research. In *Presentation at workshop to design future air quality field measurement campaigns in the United Arab Emirates*. Abu Dhabi, April 26–28.

Slaughter, J.C., E. Kim, L. Sheppard, J.H. Sullivan, T.V. Larson, and C. Claiborn. 2005. Association between particulate matter and emergency room visits, hospital admissions and mortality in Spokane, Washington. *Journal of Exposure Analysis and Environmental Epidemiology* 15(2): 153–159.

U.S. Environmental Protection Agency (EPA). 1999. The benefits and costs of the Clean Air Act, 1990–2010. EPA-410-R-99-001. U.S. Environmental Protection Agency Office of Air and Radiation. EPA Report to Congress, November.

UAE Ministry of Health. 2008. *2007 annual statistical report*. Abu Dhabi: Preventive medicine sector.

Whitford, J. 2008. Environment Agency—Abu Dhabi ambient air quality monitoring network 2007 Annual Report.

World Health Organization (WHO). 2009a. Country profiles of environmental burden of disease: United Arab Emirates. Geneva: World Health Organization. http://www.who.int/quantifying_ehimpacts/national/countryprofile/unitedarabemirates.pdf.

World Health Organization (WHO). 2009b. International Classification of Diseases (ICD). http://www.who.int/classifications/icd/en/

Zeka, A., A. Zanobetti, and J. Schwartz. 2005. Short term effects of particulate matter on cause specific mortality: Effects of lags and modification by city characteristics. *Occupational and Environmental Medicine* 62(10): 718–725.

Chapter 5
Burden of Disease from Indoor Air Pollution

Abstract Indoor air pollution has evolved into a high-priority risk across the globe, with various organizations ranking indoor air pollution in the top category of environmental risks. Indoor air pollutant concentrations are a function of indoor source emissions, the infiltration of ambient pollution via building leakage, and the air exchange rate (ventilation) in the building. Health effects range from acute conditions such as sensory irritation to chronic, potentially life-threatening conditions such as cancer and cardiovascular disease. The three primary factors that affect indoor air quality are the nature of indoor pollutant sources, ventilation of the building, and occupant behaviors. This initial modeling effort focuses on the residential environment because people spend the majority of their time indoors in residential dwellings. Deficient air quality can exist in all types of enclosed buildings and structures. In the future, the methods and models developed here could be applied to other indoor environments. The burden of disease due to a particular pollutant was calculated by multiplying the attributable fraction by the observed number of cases of the relevant health outcome in the population. The leading source of indoor air pollution contributing to excess cases of illness is environmental tobacco smoke. Altogether, it appears to cause more than 80% of the health-care facility visits attributed to indoor air pollution. The leading health outcomes attributed to indoor air pollution are cardiovascular disease and lower respiratory tract infections. An estimated 280 deaths result from those diseases, with approximately 88% of those deaths attributed to cardiovascular disease caused by environmental tobacco smoke. Our analyses suggest that indoor air pollution is a considerable risk to public health in the United Arab Emirates (UAE), accounting for at least 77,000 excess visits to health-care facilities in 2008 in addition to the 280 excess deaths. In terms of mortality, indoor air quality ranks second only to outdoor air pollution as a cause of environmentally related diseases in the UAE.

Keywords Indoor air pollution • Environmental burden of disease • Contaminant concentrations • Air exchange rates • Concentration-response function • Relative risk • Attributable fraction • Premature deaths and health-care facility visits • Environmental tobacco smoke • United Arab Emirates

J. MacDonald Gibson et al., *Environmental Burden of Disease Assessment*,
Environmental Science and Technology Library 24, DOI 10.1007/978-94-007-5925-1_5,

Overview: Nature and Sources of Indoor Air Pollution

Air pollution exposures have long been identified as a controllable cause of preventable diseases. The bulk of the research on air pollution has focused on the outdoor component, with results repeatedly showing an association between air pollution and adverse health effects (Dockery et al. 1993; Cohen et al. 2005; Pope 1991). In more recent years, however, indoor air pollution has evolved into a high-priority risk across the globe, with various organizations ranking indoor air pollution in the top category of environmental risks (California Air Resources Board 2005; DeBrouwere et al. 2007). In the national-scale ranking of environmental risks in the United Arab Emirates (UAE) described in Chap. 2, indoor air pollution ranked as the second highest priority, after outdoor air pollution.

Indoor air pollution was once thought to be primarily an issue in developing countries where the use of solid fuel indoors results in dangerously high indoor concentrations of numerous pollutants (World Health Organization 2002), but it is also a concern in developed countries where solid fuel is not used. This relatively recent increase in awareness is due to several factors, including the recognition that people spend the majority of their time indoors; the identification of many potential sources of pollutants in indoor environments; a growing body of scientific evidence documenting that indoor concentrations of pollutants can exceed health thresholds and present increased risks; and the realization that energy-efficient buildings are potentially contributing to the increased risk due to accumulation of pollutants inside tight building envelopes.

Indoor air pollutant concentrations are a function of indoor source emissions, the infiltration of ambient pollution via building leakage, and the air exchange rate (ventilation) in the building. Many indoor air pollutants are identical to outdoor air pollutants, with comparable health effects, regardless of whether they are derived from indoor or outdoor sources. These health effects range from acute conditions such as sensory irritation to chronic, potentially life-threatening conditions such as cancer and cardiovascular disease. Nazaroff and Weschler (2001) noted that the acceptable level of lifetime risk of premature mortality resulting from carcinogenic compounds in outdoor air is typically capped at one excess death per 100,000–1,000,000 people, while the lifetime risks from indoor pollutants average much higher, in the range of one excess death per 1,000–10,000 people. When indoor pollutant concentration and time spent in the indoor environment are taken into account, personal exposure to a given pollutant usually is much higher indoors than outdoors. Numerous studies have supported this fact, showing personal exposures that are highly correlated with indoor concentrations but not with outdoor concentrations (Kim et al. 2006; Liu et al. 2007). This high correlation can be attributed to the close proximity of occupants to many indoor pollutants and the increased chance of inhaling pollutants indoors, which Bennett et al. (2002) reported as a thousandfold higher than for pollutants emitted outdoors. All of these findings support the conclusion that indoor environments have the potential to present a very real risk to human health.

As mentioned previously, the extended time spent in indoor environments affects human exposures. Klepis et al. (2001) reported that people in the United States spend approximately 87% of their time indoors, with almost 69% of time spent in a residence. The authors reported similar results for the state of California, where the climate on average can be described as moderate. In more extreme climates, such as in the UAE, the percentage of time spent indoors may be expected to be higher, and the health risk from exposure to indoor pollutants also will be higher as a result (Dales et al. 2008).

Factors Affecting Indoor Air Quality

The three primary factors that affect indoor air quality are the nature of indoor pollutant sources, ventilation of the building, and occupant behaviors (U.S. Environmental Protection Agency 1991). All are important and should be considered when developing programs to reduce exposures to indoor air pollutants.

Numerous sources of indoor pollutants can be found in any home, and concentrations may vary considerably in different homes. Frequently identified sources include but are not limited to combustion products (from burning gas, wood, or kerosene); tobacco products in any form; air fresheners, incense, and candles; damp building materials or highly humid environments that result in surface condensation and mold growth; certain types of cooking; manufactured wood products; furnishings such as carpets and upholstery; cleaning products; and deteriorating or damaged lead-based paint and asbestos-containing building materials. Other sources contributing to poor indoor air quality but not necessarily emitted within the building envelope include outdoor air pollution, which can infiltrate through open windows or other points of entry, and radon gas. Radon gas may permeate through the soil and enter the home through cracks and other breaches in the foundation. The type of soil affects the rate at which radon can infiltrate a building, with sandy soil types like those found in the UAE having the greatest permeability.

Indoor pollutants may be emitted within the home from a single source such as a candle or from a source that is less containable such as new carpet, which emits volatile organic compounds for a period of time after installation. In addition, a single source may contain multiple different types of pollutants. Candles and incense emit particulate matter, often at concentrations that are a public health concern, and sometimes contain heavy metals such as lead, known to be toxic to humans. Consequently, one type of source, such as a candle, can present dramatically different health risks depending on the composition of the source.

Ventilation is an extremely important factor because it directly influences the buildup and removal of contaminants within indoor environments. It can be used to reduce the level of indoor air pollution if the source cannot be controlled. Ventilation may occur via infiltration through cracks and other openings in the building envelope, natural ventilation that occurs when windows and doors are opened, and mechanical

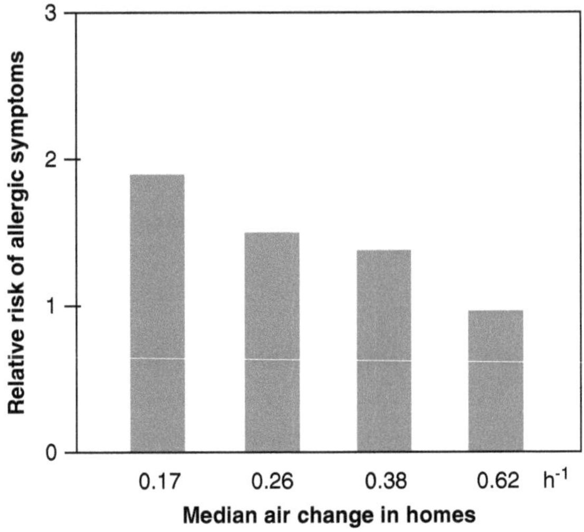

Fig. 5.1 Relationship between air exchange rates and allergic symptoms in children (developed from data in Bornehag et al. 2005)

ventilation (for example, from fans or air conditioners). The effectiveness of a building's ventilation system may depend on several factors, including:

- The age of the building. Older buildings may not have a building envelope as tight as newer buildings, which could increase the air exchange rate but also increase infiltration of outdoor air pollution.
- System design. Ventilation systems that are not designed for the space they are ventilating may not effectively remove polluted air.
- Operation and maintenance of the ventilation system. Poorly maintained systems may not perform to their specifications to reduce pollutant accumulation.

The rate at which the volume of air in a building is replaced is called the air exchange rate. Lower air exchange rates may result in higher concentrations of indoor air pollutants. Studies have shown that the air exchange rate affects not only the concentration of indoor air pollutants but also the attendant health risks. Bornehag et al. (2005) demonstrated a clear relationship between air exchange rates and risk of allergic symptoms in children, as illustrated in Fig. 5.1.

Occupant activity also can substantially affect indoor air quality. Environmental tobacco smoke (ETS) is an obvious, and perhaps the most common, example. Smoking by individuals in indoor environments such as residential dwellings not only affects the health of the active smoker but other individuals in the vicinity, too, particularly children and other susceptible populations, through the generation of secondhand smoke. Increased numbers of people in a home also can impact the quality of indoor air and potentially present an increased risk. A study conducted in the UAE on microbiological pollutants in homes reported that smaller homes with high numbers of residents had elevated concentrations of microbiological contaminants (Jaffal et al. 1997).

High-Priority Pollutants

Eight pollutants or sources of pollution were selected for inclusion in the indoor air environmental burden of disease modeling. Most of the pollutants and sources evaluated have been identified by other groups as priority pollutants due to their frequent presence in indoor air from a variety of sources and their well-characterized adverse health effects (California Air Resources Board 2005; De Brouwere et al. 2007; World Health Organization 2002). The eight priority pollutants are:

1. Environmental tobacco smoke
2. Incense combustion products
3. Coarse particulate matter (known as PM_{10}; see Chap. 4 for more information)
4. Fine particulate matter ($PM_{2.5}$)
5. Radon
6. Benzene
7. Formaldehyde
8. Microbiological pollutants (mold)

Key Health Effects of Indoor Air Pollution

Table 5.1 lists the key health effects associated with the pollutants considered in this analysis along with potential sources of the pollutants.

While Table 5.1 lists pollutants and health effects individually, contaminants may interact to cause new or exacerbated effects. The indoor environment is no different from the outdoor environment in that the air is a mixture and not simply composed of single pollutants. Multiple sources (e.g., stoves, furniture glues, candles, incense, and even the ventilation system) emit pollutants into the indoor environment. This complexity of the air matrix creates challenges when assessing health effects and risks due to air pollution (whether indoor or outdoor) because current health thresholds are typically reported for individual pollutants. For a few pollutants, scientific data have demonstrated that concomitant exposure has the potential to cause a more-than-additive health response. For example, active smoking and radon exposure are known to interact synergistically with the potential to cause a multiplicative rather than additive risk for lung cancer (U.S. Environmental Protection Agency 2003). Mixtures and potential interactions may increase the occurrence of adverse health effects from indoor exposures, as compared with estimates based on the presence of single contaminants. Studying the effects of pollutant mixtures on health is at the frontier of research, and not enough information is yet available to assess these mixture effects on the environmental burden of disease in the UAE.

A further important complication of assessing the health effects of indoor air pollutants is that children, the elderly, and people in poor health face a greater risk. Children under school age typically spend more time at home than do school-age

Table 5.1 Indoor pollutants, selected health effects, and potential sources

Pollutant	Selected health effects	Potential indoor sources
Environmental tobacco smoke (ETS)	Lung cancer; respiratory irritation; respiratory illnesses; cardiovascular disease; asthma	Tobacco products
Incense combustion products	Respiratory tract cancer	Oud; bakhoor; frankincense and other resins
Particulate matter (PM_{10}, $PM_{2.5}$)	Respiratory illnesses; cardiovascular disease; increased mortality; irritation; lung cancer	Combustion processes (including smoking, cooking, and the burning of candles and incense); house dust; pets
Radon	Lung cancer	Construction materials; granite bedrock beneath buildings; cracks in building foundation
Benzene	Leukemia; anemia	Environmental tobacco smoke; outdoor air infiltration; consumer products (e.g., glues, paints, cleaning products)
Formaldehyde	Nasopharyngeal cancer; Hodgkin's lymphoma; leukemia; irritation of eyes, nose, throat, and skin; exacerbation of asthma	Environmental tobacco smoke; manufactured wood products; consumer products; furnishings
Mold	Allergic reactions; asthma; respiratory irritation; infections	House dust; pets; humid conditions and dampness; poorly maintained heating, ventilating, and air-conditioning equipment

children. Additionally, children of all ages tend to be more active than adults, resulting in an increased breathing rate and a potentially higher dose of pollutants on a per-body-weight basis. The elderly and people in poor health are also at increased risk due to increased time spent in the home and compromised health defense systems. The existing health conditions of those who are already in poor health may be exacerbated by peak, prolonged, and repeated exposures to indoor air pollutants.

This analysis accounts for differences between adult and child susceptibilities for some, but not all, health outcomes. Data limitations prevented the examination of special risks to elderly members of the population. The relative vulnerability of sensitive subgroups such as children and the elderly to environmental pollution in the UAE is an important topic for future research.

Method for Estimating the Burden of Disease from Indoor Air Pollution

As previously mentioned, people spend the majority of their time indoors in residential dwellings. Therefore, this initial modeling effort focuses on the residential environment. Deficient air quality can exist in all types of enclosed buildings and structures. In the future, the methods and models developed here could be applied to other indoor environments (e.g., commercial buildings, schools) with the appropriate data. The model does not account for temporal variation in exposure or competing risks due to cumulative or aggregate exposures.

The analysis of the burden of disease due to indoor air pollution uses the attributable fraction approach advocated by the World Health Organization (WHO), described in detail in Chap. 3. Some previously published epidemiologic studies report relative-risk values based on measured levels of pollutants in the environment. In other cases, the studies indicated the relative risk based only on a categorical categorization of whether an individual was exposed to the pollutant or not, resulting in exposure levels such as "exposed" and "unexposed," or "high," "medium," and "low." When sufficient data exist to estimate the relative risk for a specific concentration of a pollutant and to estimate the fraction of the population exposed to that concentration, an approach WHO calls "exposure-based" can be used to estimate the attributable fraction. Otherwise, a "scenario-based" approach—categorizing the population into exposed and unexposed groups—must be used. As explained below, the exposure-based approach was used to estimate risks attributable to particulate matter, benzene, formaldehyde, and radon. The scenario-based approach was used to estimate risks due to environmental tobacco smoke, incense combustion, and mold.

Regardless of the approach used, the burden of disease due to a particular pollutant was calculated by multiplying the attributable fraction by the observed number of cases of the relevant health outcome in the population. The observed number of cases in the UAE was estimated using data from the Health Authority–Abu Dhabi (HAAD) (see Chap. 3).

PM_{10}, $PM_{2.5}$, Benzene, Formaldehyde, and Radon

The burdens of disease for PM_{10}, $PM_{2.5}$, benzene, formaldehyde, and radon were estimated using the exposure-based approach. Table 5.2 lists the health endpoints analyzed for each of these pollutants.

Two types of information are needed to implement the exposure-based approach:

1. A probability distribution that describes the variation in the observed concentration of the contaminant in indoor air
2. A concentration-response function that describes the relative risk of incurring the health outcome as a function of concentration

Ideally, information about contaminant concentrations would be derived from indoor air measurements taken for in UAE homes. However, except for radon, such data were not available at the time this analysis was conducted. Therefore, we conducted an extensive search for all previously published studies of residential indoor air quality. Probability distributions of contaminant concentrations were drawn from a meta-analysis of this literature, as described below. Since the completion of

Table 5.2 Health endpoints included in the model for particulate matter, benzene, formaldehyde, and radon

Pollutant	Health endpoint(s)
PM_{10}	Childhood asthma
$PM_{2.5}$	Childhood asthma
Benzene	Childhood asthma
Formaldehyde	Childhood asthma
Radon	Lung cancer and lung cancer mortality (adults)

this analysis, a research team affiliated with the project described in this book has collected indoor air samples from 628 randomly selected Emirati homes and has analyzed these samples for two pollutants included in the analysis in this chapter: PM and formaldehyde (Yeatts et al. 2012). In the future, the literature-derived estimates for these two contaminants should be updated based on the results of the UAE indoor air study.

Relative risk values (needed for the concentration-response functions) were drawn from an extensive survey of the international epidemiologic literature on contaminant exposure, described in more detail below.

The following sections describe the sources of information used to characterize the probability distributions of contaminant concentrations and the relative risks associated with different pollutants.

Contaminant Concentrations

Literature searches revealed no UAE-specific data on indoor air quality for pollutants other than radon and very few relevant studies from the larger Gulf region. Consequently, the global literature base provided the PM_{10}, $PM_{2.5}$, benzene, and formaldehyde concentration data that were ultimately used in this analysis. The literature reviews for those pollutants were thorough and focused on other developed countries in an attempt to approximate conditions in residential indoor environments in the UAE. The abstracted data are meant to be generally representative of indoor concentrations around the world, but not exhaustive. Without detailed information on the activity patterns of people living in the UAE, we collected data from around the world, as feasible, in an attempt to capture the variability that often exists due to culture, lifestyle, and geography. Occupant activities and basic housing characteristics in the UAE were assumed to be, in many respects, similar to those in other developed countries. Incense burning, an activity more specific to the UAE or the region, was included in preliminary literature reviews. For incense burning, we identified one epidemiologic study relating exposure to incense and respiratory tract cancer. As such, a separate module for incense burning is included in the model.

Using a meta-analysis approach based on global indoor air quality data in developed nations should provide a reasonable range of normally distributed concentration values that might be observed in the UAE. Mean and standard deviation values were

Table 5.3 Summary indoor concentrations from previous studies

Pollutant	Concentration mean (Standard deviation)
PM_{10}	92.8 μg/m³ (144.9)
$PM_{2.5}$	30.6 μg/m³ (34.36)
Benzene	9.5 μg/m³ (9.46)
Formaldehyde	47.4 μg/m³ (36.2)
Radon, city of Abu Dhabi	14.4 Bq/m³ (7.37)
Radon, emirate of Sharjah	50.3 Bq/m³ (range 8–164)[a]

[a]No standard deviation on radon concentration was provided. Instead, only minimum, mean, and maximum values were reported.

compiled from selected studies of indoor air quality. The means and variances of the pollutant concentrations observed across all these studies were then estimated. Minimum and maximum data points were also recorded when available.

Table 5.3 summarizes the concentration data abstracted from the literature. Appendix D provides detailed lists of concentration data. Studies on indoor air pollution resulting from the indoor use of solid fuel were not included in this modeling effort. Indoor air problems associated with solid fuel use are well documented and relevant to parts of the WHO's Eastern Mediterranean Region, which includes the UAE. Nevertheless, solid-fuel-use scenarios are not applicable to most of the UAE, where the use of solid fuel was reported as being less than 5% in the 2000 UAE Health and Lifestyle Survey (UAE University 2002). After more than a decade of fast-paced development and growth since the publication of that survey, that percentage should be even closer to 0.

Radon gas concentrations are geographically dependent. As such, we did not estimate radon concentrations for the UAE using literature-based data. Limited radon measurement data were obtained only for the city of Abu Dhabi and the emirate of Sharjah. Consequently, we were able to calculate the risks of residential radon exposure only in those two locations.

Epidemiologic Data

A thorough review of the epidemiologic literature was conducted in order to estimate relative risks associated with indoor air pollution exposures. Although indoor air quality has been gaining more prominence, especially during the last decade, the bulk of the research is still on outdoor air pollution. The limitations on available epidemiologic data affect exercises such as this environmental burden of disease analysis by restricting the breadth of the project and the pollutants that can be included. In addition, considerably less epidemiologic data on indoor air pollution was available for mortality than for morbidity, again limiting the scope of this project. Nonetheless, we were able to identify a number of studies we could use in our analysis. We gave preference to studies with pooled estimates from meta-analyses.

Table 5.4 Relative-risk estimates for $PM_{2.5}$, PM_{10}, radon, benzene, and formaldehyde exposures

Pollutant	Reference	Population	Health endpoint	Relative risk[a] (RR)	95% Confidence interval	Unit exposure to which RR applies (ΔC)
$PM_{2.5}$	McCormack et al. (2009)	2–5 years old	Asthma	1.03	(0.99, 1.07)	10 μg/m³
PM_{10}	McCormack et al. (2009)	2–5 years old	Asthma	1.06	(1.01, 1.12)	10 μg/m³
Radon	Darby et al. (2006)	Adult	Lung cancer	1.08	(1.03, 1.16)	100 Bq/m³
Benzene	Rumchev et al. (2004)	6 months–3 years old	Asthma	1.09[b]	(1.06, 1.12)	10 μg/m³
Formaldehyde	Rumchev et al. (2002)	6 months–3 years old	Asthma	1.003[b]	(1.002, 1.004)	10 μg/m³

[a]Relative risks for different concentrations C_i were estimated using the reported relative risks as shown and the unit exposure information. For all pollutants except radon, a log-linear concentration-response function was used, and the conversion was as follows: $RR = e^{\left(\frac{\ln(RR)C_i}{\Delta C}\right)}$. For radon, based on Darby et al. (2006), a linear concentration-response function with a slope of 0.08/100 Bq/m³ was used.

[b]These studies reported odds ratios rather than relative risks. Odds ratios were converted to relative risks. The values shown here are the odds ratios (which are very close to the calculated relative risks due to the relatively low prevalence rates and odds ratios for the health endpoints indicated).

The health endpoints selected for the UAE analysis are well recognized as significant contributors to the health burden in many countries. In all but one study, the confidence interval for the relative-risk estimate was statistically significant. Table 5.4 shows the relative-risk values used for each pollutant and health endpoint, along with the literature source from which the estimate was derived.

Mold, Environmental Tobacco Smoke, and Incense Combustion Products

For airborne mold, environmental tobacco smoke, and incense emissions, we expressed exposure as a dichotomous variable (e.g., exposed/unexposed). The epidemiologic studies selected for our analyses did not present the relative risk on a per-unit-concentration basis but rather for the exposed group relative to an unexposed group. Table 5.5 shows the health effects considered for these pollutants.

We performed an extensive literature search for epidemiologic data applicable to the pollutants shown in Table 5.5. A large volume of information is available for environmental tobacco smoke, and we were able to find several studies on relevant endpoints for use in our analysis. For incense and mold, less information is available.

Table 5.5 Health endpoints included in the model for mold, environmental tobacco smoke, and incense combustion products

Pollutants/Activities	Health endpoint(s)
Mold	Childhood asthma (6–12 years old)
	Adult asthma
Environmental tobacco smoke	Lung cancer and lung cancer mortality (adults)
	Chronic lymphocytic leukemia (adults)
	Cardiovascular disease and cardiovascular disease mortality (adults)
	Lower respiratory tract infections (<6 years old)
	Childhood asthma (<18 years old)
Incense combustion	Respiratory tract cancer and respiratory tract cancer mortality (adults)

Table 5.6 shows the studies and reported relative-risk estimates. For these pollutants, the relative risk indicates the probability of the health effect occurring in the exposed population divided by the probability of the health effect occurring in an equivalent population not exposed to the pollutant. The confidence intervals for all of the relative-risk estimates used in our analysis were statistically significant, indicating that the health endpoint of concern could be attributed to pollutant exposure. The burden of disease model represents relative risk as normally distributed in each case, with the indicated mean values and standard deviations estimated from the reported 95% confidence intervals. Relative-risk values less than zero are truncated (i.e., eliminated from the simulation, since negative relative-risk values are not possible).

Model Structure

Figure 5.2 shows the top layer of the model used to characterize the environmental burden of disease due to indoor air pollution. The top-level view in Fig. 5.2 provides an overview of the pollutants evaluated in this model; no inputs or outputs are associated with this layer.

The ***Indoor Air*** module is designed to be flexible, allowing the user to generate different scenarios and view the risks as a result of each changing scenario. For example, the user could adjust the pollutant concentration in the model or other input values if new data became available. Because the diagrams are linked, the final burden of disease estimate automatically adjusts for these changes.

Exposure-Based Modules

As described above, the exposure-based method is used to characterize risks due to five of the pollutants (shown on the left of Fig. 5.2). The structure of the submodules used for each of these pollutants is similar. This section presents the details of the ***Benzene*** module as an example of the structure of all of the modules.

Table 5.6 Epidemiologic parameters used to estimate risks of environmental tobacco smoke, mold, and incense

Reference	Population	Health endpoint	Relative risk (RR)	95% CI
Environmental tobacco smoke				
Kasim et al. (2005)	Adult	Chronic lymphocytic leukemia	2.28	(1.15, 4.53)
Boffetta (2002)	Adult	Lung cancer	1.25	(1.15, 1.37)
Cardenas et al. (1997)	Adult females	Lung cancer mortality	1.2	(0.8, 1.6)
	Adult males	Lung cancer mortality	1.1	(0.6, 1.8)
He and Whelton (1999)	Adult	Cardiovascular disease	1.25	(1.17, 1.32)
Hill et al. (2007)	Adult females	Cardiovascular disease msortality	1.35	(1.11, 1.64)
	Adult males	Cardiovascular disease mortality	1.25	(1.06, 1.47)
Vork et al. (2007)	<18 years old	Asthma	1.48	(1.32, 1.65)
Li et al. (1999)	<6 years old	Lower respiratory tract infections	1.57	(1.28, 1.91)
Mold				
Jaakkola et al. (2002)	Adult	Asthma	1.54	(1.01, 2.32)
Antova et al. (2008)	6–12 years old	Asthma	1.35	(1.20, 1.51)
Incense combustion				
Friborg et al. (2008)	Adult	Respiratory tract cancer	1.80	(1.20, 2.60)

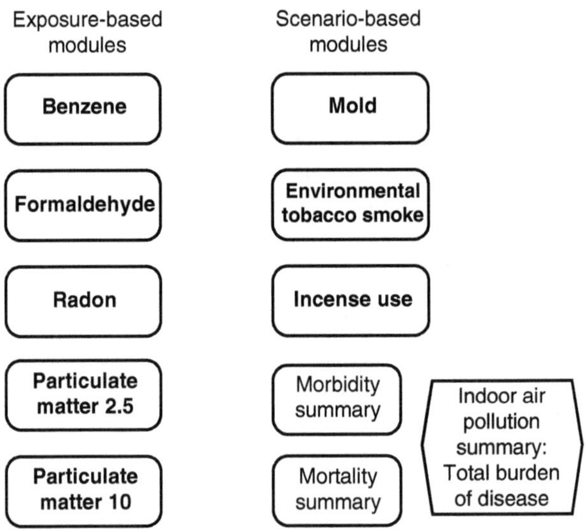

Fig. 5.2 Top-level diagram of the ***Indoor Air*** module

Table 5.7 Variables used to estimate the burden of disease due to benzene, formaldehyde, radon, PM_{10}, and $PM_{2.5}$

Variable name[a] (short-hand notation)	Type	Description	Source
[Pollutant] concentration parameters (μ_C, σ_C)	Deterministic	Mean and standard deviation of concentrations from meta-analysis of observed indoor pollutant levels; values shown in Table 5.3	See Appendix C and Table 5.3
[Pollutant] concentration (C)	Chance	Lognormal (mean = μ_C, standard deviation = σ_C)	N/A
[Pollutant] exposure levels (C_i)	Computed	Concentrations corresponding to the following values on the cumulative distribution function for C: 5%, 15%, 25%, … 95%	Computed from C
Relative risk by exposure level (RR_i)	Computed	$RR = e^{\left(\frac{\ln(RR)C_i}{\Delta C}\right)}$	Computed from RR and ΔC (shown in Table 5.4) and C_i
Attributable fraction (AF)	Computed	$AF = \frac{\sum_i P_i RR_i - 1}{\sum_i P_i RR_i}$	See Eq. 3.6 in Chap. 3
[Health endpoint] baseline (D_{total})	Deterministic	Observed number of cases of the health outcome of concern in each emirate	HAAD (2009) (see Chap. 3)
Attributable [health-care facility visits or deaths] by emirate	Computed	$(AF)(D_{total})$	See Chap. 3
Total attributable [health-care facility visits or deaths]	Computed	Sum of attributable health-care facility visits or deaths over all emirates	N/A

[a]Items in square brackets correspond to names of either the specific pollutant or the health endpoint associated with that pollutant.

Figure 5.3 shows the top layer of the ***Benzene*** module. The buttons at the top allow the user to easily view key input variable values and overall results without opening the full module. The ***Benzene Morbidity Module*** node contains the details behind the calculations; Fig. 5.4 shows the variables included in this layer. Table 5.7 describes the input variables shown in Fig. 5.4, along with all of the comparable inputs for formaldehyde, radon, $PM_{2.5}$, and PM_{10}.

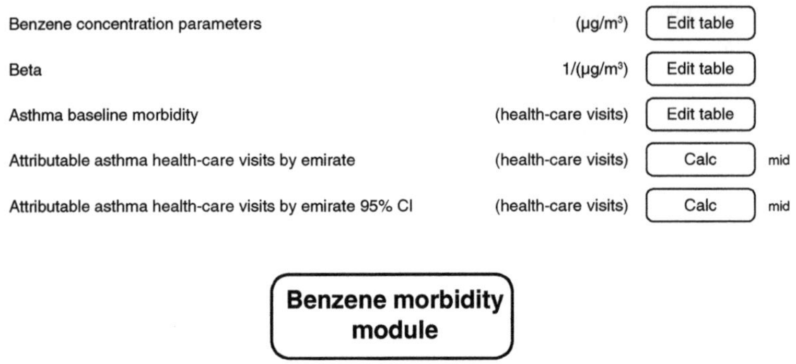

Fig. 5.3 ***Benzene*** module front page

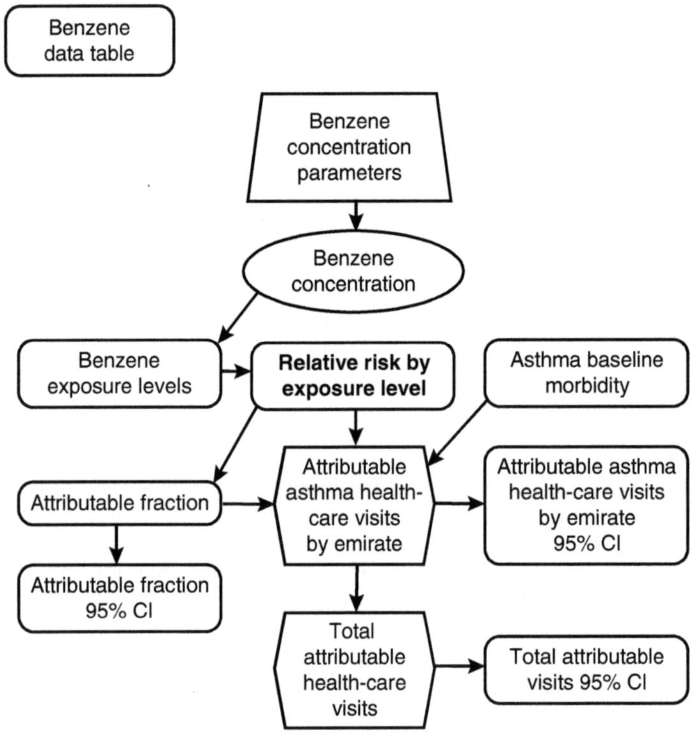

Fig. 5.4 Detailed ***Benzene Morbidity*** module

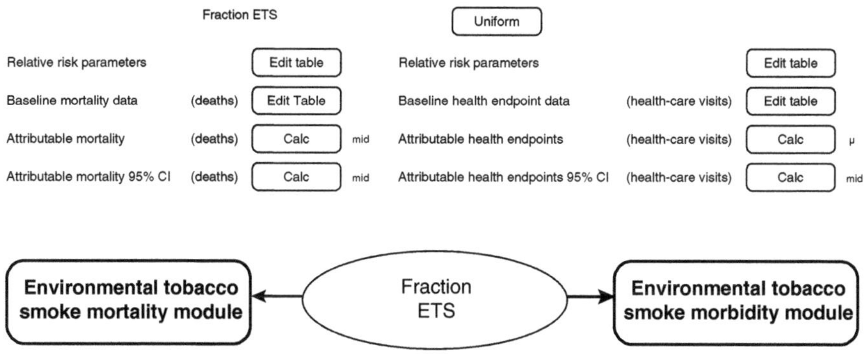

Fig. 5.5 ***Environmental Tobacco Smoke*** module front page

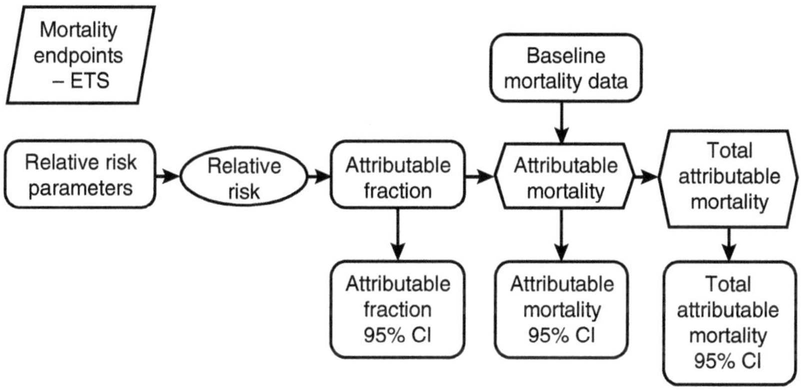

Fig. 5.6 Details of the ***Environmental Tobacco Smoke Morbidity*** module shown in Fig. 5.5

Scenario-Based Modules

Three pollutants or activities were included in the scenario-based modules: mold, environmental tobacco smoke, and incense combustion. Due to a lack of exposure-response functions in the epidemiologic literature selected for these pollutants, these modules do not include concentration data inputs. All of the pollutants analyzed in this section included at least one morbidity endpoint, while the ***Environmental Tobacco Smoke*** and ***Incense Burning*** modules also evaluated mortality.

Figures 5.5 and 5.6 show influence diagrams for the calculation of the burden of disease due to environmental tobacco smoke. Figure 5.5 shows the top level of the ***Environmental Tobacco Smoke*** module, and Fig. 5.6 shows the details for the calculation of morbidity due to environmental tobacco smoke. All of the other scenario-based modules have a similar design. Table 5.8 summarizes the input variables used in all of the modules.

Table 5.8 Variables used to estimate the burden of disease due to environmental tobacco smoke (ETS), incense, and mold

Variable name[a] (short-hand notation)	Type	Description	Source
Fraction exposed to [pollutant] ($P_{exposed}$)	Chance	Fraction of population exposed to pollutant, in each case represented as a uniform distribution, with these parameters: ETS: min = 0.15, max = 0.75 Mold: min = 0.05, max = 0.5 Incense, UAE citizens: min = 0, max = 0.9 Incense, noncitizens: min = 0, max = 0.5	ETS: Estimated from UAE University (2002); mold and incense use: assumption
Relative risk mean and standard deviation (μ_{RR}, σ_{RR})	Deterministic	Mean and standard deviation of relative risk associated with exposure (from Table 5.6)	See Table 5.6
Relative risk (RR)	Chance	Normal(μ_{RR}, σ_{RR}) (truncated at zero)	N/A
Attributable fraction (AF)	Computed	$\frac{P_{exposed}RR + (1 - P_{exposed}) - 1}{P_{exposed}RR + (1 - P_{exposed})}$	See Eq. 3.6 and Chap. 3
Baseline health endpoint data (D_{total})	Deterministic	Observed number of cases of the health outcome of concern in each emirate	HAAD (2009) (see Chap. 3)
Attributable [health-care facility visits or deaths] by emirate	Computed	(AF)(D_{total})	See Chap. 3
Total attributable [health-care facility visits or deaths]	Computed	Sum of attributable health-care facility visits or deaths over all emirates	N/A

[a]Items in square brackets correspond to names of either the specific pollutant or the health endpoint associated with that pollutant

Estimated Burden of Disease

Table 5.9 summarizes the complete results of the estimates of excess health-care facility visits per year in the UAE due to all the different indoor air pollutants considered, and Table 5.10 summarizes the mortality estimates. The tables show both the number of attributable health-care facility visits and deaths and the attributable fraction of the health endpoint due to indoor air pollution exposure. The attributable fraction is a measure of the reduction in a particular health endpoint that would be possible if the risk factor (i.e., pollutant or activity) were eliminated.

Table 5.9 Summary of estimated number of annual health-care facility visits due to indoor air pollutants

Pollutant	Citizenship	Health endpoint	Attributable fraction (95% CI)	Attributable health-care facility visits (95% CI)
Environmental tobacco smoke	All	Lung cancer	0.099 (0.034, 0.18)	44 (15, 78)
	All	Cardiovascular disease	0.099 (0.038, 0.17)	31,000 (11,000; 53,000)
	All	Chronic lymphocytic leukemia	0.34 (0.054, 0.59)	510 (75, 890)
	All	Lower respiratory tract infections in children	0.20 (0.071, 0.34)	35,000 (12,000; 60,000)
	All	Childhood asthma	0.18 (0, 0.52)	4,300 (0; 13,000)
Radon	All	Lung cancer	City of Abu Dhabi: 0.011 (0.006, 0.019) Emirate of Sharjah: 0.056 (0.030, 0.094)	6 (3, 10)
Benzene	All	Childhood asthma	0.072 (0.050, 0.095)	1,000 (690; 1,300)
Formaldehyde	All	Childhood asthma	0.014 (0.0092, 0.018)	190 (130, 250)
PM_{10}	All	Childhood asthma	0.47 (0.092, 0.78)	6,500 (1,300; 11,000)
$PM_{2.5}$	All	Childhood asthma	0.083 (0, 0.19)	1,200 (0; 2,600)
Mold	All	Adult asthma	0.12 (0.0015, 0.28)	4,000 (60; 9,200)
	All	Childhood asthma	0.086 (0.020, 0.17)	550 (120; 1,100)
Incense combustion products	UAE citizens	Respiratory tract cancer	0.24 (0.012, 0.49)	110 (5, 210)
	Noncitizens	Respiratory tract cancer	0.16 (0.0060, 0.35)	69 (3, 150)

Morbidity

Table 5.9 shows that tens of thousands of health-care facility visits may be attributable to the selected indoor air pollutants evaluated in this study. The estimates in the last column of Table 5.9 are not additive because several of the health endpoints overlap (e.g., multiple causes of childhood asthma). However, if one removes the overlapping health effects and considers only the maximum attributable number of illnesses due to each (for example, 6,500 health-care facility visits for childhood asthma as a result of PM_{10}), then the total number of attributable annual health-care

Table 5.10 Summary of estimated number of annual deaths due to indoor air pollutants

Pollutant	Citizenship	Health endpoint	Attributable fraction (95% CI)	Attributable annual deaths (95% CI)
Environmental tobacco smoke	All	Lung cancer	Females: 0.083 (0, 0.25) Males: 0.061 (0, 0.25)	8 (0, 29)
	All	Cardiovascular disease	Females: 0.13 (0.046, 0.24) Males: 0.098 (0.012, 0.22)	250 (47, 510)
Radon	All	Lung cancer	City of Abu Dhabi: 0.011 (0.006, 0.019) Emirate of Sharjah: 0.056 (0.030, 0.094)	2 (1, 3)
Incense combustion products	UAE citizens	Respiratory tract cancer	0.24 (0.012, 0.49)	13 (1, 27)
	Noncitizens	Respiratory tract cancer	0.16 (0.0060, 0.35)	10 (0, 22)

facility visits is approximately 77,000. This number is assumed to be an underestimation because it accounts for a relatively small number of pollutants in the universe of possible indoor air pollutants (although they are the ones we believe are potentially most important). In addition, the estimate excludes a number of health endpoints, such as adult asthma, for a number of the pollutants. It also does not consider synergistic effects, in which exposure to multiple pollutants can cause effects not observed in single pollutant exposure scenarios.

The leading source of indoor air pollution contributing to excess cases of illness is environmental tobacco smoke. Altogether, it appears to cause more than 80% of the health-care facility visits attributed to indoor air pollution. The leading health outcomes attributed to indoor air pollution are cardiovascular disease and lower respiratory tract infections.

Mortality

An estimated 280 deaths result from those diseases, with approximately 88% of those deaths attributed to cardiovascular disease caused by environmental tobacco smoke. Radon analyses included only the emirate of Sharjah and the city of Abu Dhabi due to data limitations; however, the mortality rates for those two emirates totaled only slightly more than one. If other emirates had comparable radon levels, the total lung cancer mortality from radon exposure would remain low.

Comparison with Other Estimates

The results of the indoor air burden of disease analyses compare reasonably well with the preliminary risk estimates prepared for the risk-ranking exercise described in Chap. 2. The preliminary estimate predicted 60–300 deaths due to indoor air pollution (see Appendix A), while the much more comprehensive estimate presented in this chapter is 280 for the pollutants selected for this analysis. Additionally, the preliminary risk analysis estimated between 200 and 300,000 short- and long-term illness cases, and the updated estimate of 77,000 falls reasonably well within that range.

Sensitivity Analysis

Table 5.11 shows how the estimated burden of disease would change if key input variables for the model were increased by 10 or 25%. For this sensitivity analysis, we focused only on the pollutants and health endpoints with the highest mortality and morbidity impacts. The most notable result of this sensitivity analysis is the profound effect of even a relatively small change in the estimated relative risk on the estimated burden of disease. For all of the contaminants and health endpoints, a small adjustment in the relative risk has a disproportionately large effect on the predicted burden of disease—an effect much larger than that caused by a similar change in the estimated pollutant concentration or fraction of the population exposed. For example, a 25% change in the assumed relative risk due to exposure to secondhand smoke (ETS) changes the estimated number of deaths attributed to ETS by 88%. In comparison, a change of 25% in the assumed fraction of the population exposed to ETS changes the estimated number of deaths by 20%. Improving estimates of the pollutant concentration and fraction of the population exposed to pollutants is important, as they also have a substantial effect on the burden of disease estimate. However, the burden of disease estimates are much more sensitive to relative risk assumptions than to pollutant concentration assumptions. Unfortunately, relative risk data are much more difficult to collect than pollutant concentration data. These estimates require large epidemiologic studies that are quite costly compared with environmental sampling alone.

Information Needed to Improve Future Burden of Disease Predictions

At the time this analysis was carried out, limited data on radon concentrations in indoor air in the UAE were available, and data on the prevalence of smoking also were available. However, UAE-specific data were not available for the other pollutants included in the risk model, necessitating an alternative approach that employed

Table 5.11 Sensitivity of the estimated burden of disease due to indoor air pollution to changes in key input variables

Pollutant	Health endpoint	Estimated attributable visits	Parameter increased	Attributable health outcomes with 10% increase	Percent change (%)	Attributable health outcomes with 25% increase	Percent change (%)
Morbidity Endpoints							
ETS	Cardiovascular disease	31,000	RR	44,000	42	61,000	97
			Fraction exposed	33,000	6	37,000	19
	Lower respiratory tract infections in children	35,000	RR	42,000	20	51,000	46
			Fraction exposed	37,000	6	41,000	17
	Childhood asthma	4,300	RR	5,100	19	6,000	40
			Fraction exposed	4,600	7	5,000	16
Benzene	Childhood asthma	1,000	RR	2,200	120	3,600	260
			Concentration	1,100	10	1,300	30
PM_{10}	Childhood asthma	6,500	RR	7,200	11	8,000	23
			Concentration	7,000	8	7,800	20
$PM_{2.5}$	Childhood asthma	1,200	RR	2,300	92	3,700	208
			Concentration	1,300	8	1,500	25
Mold	Adult asthma	4,000	RR	5,200	30	6,600	65
			Fraction exposed	4,500	13	5,100	28
	Childhood asthma	550	RR	750	36	1,000	82
			Fraction exposed	600	9	700	27
Mortality Endpoints							
ETS	Cardiovascular disease	250	RR	350	40	470	88
			Fraction exposed	270	8	300	20
	Lung cancer	8	RR	12	50	18	125
			Fraction exposed	9	13	10	25
Incense	Respiratory tract cancer	23	RR	27	17	32	39
			Fraction exposed	26	13	29	26

global literature-based concentration data to estimate the potential range of indoor air pollutant concentrations that might be expected in the UAE. For PM and formaldehyde, the burden of disease estimates can be updated with recently collected pollutant concentration data (Yeatts et al. 2012). Also available from this recent indoor air quality study are data on the frequency of incense use in Emirati homes, which could be used to update the assumptions about incense in this chapter. A remaining limitation is that the recent UAE indoor air study included only Emirati homes and hence did not account for potential differences in indoor air quality among homes occupied by expatriates (more than 80% of the population).

Another source of uncertainty in our analysis arises from the baseline health data. First, a centralized database of health data for the entire UAE does not exist. As Chap. 3 explains, we extrapolated data for 73% of the Abu Dhabi population to the rest of the country, assuming that incidence rates in the other emirates are comparable. Furthermore, our data were presented in terms of health-care facility visits for a particular health endpoint. We were not able to distinguish whether individuals made more than one visit for a given adverse health episode. The metric we used still offers value, but it does not present a true incidence rate, and that potential error perpetuates to the final burden of disease estimates as well.

The estimates of relative risks due to exposure to indoor air pollutants are a critical source of uncertainty as well. For these relative-risk estimates, we relied on the most current global epidemiologic studies of the relationship between pollutant exposures and health effects. However, as illustrated in the sensitivity analysis, the results of our predictions are highly sensitive to the assumed relative risk.

Finally, our estimates for the disease burden due to indoor air pollutants are more than likely understated. We evaluated a few select indoor pollutants, but potentially thousands of chemicals found in indoor environments could adversely affect human health. We would expect morbidity and mortality numbers to be larger if additional indoor pollutants were assessed.

Conclusions

Our analyses suggest that indoor air pollution is a considerable risk to public health in the UAE, accounting for 280 excess deaths and at least 77,000 excess visits to health-care facilities in 2008. In terms of mortality, indoor air quality ranks second only to outdoor air pollution as a cause of environmentally related diseases in the UAE.

This analysis is important not just for the UAE but also for other wealthy nations. Previous estimates of the burden of disease due to indoor air pollution have addressed mainly the use of solid fuel within homes in developing countries. Results have demonstrated a significant public health risk from this activity; however, those studies are not applicable to nations where solid fuels are no longer a primary source of fuel for home cooking. This analysis therefore focuses primarily on pollutants identified in industrialized countries where solid fuel is not commonly employed for indoor cooking, while also considering cultural activities (such as the burning of incense) that potentially could introduce alternative pollution sources into indoor environments that are not as prevalent in other parts of the world. Our final analysis includes pollutants that generally have been well characterized with regard to their adverse health effects and ubiquitous presence in indoor environments.

Several actions are possible to advance the environmental health policies of the UAE and promote healthier indoor environments. One of the key steps is to address potential problems with mechanical ventilation systems in homes. In extreme climates, such as in the UAE, the use of mechanical ventilation systems is more extensive than in temperate climates, which necessitates more oversight through measures such as building code requirements. Requiring compliance with recognized ventilation guidelines and standards for these systems is critical to controlling indoor contaminants and reducing human exposures.

In addition, actions to educate the public about indoor air quality could lead to substantial gains. Key concepts to convey to the public include risks to nonsmoking family members due to environmental tobacco smoke; risks due to incense combustion; and the need to maintain ventilation systems. Further, the UAE could initiate product-labeling or building-material-substitution campaigns in addition to campaigns aimed at behaviors that can be modified to reduce indoor air pollution and exposure to pollutants. Other countries and regions have successfully implemented similar programs and could serve as models for the UAE (Bluyssen 2009; European Chemicals Agency 2007).

As an additional important step, the UAE government could continue to sponsor research to identify factors that influence indoor air pollutant concentrations and occupant exposures. For example, surveys to measure air exchange rates in residential areas could be conducted to identify the effects different types of housing have on exposure. Also, these surveys could assess the effect of geographical differences on exposures to indoor air pollutants. The UAE also could consider increasing the availability of educational opportunities for professional personnel concerned with indoor air quality. Educational needs include training programs for building engineers as well as graduate level education for those conducting field assessments or managing larger indoor air programs at the federal, emirate, or municipal level.

References

Antova, T., S. Pattenden, B. Brunekreef, J. Heinrich, P. Rudnai, F. Forastiere, H. Luttmann-Gibson, et al. 2008. Exposure to indoor mould and children's respiratory health in the PATY study. *Journal of Epidemiology and Community Health* 62(8): 708–714.

Bennett, D.H., T.E. McKone, J.S. Evans, W.W. Nazaroff, M.D. Margni, O. Jolliet, and K.R. Smith. 2002. Defining intake fraction. *Environmental Science and Technology* 36(9): 207A–211A.

Bluyssen, P.M. 2009. Towards an integrative approach of improving indoor air quality. *Building and Environment* 44(9): 1980–1989.

Boffetta, P. 2002. Involuntary smoking and lung cancer. *Scandinavian Journal of Work, Environment and Health* 28(Supplement 2): 30–40.

Bornehag, C.G., J. Sundell, L. Hägerhed-Engman, and T. Sigsgaard. 2005. Association between ventilation rates in 390 Swedish homes and allergic symptoms in children. *Indoor Air* 15(4): 275–280.

California Air Resources Board. 2005. Report to the California legislature: Indoor air pollution in California. July. http://www.arb.ca.gov/research/indoor/ab1173/rpt0705.pdf

Cardenas, V.M., M.J. Thun, H. Austin, C.A. Lally, W.S. Clark, R.S. Greenberg, and C.W. Heath Jr. 1997. Environmental tobacco smoke and lung cancer mortality in the American Cancer Society's Cancer Prevention Study II. *Cancer Causes & Control* 8(1): 57–64.

Cohen, A.J., H.R. Anderson, B. Ostro, K.D. Pandey, M. Krzyzanowski, N. Kunzli, K. Gutschmidt, et al. 2005. The global burden of disease due to outdoor air pollution. *Journal of Toxicology and Environmental Health. Part A* 68: 1–7.

Dales, R., L. Liu, A.J. Wheeler, and N.L. Gilbert. 2008. Quality of indoor residential air and health. *Canadian Medical Association Journal* 179(2): 147–152.

Darby, S., D. Hill, H. Deo, A. Auvinen, J.M. Barros-Dios, H. Baysson, F. Bochicchio, et al. 2006. Residential radon and lung cancer: Detailed results of a collaborative analysis of individual data on 7,148 persons with lung cancer and 14,208 persons without lung cancer from 13 epidemiologic studies in Europe. *Scandinavian Journal of Work, Environment and Health* 32(Supplement 1): 1–83.

DeBrouwere, K., E. Goelen, M. Spruyt, and R. Torfs. 2007. Ranking indoor air health problems using health impact assessment: Final report. Service contract for the European Commission. *DG Environment*: Contract 061651. 2007/IMS/R/394.

Dockery, D.W., C.A. Pope, X. Xu, J.D. Spengler, J.H. Ware, M.E. Fay, B.G. Ferris, and F.E. Speizer. 1993. An association between air pollution and mortality in six U.S. cities. *The New England Journal of Medicine* 329: 1753–1759.

European Chemicals Agency. 2007. Regulation on Registration, Evaluation, Authorisation and Restriction of Chemicals (REACH). http://echa.europa.eu/reach_en.asp

Friborg, J.T., J.M. Yuan, R. Wang, W.P. Koh, H.P. Lee, and M.C. Yu. 2008. Incense use and respiratory tract carcinomas: A prospective cohort study. *Cancer* 113(7): 1676–1684.

He, J., and P.K. Whelton. 1999. Passive cigarette smoking increases risk of coronary heart disease. *European Heart Journal* 20(24): 1764–1765.

Health Authority–Abu Dhabi (HAAD). 2009. 2008 health data for Abu Dhabi emirate.

Hill, S.E., T. Blakely, I. Kawachi, and A. Woodward. 2007. Mortality among lifelong nonsmokers exposed to secondhand smoke at home: Cohort data and sensitivity analyses. *American Journal of Epidemiology* 165(5): 530–540.

Jaakkola, M.S., H. Nordman, R. Piipari, J. Uitti, J. Laitinen, A. Karjalainen, P. Hahtola, and J.J. Jaakkola. 2002. Indoor dampness and molds and development of adult-onset asthma: A population-based incident case–control study. *Environmental Health Perspectives* 110(5): 543–547.

Jaffal, A.A., I.M. Banat, A.A. El Mogheth, H. Nsanze, A. Bener, and A.S. Ameen. 1997. Residential indoor airborne microbial populations in the United Arab Emirates. *Environment International* 23(4): 529–533.

Kasim, K., P. Levallois, B. Abdous, P. Auger, and K.C. Johnson. 2005. Environmental tobacco smoke and risk of adult leukemia. *Epidemiology* 16(5): 672–680.

Kim, D., A. Sass-Kortsak, J.T. Purdham, R.E. Dales, and J.R. Brook. 2006. Sources of personal exposure to fine particles in Toronto, Ontario, Canada. *Journal of the Air & Waste Management Association* 55: 1134–1146.

Klepis, N.E., W.C. Nelson, W.R. Ott, J.P. Robinson, A.M. Tsang, P. Switzer, J.V. Behar, S.C. Hern, and W.H. Engelmann. 2001. The National Human Activity Pattern Survey (NHAPS): A resource for assessing exposure to environmental pollutants. *Journal of Exposure Analysis and Environmental Epidemiology* 11(3): 231–252.

Li, J.S., J.K. Peat, W. Xuan, and G. Berry. 1999. Meta-analysis on the association between environmental tobacco smoke (ETS) exposure and the prevalence of lower respiratory tract infection in early childhood. *Pediatric Pulmonology* 27(1): 5–13.

Liu, W., J. Zhang, L.R. Korn, L. Zhang, C.P. Wiesel, B. Turpin, M. Morandi, T. Stock, and S. Colome. 2007. Predicting personal exposure to airborne carbonyls using residential measurements and time/activity data. *Atmospheric Environment* 41: 5280–5288.

McCormack, M.C., P.N. Breysse, E.C. Matsui, N.N. Hansel, D. Williams, J. Curtin-Brosnan, P. Eggleston, and G.B. Diette. 2009. In-home particle concentrations and childhood asthma morbidity. *Environmental Health Perspectives* 117(2): 294–298.

Nazaroff, W.W., and C.J. Weschler. 2001. Indoor air and the public good. *Indoor Air* 11(3): 143–144.

Pope, C.A.III. 1991. Respiratory hospital admissions associated with PM_{10} pollution in Utah, Salt Lake, and Cache valleys. *Archives of Environmental Health* 46: 90–97.

Rumchev, K., J. Spickett, M. Bulsara, et al. 2002. Domestic exposure to formaldehyde significantly increases the risk of asthma in young children. *European Respiratory Journal* 20: 403–408.

Rumchev, K.B., J.T. Spickett, M.K. Bulsara, M.R. Phillips, and S.M. Stick. 2004. Association of domestic exposure to volatile organic compounds with asthma in young children. *Thorax* 59(9): 746–751.

U.S. Environmental Protection Agency (EPA). 1991. Building air quality: A guide for building owners and facility managers. 402-F-91-102.

U.S. Environmental Protection Agency (EPA). 2003. EPA assessment of risks from radon in homes. EPA 402-R-63-603. Washington, D.C.: EPA, Office of Air and Radiation.

UAE University. 2002. UAE health and lifestyle survey. UAEU Faculty of Medicine and Faculty of Health Sciences and College of Business and Economics.

Vork, K.L., R.L. Broadwin, and R.J. Blaisdell. 2007. Developing asthma in childhood from exposure to secondhand tobacco smoke: Insights from a meta-regression. *Environmental Health Perspectives* 115(10): 1394–1400.

World Health Organization (WHO). 2002. The world health report 2002: Reducing risks, promoting healthy life. Geneva: World Health Organization. http://www.who.int/entity/whr/2002/en/whr02_en.pdf

Yeatts, K.B., M. El-Sadig, D. Leith, W. Kalsbeek, F. Al-Maskari, D. Couper, W.E. Funk, T. Zoubeidi, R.L. Chan, C.B. Trent, C.A. Davidson, M.G. Boundy, M.M. Kassab, M.Y. Hasan, I. Rusyn, J. MacDonald Gibson, and A. Olshan. 2012. Indoor air pollutants and health in the United Arab Emirates. *Environmental Health Perspectives* 120(5): 687–694.

Chapter 6
Burden of Disease from Occupational Exposures

Abstract Workers may be exposed to physical, chemical, and biological hazards at work that may lead to occupational illness. Hazardous substance exposure routes include dermal and inhalation exposure and ingestion. Families of workers also can face risks from toxic substances brought home on contaminated work clothes or vehicles. This chapter estimates occupational exposures to harmful chemicals and noise in the United Arab Emirates (UAE) and calculates the burden of disease related to selected occupational hazards. Occupational health studies conducted in the UAE have revealed unsafe work practices and unhealthy working conditions in many industry sectors, but the majority of UAE workers who are potentially exposed to hazardous substances and noise are employed in construction, agriculture, or manufacturing. The exposures covered in this study were selected following the approach by the World Health Organization, covering common occupational carcinogens, occupational airborne particulates, and noise, excluding occupational injuries and ergonomic stressors. The estimated total number of annual deaths due to health outcomes included in this study is 47, and the total number of health-care facility visits is 17,160. In addition, the model estimates that 4,770 cases of noise-induced hearing loss occur due to occupational exposures each year. Of the health outcomes covered in the study, lung cancer and leukemia were responsible for the highest number of deaths (25 and 12, respectively). For health-care facility visits, asthma and chronic obstructive pulmonary disease contributed most to the disease burden with 11,854 and 5,012 visits, respectively. It is likely that the UAE could reduce the amount it spends on medical care by reducing exposure to respiratory irritants, carcinogens, and noise in workplaces. These numbers should not be considered to represent the total disease burden arising from all occupational exposures. Many prevalent occupational hazards, such as injuries and ergonomic stressors, were excluded because this study focuses on health risks due to releases of hazardous physical, chemical, and biological agents into the environment as a result of human activities.

J. MacDonald Gibson et al., *Environmental Burden of Disease Assessment*,

Environmental Science and Technology Library 24, DOI 10.1007/978-94-007-5925-1_6,

Keywords Occupational exposures • Environmental burden of disease • Relative risk • Attributable fraction • Premature deaths and health-care facility visits • United Arab Emirates • Occupational hygiene monitoring • Personal protective equipment • Pneumoconioses • Malignant mesothelioma • Asthma • Chronic obstructive pulmonary disease • Noise-induced hearing loss • Occupational carcinogens • Occupational airborne particulate matter • Occupational exposure to noise

Overview: Nature and Sources of Occupational Exposures

Occupational illness is an important contributor to burden of disease in any country, but nations with advanced occupational health protection systems benefit over countries with less stringent occupational hygiene monitoring and legislative enforcement. Approximately 11,000,000 occupational diseases occur in the world each year, 700,000 of which are fatal. The estimated number of occupational injuries occurring in the world annually is even higher: 100,000,000, with approximately 100,000 deaths. The leading causes of occupational illness in the world include injuries, musculoskeletal disorders, chronic respiratory diseases, skin disorders, noise-induced hearing loss, pneumoconioses, mental disorders, cancers, and pesticide and other poisonings (Leigh et al. 1999).

In addition to causing human suffering for both the worker and the worker's family, occupational illnesses strain society as a whole via direct and indirect costs. The estimated cost of treating occupational diseases and injuries in different countries ranges from 1.2 to 10.1% of gross domestic product (GDP), which for the United Arab Emirates means a potential cost of 6.4 billion to 54.1 billion dirhams, based on 2008 GDP (UAE Interact 2009; Beatson and Coleman 1997). Estimating the magnitude of occupational morbidity and mortality is difficult because most countries do not have a reliable information gathering system, such as a national occupational disease mortality surveillance system (Steenland et al. 2003; Serinken et al. 2008). Additionally, many occupational diseases have long latency periods and multiple potential causes, and primary health care providers often lack sufficient training to recognize and report occupational illnesses. These factors lead to gross underestimation of occupational injuries, diseases, and deaths. Underreporting is most significant for chronic, nonfatal occupational diseases. These illnesses, unlike fatal or severe injuries or diseases with a rapid onset, are not always easily recognized as having occupational origins (Leigh et al. 1999).

Factors Affecting Occupational Risks

Workers may be exposed to physical, chemical, and biological hazards at work that may lead to occupational illness. Hazardous substance exposure routes include dermal and inhalation exposure and ingestion. Depending on the dose, route and duration of exposure, and chemical properties of the hazardous substance, exposure symptoms may range from mild discomfort to debilitating conditions or even death. Sudden exposure to large amounts of a toxic compound may lead to acute poisoning, causing life-threatening symptoms. Chronic exposure to even trace amounts of certain substances may cause serious health effects such as cancer or reproductive and teratogenic effects years after the exposure has ceased. Families of workers also can face risks from toxic substances brought home on contaminated work clothes or vehicles (Curl et al. 2002; Thompson et al. 2003).

General factors that affect the extent of exposure to occupational hazards include administrative controls, engineering solutions, and the use of personal protective equipment. Administrative controls include replacing harmful substances with less dangerous ones, reducing amounts of chemicals used, limiting exposure times through practices such as task rotation, and promoting safe work practices. Engineering solutions include exposure-reducing systems such as air blowers, ventilation hoods, mufflers, and barriers. Personal protective equipment such as protective clothing, ear plugs, and respirators should be used if administrative or engineering controls are not enough to reduce exposures to an acceptable level. Employers should provide personal protective equipment and train workers on its proper use and maintenance. Other factors crucial to improving occupational health and safety include appropriate monitoring of contaminants in the work environment, medical monitoring of workers, proper data collection and reporting, and strict enforcement of occupational laws and regulations, which should meet the latest international exposure limits and require employers to minimize exposures through comprehensive occupational safety programs. Raising awareness of occupational risks is also a key issue, especially since many workers in the UAE are unskilled expatriate laborers unaware of safe working techniques and risks associated with their tasks.

Scope of This Analysis

This chapter estimates occupational exposures to harmful chemicals, noise, and heat in the UAE and calculates the burden of disease related to selected occupational hazards. Because the occupational health field is in its early stages in the UAE, very limited information is available on levels of harmful substances at workplaces or individual worker exposure during the workday. Depending on occupation and industrial sector, workers may be exposed to much higher levels of chemicals or noise than the general population. In addition, a wide range of harmful chemicals not typically present in outdoor air or indoor household air may be present in work environments. No UAE-specific estimates are available on the number of workers exposed to various hazards or on the potential exposure levels in different industries. Also, reliable information on the prevalence of diseases in the workforce or the general population is limited. Due to a lack of information, burden of disease calculations are based on estimations and crude assumptions extrapolated from data

Table 6.1 Distribution of workforce in each emirate by gender, 2005

	Male	%	Female	%	Total	%
Abu Dhabi	697,544	32	117,767	34	815,311	32
Dubai	875,459	40	124,159	36	999,618	39
Sharjah	385,208	17	56,742	16	441,950	17
Ajman	90,789	4	11,995	3	102,784	4
Ras al Khaimah	89,452	4	19,078	6	108,530	4
Fujairah	54,628	2	10,764	3	65,392	3
Umm al Quwain	21,008	1	5,075	1	26,083	1
Total	**2,214,088**	**100**	**345,580**	**100**	**2,559,668**	**100**

UAE Ministry of Economy (2008)

based on other countries, and the results will only provide a preliminary framework to guide further in-depth investigation in the UAE. Importantly, even though occupational injuries resulting from accidents, such as falling from a height or being hit by an object, are responsible for a large part of the occupational disease burden, work-related accidents, ergonomic problems, and musculoskeletal disorders are outside the scope of this chapter, which instead focuses on health risks due to releases of hazardous physical, chemical, and biological substances into the environment resulting from human activities.

The UAE Workforce

The number of employees in the UAE was estimated at 3,100,000 in 2007, an increase of 230,000, or 8%, from 2006 (Dubai Chamber of Commerce and Industry 2007). In the past few years the proportion of the labor force in different emirates has changed considerably. In 2001, most of the workforce was employed by Abu Dhabi. However, in 2005 Dubai employed approximately 40% and Abu Dhabi 32% of the total labor force in the UAE (UAE Ministry of Economy 2008; Dubai Chamber of Commerce and Industry 2007). The rest of the emirates together employed the remaining 28% (Sharjah 17%, Ajman 4%, Ras al Khaimah 4%, Fujairah 2%, and Umm al Quwain 1%) (UAE Ministry of Economy 2008). The proportion of expatriates in the UAE labor force is unusually high. In 2006, an estimated 90.7% of the labor force consisted of foreign workers (Economist Intelligence Unit 2007). More than 88% of Emiratis work in the public sector, with over half of the men being members of either the police or the military (Economist Intelligence Unit 2007). Table 6.1 presents the number of workers in each emirate by gender in 2005.

The sectors experiencing the fastest growth in recent years, and in 2005 employing a major part of the total workforce, include construction, trade, manufacturing, and government (UAE Ministry of Economy 2008). Table 6.2 shows the distribution of the workforce into different economic sectors by gender in 2005. The distribution of the workforce into various groups of occupations by gender in 2005 is shown in Table 6.3.

Table 6.2 Distribution of workforce into economic sectors by gender, 2005

Economic activity group	Male	%	Female	%	Total	%
Agriculture, hunting, and forestry	116,333	5	320	0	116,653	5
Fishing	5,490	0	10	0	5,500	0
Mining and quarrying	43,618	2	1,764	1	45,382	2
Manufacturing	189,337	9	11,158	3	200,495	8
Electricity, gas, and water supply	23,939	1	1,169	0	25,108	1
Construction	713,988	32	6,108	2	720,096	28
Wholesale and retail trade; repair of motor vehicles, motorcycles, and personal and household goods	306,920	14	29,582	9	336,502	13
Hotels and restaurants	88,357	4	12,400	4	100,757	4
Transport, storage, and communications	152,442	7	16,264	5	168,706	7
Financial intermediation	27,353	1	9,676	3	37,029	1
Real estate, renting, and business services	115,956	5	12,670	4	128,626	5
Public administration, defense, and compulsory social security	213,621	10	17,946	5	231,567	9
Education	33,496	2	38,167	11	71,663	3
Health and social work	23,929	1	21,537	6	45,466	2
Other community, social, and personal service activities	49,862	2	7,472	2	57,334	2
Activities of private households as employers	60,637	3	133,879	39	194,516	8
Extraterritorial organizations and bodies	1,688	0	561	0	2,249	0
Not stated activities	15,009	1	4,851	1	19,860	1
Unemployed, never worked before	32,113	1	20,046	6	52,159	2
Total	**2,214,088**	**100**	**345,580**	**100**	**2,559,668**	**100**

UAE Ministry of Economy (2008)

Occupational Exposures in UAE Industries

Occupational health studies conducted in the UAE have revealed unsafe work practices and unhealthy working conditions in many industry sectors (Gomes et al. 1999, 2001, 2002; Al Kaabi and Hadipriono 2003; Al Neaimi et al. 2001). Examples of UAE industries in which workers are at risk of being exposed to harmful substances include construction; agriculture; manufacturing; oil and gas; chemical; metal; small industries such as dry cleaning, car repair, and carpentry shops; food; and mining and quarrying. Common exposures in construction, manufacturing, and agriculture, are discussed below. Discussion is limited to these three main areas because the majority of UAE workers who are potentially exposed to hazardous substances and noise are employed in these economic sectors.

Table 6.3 Distribution of workforce into occupation groups by gender, 2005

Occupation	Male	%	Female	%	Total	%
Legislators, senior officers, and managers	86,660	4	9,874	3	96,534	4
Professionals	188,105	8	61,464	18	249,569	10
Technicians and associate professionals	171,115	8	35,844	10	206,959	8
Clerks	70,458	3	28,561	8	99,019	4
Service workers, shop and market sales workers	252,868	11	155,782	45	408,650	16
Skilled agricultural and fishery workers	46,150	2	112	0	46,262	2
Craft and related trade workers	680,205	31	6,741	2	686,946	27
Plant and machine operators and assemblers	225,194	10	5,759	2	230,953	9
Elementary occupations	386,939	17	15,983	5	402,922	16
Armed forces	58,123	3	1,528	0	59,651	2
Not stated	16,158	1	3,886	1	20,044	1
Unemployed, never worked before	32,113	1	20,046	6	52,159	2
Total	**2,214,088**	**100**	**345,580**	**100**	**2,559,668**	**100**

UAE Ministry of Economy (2008)

Construction

Driven by rapid economic growth, the construction sector is one of the fastest-growing industries in the UAE. Most of the building activities are taking place in Dubai, Abu Dhabi, and Sharjah (Dubai Chamber of Commerce and Industry 2007). The construction industry is the largest single employer in the UAE. In 2006, approximately 650,000 people worked in construction, making up 23% of the total workforce (UAE Ministry of Economy 2006). Most construction employees work for large companies employing more than 200 people. The number of small companies employing fewer than 20 workers is low (Rettab 2003). Construction is considered one of the most dangerous occupations in the world, especially in Middle Eastern countries where legislation and safety inspections are struggling to keep up with the large number of projects in progress (Al Kaabi and Hadipriono 2003). These issues are worsened by the fact that the majority of construction workers are foreign nationals who come from different cultures, seldom speak the local language, and are usually undertrained and inexperienced in construction work. A recent survey in the UAE revealed that accidents in this industry are common (Al Kaabi and Hadipriono 2003).

Construction workers are exposed to a wide range of chemical hazards, including asbestos, silica, cement, wood dusts, acids, organic solvents, and isocyanates (during the spraying of polyurethane foam); metals such as lead, nickel, manganese, and hexavalent chromium; skin irritants; and various fumes such as asphalt fumes, diesel exhaust, and welding fumes. Various cancers and other diseases such as asthma and chronic obstructive pulmonary disease have been linked to exposures occurring during construction work (Stern et al. 1995; Wang et al. 1999; Verma et al. 2003; Rushton 2007). Exposed workers include painters, carpenters, roofers, insulators, welders, demolition crews, cement workers, and highway construction workers. Construction workers who work in tunneling, masonry, foundry operations, demolition, sand-blasting, or abrasive blasting of concrete may be exposed to respirable crystalline silica dust during processing of these materials (Mazurek and Attfield 2008). On construction sites, asbestos-containing materials may release asbestos fibers when they are in poor condition, cut, or disturbed by nearby work. Since the import, production, and use of asbestos boards are now banned in the UAE (UAE Federal Government 2006), exposure is most likely to occur when renovating or demolishing older buildings. Noise is also an important hazard in construction work. These workers are also exposed to high temperatures in the summer, but this analysis does not consider the effects of exposure to excess heat.

Manufacturing

In terms of GDP, manufacturing is the largest nonoil economic sector in the UAE, making up 20% of the nation's oil-excluded GDP in 2006 (Dubai Chamber of Commerce and Industry 2007). The manufacturing sector has grown steadily in recent years due to a growing population, increasing demand for consumer goods and construction materials, and rising foreign direct investments in the UAE, particularly within Free Zones, which offer tax and economic incentives to foreign companies. In addition to the petrochemical industry, other areas that have grown remarkably include chemical fertilizers, aluminum, construction products, air conditioning and refrigerating equipment, foodstuffs, pharmaceuticals, and various small and medium businesses established in the Free Zones across the emirates (UAE Ministry of Economy 2006; UAE Interact 2007). In 2005, the Ministry of Finance and Industry registered 3,294 manufacturing establishments (Emirates Industrial Bank 2006), and in 2006 the manufacturing sector employed approximately 362,000 people (UAE Ministry of Economy 2006).

The most important manufacturing sectors in the UAE are metal fabrication, nonmetal mineral products, garments, food and beverages, plastics, furniture, chemicals, printing, basic metals, and wood production. Together these 10 sectors accounted for almost 90% of total manufacturing employment and 80% of total manufacturing establishments in 2004 (Emirates Industrial Bank 2005). In terms of employment, the two largest sectors in manufacturing are metal fabrication and nonmetal mineral products such as cement, blocks, and stones. The main exposure sources in steel manufacturing plants and iron foundries are dusts from iron ore, coal, and silica, as well as fumes and gases that include coke furnace emissions, metal fumes, iron oxides, and oxides of carbon, sulfur, and nitrogen (Gomes et al. 2001). Due to chronic exposure to dusts, fumes, and manganese, workers at these plants are at risk of impaired lung function (Gomes et al. 2001; Wang et al. 1996). Exposure to noise and heat are also common occupational problems in the metal industry. In the cement industry, workers are exposed to cement dust during many parts of the manufacturing process, including extraction, crushing, grinding, and packing the final product, and are at risk of respiratory diseases (Al Neaimi et al. 2001). In the oil, gas, and related petrochemical industries, workers are potentially exposed to a wide range of chemicals, including hydrocarbons (e.g., volatile organic compounds, polycyclic aromatic hydrocarbons), hydrogen sulfide, chlorine, styrene, and organic solvents, as well as noise and heat. Exposures in the other manufacturing industries depend on type of industry, but noise and various kinds of dusts are common occupational hazards in many manufacturing sectors.

Agriculture

The main concern among agricultural workers in the UAE is exposure to pesticides. At the federal level, pesticide use is regulated by the Ministry of Environment and Water (MOEW), assisted by local authorities in different emirates (e.g., Environment Agency–Abu Dhabi). The manufacture and formulation of pesticides is prohibited in the UAE, and only pesticides registered by MOEW can be imported and used (Al Asram 2006). Farm workers may be exposed to pesticides when preparing or applying pesticides, weeding, or hand-harvesting crops. High-risk groups in agriculture include mixers, loaders, and application operatives (Coye et al. 1986). The most significant exposure route is through skin (Zhang et al. 1991). Behaviors that increase dermal exposure include accidental splashing and spilling, spraying pesticides when skin is unprotected, wearing inappropriate personal protective equipment, failing to wash after handling pesticides, wearing clothing contaminated with pesticides, or touching pesticide-contaminated surfaces (Geer et al. 2004; Oestreich et al. 1997). Other exposure routes include inhalation, oral exposure, and contact with eyes and mucous membranes. Agriculture employs approximately 193,000 workers in the UAE (UAE Ministry of Economy 2006).

Morbidity related to pesticide exposure on UAE farms has been evaluated in several previous studies (Gomes et al. 1997, 1998, 1999; Bener et al. 1999; Beshwari et al. 1999). According to the studies, UAE farm workers are at great risk of developing adverse health effects, both acute and chronic, due to occupational exposure to pesticides (Gomes et al. 1999; Beshwari et al. 1999). The majority of UAE farm workers are uneducated immigrants unable to read complete container label instructions. Many farm workers do not believe that pesticides are harmful (Gomes et al. 1999; McDougall et al. 1993; Baker 1992; Forget 1991) and thus are not motivated to attend safety training sessions when offered. Other factors that potentially increase exposure include lack of personal protection equipment, lack of proper equipment for the preparation of chemicals for spraying, and unsafe pesticide-handling practices (Gomes et al. 1999).

Key Health Effects of Common Occupational Exposures

Carcinogens

Several cancers are known to have occupational causes, but only the three main well-documented occupational cancers are covered in this report: lung cancer, leukemia, and malignant mesothelioma. Other cancers, such as bladder cancer (resulting from exposure to aromatic amines and benzidine dyes, for example) or liver cancer (caused by exposure to vinyl chloride), are not included due to a lack of data on risks and exposures. The most significant causative agents for lung cancer in work environments include asbestos, arsenic, chromium, silica, beryllium, nickel, cadmium, and diesel exhaust. Primary leukemia-causing chemicals are benzene (a common solvent used in oil refineries, chemical plants, and gasoline related industries) and ethylene oxide (used in the production of industrial chemicals) (Driscoll et al. 2004b). Malignant mesothelioma is a form of cancer almost exclusively caused by asbestos exposure. Exposure to asbestos may also lead to asbestosis, as discussed below.

Particulate Matter

Asthma and Chronic Obstructive Pulmonary Disease

The most common occupational health effects associated with exposure to airborne particulates include asthma, chronic obstructive pulmonary disease (COPD), and pneumoconioses, particularly asbestosis and silicosis (Driscoll et al. 2004a). In fact, occupational asthma is estimated to be the most common work-related respiratory disorder in industrialized countries (Kogevinas et al. 1999). Hundreds of biological and nonbiological agents present at workplaces have been associated with occupational asthma (Chan-Yeung and Malo 1994; Venables and Chan-Yeung 1997). Examples of chemical agents include isocyanates, chlorofluorocarbons, alcohols, acrylates, metals and their salts, and welding fumes. Biological agents include grains, flours, plants, woods, insects, fungi, feathers and other animal parts, drugs, and enzymes (Canadian Centre for Occupational Health and Safety 2005). These agents are found in numerous workplaces, such as facilities processing food and natural products, manufacturing and construction sites, and animal-handling facilities (Driscoll et al. 2004a). Agents causing COPD include various kinds of nonspecific dusts and fumes, present in a wide range of industries. COPD is expected to be the fifth-highest cause of disability in the world by 2020 (Murray and Lopez 1996).

Asbestosis and Silicosis

Occupational exposure to silica is associated with a serious occupational lung disease called silicosis and other conditions such as chronic obstructive lung disease, lung cancer, pulmonary tuberculosis, chronic renal disease, rheumatoid arthritis, and autoimmune diseases (Verma et al. 2003; Rushton 2007). Activities in which occupational exposure to silica is most common include mining, quarrying and tunneling, paint and chemical manufacturing, and processes that involve sand, such as sandblasting, foundry work, cement work, brickwork, pottery-making, and glassmaking (Driscoll et al. 2004a).

Asbestos is a fibrous material found in insulation and fireproofing materials, automotive brakes, cement pipes, wallboard, and other materials. Previously, exposures to asbestos occurred mostly in mining, manufacturing, and construction, but removal of asbestos from buildings and structures is now the main cause of exposure in developed countries (Driscoll et al. 2004a). Exposure to asbestos causes lung diseases such as asbestosis and mesothelioma and increases the risk of lung cancer. Both asbestosis and silicosis are irreversible but preventable conditions.

Noise-Induced Hearing Loss

Noise-induced hearing loss is one of the most common occupational illnesses worldwide, and it has been named one of the most important research priorities of the century by the U.S. National Institute for Occupational Safety and Health (2008). Hearing loss is still often overlooked, however, because it occurs gradually over time and without pain. Workers may therefore be unaware of it until the problem grows severe. Short-term symptoms of excess noise include temporary changes in hearing or a temporary ringing in the ears (tinnitus). These problems usually vanish soon after leaving a noisy area. Repeated exposures to loud noise can cause permanent hearing loss or tinnitus that can lead to disability through loss of communication, socialization, and responsiveness to the environment. Importantly, noise-induced hearing loss is completely preventable, but once it has been acquired, it is permanent and irreversible (U.S. Occupational Safety and Health Administration 2007).

Even though noise is a common hazard in almost any workplace, workers in certain industries have more frequent exposures to high levels of noise. Sectors in which high numbers of workers are exposed to noise include agriculture, mining, construction, manufacturing and utilities, transportation, and the military (U.S. National Institute for Occupational Safety and Health 2008). Industrial workers are especially at risk of hearing loss because organic solvents such as styrene and toluene may cause hearing loss themselves or promote noise-induced hearing loss (Rabinowitz et al. 2008; Śliwinska-Kowalska 2007). There are several industries and operations in the UAE in which people are typically exposed to noise, including construction, various kinds of factories, and metal industries.

Heat Stress

Heat stress typically occurs when workers are exposed to high air temperatures, high humidity, heat radiating from a hot surface, direct physical contact with hot objects, or strenuous physical activity. In the UAE, the climate is very hot for most of the year. Outdoor activities where workers are at risk of heat stress in hot weather include farming, surface mining, roofing, road work, and other construction activities (U.S. National Institute for Occupational Safety and Health 1992). Workers are exposed to high temperatures in several indoor operations as well, including iron, steel, and nonferrous foundries; brick-firing and ceramic operations; glass-products manufacturing plants; rubber-products factories; electrical utilities (particularly boiler rooms); bakeries; confectioneries; commercial kitchens; laundries; food canneries; chemical manufacturing plants; mines; and smelters (U.S. Occupational Safety and Health Administration 2008). Working in a hot environment may cause dizziness, fainting, heat rash, and muscle cramps. More serious heat-related conditions include heat exhaustion and heat stroke, which can be fatal. Although heat related illnesses are an important concern in the UAE, particularly in the agriculture and construction sectors, quantifying the burden of disease associated with occupational exposure to excess heat is beyond the scope of this report.

Method for Estimating the Burden of Disease from Occupational Exposures

Risk Factors Covered in the Burden of Disease Calculations

Due to the large number of harmful substances present in work environments, it would be impossible to calculate the disease burden resulting from exposure to all agents. In fact, workers are exposed to hundreds of harmful chemicals and other hazards in occupational settings, including carcinogenic chemicals, heavy metals, dusts, and so on. Therefore, the exposures covered in this study were selected following the approach by the World Health Organization (WHO), covering common occupational carcinogens (Driscoll et al. 2004b), occupational airborne particulates (Driscoll et al. 2004a), and noise (Concha-Barrientos et al. 2004a), excluding occupational injuries, ergonomic stressors, and heat exposure. The three cancers included in this study, lung cancer, leukemia, and malignant mesothelioma, account for the majority of cancers resulting from occupational exposures (Driscoll et al. 2005a). For airborne particulates, health outcomes covered include asthma, chronic obstructive pulmonary disease (COPD), and two pneumoconioses: asbestosis and silicosis. For occupational exposure to noise the resulting health outcome is noise-induced hearing loss (NIHL).

Table 6.4 Occupational hazards and related health outcomes included in this study

Occupational hazard	Health outcome
Occupational carcinogens	
Arsenic, asbestos, beryllium, cadmium, chromium, diesel exhaust, nickel, silica	Lung cancer[a]
Benzene, ethylene oxide	Leukemia
Asbestos	Malignant mesothelioma
Airborne particulate matter	
Unspecified dusts and fumes	Asthma, chronic obstructive pulmonary disease (COPD)
Asbestos	Asbestosis
Silica	Silicosis
Noise	
Noise exposure above 85 dB(A)	Noise-induced hearing loss

[a]Includes cancers of the bronchus, trachea, and lung

Approach for Estimating Disease Burden Due to Occupational Exposures

The disease burden resulting from exposure to occupational hazards is estimated following the attributable fraction approach developed by the WHO (Driscoll et al. 2004a, b; Concha-Barrientos et al. 2004a, b). This approach involves calculating the fraction of disease burden attributable to occupational exposures and using the resulting attributable fraction to estimate the magnitude of death and disability resulting from exposure to occupational hazards. The general method is described in more detail in Chap. 3.

The occupational disease burden is estimated for the year 2008. Source data for 2008 are used whenever possible. When data for that year are not available, data for a year as close as possible to 2008 are used. In this study, mortality is expressed as estimated number of deaths, and morbidity is given as estimated number of health-care facility visits since no pertinent disease incidence or prevalence data for the UAE were available. The occupational hazards and related health outcomes included in the burden of disease calculations are summarized in Table 6.4.

Attributable Fraction

Even though many diseases have been associated with exposure to occupational hazards, only a few diseases are caused exclusively by occupational exposures. Examples of illnesses that are thought to be fully caused by work-related exposures are pneumoconioses such as asbestosis and silicosis. For other illnesses, the proportion of the disease burden caused by work-related exposure has to be estimated. The percentage of deaths or disabilities attributable to occupational factors can be expressed as the attributable fraction (AF). By definition, the attributable fraction is the proportion of cases related to a certain exposure, or the fraction of disease in a population that might be prevented if exposure to a causative agent were reduced or eliminated (Coughlin et al. 1994). The attributable fraction can be calculated as:

$$AF = \frac{\sum_i P_i RR_i - 1}{\sum_i P_i RR_i} \quad (6.1)$$

Where:

AF = Attributable fraction
P_i = Proportion of population at exposure category i
RR_i = Relative risk at exposure category i, compared with reference level

Relative risks (RR) for various diseases can be obtained from international epidemiologic literature. After calculating AF, the mortality and morbidity resulting from a particular exposure can be estimated by multiplying the number of deaths and disease incidences in the general population by AF. Modeling the burden of disease related to exposure to occupational carcinogens (lung cancer, leukemia, and malignant mesothelioma), airborne particulate matter (asthma, COPD, asbestosis, and silicosis), and noise (noise-induced hearing loss) is discussed in more detail in subsequent sections.

Quantitative Modeling

The occupational exposures model consists of three individual modules, one for each of the primary exposures: carcinogens, particulate matter, and noise. Related diseases are handled within each exposure module. To facilitate data entry and analysis, the main page includes all input variables organized by exposure module. Results and model details are also linked to the main page. Figures 6.1 and 6.2 depict the structure of the occupational exposures model, and Table 6.5 details variables shown in the figures. Influence diagrams describing the detailed structure of each module, as well as variables used in each individual exposure module, are presented in the corresponding sections below.

Occupational Carcinogens

Proportion of Workforce Exposed to Carcinogens

The first step in calculating the attributable fraction for occupational carcinogens is to determine the proportion of workers in different economic sectors. Then, since not all workers within an economic sector are exposed to carcinogens, the proportion of workers exposed to a given carcinogen within each sector must be estimated. The proportion of the UAE workforce employed in each economic sector was derived from the UAE Ministry of Economy (2008). Data are classified according to

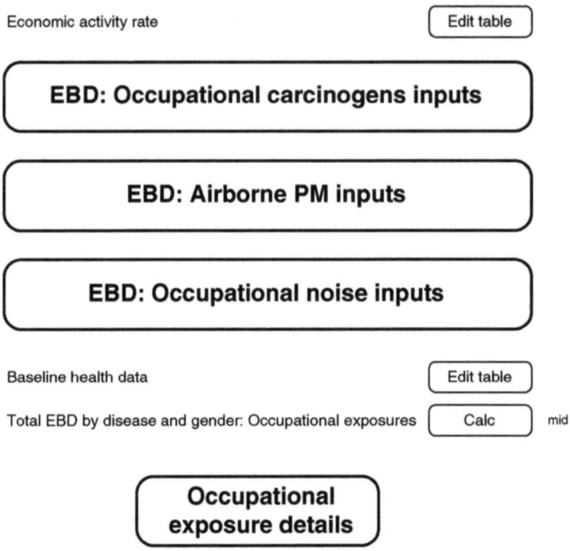

Fig. 6.1 Top-level diagram of the ***Occupational Exposures*** module

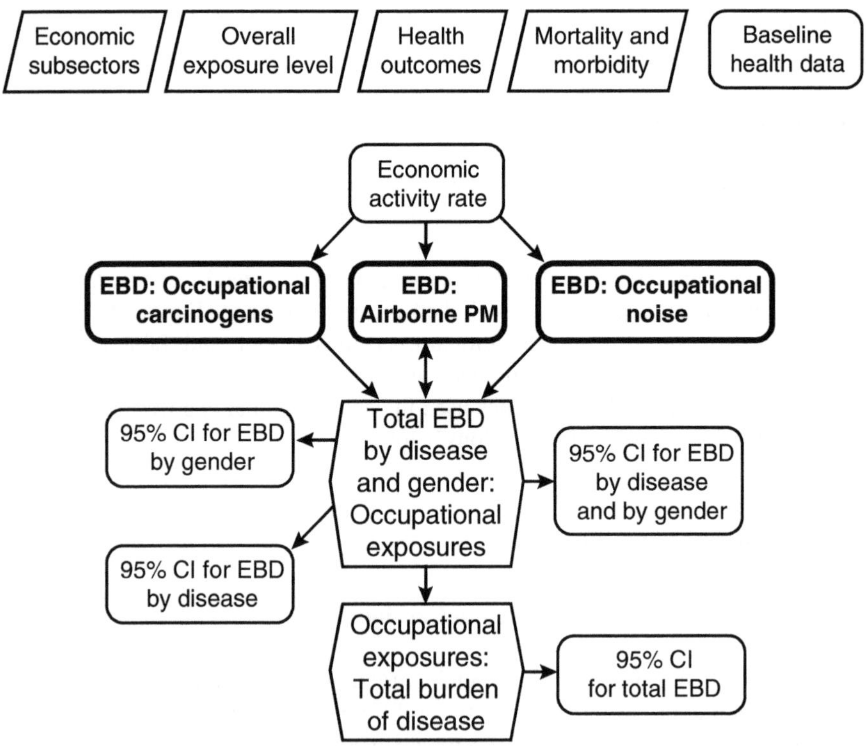

Fig. 6.2 Influence diagram of the ***Occupational Exposure Details*** node

Table 6.5 Indexes and variables used in the ***Occupational Exposures*** module

Index/variable	Description	Definition	Source
Economic subsectors	Proportion of UAE workforce classified into various sectors following the International Standard of Industrial Classification of All Economic Activities	Agriculture, mining, manufacturing, electrical, construction, trade, transportation, finance, services	UAE Ministry of Economy (2008)
Overall exposure level	Exposure levels at which the population is exposed to occupational hazards (carcinogens and leukemogens)	Background: People outside the workforce Low: Workers exposed below the appropriate U.S. Permissible Exposure Limit (PEL) High: Workers exposed above the PEL	Driscoll et al. (2004b)
Health outcomes	Diseases included in the model	Listed in Table 6.4	
Mortality and morbidity	Health endpoints considered in the model	Number of deaths and health-care facility visits	
Fractiles	Fractiles used to calculate the 95% confidence intervals	0.025 and 0.975	
Baseline health data	Number of deaths and health-care facility visits in the general UAE population, by disease and by gender	Listed in Tables 6.10, 6.14, 6.23, and 6.32	Health Authority–Abu Dhabi (HAAD) (2009), Mathers et al. (2000)
Economic activity rate	Percentage of population working or seeking work, including people in paid employment, the self-employed, those who produce goods and services for their own household consumption, and unemployed persons seeking work	Male: 0.891 Female: 0.418	UAE Ministry of Economy (2008)
Total EBD by disease and gender: Occupational exposures	Disease burden due to occupational exposures by disease and gender	Number of deaths and health-care facility visits (Table 6.37)	
Total EBD: Occupational exposures	Total disease burden due to occupational exposures	Number of deaths and health-care facility visits (Table 6.36)	
95% CI for EBD	95% confidence interval for the disease burden by disease and/or gender, or the total disease burden	2.5 and 97.5% fractiles of the corresponding probability distribution	

the International Standard of Industrial Classification of All Economic Activities (ISIC), a widely used industrial classification scheme developed by the United Nations.

Because no carcinogen exposure database exists for the Middle East, a large international Carcinogen Exposure (CAREX) database covering more than 32 million workers is used. The CAREX database provides information on the proportion of workers exposed to higher than background levels of 139 carcinogens in 19 European Union countries. For 15 countries the exposure data are from 1991 to 1993; for the remaining four countries, the data are from 1997. The CAREX exposure estimates were originally constructed in two phases. Preliminary estimates for the proportion of workers exposed to carcinogens were generated in the first phase by combining national labor force data for each country with estimates of carcinogen exposure. Because available country-specific exposure data were limited, two reference countries for which relatively comprehensive exposure data had been collected were used: Finland and the United States. Estimates of exposure prevalence were constructed based on the reference countries, using the most valid data as a default value (the Finnish value, the U.S. value, or a mean of the two). In the second phase, a panel of national experts in each European Union country reviewed the preliminary estimates and either approved the suggested default estimates or modified them to better represent their country. CAREX does not provide information on the level of exposure, and it is assumed that the exposure patterns are the same for male and female workers. When the CAREX database is used, it also has to be assumed that the proportion of exposed workers in the UAE for each carcinogen is similar to that in the database.

The categories used in the employment data (i.e., the proportion of workers in each economic sector) should match the classes used in the carcinogen exposure data as closely as possible. The CAREX database is categorized according to the 2nd Review of the ISIC scheme, whereas the UAE employment data are classified by the 3rd Review, which has more categories than the 2nd. Therefore, nine employment categories in the UAE employment data (hotels and restaurants; real estate, renting, and business services; public administration, defense, and compulsory social security; education; health and social work; other community, social, and personal service activities; activities of private households as employers; extra-territorial organizations and bodies; and not-stated activities) are combined under the Services category. Table 6.6 presents the resulting proportion of workers employed in each economic sector, by gender, in 2005. Table 6.7 lists proportions of the workforce exposed to eight lung carcinogens and two leukemogens, based on the CAREX database.

The proportion of workers exposed in the UAE is then calculated by multiplying the values in Table 6.7 with the proportion of male and female workers in each economic sector (Table 6.6), and the resulting proportions are summed across all economic sectors separately for lung carcinogens and leukemogens.

Turnover of Workforce

Many diseases, such as cancers, have long latency periods. Therefore, workers continue being at risk even after moving to another job or retiring because the

Table 6.6 Proportion of UAE male and female workers in each economic sector

	Proportion of workers		
Economic sector	Male	Female	Total
Agriculture	0.055	0.001	0.048
Mining	0.020	0.005	0.018
Manufacturing	0.086	0.032	0.078
Electrical	0.011	0.003	0.001
Construction	0.332	0.018	0.281
Trade	0.139	0.086	0.131
Transportation	0.069	0.047	0.066
Finance	0.012	0.028	0.014
Services	0.287	0.780	0.353
Total	**1.00**	**1.00**	**1.00**

UAE Ministry of Economy (2008)

disease process progresses even after the exposure has ceased. This is why workers exposed in the past have to be taken into consideration and treated as currently exposed. Occupational turnover (OT) can be calculated using one of these equations, as described by Concha-Barrientos et al. in 2004:

$$OT = \frac{P_t}{P_0} \tag{6.2a}$$

$$OT = \frac{P_0 + P_0 tATR - x(P_0 + P_0 tATR)}{P_0} \tag{6.2b}$$

Where:

P_t = proportion of workers who have ever been occupationally exposed, during time period t, who are still living
P_0 = proportion of workers occupationally exposed at $t=0$ (original workers)
ATR = annual turnover rate
t = working time period (typically considered 40 years in many countries)
x = estimated death rate over time period t

Annual turnover rates within various occupations in several countries presented in the literature were compiled by Concha-Barrientos et al. in 2004. In these data, the variation in annual turnover was high, ranging from 2% in the informatics industry in the Eastern Caribbean area to 500% in the U.S. restaurant industry. Annual turnover varied from 3 to 40% within manufacturing industries in this data set. Due to the large number of expatriate workers in the UAE, the annual turnover rate is estimated to be high, and a range of 10–40% is used in the model. A typical working lifetime is considered 40 years in many countries, but due to the unique composition of the UAE workforce, this is likely to be shorter for the expatriate workers. Thus, a

Table 6.7 Proportion of workers exposed to carcinogens in each industry sector in the CAREX survey

Carcinogen	Agriculture	Mining	Manufacturing	Electrical	Construction	Trade	Transportation	Finance	Services
Lung carcinogens									
Silica	0.004	0.230	0.023	0.014	0.189	0.000	0.00476	0.000	0.001
Cadmium	0.000	0.000	0.005	0.003	0.003	0.000	0.00065	0.000	0.000
Nickel	0.000	0.020	0.017	0.004	0.000	0.000	0.00003	0.000	0.000
Arsenic	0.001	0.001	0.004	0.001	0.001	0.000	0.00000	0.000	0.000
Chromium	0.000	0.003	0.021	0.004	0.002	0.000	0.00370	0.000	0.002
Diesel exhaust	0.006	0.220	0.011	0.034	0.058	0.005	0.13438	0.000	0.009
Beryllium	0.000	0.001	0.002	0.001	0.000	0.000	0.00011	0.000	0.000
Asbestos	0.012	0.102	0.006	0.017	0.052	0.003	0.00684	0.000	0.003
Leukemogens									
Benzene	0.001	0.002	0.003	0.001	0.001	0.01	0.00500	0	0.02
Ethylene oxide	0.00012	0.00137	0.0006	0.00006	0.00027	0	0.00002	0	0.000057

Adapted from Driscoll et al. (2004b)

range of 10–40 years is assumed in the model. For the mortality rate, an overall death rate in the working-age population (ages 15–65) is determined from a data set covering age-specific death rates in the UAE (UAE Ministry of Health 2007). A mortality rate of 0.0013 is calculated by weighting the death rate in each age group with the relative proportion of population in that age group. The proportion of the workforce currently exposed to carcinogens, determined in the previous step, is multiplied by the turnover factor to determine the proportion of the workforce ever exposed to lung carcinogens and leukemogens.

Level of Exposure to Carcinogens

Since not all workers are exposed to the same level of carcinogens within a certain occupation or economic sector, exposed workers are divided into low and high exposure groups. Following the WHO approach, two levels of exposure are chosen for lung carcinogens and leukemogens: high exposure, above the U.S. Occupational Safety and Health Administration (OSHA) Permissible Exposure Limit (PEL), and low exposure, below the PEL (Driscoll et al. 2004b; Concha-Barrientos et al. 2004b). PELs are used because risk estimates and exposure data are often reported in reference to PELs, and many PELs have remain unchanged since they were established in 1971 (Driscoll et al. 2004b). Because data on the level of carcinogens that the UAE workforce is exposed to do not exist, the proportion of workers in each exposure group is estimated following the approach by Concha-Barrientos et al. (2004b). Based on higher prevalence of occupational health and safety programs in the industrialized regions (countries in WHO subregion A), 10% of exposed workers are estimated to be exposed to high levels and 90% to low levels of carcinogens. For industrializing regions (WHO subregions B–E, including the UAE), 50% of exposed workers are partitioned into the high exposure group and 50% into the low exposure group.

To determine the fraction of the total population in each exposure group, the proportions of male and female workers in the high and low exposure groups are then multiplied by the proportion of the population in the workforce, i.e., the economic activity rate. The reported economic activity rate in 2008 was 89% for males and 42% for females (UAE Ministry of Economy 2008). The proportion of the population outside the workforce, 11% of males and 58% of females, is considered the unexposed background group.

Relative Risk

The relative risk estimates for lung carcinogens and leukemogens are based on extensive epidemiologic studies (Steenland et al. 1996, 2003; Nurminen and Karjalainen 2001; Lynge et al. 1997; International Agency for Research on Cancer 1997) and are summarized in Driscoll et al. (2004b). Following the WHO method (Driscoll et al. 2004b, 2005a), a mean relative risk of 1.49 was calculated for lung

Table 6.8 Relative risk estimates for lung cancer and leukemia

Risk factor	Relative risk[a]	95% CI	Proportion of workers exposed[b]	Weighted RR[c]
Lung carcinogens				
Arsenic	3.69	(3.06, 4.46)	0.0006	0.0190
Asbestos	2.00	(1.90, 2.11)	0.0189	0.3160
Beryllium	1.49	NA[d]	0.0002	0.0023
Cadmium	1.49	(0.96, 2.22)	0.0013	0.0160
Chromium	2.78	(2.47, 3.52)	0.0032	0.0747
Diesel exhaust	1.31	(1.13, 1.44)	0.0339	0.3716
Nickel	1.56	(1.41, 1.73)	0.0017	0.0221
Silica	1.33	(1.21, 1.45)	0.0597	0.6654
Total			*0.1195*	*1.4871*
Leukemogens				
Low exposure				
Benzene	2.0	(1.8, 2.2)	0.0093	1.9263
Ethylene oxide	1.1	NA	0.0004	0.0405
Total (low exposure)			*0.0097*	*1.9668*
High exposure				
Benzene	4.0	(3.6, 4.4)	0.0093	3.8527
Ethylene oxide	3.5	NA	0.0004	0.1289
Total (high exposure)			*0.0097*	*3.9816*

[a]Driscoll et al. (2004b). Relative risk estimates based on Steenland et al. (1996, 2003), Nurminen and Karjalainen (2001), International Agency for Research on Cancer (1997), and Lynge et al. (1997).
[b]Calculated from information in Tables 6.6 and 6.7.
[c]Calculated by weighting the relative risk with the proportion of workers exposed.
[d]NA = not available.

carcinogens by weighting each relative risk with the proportion of the population exposed to that carcinogen and summing these weighted relative risks (Table 6.8) (Driscoll et al. 2004a, b). A 95% confidence interval (1.44–1.53) for the mean relative risk is calculated by computing a variance for each individual relative risk and multiplying each variance by the square of the weight used to calculate the weighted average risk. The composite relative risk is partitioned into separate relative risks for low (1.21) and high (1.77) exposure following the approach by Driscoll et al. (2005a), where the ratios of the low and high relative risks to the average U.S. relative risk are applied to produce low and high relative risks for other regions (Driscoll et al. 2005a). The 95% confidence intervals (1.18–1.24 for low exposure and 1.70–1.83 for high exposure) are calculated as described above.

The relative risks for benzene and ethylene oxide are combined into summary relative risks following the same weighting method (Driscoll et al. 2005a). Since relative risks were available in the literature separately for low and high exposure, these values are weighted directly to produce summary relative risks for low (1.97) and high (3.98) exposures (Table 6.8). The relative risks for low and high exposure groups, listed in Table 6.9, are used with information determined in the previous steps to calculate the attributable fraction for lung cancer and leukemia, using Eq. 6.1.

Table 6.9 Summary relative risk estimates used in the burden of disease calculations for occupational carcinogens and leukemogens

	Summary relative risk (95% CI)[a]	
	Low exposure	High exposure
Lung carcinogens	1.21 (1.18, 1.24)	1.77 (1.70, 1.83)
Leukemogens	1.97 (1.8, 2.2)	3.98 (3.6, 4.4)

[a]For leukemogens, the 95% confidence interval (CI) for benzene was used directly for the summary relative risk since no 95% CI was available for ethylene oxide.

Table 6.10 Estimated number of deaths and health-care facility visits for males and females in the UAE population due to lung cancer, leukemia, and malignant mesothelioma

	Number of deaths		Number of health-care facility visits	
	Male	Female	Male	Female
Lung cancer	84	36	389	54
Leukemia	75	54	1,080	440
Malignant mesothelioma	6	0	28	0

HAAD (2009)

Malignant Mesothelioma

A different approach is used for malignant mesothelioma, which is almost solely caused by exposure to asbestos. Not all asbestos exposure is occupational, however, even though most of it is work-related, particularly for men. Calculating the AF from relative risk in the exposed population versus the nonexposed population is not feasible since mesothelioma does not occur in populations that have not been exposed to asbestos. It has been estimated in the literature that 90% of mesothelioma in males and 25% in females is related to occupational exposure to asbestos (Steenland et al. 2003; Nurminen and Karjalainen 2001). These percentages (which represent AFs) can be used to calculate the disease burden due to occupational exposures.

Baseline Mortality and Morbidity for Lung Cancer, Leukemia, and Malignant Mesothelioma

Currently, no central database exists in the UAE from which mortality and morbidity estimates for many of the diseases considered in this study could be derived. Mortality and morbidity data are extrapolated for the entire UAE from a data set covering 73% of the population in Abu Dhabi emirate (HAAD 2009). Table 6.10 presents the number of deaths and health-care facility visits due to lung cancer, leukemia, and malignant mesothelioma in the general UAE population.

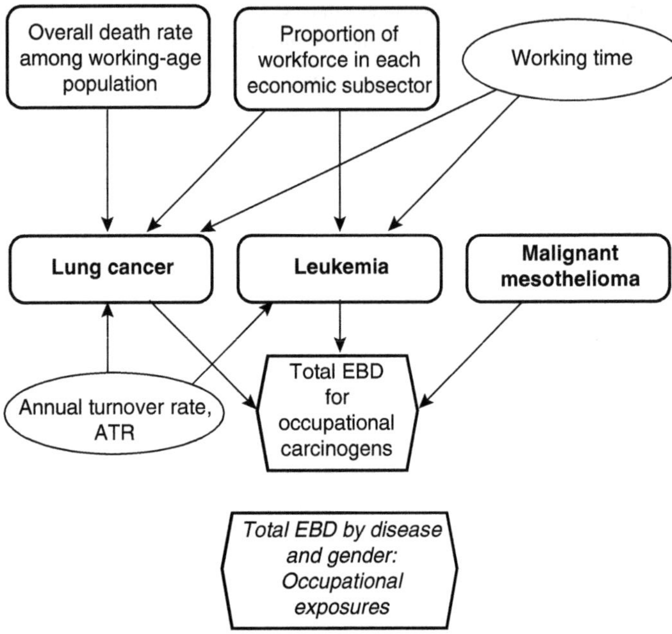

Fig. 6.3 Influence diagram of the ***Occupational Carcinogens*** module

Occupational Carcinogens Module

The occupational carcinogens module is divided into three submodules: lung cancer, leukemia, and malignant mesothelioma. The lung cancer and leukemia submodules share several input variables, as shown in the overall influence diagram of the occupational carcinogens module in Fig. 6.3. Because the lung cancer and leukemia submodules follow a similar structure, only the detailed influence diagram for the lung cancer submodule is presented here, including the embedded ***Occupational Turnover*** node (Figs. 6.4 and 6.5, respectively). Figure 6.6 illustrates the detailed structure of the malignant mesothelioma submodule. Descriptions of variables in each are presented in Tables 6.11, 6.12, and 6.13.

Occupational Airborne Particulate Matter

Exposure to various kinds of dusts is common at almost any workplace. Hundreds of particulates occur in the occupational environment, and many respiratory conditions are potentially work related. The most important nonmalignant respiratory diseases resulting from occupational exposure to airborne particulates include asthma, chronic obstructive pulmonary disease (COPD), and pneumoconioses

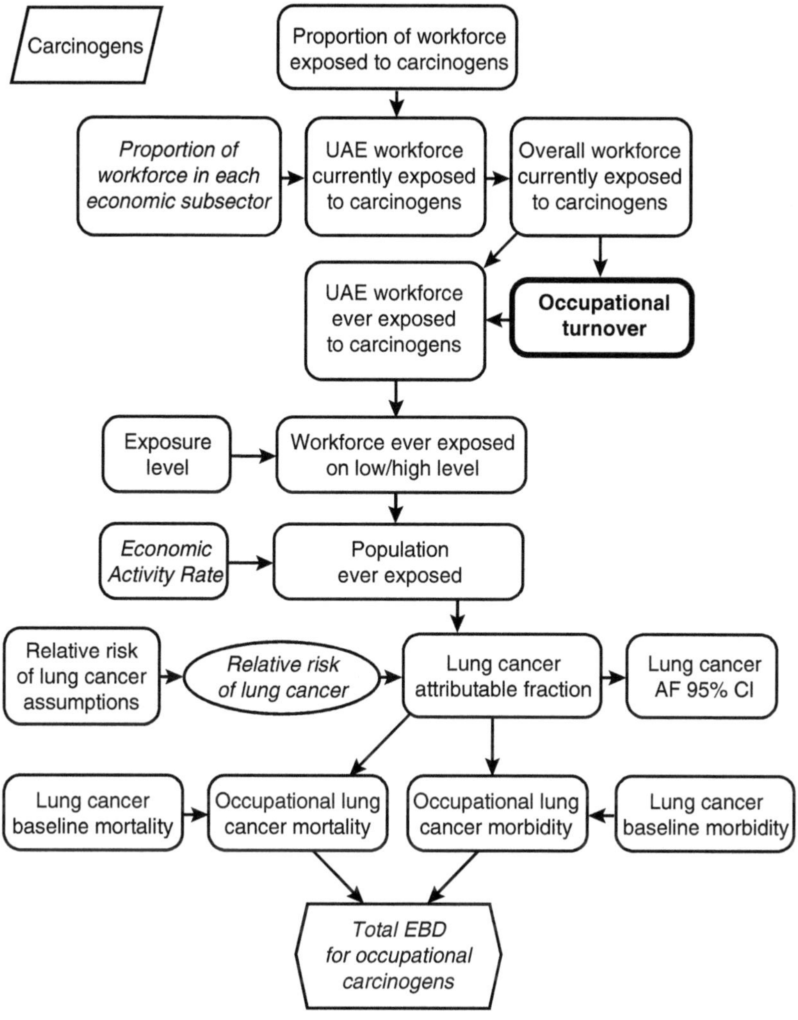

Fig. 6.4 Influence diagram of the detailed ***Lung Cancer*** submodule. The ***Leukemia*** submodule is structurally the same

(asbestosis and silicosis). Whereas asbestosis and silicosis are caused exclusively by exposure to asbestos and silica, hundreds of agents can cause occupational asthma. It is impossible to estimate relative risks for all of these substances. Consequently, instead of looking at individual agents, occupation or industrial sector can be used as a substitute for exposure to substances associated with occupational asthma or COPD (Driscoll et al. 2004a). Overall, the approach for calculating the burden of disease due to work-related asthma and COPD follows the WHO approach (Driscoll et al. 2004a) and is similar to that used for lung cancer and leukemia, discussed above.

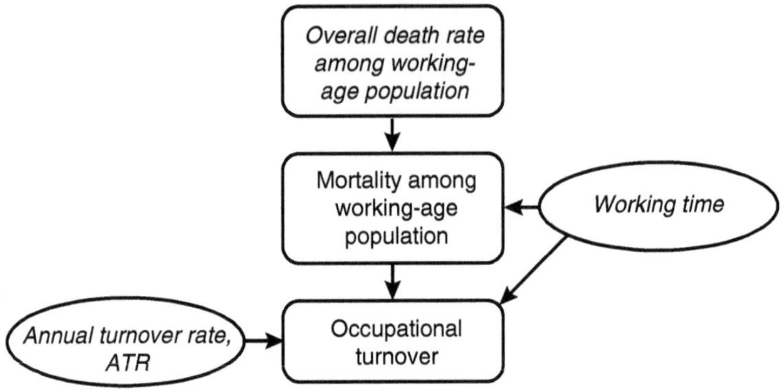

Fig. 6.5 Influence diagram of the ***Occupational Turnover*** node

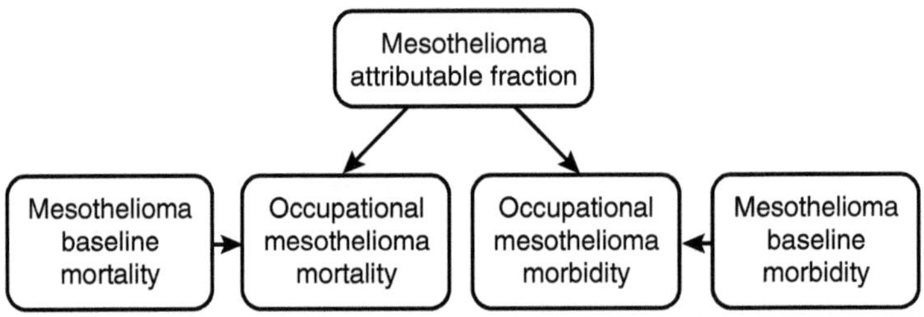

Fig. 6.6 Influence diagram of the ***Malignant Mesothelioma*** submodule

Table 6.11 Description of variables in the ***Occupational Carcinogens*** module

Index/Variable	Description	Definition	Source
Overall death rate among working-age population	Age-adjusted death rate for working-age population (15–65 years)	0.0013	Calculated from data from UAE Ministry of Health (2007)
Proportion of workforce in each economic subsector	Proportion of UAE workforce in each economic subsector, by gender, categorized according to the 2nd Review of the International Standard of Industrial Classification of All Economic Activities (ISIC) scheme	Listed in Table 6.6	UAE Ministry of Economy (2008)
Working time	Total duration of occupational exposure	10–40 years, uniform distribution	
Annual turnover rate, ATR	Occupational turnover rate (people retiring or moving to other jobs), estimated as percentage of workforce replaced each year	10–40%, uniform distribution	Estimated based on studies compiled in Concha-Barrientos et al. (2004b)

Table 6.12 Description of variables in the ***Lung Cancer*** submodule, including the ***Occupational Turnover*** node

Index/Variable	Description	Definition	Source
Carcinogens	The eight lung carcinogens included in the study	Arsenic, asbestos, beryllium, cadmium, chromium, diesel exhaust, nickel, silica	Driscoll et al. (2004b)
Proportion of workforce exposed to carcinogens	Proportion of workers exposed to selected carcinogens in each economic subsector, based on the Carcinogen Exposure (CAREX) database	Listed in Table 6.7	Driscoll et al. (2004b)
UAE workforce currently exposed to carcinogens	Proportion of UAE workforce currently exposed to carcinogens, by gender and by carcinogen	Proportion of workforce in each economic subsector multiplied by proportion of workers exposed to carcinogens	Driscoll et al. (2004b)
Overall workforce currently exposed to carcinogens	Proportion of workforce currently exposed to all eight carcinogens, by gender	Calculated by adding together proportions of workforce exposed to each carcinogen within each economic sector, and summing these totals	Driscoll et al. (2004b)
UAE workforce ever exposed to carcinogens	Proportion of UAE workforce ever exposed to carcinogens, by gender	Overall workforce currently exposed to carcinogens multiplied by the occupational turnover factor	Driscoll et al. (2004b)
Exposure level	Proportion of workforce exposed at low (below the relevant U.S. Permissible Exposure Limit, PEL) and high level (above PEL)	Background: 0 (since only workforce, but not population, is considered at this point) Low: 0.5 High: 0.5	Driscoll et al. (2004b)
Workforce ever exposed on low/high level	Proportion of workforce exposed to carcinogens at different levels	UAE workforce ever exposed to carcinogens multiplied by the exposure level	Driscoll et al. (2004b)
Population ever exposed	Proportion of the population exposed to carcinogens at different levels	Workforce ever exposed to low/high levels multiplied by the economic activity rate; proportion of UAE population never exposed to carcinogens calculated by subtracting the proportion of population exposed at low and high levels from 1	Driscoll et al. (2004b)

(continued)

Table 6.12 (continued)

Index/Variable	Description	Definition	Source
Relative risk of lung cancer assumptions	Relative risk of lung cancer indexed by overall exposure level and normal parameters	Background: mean = 1, standard deviation = 0 Low: mean = 1.21, sd = 0.0148 High: mean = 1.77, sd = 0.0316	Steenland et al. (1996, 2003), Nurminen and Karjalainen (2001), Driscoll et al. (2004b, 2005a)
Relative risk of lung cancer	Relative risk of lung cancer by exposure level	Distribution of lung cancer relative risk in each exposure group using normal parameters defined above	
Lung cancer attributable fraction	Proportion of lung cancer attributable to occupational exposure to lung carcinogens	Calculated using Eq. 6.1	Driscoll et al. (2004b)
Lung cancer AF 95% CI	95% confidence interval for lung cancer attributable fraction	2.5 and 97.5% fractiles of the probability distribution of lung cancer AF	
Lung cancer baseline mortality/morbidity	Number of deaths and health-care facility visits due to lung cancer in the general UAE population, by gender	Listed in Table 6.10	HAAD (2009)
Occupational lung cancer mortality/ morbidity	Number of deaths and health-care facility visits due to lung cancer resulting from occupational exposure to carcinogens, by gender	Lung cancer baseline mortality/morbidity multiplied by the lung cancer AF	Driscoll et al. (2004b)
Mortality among working-age population	Mortality rate over the working time period	Age-adjusted death rate among working-age population multiplied by the working time period	Concha-Barrientos et al. (2004b)
Occupational turnover	Adjustment factor to account for people previously exposed, but retired or working in another occupation (accounting for latency period between exposure and disease onset)	Calculated using Eq. 6.2b	Concha-Barrientos et al. (2004b)

Table 6.13 Description of variables in the ***Malignant Mesothelioma*** submodule

Index/Variable	Description	Definition	Source
Mesothelioma attributable fraction	Proportion of mesothelioma attributable to occupational exposure to asbestos	Male: 0.9 Female: 0.25	Nurminen and Karjalainen (2001), Steenland et al. (2003)
Mesothelioma baseline mortality/morbidity	Number of deaths and health-care facility visits due to malignant mesothelioma in the general UAE population, by gender	Listed in Table 6.10	HAAD (2009)
Occupational mesothelioma mortality/morbidity	Number of deaths and health-care facility visits due to mesothelioma resulting from occupational exposure to asbestos, by gender	Mesothelioma baseline mortality/morbidity multiplied by the mesothelioma AF	Driscoll et al. (2004b)

Table 6.14 Number of baseline deaths and health-care facility visits due to asbestosis and silicosis in the UAE

Pneumoconiosis	Deaths	Health-care facility visits
Asbestosis	0	3
Silicosis	0	8

HAAD (2009)

The method is simpler for pneumoconioses because all cases are attributable to work (AF = 100%). The number of deaths due to pneumoconioses can be assessed by counting the number of asbestosis and silicosis deaths in the country using national death records (Driscoll et al. 2004a). The number of recorded deaths and health-care facility visits due to asbestosis and silicosis in the UAE is listed in Table 6.14.

Proportion of Workforce in Each Occupation-Industry Group for Calculating AF for Asthma

For calculation of attributable fraction for occupational asthma, the proportion of workers in each occupation-industry group (i.e., the proportion of administration, services, production workers, etc., within each industry) is needed (Driscoll et al. 2004a). The cross-classification of workers into occupations and industry sectors should correspond with the grouping used in the literature from which the relative risks were derived as closely as possible. The relative risks used in this study are based on Karjalainen et al. (2001, 2002) and Kogevinas et al. (1999). The distribution of the UAE workforce into different occupations within each economic sector is based on information from the Ministry of Economy by applying the percentages

Table 6.15 Proportion of UAE male workforce in each occupational category, by industry

Industry	Technical	Administration	Clerical	Sales	Services	Agriculture	Production	Total
Agriculture	0.0018	0.0004	0.0003	0.0004	0.0004	0.0218	0.0298	0.0550
Mining	0.0101	0.0022	0.0014	0.0009	0.0009	0.0000	0.0042	0.0197
Manufacturing	0.0235	0.0107	0.0041	0.0020	0.0020	0.0003	0.0430	0.0855
Electrical	0.0046	0.0007	0.0007	0.0000	0.0000	0.0000	0.0047	0.0108
Construction	0.0994	0.0240	0.0098	0.0003	0.0003	0.0000	0.1886	0.3225
Trade	0.0290	0.0125	0.0037	0.0357	0.0357	0.0014	0.0207	0.1386
Transportation	0.0208	0.0054	0.0069	0.0012	0.0012	0.0001	0.0332	0.0689
Finance	0.0063	0.0034	0.0022	0.0000	0.0000	0.0000	0.0005	0.0124
Services	0.0813	0.0197	0.0142	0.0340	0.0340	0.0022	0.1014	0.2866
Total	**0.2768**	**0.0790**	**0.0433**	**0.0745**	**0.0745**	**0.0258**	**0.4261**	**1.0000**

Reclassified employment data based on data from the UAE Ministry of Economy (2008)

Table 6.16 Proportion of UAE female workforce in each occupational category, by industry

Industry	Technical	Administration	Clerical	Sales	Services	Agriculture	Production	Total
Agriculture	0.0004	0.0000	0.0001	0.0001	0.0001	0.0003	0.0000	0.0010
Mining	0.0029	0.0002	0.0017	0.0000	0.0000	0.0000	0.0002	0.0051
Manufacturing	0.0152	0.0032	0.0094	0.0016	0.0016	0.0000	0.0014	0.0323
Electrical	0.0010	0.0009	0.0015	0.0000	0.0000	0.0000	0.0000	0.0034
Construction	0.0109	0.0010	0.0049	0.0000	0.0000	0.0000	0.0009	0.0177
Trade	0.0310	0.0099	0.0243	0.0096	0.0096	0.0000	0.0012	0.0856
Transportation	0.0170	0.0043	0.0157	0.0039	0.0039	0.0000	0.0024	0.0471
Finance	0.0139	0.0024	0.0116	0.0000	0.0000	0.0000	0.0001	0.0280
Services	0.2168	0.0425	0.0685	0.2169	0.2169	0.0000	0.0184	0.7799
Total	**0.3091**	**0.0644**	**0.1377**	**0.2321**	**0.2321**	**0.0003**	**0.0246**	**1.0000**

Reclassified employment data based on data from the UAE Ministry of Economy (2008)

of male and female workers in different occupation-industry groups in 2008 to the actual number of workers in 2005. To match the relative risk data, the UAE data are reclassified by combining professional and technical occupations into technical workers, relabeling legislators as administration workers, dividing the "service workers and shop and market sales" category in half into separate categories for sales and services, and combining the remaining occupations (excluding workers in the clerical and agricultural category) into production workers. Economic subsectors are reclassified the same way as for carcinogens, as discussed above. Proportions of male and female workers in each occupation-industry group are presented in Tables 6.15 and 6.16, respectively.

The proportions of workers in these occupational groups then had to be adapted to the groups used in the literature for relative risks. Following the approach by Driscoll et al. (2004a), clerical workers were added to the administration group, and production workers were separated into miners (production workers in the mining industry), transportation workers (production workers in the transportation industry),

Table 6.17 Proportion of UAE male workforce in each occupational category, reclassified for calculating AF for asthma

Technical	Administration	Sales	Services	Agriculture	Mining	Transportation	Manufacturing	Total
0.277	0.122	0.075	0.075	0.026	0.004	0.033	0.389	1.000

Table 6.18 Proportion of UAE female workforce in each occupational category, reclassified for calculating AF for asthma

Technical	Administration	Sales	Services	Agriculture	Mining	Transportation	Manufacturing	Total
0.309	0.202	0.232	0.232	0.000	0.000	0.002	0.022	1.000

and manufacturers (production workers in other industries), resulting in the figures in Tables 6.17 and 6.18 for males and females, respectively.

To obtain proportions of the total male and female populations in each occupational group, the figures in Tables 6.17 and 6.18 are multiplied by the economic activity rates in the UAE, 89% of men and 42% of women (UAE Ministry of Economy 2008). The Population outside the workforce is considered to be in the background group, 11% of males and 58% of females.

Proportion of Workforce in Each Industry Group for Calculating AF for COPD

For COPD, the relative risks are determined in the epidemiologic literature for industrial sectors, and thus the same grouping can be used as was for carcinogens (Table 6.6). Next, the figures in Table 6.6 need to be adapted to the three exposure levels used for the relative risks, based on Korn et al. (1987). The three exposure groups are background (includes combined proportions of workers in trade, finance, and services from Table 6.6), low (agriculture, electricity, and transportation), and medium/high (mining, manufacturing, and construction) (Korn et al. 1987). Proportions of male and female workforce in the three exposure groups are presented in Table 6.19.

The figures in Table 6.19 need to be adjusted for the proportion of the population in the workforce by multiplying the proportion of workforce in each exposure category with the economic activity rate (89% of males and 42% of females). The remaining 11 and 58% of the male and female populations not in the workforce are considered to be exposed to background levels of dusts. Table 6.20 presents adjusted figures for proportions of male and female populations in each exposure group.

For asthma and COPD, the number of exposed workers is not adjusted for previous exposures, since people currently working in occupations with increased risk already make up the majority of the population (Karjalainen et al. 2001; Driscoll et al. 2004b).

Table 6.19 Proportion of the UAE workforce in each of three exposure groups[a]

		Proportion of workers	
Exposure group[a]	Industries	Male	Female
Background	Trade, finance, and services	0.437	0.892
Low	Agriculture, electricity, and transportation	0.135	0.051
Medium/high	Mining, manufacturing, and construction	0.429	0.055
Total	**All industries**	**1.00**	**1.00**

[a]Based on Korn et al. (1987)

Table 6.20 Proportion of the UAE population in each exposure group

		Proportion of population	
Exposure group	Industries	Male	Female
Background	Trade, finance, services, or not in workforce	0.497	0.955
Low	Agriculture, electricity, and transportation	0.121	0.021
Medium/high	Mining, manufacturing, and construction	0.384	0.023
Total	**All industries**	**1.00**	**1.00**

Table 6.21 Relative risk estimates for asthma by occupation and gender

	Male		Female	
Occupation	Relative risk	95% CI	Relative risk	95% CI
Background	1.00		1.00	
Administration	1.00		1.00	
Technical	1.05	(0.98, 1.12)	1.06	(1.03, 1.10)
Sales	1.10	(1.05, 1.23)	1.13	(1.08, 1.18)
Agricultural	1.41	(0.98, 2.02)	1.41	(0.98, 2.02)
Mining	1.95	(1.58, 2.40)	1.00	(0.25, 4.02)
Transportation	1.31	(1.22, 1.40)	1.22	(1.13, 1.31)
Manufacturing	1.56	(1.47, 1.65)	1.33	(1.27, 1.39)
Services	1.53	(1.42, 1.66)	1.41	(1.35, 1.46)

Driscoll et al. (2004a). Relative risks based on Karjalainen et al. (2001, 2002), and Kogevinas et al. (1999)

Relative Risk for Asthma and COPD

Relative risk estimates for asthma and COPD are derived from epidemiological literature because regional information was not available for the UAE (Kogevinas et al. 1999; Karjalainen et al. 2001, 2002). Because countless particulates can cause asthma and/or COPD, relative risks are presented for entire occupations or industrial sectors rather than for specific occupational agents. The relative risk estimates used in this study for asthma and COPD were derived from Driscoll et al. (2004b) and are listed in Tables 6.21 and 6.22, respectively. Risk estimates for asthma are based on studies by Karjalainen et al. (2001, 2002) and Kogevinas et al. (1999); COPD risk estimates come from Korn et al. (1987). The attributable fractions for asthma and COPD for males and females are calculated using Equation 6.1.

Table 6.22 Relative risk estimates of COPD mortality for the WHO Eastern Mediterranean B Region (including the UAE)

	Relative risk of COPD	
Exposure level	Male	Female
Unexposed	1.0	1.0
Low	1.2	1.1
High	1.8	1.4

Driscoll et al. (2004a). Relative risks based on Korn et al. (1987)

Table 6.23 Estimated number of UAE deaths and health-care facility visits related to asthma and COPD

	Deaths		Health-care facility visits	
	Male	Female	Male	Female
Asthma	4	6	49,137	23,164
COPD	9	28	19,861	7,352

Calculated from health data collected in Abu Dhabi emirate (HAAD 2009)

Baseline Asthma and COPD Morbidity and Mortality Rates

Morbidity and mortality due to asthma and COPD in the general UAE population are estimated from health data gathered in Abu Dhabi emirate (HAAD 2009) and extrapolated to cover the entire UAE population. Table 6.23 presents the number of deaths and health-care facility visits related to asthma and COPD.

Occupational Particulate Matter Module

As illustrated in Fig. 6.7, the ***Occupational Particulate Matter*** module is divided into three submodules, covering asthma, COPD, and asbestosis and silicosis. The detailed influence diagrams of the asthma and COPD submodules are presented in Figs. 6.8 and 6.9, respectively. The variables used in the asthma and COPD submodules are described in Table 6.24. Only the COPD variables different from those in the asthma submodule are included in Table 6.24. The ***Asbestosis and Silicosis*** submodule follows essentially the same structure as the malignant mesothelioma submodule described earlier, except that the variables are indexed by asbestosis and silicosis (Fig. 6.10).

Occupational Exposure to Noise

Noise is a very common risk factor in many work environments, particularly in manufacturing, transportation, mining, construction, agriculture, and the military. Hearing impairment is the best characterized consequence of noise exposure and

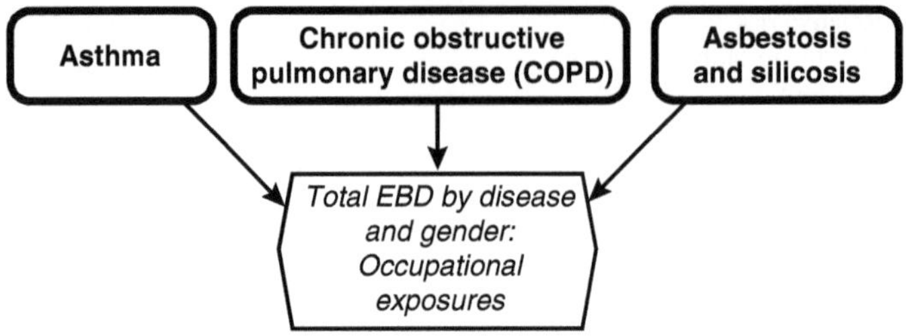

Fig. 6.7 Structure of the ***Occupational Particulate Matter*** module

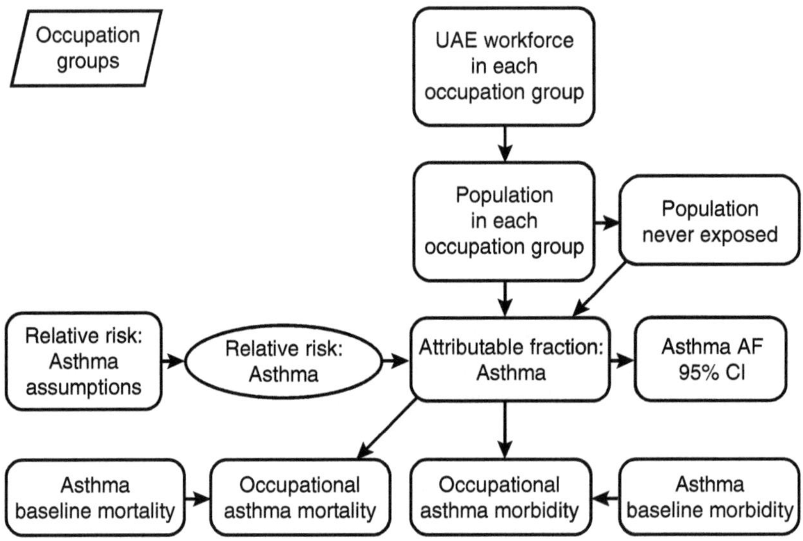

Fig. 6.8 Influence diagram of the ***Occupational Asthma*** submodule

shows the strongest epidemiologic link between exposure and health effect. Even though exposure to occupational noise has been linked to other health effects as well, including annoyance, hypertension, and disturbance of psychological well being (Concha-Barrientos et al. 2004), epidemiologic evidence linking these health effects to occupational noise is not as strong as for hearing impairment, so these other effects will not be covered in this report.

WHO defines disabling hearing loss as "permanent unaided hearing threshold level for the better ear of 41 dBHL or greater for the four frequencies of 500, 1,000, 2,000, and 4,000 Hz" (Concha-Barrientos et al. 2004a). The WHO definition for hearing impairment (WHO 2009) is described in Table 6.25, adapted from Concha-Barrientos et al. (2004a).

Exposure to noise is usually measured as A-weighted decibels, dB(A), which take into consideration the sensitivity of the human ear to sound at different frequen-

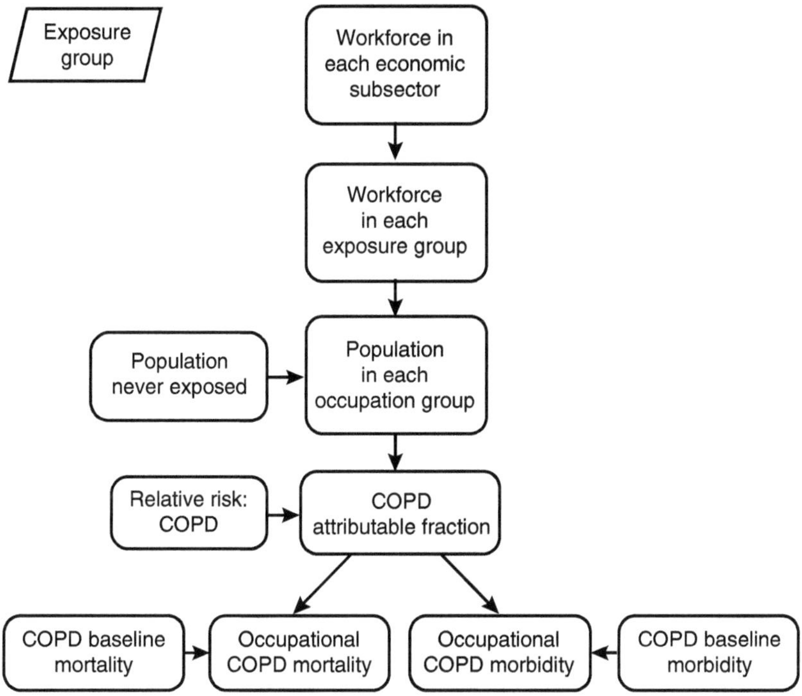

Fig. 6.9 Influence diagram of the ***Occupational COPD*** submodule

cies. Because occupational noise often fluctuates over time, the equivalent sound level is determined over a time period, typically an 8-h work shift. Exposure to occupational noise is usually divided into three categories that correspond to common regulatory limits, which are 85 dB(A) in most developed countries and 90 dB(A) in many developing nations over an 8-h work day. Exposure to sound levels less than 85 dB(A) is considered minimum, exposure to 85–90 dB(A) is considered moderately high, and exposure to higher than 90 dB(A) is considered high, as presented in Table 6.26 (Concha-Barrientos et al. 2004a). As a rule of thumb, a sound level is over 85 dB if voices must be raised to be heard.

Sources of Noise Exposure Data for the UAE

Exposure to occupational noise is usually determined for different occupations and has been shown to be the most important determinant for exposure level (Concha-Barrientos et al. 2004a). For example, workers in occupations categorized under the production category are typically exposed to higher levels of noise than workers in administrative occupations in each industrial sector. However, many factors affect the exposure levels within each occupational category, including types of processes in a facility; types of raw materials, machinery and tools used; the extent of

Table 6.24 Description of variables used in the ***Occupational Asthma*** and ***Occupational COPD*** submodules

Index/Variable	Description	Definition	Source
Asthma			
Occupation groups	Classification of workers into occupation groups that match the grouping used in the literature from which relative risks were derived	Listed in Tables 6.17 and 6.18	Driscoll et al. (2004a)
UAE workforce in each occupation group	Proportion of UAE workforce in each occupation group, by gender	Listed in Tables 6.17 and 6.18	Workforce data: UAE Ministry of Economy 2008; method for matching UAE data with occupation groups used here is described in Driscoll et al. (2004a)
Population in each occupation group	Proportion of UAE population in each occupation group, by gender	Proportion of workforce in each occupation group multiplied by the economic activity rate	Driscoll et al. (2004a)
Population never exposed	Proportion of people outside the work-force (= background), by gender	= 1-economic activity rate	Driscoll et al. (2004a); UAE Ministry of Economy 2008
Relative risk: asthma assumptions	Relative risk of asthma indexed by occupation group, gender, and normal parameters	Listed in Table 6.21; Standard deviations calculated from 95% confidence intervals	Relative risks based on Karjalainen et al. (2001, 2002) and Kogevinas et al. (1999) and summarized in Driscoll et al. (2004a)
Relative risk: asthma	Relative risk of asthma by occupation group and gender	Distribution of asthma relative risk in each occupation group defined by the normal parameters	
Attributable fraction: asthma	Proportion of asthma attributable to occupational exposure to airborne PM	Calculated using Eq. 6.1	Driscoll et al. (2004a)

(continued)

Table 6.24 (continued)

Index/Variable	Description	Definition	Source
Asthma AF 95% CI	95% confidence interval for the asthma attributable fraction	2.5 and 97.5% fractiles of the probability distribution of asthma AF	
Asthma baseline mortality/ morbidity	Number of deaths and health-care facility visits due to asthma in the general UAE population, by gender	Listed in Table 6.23	HAAD (2009)
Occupational asthma mortality/ morbidity	Number of deaths and health-care facility visits due to asthma resulting from occupational exposure to airborne PM, by gender	Asthma baseline mortality/ morbidity multiplied by asthma AF (Table 6.37)	Driscoll et al. (2004a)
Chronic obstructive pulmonary disease (COPD)			
Exposure group	Exposure groups (background, low, and high) corresponding to relative risk estimates for COPD	Economic subsectors included in each exposure group listed in Table 6.19	Korn et al. (1987), Driscoll et al. (2004a)
Workforce in each economic subsector	Proportion of UAE workforce in each economic subsector, by gender	Listed in Table 6.6	UAE Ministry of Economy 2008
Workforce in each exposure group	Proportion of workforce in each exposure group defined above	Listed in Table 6.19	Korn et al. 1987, Driscoll et al. (2004a)
Relative risk: COPD	Relative risk of COPD in each exposure group (background, low, and high)	Listed in Table 6.22	Korn et al. (1987), Driscoll et al. (2004a)

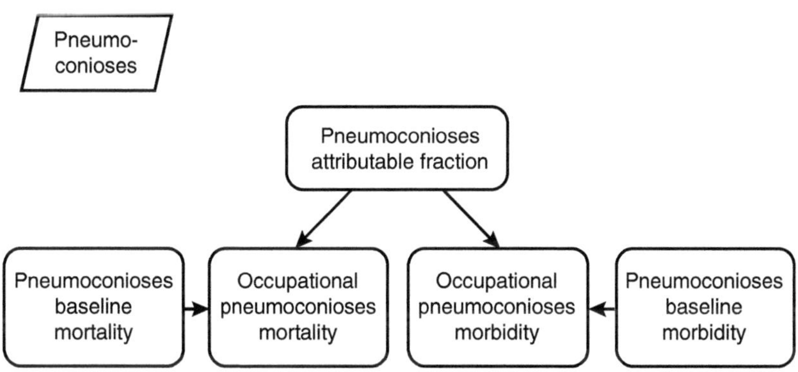

Fig. 6.10 Influence diagram of the ***Asbestosis and Silicosis*** submodule

Table 6.25 World health organization definition of hearing impairment

Grade of hearing impairment	Audiometric ISO value (better ear)[a]	Performance
0 No impairment	≤25 dB	No or very slight hearing problems; able to hear whispers
1 Slight impairment	26–40	Able to hear and repeat words spoken in normal voice at 1 m
2 Moderate impairment	41–60	Able to hear and repeat words spoken in raised voice at 1 m
3 Severe impairment	61–80	Able to hear some words when shouted into better ear
4 Profound impairment, including deafness	≥81	Unable to hear and understand even shouted words

Adapted from Concha-Barrientos et al. (2004a)
[a]Averages of values at 500, 1,000, 2,000, 4,000 Hz

Table 6.26 Categories of noise exposure used in this study

Noise exposure	dB(A)
Minimum	<85
Moderately high	85–90
High	>90

Concha-Barrientos et al. (2004a)

engineering controls and work practices used to control exposure; and availability, use and maintenance of personal protective equipment (Concha-Barrientos et al. 2004a). Ideally, noise exposure data for estimating the burden of disease due to occupational noise should come from local exposure data, such as measurements of average noise levels within major occupations in the UAE. Because these data do not currently exist, extensive data from the United States (Centers for Disease Control and Prevention 1986; U.S. National Institute for Occupational Safety and Health 1992) covering more than nine million production workers, adjusted to reflect estimated exposures in the WHO Eastern Mediterranean B region by Concha-Barrientos et al. (2004b), were used.

Estimating Exposure to Occupational Noise

The approach for assessing the disease burden due to occupational noise is based on the WHO Environmental Disease Burden Series by Concha-Barrientos et al. (2004a). The first step is to estimate the proportion of workers in each occupational category within each economic subsector that is exposed to a noise level of greater than 85 dB(A) (Centers for Disease Control and Prevention 1986; U.S. National Institute for Occupational Safety and Health 1998). Partitioning the exposure estimates into moderately high (85–90 dB(A)) and high (>90 dB(A)) noise levels, with modifications to reflect working conditions in different regions of the world, is described in Concha-Barrientos et al. (2004b). The estimated proportions of the

Table 6.27 Proportion of workforce within each occupation and industry subsector exposed to noise levels of 85–90 dB(A) in the WHO Eastern Mediterranean B region (including the UAE)

	Occupation						
Industry	Professional	Administration	Clerical	Sales	Services	Agriculture	Production
Agriculture	0.05	0.05	0.05	0.09	0.09	0.14	0.01
Mining	0.05	0.05	0.05	0.09	0.09	0.14	0.04
Manufacturing	0.05	0.05	0.05	0.09	0.09	0.14	0.01
Electrical	0.05	0.05	0.05	0.09	0.09	0.14	0.04
Construction	0.05	0.05	0.05	0.09	0.09	0.14	0.01
Trade	0.02	0.02	0.02	0.09	0.09	0.14	0.01
Transportation	0.02	0.02	0.02	0.09	0.09	0.14	0.01
Finance	0.02	0.02	0.02	0.09	0.09	0.14	0.00
Services	0.02	0.02	0.02	0.09	0.09	0.14	0.00

Concha-Barrientos et al. (2004b)

Table 6.28 Proportion of workforce within each occupation and industry subsector exposed to noise levels above 90 dB(A) in the WHO Eastern Mediterranean B region (including the UAE)

	Occupation						
Industry	Professional	Administration	Clerical	Sales	Services	Agriculture	Production
Agriculture	0.00	0.00	0.00	0.03	0.03	0.06	0.19
Mining	0.00	0.00	0.00	0.03	0.03	0.06	0.81
Manufacturing	0.00	0.00	0.00	0.03	0.03	0.06	0.21
Electrical	0.00	0.00	0.00	0.03	0.03	0.06	0.14
Construction	0.00	0.00	0.00	0.03	0.03	0.06	0.17
Trade	0.00	0.00	0.00	0.03	0.03	0.06	0.12
Transportation	0.00	0.00	0.00	0.03	0.03	0.06	0.11
Finance	0.00	0.00	0.00	0.03	0.03	0.06	0.02
Services	0.00	0.00	0.00	0.03	0.03	0.06	0.03

Concha-Barrientos et al. (2004b)

workforce exposed to moderately high and high noise levels in the WHO Eastern Mediterranean B region (including the UAE) are presented in Tables 6.27 and 6.28, respectively.

The next step is to determine the proportion of workers in each occupation-industry category in the UAE. This information is derived from UAE Ministry of Economy data and reclassified to match the occupational categories in Tables 6.27 and 6.28. Professionals, technicians, and associate professionals were combined into professional workers; legislators, senior officials, and managers are relabeled as administrative workers; service workers and shop and market sales workers are divided into two separate categories for sales workers and service workers, and the remaining four occupational categories are combined into production workers. Proportions of the workforce in each occupation-industry group are presented in Tables 6.29 and 6.30 for males and females, respectively.

Table 6.29 Proportion of UAE male workforce in each occupation, reclassified for estimating noise exposure

	Occupation						
Industry	Professional	Administration	Clerical	Sales	Services	Agriculture	Production
Agriculture	0.03	0.01	0.01	0.01	0.01	0.40	0.54
Mining	0.51	0.11	0.07	0.05	0.05	0.00	0.21
Manufacturing	0.27	0.12	0.05	0.02	0.02	0.00	0.50
Electrical	0.42	0.07	0.06	0.00	0.00	0.00	0.44
Construction	0.31	0.07	0.03	0.00	0.00	0.00	0.58
Trade	0.21	0.09	0.03	0.26	0.26	0.01	0.15
Transportation	0.30	0.08	0.10	0.02	0.02	0.00	0.48
Finance	0.51	0.28	0.18	0.00	0.00	0.00	0.04
Services	0.28	0.07	0.05	0.12	0.12	0.01	0.35

Reclassified employment data based on data from UAE Ministry of Economy 2008

Table 6.30 Proportion of UAE female workforce in each occupation, reclassified for estimating noise exposure

	Occupation						
Industry	Professional	Administration	Clerical	Sales	Services	Agriculture	Production
Agriculture	0.42	0.00	0.08	0.08	0.08	0.33	0.00
Mining	0.57	0.05	0.33	0.00	0.00	0.00	0.05
Manufacturing	0.47	0.10	0.29	0.05	0.05	0.00	0.04
Electrical	0.30	0.26	0.44	0.00	0.00	0.00	0.00
Construction	0.61	0.06	0.28	0.00	0.00	0.00	0.05
Trade	0.36	0.12	0.28	0.11	0.11	0.00	0.01
Transportation	0.36	0.09	0.33	0.08	0.08	0.00	0.05
Finance	0.50	0.09	0.41	0.00	0.00	0.00	0.00
Services	0.28	0.05	0.09	0.28	0.28	0.00	0.02

Reclassified employment data based on data from UAE Ministry of Economy 2008

The proportion of the male workforce exposed to moderately high noise levels in each economic subsector is then assessed by multiplying the numbers in Table 6.27 by the proportion of male workers employed within these occupation-industry groups (Table 6.29) and summing proportions across each industry sector. Next, the percentage distribution of the male workforce into each economic subsector is determined (Table 6.6). Then the proportion of male workforce exposed to elevated noise levels in each economic subsector is calculated by multiplying the proportions determined in the previous two steps and then summing across all sectors. This overall worker exposure estimate is adjusted using the economic activity rate in the UAE (89% for males) to determine the overall population exposure. The same calculations were then performed for females and for noise exposure of greater than 90 dB(A).

Table 6.31 Relative risks for hearing loss by age group and exposure level

Exposure level	15–29	30–44	45–59	60–69	70–79
<85 dB(A)	1.00	1.00	1.00	1.00	1.00
85–90 dB(A)	1.96	2.24	1.91	1.66	1.66
>90 dB(A)	7.96	5.62	3.83	2.82	2.82

Concha-Barrientos et al. (2004b)

Relative Risk for Developing Hearing Loss

In terms of causality, relative risks of hearing loss are not believed to differ between countries, and relative risks based on previous studies can be used. However, the WHO definition for disabling hearing loss differs from the criteria used in most of the occupational health studies, so the published relative risks have to be adjusted with a correction factor. This procedure is described in detail by Concha-Barrientos et al. (2004b). The resulting relative risks of hearing loss at different exposure levels are presented in Table 6.31. In general, when people are exposed to certain levels of noise, most hearing loss occurs within the first few years. If exposure continues, so does hearing damage, though its progress is slower. However, personal susceptibility to noise damage varies greatly between individuals, so that workers with similar noise exposures may end up with different degrees of hearing loss (Concha-Barrientos et al. 2004b). Attributable fractions for males and females by age group were calculated using Eq. 6.1.

Hearing Loss Due to Occupational Noise Exposure in the UAE Population

The number of noise-induced hearing loss (NIHL) cases resulting from noise exposure in occupational settings can be calculated by multiplying the number of NIHL incidences in each age group with the corresponding attributable fractions. In the data set available to the research team, zero cases of NIHL were recorded in Abu Dhabi emirate. However, noise-induced hearing loss is one of the most common occupational health conditions in the world, and thus an alternative method to estimate the baseline NIHL in the UAE population is used.

Estimating the prevalence of NIHL in the general population is difficult because adult-onset hearing loss has two main causes: age and exposure to noise. The majority of studies looking at effects of noise exposure include only occupational noise exposures, whereas studies describing age-related hearing loss typically include only the elderly. Hearing loss can be defined several ways, and this study uses the WHO definition (WHO 2009). How much of adult-onset hearing loss is related to noise exposure and how much is related to age, particularly in people of different ages, is unclear. The prevalence of baseline hearing loss is estimated from a WHO study by Mathers et al. (2000), which presents the prevalence rate of adult-onset

Table 6.32 Estimates of the prevalence of noise-induced hearing loss cases in the UAE general population, by age group

	15–29	30–44	45–59	60–69	70–79
Male	0	12,600	6,937	2,176	1,302
Female	0	2,723	2,023	1,059	968

hearing loss in the Eastern Mediterranean region, by gender and age group, itself based on a study conducted in Oman (Khabori et al. 1996). The UAE prevalence of adult-onset hearing loss is calculated by adjusting the Eastern Mediterranean rate (prevalence/100,000) to the UAE population.

Since hearing loss prevalence presented in Mathers et al. (2000) include both NIHL and age-related hearing loss (AHL), the proportion of NIHL out of the total prevalence must be estimated. It should be noted that in the data set by Mathers et al. (2000) used here, baseline hearing loss in the 15–29 age group was zero. For other age groups, the NIHL proportion estimates are based on studies from other countries. According to Thorne et al. (2008), studies conducted in New Zealand and Australia suggest the NIHL proportion of overall hearing loss ranges from 30 to 50%. It is assumed that age-related hearing loss becomes more important as a population ages, and the following percentages are used to adjust the prevalence of hearing loss in each age group: 50% for the 30–44 age group, 38% for the 45–59 age group, and 30% for the 60–69 and 70–79 age groups. The resulting baseline prevalence of noise-induced hearing loss in the UAE population is presented in Table 6.32. The influence diagram of the occupational noise model is presented in Fig. 6.11, and the input variables are described in Table 6.33.

Estimated Burden of Disease

Estimates of Attributable Fractions and the Disease Burden

The attributable fractions (AF) for occupational illnesses included in this study vary from 9% (leukemia) to 28% (lung cancer) for males and from 1% (COPD) to 9% (leukemia) for females. AFs for noise-induced hearing loss for males range from 11% (age groups 60–69 and 70–79) to 30% (15–29 age group). For females, AFs for NIHL range from 3% (age groups 60–69 and 70–79) to 6% (15–29 age group). Asbestosis and silicosis are caused exclusively by exposure to asbestos and silica; consequently, AFs for asbestosis and silicosis are assumed to be 100%. It is also assumed, based on the published literature, that the AF for malignant mesothelioma is 90% for males and 25% for females (Nurminen and Karjalainen 2001; Steenland et al. 2003). The AFs for lung cancer, leukemia, asthma, and COPD are presented in Table 6.34, and AFs for NIHL in each age group are given in Table 6.35.

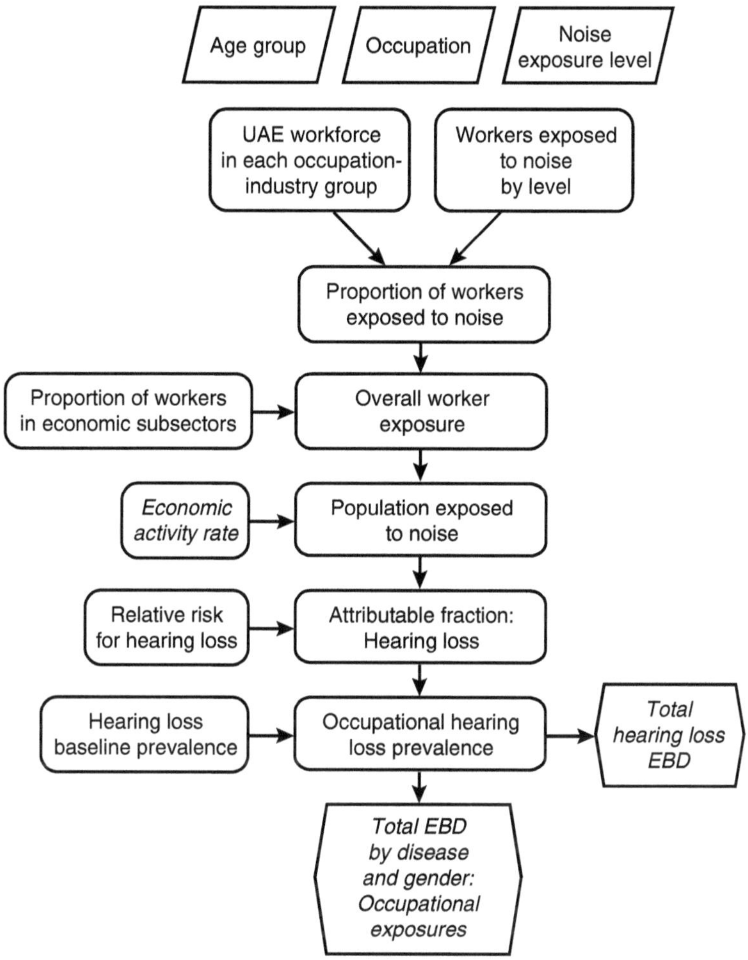

Fig. 6.11 Influence diagram of the ***Occupational Noise*** module

The estimated total number of deaths due to health outcomes included in this study is 47, out of which 40 are male and 7 are female (Table 6.36). The total number of health-care facility visits is 17,160 (15,670 male; 1,497 female). In addition, the model estimates that 4,770 cases (4,494 male; 276, female) of NIHL occur due to occupational exposures. Of the health outcomes covered in the study, lung cancer and leukemia were responsible for the highest number of deaths (25 and 12, respectively). For health-care facility visits, asthma and COPD contributed most to the disease burden with 11,854 and 5,012 visits, respectively. Table 6.37 lists these results by health outcome.

Table 6.33 Description of variables in the ***Occupational Noise*** module

Index/Variable	Description	Definition	Source
Age group	Age groups matching the relative risks	15–29, 30–44, 45–59, 60–69, 70–79	Concha-Barrientos et al. (2004b)
Occupation	Distribution of workforce into different occupations within each economic sector	Professional, administrative, clerical, sales, services, agriculture, production	Concha-Barrientos et al. (2004b)
Noise exposure level	Exposure level to noise: moderately high or high	Moderately high exposure: 85–90 dB(A); high exposure: >90 dB(A), based on Recommended Exposure Limits (REL)	Concha-Barrientos et al. (2004b)
UAE workforce in each occupation-industry group	Distribution of UAE workforce into different occupations within each economic sector	Listed in Tables 6.29 and 6.30	UAE Ministry of Economy (2008), Concha-Barrientos et al. (2004b)
Workers exposed to noise by level	Estimated proportion of workers exposed to noise at moderately high (85–90 dB(A)) or high (>90 dB(A)) levels, by gender, in the WHO Eastern Mediterranean B region	Listed in Table 6.27 for 85–90 dB(A) and in Table 6.28 for >90 dB(A)	Concha-Barrientos et al. (2004b)
Proportion of workers exposed to noise	Proportion of UAE workers exposed to noise in each occupation-industry group, by gender	Proportion of UAE workers in each occupation-industry group multiplied by the estimated proportion of workers exposed to noise in each group	Concha-Barrientos et al. (2004b)
Proportion of workers in economic subsectors	Proportion of UAE workforce in each economic subsector, by gender, categorized according to the 2nd Review of the International Standard of Industrial Classification of All Economic Activities (ISIC)	Listed in Table 6.6	UAE Ministry of Economy (2008), Concha-Barrientos et al. (2004b)

(continued)

Table 6.33 (continued)

Index/Variable	Description	Definition	Source
Overall worker exposure	Proportion of workforce exposed to noise in each economic subsector at different levels, by gender	Calculated by summing proportion of workers exposed to noise over various occupations within each economic subsector, then multiplying these totals by proportion of workforce in each economic subsector	Concha-Barrientos et al. (2004b)
Population exposed to noise	Proportion of UAE population exposed to occupational noise	Overall worker exposure multiplied by the economic activity rate	Concha-Barrientos et al. (2004b)
Relative risk for hearing loss	Relative risk of hearing loss, by age group	Listed in Table 6.31; relative risk for the unexposed background group is considered 1	Concha-Barrientos, et al. (2004a)
Attributable fraction: hearing loss	Proportion of NIHL attributable to occupational exposure to noise	Calculated using Eq. 6.1 (Table 6.35)	
Hearing loss baseline prevalence	Estimated number of NIHL cases in the general UAE population, by age group and gender	Listed in Table 6.32	Mathers et al. (2000), Thorne et al. (2008)
Occupational hearing loss prevalence	Prevalence of NIHL due to occupational exposure to noise	Hearing loss baseline prevalence multiplied by hearing loss AF	

Table 6.34 Attributable fractions for lung cancer, leukemia, malignant mesothelioma, asthma, and COPD in the UAE

	Male		Female	
	Mean (%)	95% CI (%)	Mean (%)	95% CI (%)
Lung cancer	28	(14, 45)	4	(2, 8)
Leukemia	9	(4, 17)	9	(4, 16)
Asthma	21	(19, 24)	6	(5, 7)
COPD	25	N/A[a]	1	N/A

[a]95% confidence interval not applicable since only deterministic input values used in model

Table 6.35 Attributable fractions for occupational noise-induced hearing loss in the UAE population ages 15–79 (%)

	15–29	30–44	45–59	60–69	70–79
Male	30	24	16	11	11
Female	6	5	4	3	3

Table 6.36 Total number of occupational deaths and health-care facility visits due to lung cancer, leukemia, malignant mesothelioma, asthma, COPD, asbestosis, and silicosis

	Male		Female		Total	
	Mean	95% CI	Mean	95% CI	Mean	95% CI
Deaths	40	(23, 59)	7	(3, 12)	47	(26, 72)
Health-care facility visits	15,670	(14,400; 16,900)	1,497	(1,320; 1,670)	17,160	(15,700; 18,600)

Table 6.37 Estimated mortality (number of deaths) and morbidity (number of health-care facility visits) due to lung cancer, leukemia, malignant mesothelioma, asthma, COPD, asbestosis, silicosis, and noise-induced hearing loss in the UAE resulting from occupational exposures

	Male		Female		Total[a]	
	Mean	95% CI	Mean	95% CI	Mean	95% CI
Deaths						
Lung cancer	24	(12, 37)	2	(1, 3)	25	(12, 41)
Leukemia	7	(3, 13)	5	(2, 9)	12	(5, 22)
Malignant mesothelioma	6	N/A[b]	0	N/A	6	N/A
Asthma	1	(1, 1)	0	(0, 0)	1	(1, 1)
COPD	2	N/A	0	N/A	2	N/A
Asbestosis[c]	0	N/A	0	N/A	0	N/A
Silicosis[c]	0	N/A	0	N/A	0	N/A
Health-care facility visits						
Lung cancer	110	(54, 174)	2	(1, 4)	112	(54, 180)
Leukemia	99	(41, 183)	39	(16, 72)	138	(57, 255)
Malignant mesothelioma	25	N/A	0	N/A	25	N/A
Asthma	10,475	(9,180; 11,725)	1,379	(1,200; 1,557)	11,854	(10,546; 13,107)
COPD	4,936	N/A	76	N/A	5,012	N/A
Asbestosis	3	N/A	0	N/A	3	N/A
Silicosis	8	N/A	0	N/A	8	N/A
Noise-induced hearing loss[d]	4,494	N/A	276	N/A	4,770	N/A

[a]Total number of deaths may not equal the sum of male and female deaths (Table 6.36) because of rounding.
[b]95% confidence interval not applicable since only deterministic input values used in the model.
[c]Number of deaths at baseline in the data set available was zero for asbestosis and silicosis.
[d]For noise-induced hearing loss the morbidity is expressed as prevalence instead of health-care facility visits because a different data set for baseline morbidity was used.

Comparison of Results with Previous Estimates

Attributable Fractions

Attributable fractions are considerably higher for males than females in the UAE, since males make up almost the entire workforce in areas in which exposures are potentially high, such as construction, agriculture, and mining. In addition, the economic activity rate is considerably lower for females, and thus a larger proportion of the female population is outside the workforce and considered to be in the background exposure group. An exception is leukemia, for which the attributable fraction is the same for males and females. One explanation is that a relatively large proportion of workers in the CAREX services category is estimated to be exposed to benzene, and a large percentage of female workers in the UAE is classified under the services sector.

To our knowledge, the WHO approach has not yet been used to calculate disease burden related to occupational hazards in other Middle Eastern countries individually, but the WHO has estimated the disease burden due to occupational exposures for various regions of the world in its Global Burden of Disease study (Concha-Barrientos et al. 2004b). Thus, the results for the UAE can be compared with the WHO estimates for the Eastern Mediterranean Region B (EMR-B), which includes the UAE. When comparing these numbers, it should be kept in mind that WHO estimates are calculated for the entire EMR-B region, which encompasses 13 countries. Also, the labor force in the UAE is unique in that it comprises mostly young males, a large proportion of whom work in high-risk occupations such as construction. The disease burden in this study is estimated for the year 2008, whereas the baseline year in the WHO study was 2000.

The AFs for lung cancer, leukemia, asthma, and COPD estimated in this study are compared with the WHO estimates for the EMR-B region in Fig. 6.12. Error bars represent 95% confidence intervals calculated in the model. The 95% confidence intervals were not available for the WHO estimates. The AF for malignant mesothelioma is not included in the comparison since practically all mesothelioma is due to asbestos exposure and most exposure takes place at work. The estimated AFs were derived from literature. Similarly, AFs for asbestosis and silicosis are excluded from the comparison because the AFs were assumed to be 100%.

Overall, many of the attributable fractions in this study are higher than the estimates by the WHO for the EMR-B region, particularly the AFs for male workers. AFs for lung cancer (28% for males and 4% for females) and leukemia (9% for males and females) are higher than the corresponding numbers for the EMR-B region (12% for males and 2% for females for lung cancer; 3% for males and 2% for females for leukemia) mainly because of the difference in the occupational turnover (OT) factor used. The WHO used an OT factor of 4 (which assumes annual workforce turnover of 10%), whereas in this study the OT factor is approximately 7 due to an assumption that the annual turnover is 10–40%, based on the unique composition of the UAE workforce, mostly expatriate workers. If an OT factor of 4 were used in our calculations, the lung cancer AF would be 19% for males and 3% for females, and the leukemia AF would be 6% for males and 5% for females, closer to WHO estimates.

The AFs for asthma and COPD corresponded with WHO estimates for the EMR-B region. The COPD AF for males is somewhat higher in this study (25% compared with 17%), which may be due to the high proportion of workers in the construction sector, included in the high-exposure group within the COPD model.

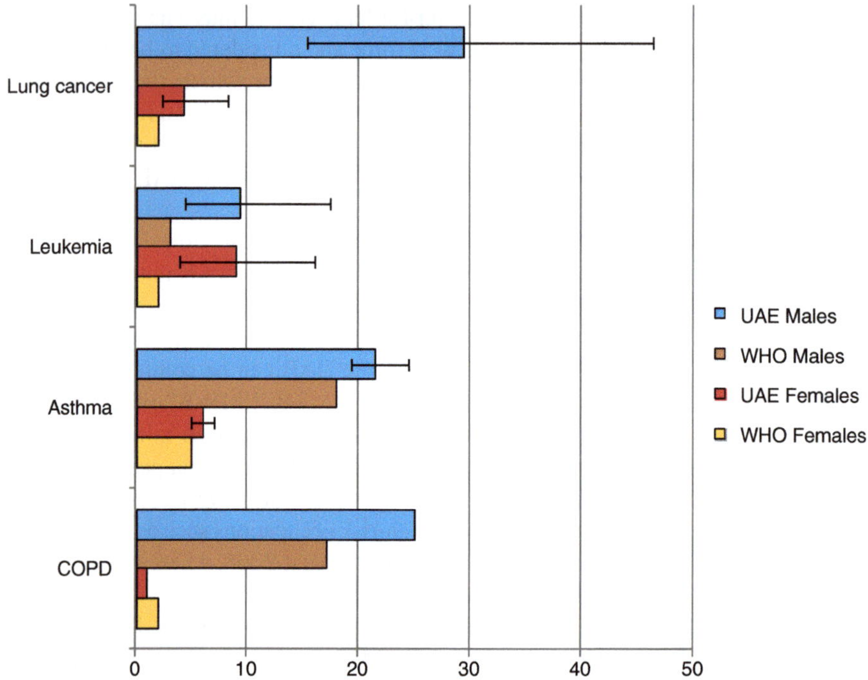

Fig. 6.12 Attributable fractions for lung cancer, leukemia, asthma, and COPD estimated in this study (UAE males and females) and by WHO for the EMR-B region (WHO EMR-B males and females). Error bars represent 95% confidence intervals calculated in the model. The 95% confidence intervals were not available for WHO estimates

AFs for noise-induced hearing loss are calculated by gender for different age groups. The AFs are considerably higher for males than females, also seen in the global estimates by the WHO (Fig. 6.13). This result reflects the much higher proportion of males in the UAE workforce compared with females, and the fact that more males work in occupations where a higher proportion of workers is estimated to be exposed to high noise levels, such as in production. The AFs decrease by age, demonstrating that occupational noise exposure is more important in younger age groups than older ones, corresponding with global WHO estimates (Concha-Barrientos et al. 2004b). Overall, the AFs for females in the UAE are smaller than the global estimates. This may be due to different exposure patterns, since females in the UAE do not commonly work in occupations with potential noise exposure, and the economic activity rate for females in the UAE is relatively low.

Mortality and Morbidity

The occupational burden of disease calculated in this study represents morbidity and mortality related only to exposure to selected carcinogens, particulate matter, and noise. Disease burden resulting from occupational exposures is considerably higher for males

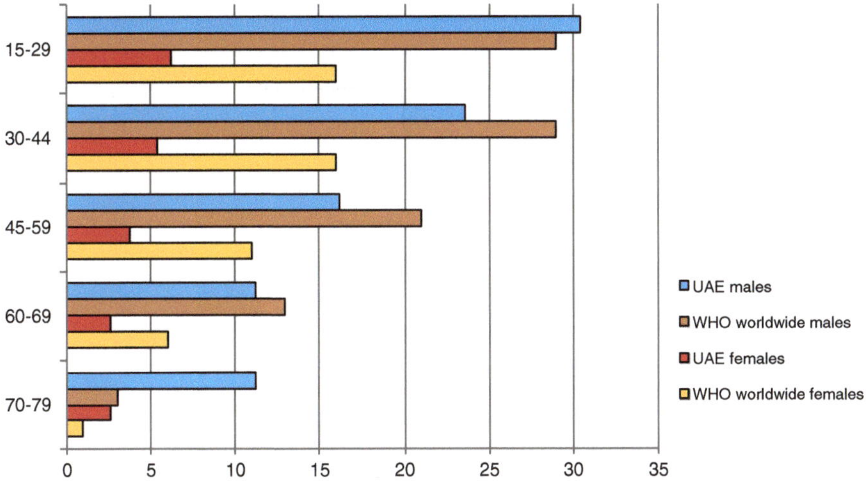

Fig. 6.13 Attributable fractions for noise-induced hearing loss, by age group and gender. The UAE males and females represent results from this study, and WHO global males and females represent global estimates by WHO

than females, as could be expected from the higher attributable fractions. For males, lung cancer is responsible for the highest number of deaths (24), and for females the highest number of deaths result from leukemia (5). These results correspond with the highest attributable fractions for males and females, respectively, and with the fact that lung cancer and leukemia are responsible for the highest number of baseline deaths.

The estimated total number of deaths due to occupational exposures calculated in this study (47) corresponds well with previous WHO approximations and with preliminary risk estimates compiled by the RAND Corp. for the risk ranking exercise described in Chap. 2 (see Appendix A). In the preliminary (RAND) analysis used to inform the ranking exercise, the number of deaths in the UAE resulting from occupational exposures was 90, with the minimum number of deaths estimated at 0 and the maximum at 100. WHO mortality estimates for the Eastern Mediterranean B region are extrapolated to the UAE by comparing the UAE population with the population of the EMR-B region. The total number of deaths estimated in this study is very close to the number extrapolated for the UAE from the WHO estimates for the EMR-B region (Driscoll et al. 2004b, 2005a, b). Figure 6.14 compares the mortality results with the preliminary estimates in Appendix A and with the previous WHO estimates.

The results of this study were also compared with the mortality estimates by WHO by disease (Fig. 6.15) and by disease and gender (Fig. 6.16). For the number of deaths categorized by disease, UAE estimates are higher for cancers but lower for asthma and COPD. The WHO estimate for COPD deaths for males (22) is particularly high compared to the estimated deaths in this study (2). This result may be due to the lower AFs in this study and the lower number of asthma and COPD deaths in the baseline health data. In general, however, estimated numbers of deaths due to occupational exposures in this study are of the same magnitude as estimates by WHO and RAND.

The estimates of morbidity resulting from occupational exposures are expressed as health-care facility visits instead of disease incidence or prevalence since only the

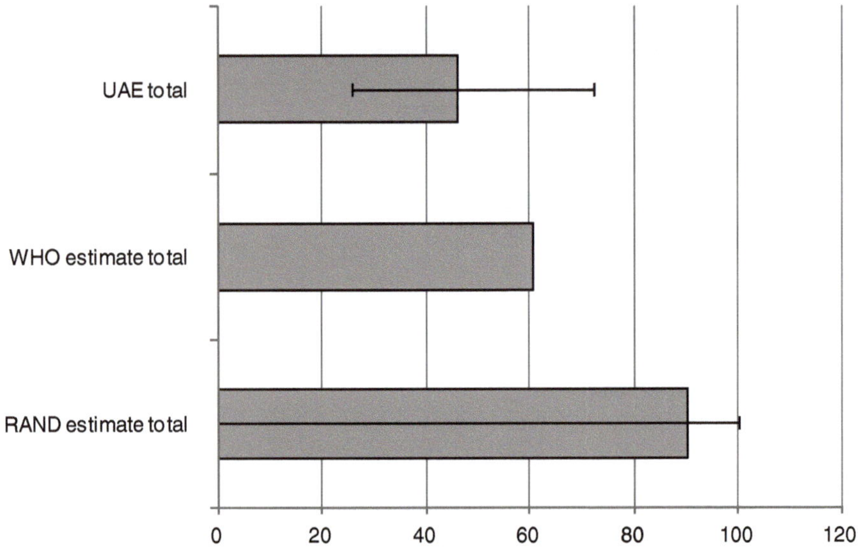

Fig. 6.14 Comparison of total mortality estimated in this chapter (UAE total) with WHO and Appendix A estimates. The WHO mortality estimate for the Eastern Mediterranean B region was extrapolated to the UAE by comparing populations of the UAE and the EMR-B region. The error bars in the UAE estimates represents 95% confidence intervals calculated in the model, whereas the error bars in Appendix A represent estimated minimum and maximum number of deaths. The 95% confidence interval was not available for the WHO estimate

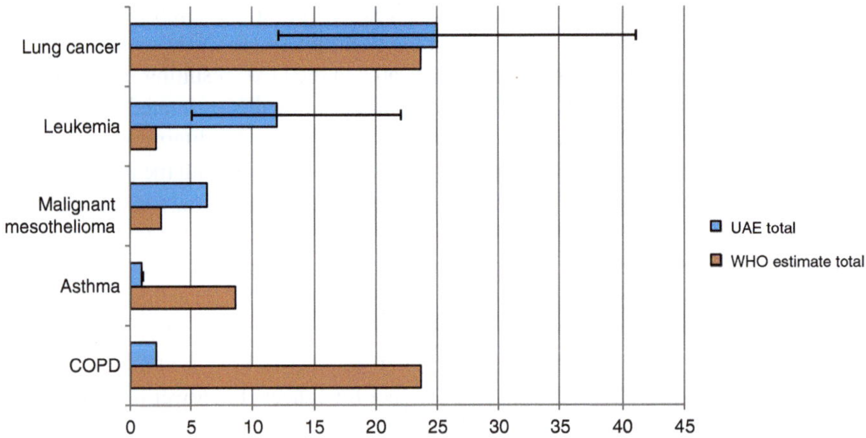

Fig. 6.15 Comparison of total mortality estimated in this study (UAE total) with WHO estimates, by disease. WHO mortality estimates for the Eastern Mediterranean B region were extrapolated to the UAE by comparing populations of the UAE and EMR-B region. Error bars for UAE estimates represent 95% confidence intervals calculated in the model

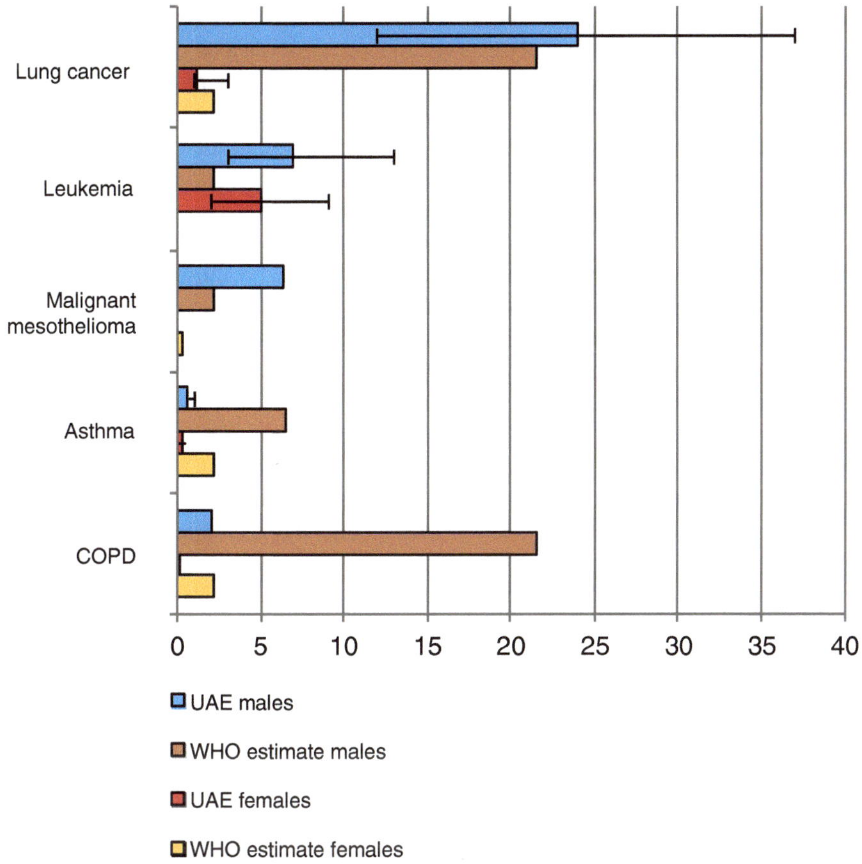

Fig. 6.16 Comparison of total mortality estimated in this study (UAE total) with WHO estimates, by disease and gender. WHO mortality estimates for the Eastern Mediterranean B region were extrapolated to the UAE by comparing populations of the UAE and the EMR-B region. Error bars for UAE estimates represent 95% confidence intervals calculated in the model

number of visits to hospitals and clinics in 2008 was available to the research team. Due to a lack of further information, we could not convert the morbidity data to disability-adjusted life years (DALYs). Thus, the morbidity results cannot be compared directly with WHO estimates, expressed as DALYs. In this study, diseases that contribute most to overall morbidity are asthma, COPD, and NIHL. These are also responsible for the highest number of DALYs in WHO estimates for the EMR-B region (Concha-Barrientos et al. 2004b).

In the baseline health data available from the UAE health authorities, zero cases of noise-induced hearing loss were recorded. Since NIHL is one of the most prevalent occupational health conditions around the world, an alternative data set was used (Mathers et al. 2000). In this data set, however, the number of hearing loss cases was zero in the 15–29 age group. This figure is likely to underestimate the disease burden resulting from noise exposure in the UAE since most of the expatriate workers are young males and are potentially exposed to loud noise when working, for example, in construction. In WHO global estimates, approximately 15% of the NIHL disease burden occurred in the 15–29 age group (Concha-Barrientos et al. 2004b). Most cases of occupational NIHL in this study were recorded in the 30–44 age group, after which the prevalence fell as age increased. The same trend can be seen in the WHO estimates of NIHL (Concha-Barrientos et al. 2004b).

It should be noted that the estimated total number of deaths (47) and health-care facility visits (17,160) in this study include only selected occupational health outcomes. Consequently, these numbers should not be considered to represent the total disease burden arising from all occupational exposures. Many prevalent occupational hazards, such as injuries and ergonomic stressors, were excluded because this study focuses on health risks due to releases of hazardous physical, chemical, and biological agents into the environment as a result of human activities. Due to a lack of data, exposure to several chemicals was also excluded from the quantitative assessment of this study. In addition, the calculations rely on the baseline health data available from the UAE at the time of the study. The results should be regarded as preliminary estimates because several assumptions and estimations had to be made when calculating the disease burden. Nevertheless, the numbers of deaths and health-care facility visits resulting from occupational exposures estimated in this study are substantial and indicate that occupational illnesses are an important source of disease burden in the UAE.

Sources of Uncertainty

General Sources of Uncertainty

Because limited information was available from the UAE, in many instances data had to be derived from studies conducted in other countries, and it was necessary to make assumptions based on best estimates. Several general sources of uncertainty exist in the disease burden estimates for occupational illnesses, including uncertainty related to the exposure and relative risk estimates, gender and age, smoking, latency of illnesses, excluded exposures and conditions, and occupational turnover (Driscoll et al. 2004b). Most of these uncertainties apply to occupational lung cancer, asthma, and COPD. The two main sources of uncertainty in the NIHL calculations are exposure and relative risk estimates (Concha-Barrientos et al. 2004a). Assumptions made during each step of the disease burden calculations and the potential uncertainty they introduce into the model are discussed below.

Employment Data

Even though recent data were available from the UAE Ministry of Economy on the distribution of the UAE workforce into different economic subsectors and occupations, the data were categorized differently than the data from which relative risk estimates were derived. Consequently, assumptions had to be made when matching UAE employment data with exposure and relative risk data. In addition, the baseline year for the calculations was 2008, but employment data were from the latest census in 2005 and the latest labor survey, conducted in 2008 (UAE Ministry of Economy 2008). However, the uncertainty resulting from these factors is not likely to be significant.

CAREX Data

Because no other source of data was available, UAE carcinogen exposure patterns were assumed to be the same as in the Carcinogen Exposure (CAREX) database. This database assumes that exposure patterns are the same for males and females, and for different age groups. In the UAE, it is likely that within the economic sectors used in the CAREX database, a larger proportion of females work in occupations with potentially lower risks of carcinogen exposure, such as professional, service, and trade workers. Thus, a lower proportion of females than males may be exposed to carcinogens.

Because the exposure data in the CAREX database are estimated for European countries based on two reference countries (Finland and the United States), the exposures in the UAE may differ from these estimates. Overall, proportions of workers exposed to carcinogens in the UAE could be slightly higher than those in the CAREX database, because programs to reduce occupational exposures are not yet widespread in UAE workplaces. Exposure measurements conducted at various UAE workplaces would give more information on carcinogen exposures in the UAE.

Exposure Estimates

Due to a lack of country-specific data, it was assumed that 50% of the workforce exposed to carcinogens was exposed at a low level and 50% at a high level in reference to the relative U.S. PEL. This assumption was based on the WHO approach, in which these percentages were used for the B, C, D, and E regions of the world, but at the moment this UAE assumption is not based on any exposure data collected in the country. If percentages in the model were changed to values used by WHO for the A region (90% of the workforce exposed at low level and 10% at high level), the number of deaths due to occupational lung cancer and leukemia would be reduced to approximately half the current estimates. This reduction indicates the importance of collecting exposure data in the UAE and using this information in the model to increase the accuracy of the disease burden estimates.

Relative Risk Estimates

Following the WHO approach, relative risk estimates were assumed to be the same for disease incidence and mortality, even though relative risks were based on disease incidence studies. This is most directly applicable to lung cancer and malignant mesothelioma, but the number of deaths may not be the same as the number of incidence cases for other diseases such as leukemia, asthma, and COPD. However, the relative rate can be assumed to be comparable in many cases (Driscoll et al. 2004a, b).

For carcinogens, the same mean relative risks were used for males and females. Even though this is likely to be true for most occupational carcinogen exposures (Stellman 1994; Jahn et al. 1999), there may be gender differences in exposure patterns (Setlow et al. 1998), which may result in different risks based on gender (Driscoll et al. 2004b). Relative risk estimates were also assumed to be the same for all age groups (except for NIHL), since exposure or relative risk data are only rarely available for different age groups. Even though older age groups can be expected to have a higher absolute risk of disease based on more years of cumulative exposure, the attributable fraction approach using relative risk will result in a lower disease burden in the younger age groups because the baseline disease incidence is lower. Smoking is the most important confounder for lung cancer and respiratory disease estimates (Driscoll et al. 2004a, b). Relative risk estimates were, whenever possible, derived from studies that controlled for smoking, and thus effects of smoking are accounted for in the relative risk estimates.

Baseline Health Data

The only available baseline mortality and morbidity data collected in the UAE were based on information gathered in Abu Dhabi emirate and had to be extrapolated to cover the other emirates. Information from Abu Dhabi emirate may not accurately represent disease patterns in more rural areas of the UAE because of potential differences in access to health care. In addition, diagnosing certain conditions, such as asbestosis and silicosis, can be difficult and requires expertise that may not be available in all health-care facilities. In the data set available to the research team, zero deaths and very few health-care facility visits due to asbestosis and silicosis were recorded. Even though the import, production and use of asbestos boards have been banned in the UAE, mixing asbestos with cement is still legal and at least one UAE factory produces asbestos-containing cement pipes (Landais 2009). It is likely that the baseline health data underestimate the true incidence of asbestosis and silicosis in the UAE. Similarly, zero health-care facility visits related to NIHL were recorded, likely a gross underestimation of the condition's prevalence. Lack of data covering mortality and morbidity across different emirates and possible difficulties in diagnosis, correct coding, and systematic recording of diseases suggest the baseline health estimates for other diseases covered in this study may be underestimated as well.

Excluded Exposures and Illnesses

Within occupational carcinogens and occupational airborne particulate matter, some exposures were excluded because levels of exposure were expected to be very low, insufficient evidence linked exposure with outcome, or data on exposure patterns were lacking. Similarly, some cancers and nonmalignant respiratory diseases were omitted due to a lack of information on relevant exposures and risks. Omitting some exposures and health outcomes related to carcinogens and airborne particulate matter results in underestimation of the total disease burden (Driscoll et al. 2004a, b).

Sensitivity Analysis

The sensitivity of the model to changes made in the input variables was tested with a sensitivity analysis. The purpose was to identify variables responsible for the largest changes in the disease burden estimates. The sensitivity analysis was performed by selecting input variables that included ranges of uncertain values (e.g., annual turnover rate) or that were based on assumptions (e.g., proportion of workers exposed at various levels). Only input variables meaningful for the improvement of the model were included. For example, the proportion of workforce in each economic subsector and the economic activity rate were excluded from the sensitivity analysis.

The analysis was performed by reducing the value of each input variable by 10% one at a time, while keeping other variables unchanged, and seeing which variable produced the largest change in the final output numbers. When the CAREX data, working time, annual turnover rate, exposure level, and relative risk were lowered one at a time by 10% of their original values, the relative risk variable was responsible for the highest change in the number of deaths and health-care facility visits.

However, the occupational turnover factor and exposure level variables in the carcinogen exposure model are solely based on assumptions and would benefit from UAE-specific data the most. In a hypothetical best-case scenario, for instance, in which 90% of workers (instead of 50%) were exposed at a low level, the number of lung cancer deaths in males would drop from 24 to 15, a larger change than that produced by lowering the relative risk by 10%, which would decrease the number of deaths from 24 to 18. Conversely, in a worst-case scenario in which 90% of workers were exposed at a high level, the number of lung cancer deaths in males would increase to 26. Similarly, the inputs in the occupational turnover module (working time and annual turnover rate) are not based on UAE data and have large effects on the results if changed by more than the 10% used in this sensitivity analysis. These examples show that it is crucial to collect more local data in the UAE to improve the accuracy of the occupational exposures model.

Information Needed to Improve Future Burden of Disease Predictions

All three parts of the occupational exposures model—the carcinogen model, the particulate matter model, and the noise model—are based on determining the proportion of UAE population working in each economic subsector or occupation. This information was available from the Ministry of Economy, which regularly surveys the distribution of the workforce in the country. The most pressing data needed to improve the occupational exposures model arise from the limited exposure and baseline health data. To improve the model's estimates regarding occupational exposures, levels of contaminants at workplaces within different UAE industries need to be determined. This information can be used to estimate more accurately the proportion of people who are exposed as well as the magnitude of exposure to occupational hazards. Comprehensive assessment of occupational exposures in UAE industries is a huge undertaking that will require strong, established occupational health and safety capacity. However, even small-scale assessments and pilot studies would provide important information on model assumptions and potential adjustments needed in estimates now based on data derived from other populations and countries. Information on occupational turnover, including annual turnover rate (percentage of workers replaced each year) and working time, is needed in the carcinogen model. Typical working time is usually assumed to be 40 years, but due to the unique composition of the UAE workforce, the range of values is likely to be wide for both working time and annual turnover rate. The model would thus benefit from local worker data gathered for various industries and occupations.

Comprehensive baseline health data are also needed. As in many other countries, mortality data can be derived from death registries in the UAE. Morbidity data are not recorded systematically, however. Based on the fact that zero cases of noise-induced hearing loss were recorded in the data set available to the research team, it is likely that this condition and possibly others are underreported in the UAE. Further, a lack of incidence data for nonfatal health conditions prevented the calculation of disability-adjusted life years (DALYs) in this study. Expressing the disease burden in DALYs would allow a better comparison of the burden of disease related to various risk factors within the UAE and in other countries. Future estimates would also be more accurate if baseline health data were available for all emirates, not just Abu Dhabi.

Overall, characterization of occupational exposures and the prevalence of occupational diseases can significantly improve model predictions as well as provide critical information on conditions at various working environments in the UAE. This information can be used to choose the most appropriate methods to reduce exposures, whether by applying engineering controls, reducing exposure time, selecting correct personal protective equipment, or promoting safer working techniques through education.

Conclusions

Results of this study show that occupational exposures are an important source of the disease burden in the UAE. Risk factors covered include selected carcinogenic chemicals, particulate matter, and noise. We estimate that 47 deaths and 17,160 health-care facility visits were attributable to these occupational hazards in 2008.

Of the risk factors and health conditions considered in this study, noise-induced hearing loss in the 15–29 age group and lung cancer related to carcinogen exposure had the highest attributable fractions in males, 29 and 28%, respectively, suggesting that more than one-quarter of NIHL cases and lung cancers in males of working age could be prevented by reducing exposure to hazardous substances and noise in workplaces. Lung cancer and leukemia appear to be associated with the highest number of deaths (38), whereas asthma and COPD contribute most to occupationally attributable health-care facility visits (approximately 16,900). Thus it is likely that the UAE could reduce the amount it spends on medical care by reducing exposure to respiratory irritants, carcinogens, and noise in workplaces.

Importantly, the estimates in this chapter represent occupational morbidity and mortality related only to exposure to selected pollutants and health effects and should not be considered representative of the total disease burden arising from all occupational hazards. Occupational morbidity and mortality estimated in this chapter is a gross underestimation of the total occupational disease burden in the UAE because injuries, ergonomic stressors, heat and other potentially important risk factors are outside the scope of this work and because the baseline health data are believed to underestimate the true disease burden in the UAE population.

Due to the rapid rate of development in the UAE, environmental and occupational risk factors have the potential to become serious problems, as seen in other countries during similar industrialization booms. By evaluating these hazards proactively, the UAE is taking important steps toward becoming a leading example in the Gulf region and in the world.

Overall, a considerable need for information exists in order to improve future burden of disease estimates related to occupational exposures in the UAE. To facilitate the collection and use of such data, government agencies should share information more efficiently and work together in reducing the disease burden from occupational exposures. The progressive approach taken by the UAE to assess the national burden of disease is important for successfully addressing the current and potential forthcoming occupational hazards and risks.

References

Al Asram, O. 2006. Wastes and pollution sources of Abu Dhabi Emirate. AGEDI Sector Paper. Environment Agency–Abu Dhabi. http://www.agedi.ae/sectorpapers/default.aspx

Al Kaabi, N., and F.C. Hadipriono. 2003. Construction safety performance in the United Arab Emirates. *Civil Engineering and Environmental Systems* 20(3): 197–212.

Al Neaimi, Y.I., J. Gomes, and O.L. Lloyd. 2001. Respiratory illnesses and ventilatory function among workers at a cement factory in a rapidly developing country. *Occupational Medicine* 51(6): 367–373.

Baker, J.E. 1992. Primary, secondary, and tertiary prevention in reducing pesticide-related illness in farmers. *Journal of Community Health Nursing* 9(4): 245–254.

Beatson, M., and M. Coleman (eds.) 1997. International comparisons of the economic costs of work accidents and work-related ill health. In *Costs and benefits of occupational safety and health: Proceedings of the European conference on costs and benefits of occupational safety and health, The Hague, 28–30 May.*

Bener, A., G.G. Lestringant, M.M. Beshwari, and M.A. Pasha. 1999. Respiratory symptoms, skin disorders and serum IgE levels in farm workers. *Allergy and Immunology* 31(2): 52–56.

Beshwari, M.M., A. Bener, A. Ameen, A.M. Al Mehdi, H.Z. Ouda, and M.A. Pasha. 1999. Pesticide-related health problems and diseases among farmers in the United Arab Emirates. *International Journal of Environmental Health Research* 9: 213–221.

Canadian Centre for Occupational Health and Safety. 2005. Asthma. *OSH answers*. http://www.ccohs.ca/oshanswers/diseases/asthma.html#_1_5

Centers for Disease Control and Prevention. 1986. Perspectives in disease prevention and health promotion, leading work-related diseases and injuries—United States. *Morbidity and Mortality Weekly Report* 35(12): 185–188.

Chan-Yeung, M., and J.L. Malo. 1994. Aetiological agents in occupational asthma. *European Respiration Journal* 7(2): 346–371.

Concha-Barrientos, M., D. Campbell-Lendrum, and K. Steenland. 2004. Occupational noise: Assessing the burden of disease from work-related hearing impairment at national and local levels. Environmental Burden of Disease Series, No. 9. Geneva: World Health Organization.

Concha-Barrientos, M., D.I. Nelson, T. Driscoll, N.K. Steenland, L. Punnett, M. Fingerhut, A. Prüss-Üstün, C. Corvalán, J. Leigh, and S.W. Tak. 2004b. Selected occupational risk factors. In *Comparative quantification of health risks*, ed. M. Ezzati, A.D. Lopez, A. Rodgers, and C.J.L. Murray, 1651–1801. Geneva: World Health Organization.

Coughlin, S.S., J. Benichou, and D.L. Weed. 1994. Attributable risk estimation in case-control studies. *Epidemiologic Reviews* 16(1): 51–64.

Coye, M.J., J.A. Lowe, and K.T. Maddy. 1986. Biological monitoring of agricultural workers exposed to pesticides: I. Cholinesterase activity determinations. *Journal of Occupational Medicine* 28(8): 619–627.

Curl, C.L., R.A. Fenske, J.C. Kissel, J.H. Shirai, T.F. Moate, W. Griffith, G. Coronado, and B. Thompson. 2002. Evaluation of take-home organophosphorus pesticide exposure among agricultural workers and their children. *Environmental Health Perspectives* 110(12): A787–A792.

Driscoll, T., D.I. Nelson, K. Steenland, J. Leigh, M. Concha-Barrientos, M. Fingerhut, and A. Prüss-Üstün. 2005a. The global burden of disease due to occupational carcinogens. *American Journal of Industrial Medicine* 48(6): 419–431.

Driscoll, T., D.I. Nelson, K. Steenland, J. Leigh, M. Concha-Barrientos, M. Fingerhut, and A. Prüss-Üstün. 2005b. The global burden of non-malignant respiratory disease due to occupational airborne exposures. *American Journal of Industrial Medicine* 48(6): 432–445.

Driscoll, T., K. Steenland, D.I. Nelson, and J. Leigh. 2004. Occupational airborne particulates: Assessing the environmental burden of disease at national and local levels. Environmental Burden of Disease Series, No. 7. Geneva: World Health Organization.

Driscoll, T., K. Steenland, A. Prüss-Üstün, D.I. Nelson, and J. Leigh. 2004. Occupational carcinogens: Assessing the environmental burden of disease at national and local levels. Environmental Burden of Disease Series, No. 6. Geneva: World Health Organization.

Dubai Chamber of Commerce and Industry. 2007. UAE macroeconomic report. Macroeconomic report series. DCCI—Data Management and Business Research Department. http://web2.dubaichamber.ae/pdf/reports/UAE-Macroeconomic-Report.pdf

Economist Intelligence Unit. 2007. United Arab Emirates Country Profile. London: Economist Intelligence Unit.

Emirates Industrial Bank. 2005. Characteristics of the UAE manufacturing industries. *Journal of the Emirates Industrial Bank* 20(9). http://www.emiratesindustrialbank.net/archives.htm

Emirates Industrial Bank. 2006. Investment profile of the UAE manufacturing industries. *Journal of the Emirates Industrial Bank* 21(4). http://www.emiratesindustrialbank.net/archives.htm

Forget, G. 1991. Pesticides and the third world. *Journal of Toxicology and Environmental Health* 32(1): 11–31.

Geer, L.A., N. Cardello, M.J. Dellarco, T.J. Leighton, R.P. Zendzian, J.D. Roberts, and T.J. Buckley. 2004. Comparative analysis of passive dosimetry and biomonitoring for assessing chlorpyrifos exposure in pesticide workers. *Annals of Occupational Hygiene* 48: 683–695.

Gomes, J., O. Lloyd, and N. Norman. 2002. The health of the workers in a rapidly developing country: Effects of occupational exposure to noise and heat. *Occupational Medicine* 52(3): 121–128.

Gomes, J., O.L. Lloyd, N.J. Norman, and P. Pahwa. 2001. Dust exposure and impairment of lung function at a small iron foundry in a rapidly developing country. *Occupational and Environmental Medicine* 58(10): 656–662.

Gomes, J., O.L. Lloyd, and D.M. Revitt. 1999. The influence of personal protection, environmental hygiene and exposure to pesticides on the health of immigrant farm workers in a desert country. *International Archives of Occupational and Environmental Health* 72(1): 40–45.

Gomes, J., O. Lloyd, D.M. Revitt, and J.N. Norman. 1997. Erythrocyte cholinesterase activity levels in desert farm workers. *Occupational Medicine* 47(2): 90–94.

Gomes, J., O. Lloyd, M.D. Revitt, and M. Basha. 1998. Morbidity among farm workers in a desert country in relation to long-term exposure to pesticides. *Scandinavian Journal of Work, Environment, and Health* 24(3): 213–219.

Health Authority–Abu Dhabi (HAAD). 2009. *2008 health data for Abu Dhabi emirate*. Abu Dhabi: Health Authority–Abu Dhabi (HAAD).

International Agency for Research on Cancer. 1997. Silica, some silicates, coal dust and para-aramid fibrils. In *IARC monographs on the evaluation of carcinogenic risks to humans*, vol. 68. Lyon: World Health Organization. http://monographs.iarc.fr/ENG/Monographs/vol68/index.php

Jahn, I., W. Ahrens, I. Brüske-Hohlfeld, M. Kreuzer, M. Möhner, H. Pohlabeln, H.E. Wichmann, and K.H. Jöckel. 1999. Occupational risk factors for lung cancer in women: Results of a case-control study in Germany. *American Journal of Industrial Medicine* 36(1): 90–100.

Karjalainen, A., K. Kurppa, R. Martikainen, T. Klaukka, and J. Karjalainen. 2001. Work is related to a substantial portion of adult-onset asthma incidence in the Finnish population. *American Journal of Respiratory Critical Care Medicine* 164(4): 565–568.

Karjalainen, A., K. Kurppa, R. Martikainen, J. Karjalainen, and T. Klaukka. 2002. Exploration of asthma risk by occupation—Extended analysis of an incidence study of the Finnish population. *Scandinavian Journal of Work, Environment, and Health* 28(1): 49–57.

Khabori, M., A.J. Mohammed, R. Khandekar, and N. Prakesh. 1996. National survey for causes of deafness and common ear disorders in Oman. Oman Ear Study (OES '96) survey report. Sultanate of Oman Ministry of Health. Geneva: World Health Organization.

Kogevinas, M., J.M. Anto, J. Sunyer, A. Tobias, H. Kromhout, and P. Burney. 1999. Occupational asthma in Europe and other industrialised areas: A population-based study. European Community Respiratory Health Survey Study Group. *Lancet* 353(9166): 1750–1754.

Korn, R.J., D.W. Dockery, F.E. Speizer, J.H. Ware, and B.G. Ferris Jr. 1987. Occupational exposures and chronic respiratory symptoms: A population-based study. *American Review of Respiratory Disease* 136(2): 298–304.

Landais, E. 2009. Environment expert calls for asbestos ban. *Gulf News*, September 30. http://gulfnews.com/news/gulf/uae/environment/environment-expert-calls-for-asbestos-ban-1.511828.

Leigh, J., P. Macaskill, E. Kuosma, and J. Mandryk. 1999. Global burden of disease and injury due to occupational factors. *Epidemiology* 10(5): 626–631.

Lynge, E., A. Anttila, and K. Hemminki. 1997. Organic solvents and cancer. *Cancer Causes and Control* 8(3): 406–419.

Mathers, C., A. Smith, and M. Concha. 2000. Global burden of hearing loss in the year 2000. *Global Burden of Disease 2000*. World Health Organization. http://www.who.int/healthinfo/statistics/bod_hearingloss.pdf

Mazurek, J.M., and M.D. Attfield. 2008. Silicosis mortality among young adults in the United States, 1968–2004. *American Journal of Industrial Medicine* 51(8): 568–578.

McDougall, L., L. Magloire, C.J. Hospedales, J.E. Tollefson, M. Ooms, N.C. Singh, and F.M. White. 1993. Attitudes and practices of pesticide users in Saint Lucia, West Indies. *Bulletin of the Pan American Health Organization* 27(1): 43–51.

Murray, C.J.L., and A.D. Lopez. 1996. *The global burden of disease: A comprehensive assessment of mortality and disability from diseases, injuries and risk factors in 1990 and projected to 2020*. The Global Burden of Disease and Injury series. Cambridge, MA: Harvard University Press.

Nurminen, M., and A. Karjalainen. 2001. Epidemiologic estimate of the proportion of fatalities related to occupational factors in Finland. *Scandinavian Journal of Work, Environment, and Health* 27(3): 161–213.

Oestreich, A., P. Schmid, and C. Schlatter. 1997. Biological monitoring of the fungicide epoxiconazol during application. *Archives of Environmental Contamination and Toxicology* 33: 329–335.

Rabinowitz, P.M., D. Galusha, M.D. Slade, C. Dixon-Ernst, A. O'Neill, M. Fiellin, and M.R. Cullen. 2008. Organic solvent exposure and hearing loss in a cohort of aluminium workers. *Occupational and Environmental Medicine* 65(4): 230–235.

Rettab, B. 2003. *Dubai Chamber of Commerce and Industry 2003 survey: Structure and performance of Dubai major economic sectors, in sector monitor series*. Abu Dhabi: Dubai Chamber of Commerce and Industry.

Rushton, L. 2007. Occupational causes of chronic obstructive pulmonary disease. *Reviews on Environmental Health* 22(3): 195–212.

Serinken, M., O. Karcioglu, M. Zencir, and I. Turkcuer. 2008. Direct medical costs and working days lost due to nonfatal occupational injuries in Denizli, Turkey. *Journal of Occupational Health* 50(1): 70–74.

Setlow, V., C. Lawson, and N. Woods. 1998. *Gender differences in susceptibility to environmental factors: A priority assessment*, 1998. Washington, DC: The National Academies Press.

Sliwinska-Kowalska, M. 2007. Exposure to organic solvent mixture and hearing loss: Literature overview. *International Journal of Occupational Medicine and Environmental Health* 20(4): 309–314.

Steenland, K., C. Burnett, N. Lalich, E. Ward, and J. Hurrell. 2003. Dying for work: The magnitude of U.S. mortality from selected causes of death associated with occupation. *American Journal of Industrial Medicine* 43(5): 461–82.

Steenland, K., D. Loomis, C. Shy, and N. Simonsen. 1996. Review of occupational lung carcinogens. *American Journal of Industrial Medicine* 29(5): 474–490.

Stellman, J.M. 1994. Where women work and the hazards they may face on the job. *Journal of Occupational Medicine* 36(8): 814–825.

Stern, F., P. Schulte, M.H. Sweeney, M. Fingerhut, P. Vossenas, G. Burkhardt, and M.F. Kornak. 1995. Proportionate mortality among construction laborers. *American Journal of Industrial Medicine* 27(4): 485–509.

Thompson, B., G.D. Coronado, J.E. Grossman, K. Puschel, C.C. Solomon, I. Islas, C.L. Curl, J.H. Shirai, J.C. Kissel, and R.A. Fenske. 2003. Pesticide take-home pathway among children of agricultural workers: Study design, methods, and baseline findings. *Journal of Occupational and Environmental Medicine* 45(1): 42–53.

Thorne, P.R., S.N. Ameratunga, J. Stewart, N. Reid, W. Williams, S.C. Purdy, G. Dodd, and J. Wallaart. 2008. Epidemiology of noise-induced hearing loss in New Zealand. *The New Zealand Medical Journal* 121(1280): 33–44.

UAE Federal Government. 2006. Federal Cabinet Resolution No. (39) of 2006 on banning the import and production of asbestos. Abu Dhabi, UAE: Cabinet of the UAE.

UAE Interact. 2007. *UAE Yearbook 2007*. http://www.uaeinteract.com/uaeint_misc/pdf_2007/index.asp

UAE Interact. 2009. UAE economy records 7.4% growth in 2008. http://uaeinteract.com/docs/UAE_economy_records_7.4_growth_in_2008/34951.htm

UAE Ministry of Economy. 2006. *The Annual Economic and Social Report*.

UAE Ministry of Economy. 2008. Labour Force Survey. Statistic Abstract 2008 – Employment 2008. http://www.economy.ae/English/EconomicAndStatisticReports/StatisticReports/StatisticAbstract/Pages/sa2008.aspx

UAE Ministry of Health. 2007. *Summary Statistics 2007*. Abu Dhabi, UAE: Ministry of Health.

U.S. National Institute for Occupational Safety and Health. 1992. Working in hot environments. http://www.cdc.gov/niosh/hotenvt.html#intro

U.S. National Institute for Occupational Safety and Health. 1998. Criteria for recommended standard: Occupational noise exposure. http://www.cdc.gov/niosh/docs/98-126/

U.S. National Institute for Occupational Safety and Health. 2008. Noise and hearing loss prevention. NIOSH Safety and Health Topic. http://www.cdc.gov/niosh/topics/noise/abouthlp/abouthlp.html

U.S. Occupational Safety and Health Administration. 2007. Noise and Hearing conservation—Health effects. http://www.osha.gov/SLTC/noisehearingconservation/health_effects.html

U.S. Occupational Safety and Health Administration. 2008. Heat stress—Safety and health topics. http://www.osha.gov/SLTC/heatstress/index.html

Venables, K.M., and M. Chan-Yeung. 1997. Occupational asthma. *Lancet* 349(9063): 1465–1469.

Verma, D.K., L.A. Kurtz, D. Sahai, and M.M. Finkelstein. 2003. Current chemical exposures among Ontario construction workers. *Applied Occupational and Environmental Hygiene* 18(12): 1031–1047.

Wang, E., J.M. Dement, and H. Lipscomb. 1999. Mortality among North Carolina construction workers, 1988–1994. *Applied Occupational Environmental Hygiene* 14(1): 45–58.

Wang, M.L., L. McCabe, J.L. Hankinson, M.H. Shamssain, E. Gunel, N.L. Lapp, and D.E. Banks. 1996. Longitudinal and cross-sectional analyses of lung function in steelworkers. *American Journal of Respiratory Critical Care Medicine* 153: 1907–1913.

World Health Organization (WHO). 2009. Prevention of deafness and hearing impairment. Grades of hearing impairment. http://www.who.int/pbd/deafness/hearing_impairment_grades/en/index.html

Zhang, Z.W., J.X. Sun, S.Y. Chen, Y.Q. Wu, and F.S. He. 1991. Levels of exposure and biological monitoring of pyrethroids in spraymen. *British Journal of Industrial Medicine* 48(2): 82–86.

Chapter 7
Burden of Disease from Climate Change

Abstract Expected climate change may be particularly important in the United Arab Emirates (UAE) due to its already hot and arid climate. Compared with other nations, the UAE has a relatively low level of total greenhouse gas (GHG) emissions, with an estimated 0.31–0.42% of global emissions since 1994. However, the UAE has one of the highest levels of GHG emissions per capita, consistently ranking second or third in the world over the past two decades. Climate change is likely to have only limited impacts on infectious and diarrheal diseases in the UAE due to relatively low baseline levels of these climate-sensitive diseases. The major impacts of climate change in the UAE are expected to be increased heat stress and possibly increased water- and vector-borne diseases, as well as decreased water availability and food production. The total burden of disease from climate change is inherently difficult to determine due to the many mechanisms through which climate can affect public health and the high level of uncertainty with future climate scenarios, GHG emission levels, and human adaptation measures. Our model includes only the effect of climate change on cardiovascular disease. The results show that climate change currently has minimal effects on human health relative to the other modeled priority areas. There were approximately 410 additional health-care facility visits and three additional deaths from cardiovascular disease in the UAE in 2008 due to the added risks of climate change.

Keywords Climate change • Environmental burden of disease • Relative risk • Attributable fraction • Premature deaths and health-care facility visits • United Arab Emirates • Per-capita greenhouse gas emissions • Extreme heat events • Inland and coastal flooding • Emissions-reduction scenarios • Heat-related cardiovascular disease • Climate change mitigation and adaptation

J. MacDonald Gibson et al., *Environmental Burden of Disease Assessment*, Environmental Science and Technology Library 24, DOI 10.1007/978-94-007-5925-1_7,

Overview: Nature and Causes of Climate Change

Climate change is a complex global environmental problem with the potential for significant long-term impacts on ecosystems, coastal areas, water resources, and human health. According to the Intergovernmental Panel on Climate Change (IPCC), "warming of the climate system is unequivocal, as is now evident from observations of increases in global average air and ocean temperatures, widespread melting of snow and ice and rising global average sea level" (IPCC 2007a). While there are large variations in its regional sources and impacts, current climate change is primarily associated with anthropogenic emissions of greenhouse gases (GHG) from energy use, industrial processes, transportation, agriculture, land-use change, and waste management. Despite the scientific consensus that climate change will alter the distribution, intensity, and frequency of precipitation and other weather phenomena, considerable difficulties persist in attempting to quantify the precise regional changes in climate that will occur and their impact on the environment.

Expected climate change may be particularly important in the United Arab Emirates due to its already hot and arid climate. The IPCC projects that average surface temperatures in the Arabian Peninsula may increase by 1–2°C by 2030–2050, with projected temperature increases by 2100 of 2.3–5.9°C relative to 1961–1990 levels (IPCC 1998; UAE Ministry of Energy 2006). Changes are expected worldwide in the intensity and frequency of precipitation, leading to a change in the total amount of rainfall received in many regions and an increased risk of flash flooding in some areas. With respect to the UAE, potentially significant declines in regional precipitation levels are projected by the IPCC, although the magnitude of change is difficult to project (IPCC 1998; UAE Ministry of Health 2007). Less total precipitation may result in further depletion of groundwater reserves, elevating the need for desalinated water in order to provide potable water as well as water for irrigation in the UAE.

Climate change will likely have many effects on economic and human well-being in the UAE that will require planning for adaptation. However, the main focus of this document is the effect on health.

Greenhouse Gas Emissions in the UAE

Compared with other nations, the UAE has a relatively low level of total greenhouse gas emissions, with an estimated 0.31–0.42% of global emissions since 1994 (World Resources Institute 2009). However, the UAE has one of the highest levels of GHG emissions per capita, consistently ranking second or third in the world over the past two decades (Kazim 2007; World Resources Institute 2009). The only official GHG inventory conducted by the UAE is for the year 1994 (Table 7.1) and was included in their initial national communication to the United Nations Framework Convention on Climate Change (UNFCCC) (UAE Ministry of Energy 2006). An updated official emissions inventory is planned for inclusion in the second national communication

Table 7.1 Total greenhouse gas emissions in the UAE in 1994 in million metric tons (Mt) from the UAE's Initial National Communication to the UNFCCC

Sector	$CO_{2\text{-eq}}$	Percent of total $CO_{2\text{-eq}}$ emissions (%)	CO_2	CH_4	N_2O	NO_x	CO	NM-VOC[a]	SO_2
Energy	70.879	95.22	60.246	0.396	0.005	0.162	0.836	0.095	18.310
Industrial processes	3.455	4.64	3.443	0.001	0	0.001	0.138	0.006	0.005
Waste management	2.552	3.43	0	0.108	0	0	0	0	0
Agriculture	1.777	2.39	0	0.048	0.002	0	0	0	0
Land-use change and forestry	−4.227	−5.68	−4.227	0	0	0	0	0	0
Total	**74.436**	**100**	**59.462**	**0.553**	**0.007**	**0.163**	**0.974**	**0.101**	**18.315**

[a] Nonmethane volatile organic compounds

Table 7.2 Estimates of the UAE's total GHG emissions ($CO_{2\text{-eq}}$ and CO_2)

Source	Year	Total GHG emissions ($CO_{2\text{-eq}}$)	Total CO_2-only emissions	Includes land-use change and forestry?
Kazim (2007)	1980–2003 average	86.1 Mt	Not specified	Not specified
UAE Ministry of Energy (2006)	1994	74.436 Mt	59.462 Mt	Yes
Earth Trends Country Profile	1998	Not specified	88.198 Mt	Not specified
World Resources Institute (2009)	2000	124.6 Mt	88.6 Mt	No
UAE Embassy[a]	2002	Not specified	94.163 Mt	Not specified
CDIAC[b]	2006	Not specified	38.060 Mt[c]	No

[a]http://www.uae-embassy.org/uae/energy/climate-change
[b]Carbon Dioxide Information Analysis Center (2009), Oak Ridge National Laboratory, http://cdiac.ornl.gov/trends/emis/meth_reg.html
[c]Only includes emissions from fossil fuel use

to the UNFCCC, to be completed by 2013. Numerous more recent estimates of the UAE's total and per capita GHG emissions are available from a variety of international sources (Tables 7.2 and 7.3), but many of these have diverse scopes and underlying assumptions that make comparison among estimates somewhat challenging.

The rapid development that has occurred in the UAE since 1994, both in urban expansion and monetary wealth, has likely resulted in significantly higher annual GHG emissions than those reported in the 1994 inventory. According to Kazim (2007), the three factors influencing the rates of energy use in the UAE are

Table 7.3 Estimates of the UAE's per capita GHG emissions

Source	Year	Annual per capita GHG emissions (tons $CO_{2\text{-eq}}$ per person)	Annual per capita CO_2 emissions (tons CO_2 per person)	World rank per capita	Includes land-use change and forestry?
Kazim (2007)	1980–2003 average	10.5	Not specified	1	Not specified
Navigating the numbers[a]	2000	36.1	25.2	2 (GHG), 3 (CO_2)	No
World Resources Institute (2009)	2000	38.4	27.3	2	No
UAE Embassy[b]	2002	Not specified	25.1	4	Not specified

[a]Baumert et al. (2005)

[b]http://www.uae-embassy.org/uae/energy/climate-change

population growth, economic growth, and high urbanization, with the last two factors playing the most critical role. For example, in 2006 the UAE's GDP per capita was 141,700 dirhams, or \$38,600, whereas in 1995 it was 65,000 dirhams, or \$17,700 (UAE Ministry of Economy 2007). It is highly probable that GHG emissions have increased substantially since 1994, but the exact amount of change cannot be known without a current inventory. An updated inventory of emissions, scheduled to be completed by 2013, is a necessary first step in determining the scope of the GHG emissions challenge facing the UAE.

Energy Sector

Extraction and combustion of carbonaceous fossil fuels for energy use is the greatest source of GHG emissions in the world, and the same is true in the UAE (IPCC 2007a; UAE Ministry of Energy 2006; Radhi 2009). In 1994, 95% of GHG emissions in the UAE originated from the energy sector (UAE Ministry of Energy 2006). Natural gas is the primary source of energy in the UAE. Natural gas powers the petrochemical industry and plants that produce both electricity and desalinated water for residential, commercial, and industrial uses. Due to the high demand for air conditioning, in summertime diesel generators provide additional power to meet peak demand. In addition, about 13% of the oil produced in the UAE is used domestically, primarily to fuel the transportation sector (Kazim 2007). Since the 1980s, the average per-capita energy consumption in the UAE has grown at a rate of about 1.4% a year—among the highest rates in the world. As a result, from 1980 to 2003 the UAE's annual per-capita energy consumption was on average nine times greater than that of the world, and twice that of the United States (Fig. 7.1) (Kazim 2007). If current consumption trends continue in the UAE, rising energy use will translate into even higher GHG emissions in the future.

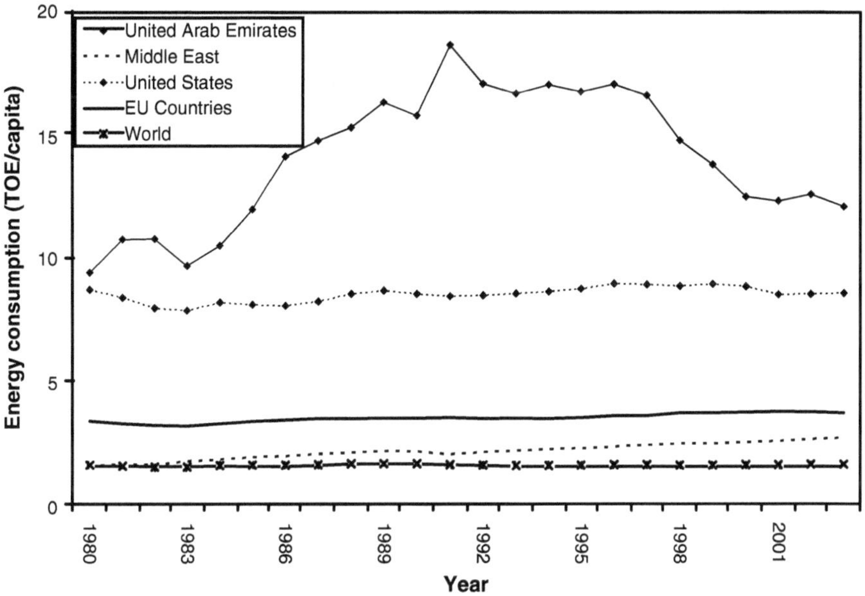

Fig. 7.1 A comparison of the UAE's energy consumption in tons of oil equivalent (TOE) per capita, 1980–2003 (Kazim 2007)

Transportation

Transportation accounted for almost 30% of energy-related GHG emissions in the UAE in 1994, with an estimated 17,683 Gg of CO_2 emissions (UAE Ministry of Energy 2006). The transportation sector in the UAE is steadily expanding and is likely to be an even larger source of GHG emissions in the future. Major factors contributing to the increase in transportation emissions are the increased level of vehicle ownership; the growing percentage of larger, less-fuel-efficient vehicles in the personal vehicle fleet; and growth in annual vehicle miles traveled. The number of vehicles in the UAE rose from 792,000 in 2003 to 1,078,000 in 2006, an increase of more than 36% (UAE Ministry of Economy 2007).

Buildings

Another important component of the energy sector is building construction and operation, although emissions from those sources are often implicitly included in the energy-use estimates of the residential, industrial, and commercial sectors. The residential, commercial, and industrial buildings of the UAE accounted for nearly 46% of electricity consumption in the UAE in 2005 (Radhi 2009). Some of the major energy uses in buildings are heating, ventilation and air-conditioning systems; lighting; electronic appliances and equipment; and water heating (Levine et al. 2007).

Other Sectors

The remaining sectors contributing to the UAE's GHG emissions are industrial processes, waste management, agriculture, and irrigation required for programs intended to "green the desert." The major source of emissions from industry is from the production of cement. Waste-management emissions originate from municipal landfills and wastewater treatment processes and represent the dominant source of methane emissions in the UAE. Agricultural emissions are composed of enteric fermentation products from livestock, manure management, and agricultural soils. These three sectors together account for a little more than 10% of the UAE's gross emissions. The land-use change (i.e., desert greening) sector, while offsetting some GHG emissions due to the planting of trees, also consumes power because it relies exclusively on irrigation. Irrigation water comes either from pumping scarce groundwater or from treating sewage, both of which consume energy.

Key Health Effects of Climate Change in the UAE

The health effects associated with climate change can be indirectly influenced by changing climate patterns or can be a direct result of extreme weather events. The major health outcomes considered in global climate change burden-of-disease studies are mortality from increased flooding (especially inland), morbidity and mortality due to increased frequency and intensity of extreme heat events, population displacement and morbidity due to sea level rise, increased malnutrition rates where agriculture is negatively impacted, potential changes in disease vector ecology leading to increased risk of infectious diseases such as malaria and dengue fever, and water- or foodborne diarrheal diseases. Climate change is likely to have only limited impacts on infectious and diarrheal diseases in the UAE due to relatively low baseline levels of climate-sensitive diseases. In a special report on regional impacts of climate change, the IPCC projects that the impacts of climate change in the UAE's region "are likely to be detrimental to the health of the population, mainly through heat stress and possible increases in vector-borne (e.g., dengue fever and malaria) and waterborne diseases. Decreases in water availability and food production (especially if there is a shortage of water for irrigation) would indirectly affect the health of the population" (IPCC 1998). Nonetheless, compared to many other countries in the Eastern Mediterranean region, the UAE, with its high income and an infrastructure equipped to handle extremely high temperatures, is well positioned to handle the heat and food supply effects of climate change.

The total burden of disease from climate change is inherently difficult to determine due to the many mechanisms through which climate can affect public health and the high level of uncertainty with future climate scenarios, GHG emission levels, and human adaptation measures. Climate variability occurs naturally over time and makes demonstrating causality between climate change and a certain human health response extremely challenging. Ultimately, the effects of climate change on health will depend upon socioeconomic factors and the ability of the public health system to manage additional health threats. Despite these challenges, a 2002 study of major global health risk factors by Ezzati et al. (2002) calculated that the worldwide burden of disease due to climate change in 2000 was 5,517,000 disability-adjusted life years (DALYs), a common metric for combining morbidity and mortality into a single number to represent human health impact. In comparison with DALY estimates for other environmental risks in this study, for example 38,539,000 DALYs attributable to indoor smoke from solid fuels and 54,158,000 DALYs attributable to inadequate sanitation, the DALY estimate for climate change is relatively low; however, this study demonstrates that there is a measurable effect of climate change on human health (Zhang et al. 2007; Ezzati et al. 2002).

Extreme Heat Events

Extreme heat events or "heat waves" occur when stagnant warm air masses cause consecutive nights with high minimum temperatures (Luber and McGeehin 2008). The heat events with the most severe consequences tend to occur early in the summer, before populations are accustomed to hot weather, and generally exhibit high nighttime temperatures and long durations. Another phenomenon contributing to the severity of heat waves is the urban heat island effect, in which cities experience higher temperatures than surrounding nonurban areas due to the different radiative properties of urban landscapes. Because the majority of the world's population now resides in urban centers, a large number of people are affected by the urban heat island effect. In the UAE, more than 75% of the population resides in urban areas (UAE Ministry of Energy 2006). With average summer temperatures often exceeding 38°C, extreme heat conditions are not uncommon in the UAE. With changes in the global climate, however, extreme heat events have been occurring with more frequency and severity worldwide over the past several decades, and this trend is likely to continue (Trenberth et al. 2007).

Cardiovascular disease is the leading cause of death during extreme heat events, and, therefore, death from cardiovascular disease is the health outcome generally used to measure the health impacts of heat waves (McMichael et al. 2004). Other illnesses associated with heat waves are heat cramps, heat syncope, heat exhaustion, heat stroke, and exacerbation of respiratory disease; however, these are often not reported to public health services and are therefore difficult to quantify (Luber and McGeehin 2008). The majority of heat-related deaths occur in populations older than 65; populations with chronic illness or preexisting medical conditions are particularly vulnerable as well. One of the deadliest extreme heat events in recent history occurred in Europe in August 2003. During this event, 14,802 people died in France alone in a 20-day period, with 60% of deaths occurring in people age 75 or older; across Europe, the total death toll was estimated to be around 35,000 (Kovats and Hajat 2008; Confalonieri et al. 2007). Some of the deaths associated with heat waves can be attributed to short-term mortality displacement, where they likely would have occurred in the near term regardless of the heat wave. This phenomenon is demonstrated by a decrease in mortality following some extreme heat events (Confalonieri et al. 2007).

Inland and Coastal Flooding

Another direct health impact expected to be associated with climate change is an increased risk of drowning and morbidity due to inland flooding. In the Middle East the number of precipitation events is likely to decrease, but those that do occur in the less arid regions may become more intense (UAE Ministry of Energy 2006). In general, the significant nonpermeable surface cover in many major cities makes them more susceptible to increased risk of injuries, property damage, and drowning by contributing to inland flash flooding associated with more intense precipitation events. The UAE is primarily a very arid nation, however, with only 120 mm of rainfall per year on average. Therefore, the risk of inland flooding due to climate change is unlikely to have a significant impact on health in the UAE.

Coastal flooding will be associated with both a rise in sea level and the possible intensification of coastal storms. The IPCC projects global average sea level to rise 0.18–0.59 m by 2090–2099, relative to 1980–1999 levels (IPCC 2007b). In the long term, the UAE will be particularly susceptible to rising sea levels because it has 1,318 km of coastline and a significant portion of its population residing in major urban centers on the coast (UAE Ministry of Energy 2006). According to an assessment of climate-related coastal impacts in Abu Dhabi by Fencl and Klein (2008), 85% of the UAE population and 90% of its infrastructure are in coastal areas vulnerable to rising sea level. The major impacts associated with coastal flooding will be population displacement and destruction of coastal infrastructure. It will likely take a century or more for a significant amount of sea-level rise to occur, allowing the UAE to adequately plan for and adapt to this gradual but inevitable threat. On the other hand, climate change could alter the frequency and severity of coastal storms, which could be exacerbated by a rise in sea level. Severe coastal storms are relatively infrequent in the Arabian Gulf and are therefore unlikely to be a major risk factor for the UAE, but they do occur; the strongest recorded tropical storm in the Arabian Sea made landfall in Oman in 2007 (Fencl and Klein 2008). Long-term increases in sea level are expected to reduce the amount of available land, but storms will damage property and may pose the greater health risk.

Vector-Borne Disease

Climate change can alter the transmission of infectious vector-borne diseases. Increased variability in precipitation and temperature levels likely will alter the development, reproduction, behavior, and population dynamics of disease-carrying vectors. Malaria and dengue fever are two of the major vector-borne diseases related to climate change, and the impacts of climate on malaria have been extensively studied (Kovats et al. 2000; Rogers and Randolph 2000; Mouchet and Manguin 1999; Reiter 2001). In some areas, the endemic range of these vectors may increase, whereas in other regions, changes in precipitation or temperature may reduce the risk of vector-borne disease transmission. The UAE is situated in an endemic malarious region, with climate conditions amenable for both *Plasmodium falciparum* and *P. vivax* malaria transmission through the *Anopheles stephensi* and *A. culicifacies* mosquito vectors, two of the 12 recorded *Anopheles* species in the country (UAE Ministry of Health 2003; Beljaev 2002). After a long, successful campaign against malaria, the UAE was officially declared malaria-free by the World Health Organization in 2007, with the last endemic case reported in 1997 (Meleigy 2007). This certification means there is no evidence of ongoing malaria transmission within the UAE and that the health service is proactively managing malaria infections that do occur. This does not mean malaria is totally absent from the UAE. Most of the malaria cases that do occur in the UAE originate from the immigrant worker population, primarily from the Indian subcontinent (*Weekly Epidemiological Record* 2007). The ability of the UAE's current control program and public health system to meet any possible increased risk will ultimately determine the future additional malarial burden due to climate change.

Water- and Foodborne Illness

Water- and foodborne pathogens will also likely be affected by climate change, leading to an increased incidence of diarrheal diseases in some areas. A wide variety of bacterial and viral pathogens are transmitted through food and water sources, and their impact is generally measured together under the umbrella of diarrheal disease. According to Miraglia et al. (2009), it is difficult to predict actual increases in climate-change-related water- and foodborne bacterial diseases. But many of these pathogens exhibit a seasonal dependence on high temperatures, and, therefore, increased temperatures projected for the future will likely lead to an increased presence of pathogens (Miraglia et al. 2009). For example, the bacteria responsible for many common cases of food poisoning, S*almonella,* are temperature-dependent and have been shown to increase almost linearly with each additional degree of average monthly temperature (Confalonieri et al. 2007; D'Souza et al. 2004). Harmful algal blooms, which contribute to certain types of shellfish poisoning such as ciguatera, have also shown a correlation with increases in ocean temperature and may undergo a poleward expansion with rising average sea temperatures (Confalonieri et al. 2007). Additionally, extreme weather events may impact the spread of waterborne diseases by increasing contaminant concentration in times of drought and contributing to overflows of untreated effluent waters during flood events (Confalonieri et al. 2007).

Nutrition

Changes in precipitation and temperature variability will impact the global distribution of productive agricultural land, with increased productivity projected in higher latitudes and decreased productivity primarily in lower latitudes. In the short term, agricultural productivity will likely increase in most areas due to the fertilizing effect of increased atmospheric concentrations of CO_2 (Easterling et al. 2007). In the Middle East, agricultural productivity is projected to decline due to soil degradation from climate change (UAE Ministry of Energy 2006). In some regions of the world, the decreased availability of agricultural products associated with lowered productivity will increase the incidence of malnutrition. Thus, malnutrition is often included as a health indicator in global burden-of-disease studies of climate change. Because the UAE is a relatively wealthy nation, malnutrition is unlikely to emerge as a serious health issue due to climate change.

Air Pollution

Many air pollutants are affected by changes in climate. For example, the production of tropospheric ozone increases with higher temperatures, as do emissions of volatile organic compounds (VOCs) from biogenic sources (Kinney 2008). Modeling studies in the United States have shown that climate change is generally expected to increase ozone (U.S. Environmental Protection Agency 2009). Additional air quality impacts associated with climate change include an increased range of aeroallergens, higher probability of dust storms due to drought, and increased wildfires leading to emissions of particulate matter. All of these factors contribute to the incidence of asthma, allergic reactions, and other respiratory ailments (Kinney 2008). The UAE is influenced by emissions from a large region. Climate effects on dust and smoke in Central Asia and North Africa will also influence the UAE. Because air quality is treated extensively elsewhere in this report, it will be excluded from further discussion in this chapter.

Method for Estimating the Burden of Disease from Climate Change

Our objective for modeling climate change in the UAE was to determine the consequences of climate change for human health, at present and in the near future, expressed as total premature deaths as well as additional health-care facility visits. Constructing this climate change environmental-burden-of-disease model involved a series of steps. First, the objective of the model was identified and assumptions clearly specified. Second, we reviewed the different mechanisms by which climate change could affect human health and evaluated their applicability to the UAE, selecting the best indicators to include in the model. Third, we established the baseline rates of relevant health outcomes reflecting the normal occurrence of these health indicators without accounting for climate change. Baseline rates of mortality and morbidity rely on health data from the Health Authority–Abu Dhabi (HAAD). Fourth, we input the relative-risk estimates developed by McMichael et al. (2004) for the EMR-B[1] subregion (Bahrain, Cyprus, Iran, Jordan, Kuwait, Lebanon, Libya, Oman, Qatar, Saudi Arabia, Syria, Tunisia, and the UAE) in order to calculate the additional mortality and morbidity from the added risks of climate change. Finally, we extrapolated the additional environmental burden of disease due to climate change by comparing baseline mortality and incidence rates with the new projected rates under two climate change scenarios.

The World Health Organization's Climate Change Burden of Disease Study

It is important to note that determining the relative risk of adverse health effects due to climate change is extremely complex and must incorporate data on the climate-health response for various climate change predictions over multiple time scales. Due to the complexity of information needed (i.e., detailed temperature records and daily location-specific health records for cardiovascular disease), the *UAE Burden of Disease Model* uses relative-risk estimates from a prior global study of the climate change environmental burden of disease. As such, many of the parameters in our model were selected to be compatible with this global study's parameters. The World Health Organization's (WHO) *Comparative Quantification of Health Risks: Global and Regional Burden of Disease Attributable to Selected Major Risk Factors* (McMichael et al. 2004) includes a climate change chapter written by McMichael et al. We chose to use the relative-risk estimates from this WHO climate change chapter because it is the most comprehensive assessment of the global burden of disease from climate change published to date. McMichael et al. (2004) studied the health effects of climate change for 14 subregions of the world, using five indicators: heat exposure, inland and coastal flooding, malaria, diarrhea, and malnutrition. The UAE model uses nine relative-risk estimates, developed by McMichael et al. for the EMR-B subregion for three climate change scenarios over three time periods: 2007, 2020, and 2030.

[1] A World Health Organization (WHO) coding system classifies member states within its six major regions into subregions according to mortality rates (very low, low, or high) in two groups: children 5 years or younger and adult males ages 15–59. The "EMR-B" subregion includes nations in the WHO's Eastern Mediterranean ("EMR") region that have low mortality in both demographics.

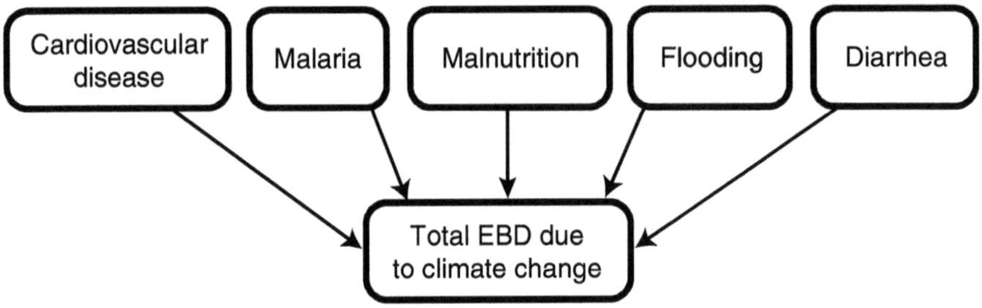

Fig. 7.2 Influence diagram for the ***Climate Change*** module

Model Assumptions

Many of the assumptions of the UAE climate change burden of disease model mirror those made by McMichael et al. (2004) in order to apply their relative-risk estimates. The model constrains its results for relative risks to human-induced climate change, omitting natural climate variability. The baseline estimate for the model was developed by assuming that the world before 1990 had only nominal effects of climate change. The years 1960–1990 were used to construct baseline rates for the relative risks. In estimating health effects, the model assumes that the UAE population will incur effects evenly over time. The model also assumes that relative risks for the EMR-B subregion can be directly applied to the UAE as a single country, so that relative risk in the UAE is the same as for the EMR-B subregion. In addition, the mortality relative-risk estimate taken from McMichael et al. (2004) is presumed to be the same as the morbidity relative-risk estimate. Lastly, the climate change model assumes the relative risk for 2008 is the same as for 2005 in order to estimate the fraction of observed illnesses reported in 2008 (the study year for this book) that could be attributable to climate change.

Indicators of Climate Change Effects on Human Health

The WHO study conducted by McMichael et al. used five indicators to calculate the overall environmental burden of disease due to climate change worldwide, as shown in Fig. 7.2. We chose to include only the effect of climate change on cardiovascular disease. The other four indicators used in the worldwide study—malaria, malnutrition, flooding, and diarrhea—were excluded.

Malaria was excluded from the model because the relative risk for malaria in the EMR-B subregion is 1, indicating no effect due to climate change (McMichael et al. 2004). In addition, the UAE has an extremely low incidence rate of malaria with only 1,663 cases and two fatalities reported in 2006 (UAE Ministry of

Health 2007). Furthermore, as mentioned previously, the UAE was declared "malaria free" by the WHO in 2007, indicating the success of its malaria control program. It is assumed that this program will continue in the future, eliminating the need to model future climate-change-related malaria incidence in the UAE.

Malnutrition was excluded from the UAE model for a number of reasons. First, the UAE had only eight reported deaths in 2006 from protein-calorie malnutrition (UAE Ministry of Health 2007), so the effect of climate change on malnutrition would be minimal given its extremely low baseline rate. Second, in the UAE, all agriculture is irrigated and a great portion of the food supply is imported. The UAE is even buying land in arable regions of the world to increase food production. With continuing irrigation of domestic lands, development of land in climatic regions more suitable to agriculture, and continued importation of domestic food, malnutrition is not likely to become a serious problem in the near future for the UAE, unless the entire developed world begins to face effects of malnutrition. Third, the UAE's wealth is projected to continue increasing due to its abundant oil reserves and increasing investments in nonpetroleum assets around the world, and it is therefore likely that the UAE will continue to be able to afford to provide food for its population. This point is supported by the assumption by McMichael et al. (2004) that developed countries are immune to climate change effects on nutrition.

Fatalities and injuries from flooding were also excluded from the UAE model for a number of reasons. First, the UAE has virtually no surface freshwater bodies, so inland flooding is very unlikely to occur. The main flood-related risk is the potential for sea-level rise. However, the Arabian Gulf is relatively calm and severe coastal storms are rare, so a rise in sea level will occur gradually over time, not posing the same human health risk as a rapid increase in sea level or coastal storm surges (Fencl and Klein 2008). Moreover, without knowing how the UAE might shift its coastal infrastructure in the future in response to climate change, it seems premature to speculate on the burden of disease due to flooding. Thus, although flooding and sea level rise are important and vast risks for coastal infrastructure, they are only nominal risks for human health in the UAE.

Diarrhea incidence rates were measured in the worldwide WHO study by McMichael et al. (2004) to represent the overall effect of climate change on waterborne disease. However, the UAE provides highly treated potable water to nearly all residents, with 95% of the supply coming from desalination facilities and the remaining 5% from groundwater treated with reverse osmosis and disinfection (Environment Agency–Abu Dhabi 2006). In addition, a large portion of the population drinks highly treated bottled water. Thus, it is very likely that none of the population is drinking untreated water, which means there will be little to no effect on waterborne diarrhea rates in the UAE from climate change. Furthermore, McMichael et al. (2004) assume that the effect of climate change on waterborne diarrhea rates is nominal in countries where per capita GDP is greater than $6,000. The per capita GDP of the UAE was an estimated $44,600 in 2008, one of the highest in the world and well above the McMichael threshold (Central Intelligence Agency 2009). Thus, waterborne diarrhea was excluded from the UAE climate change model.

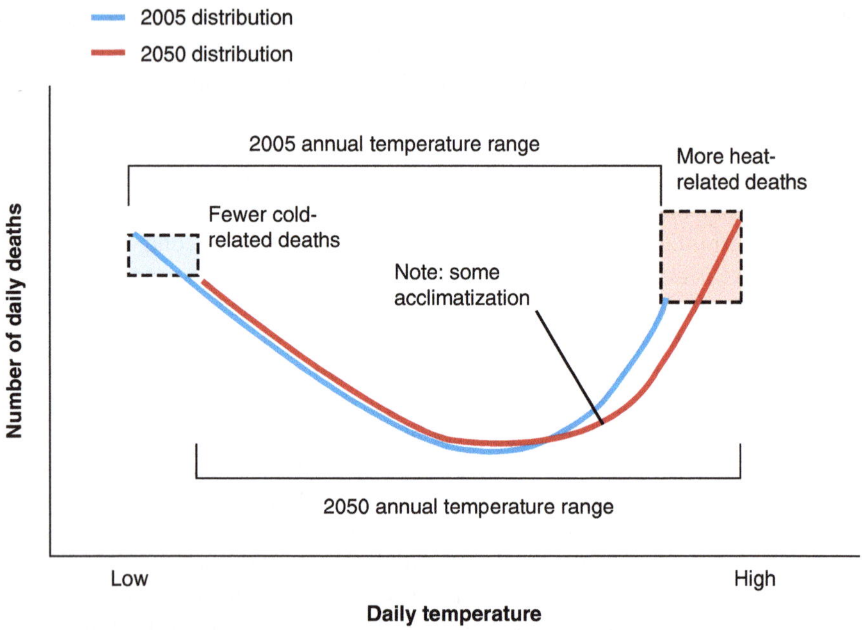

Fig. 7.3 A schematic representation of the effect of climate change on temperature-related mortality rates. 2050 temperature distribution accounts for higher global temperatures due to climate change (McMichael et al. 2006)

The UAE model does include the effect of climate change on cardiovascular disease rates because cardiovascular disease is expected to respond to changes in temperature, and this impact is believed to be the most pertinent to the UAE of all potential health effects from climate change. Many previously published studies (Kalkstein and Smoyer 1993; Kalkstein and Greene 1997; Ballester et al. 1997; Gouvenia et al. 2003; McGeehin and Mirabelli 2001; McMichael et al. 2006) have demonstrated a direct relationship between increases in temperature and cardiovascular disease rates. Higher temperatures can cause heat stress and heat strokes, and these events can lead to cardiovascular failure and death. The magnitude of heat-related mortality is difficult to determine because many factors affect this relationship, including regional heat tolerance, city heat-wave warning programs, heat duration, and air-conditioning availability, among others (McGeehin and Mirabelli 2001). Figure 7.3 shows the likely change in future cardiovascular disease rates with higher ambient temperatures from climate change. These rates may decrease in high latitude regions where extreme cold weather deaths from cardiovascular disease may decrease over time due to fewer very cold winter days in the future. Rates will increase in warmer regions where extreme hot days above a certain temperature threshold will increase over time. In the UAE, climate change will increase the incidence of cardiovascular disease as seen by McMichael's relative-risk estimates, which are greater than 1 (Table 7.4). The UAE's hot and arid climate makes the effects of fewer cold days irrelevant.

Table 7.4 Relative risks of cardiovascular disease in the EMR-B subregion due to climate change for three time periods (2005, 2020, and 2030) and three climate scenarios[a]

	2005			2020			2030		
	Mid	Low	High	Mid	Low	High	Mid	Low	High
S550	1.001	1.00	1.001	1.001	1.00	1.003	1.002	1.00	1.004
S570	1.001	1.00	1.002	1.002	1.00	1.004	1.002	1.00	1.004
Unmitigated	1.001	1.00	1.003	1.003	1.00	1.005	1.003	1.00	1.007

[a]Stabilization at 550 ppm, 750 ppm, and an unmitigated "business-as-usual" scenario (McMichael et al. 2004)

Relative-Risk Estimate for Cardiovascular Disease

McMichael et al. (2004) examined temperature-related mortality in five climate zones: hot/dry, warm/humid, temperate, cold, and polar. The hot/dry climate estimates, most applicable to the UAE, were extrapolated from data from Delhi, India. The quantitative estimates of the health impacts of climate change were determined by examining the short-term relationship between high temperatures above a threshold and increases in mortality from cardiovascular disease, as well as lower temperatures below a low threshold. It is important to note that the McMichael study "made the important assumption that these [short-term] relationships are also relevant to long-term climate change." An exposure-response relationship was calculated with a temperature threshold for each climate zone, which found the total deaths due to climate change from the change in both heat-attributable and cold-attributable deaths. Each climate change scenario has different predicted amounts of temperatures above the heat threshold. The temperature-mortality relationship was coupled with the temperature information for each climate change scenario to generate the relative risk of cardiovascular disease. These relative risks list a low, midrange, and high estimate for each scenario. The midrange estimate includes an adjustment for biological adaptation, which assumes that humans can acclimate over time, becoming less vulnerable to gradual increases in the mean temperature. McMichael accounted for this human acclimatization by assuming that the threshold temperature increases over time. The high estimate assumes that there is no human adaption, while the low estimate assumes humans will adapt completely to climate change, resulting in no risk from climate change, represented by a relative-risk value of 1.

The relative-risk estimates formulated by McMichael et al. (2004) are represented as a triangular distribution in the model, with the midrange estimate as its mode (see Table 7.4). McMichael formulated the cardiovascular disease relative-risk estimates by first choosing three climate change scenarios to study: (1) "Unmit," an unmitigated emissions trend approximately equal to the IPCC IS92a scenario, widely used as the standard scenario for impact assessments (IPCC 2008); (2) "s750," an emissions-reduction scenario with stabilization at 750 ppm CO_2 equivalent by 2210; and (3) "s550," an emissions-reduction scenario with stabilization at 550 ppm CO_2 equivalent by 2170. Because GHG emissions are highly influenced by the uncertain future, the IPCC created 40 scenarios for future GHG emissions, each assuming different levels of future technological and economic development. These scenarios are used in modeling to show the different possible outcomes that can occur for different GHG emissions scenarios. In this way, modelers do not have to model GHG emissions; they can adopt a number of the standard IPCC scenarios already delineated. The study used the HadCM2 global climate change model, also used by the IPCC, to project future temperature and precipitation data. HadCM2 is one of a number of reputable global climate models used for predicting future climate (IPCC 2008). The baseline scenario was established by using the average climate conditions from the World Meteorological Organization for 1961–1990.

Model Setup

The ***Climate Change*** module calculates the number of excess fatalities and health-care facility visits through a series of steps, shown in Fig. 7.4. Using data from HAAD, the model estimated the fatality rate from cardiovascular disease in 2008. The baseline cardiovascular disease fatality rate was multiplied by population, assumed to grow exponentially over time, to calculate the total baseline fatalities related to cardiovascular disease in the years 2008, 2020, and 2030. The total number of deaths was computed by multiplying the total baseline fatalities by the relative risk developed by McMichael et al. (2004) as explained above. The excess fatalities due to climate change were then found by subtracting the baseline fatalities from the total fatalities.

Similar methodology was used to calculate excess health-care facility visits related to cardiovascular disease from climate change. The baseline rate of morbidity was calculated by finding the sum of the number of reported cases of nonfatal incidences of cardiovascular disease in the UAE in 2008. Table 7.5 lists the model's five input variables, their source, description, and definition. Table 7.6 lists the remaining variables, their equation, and a description of what they represent.

Estimated Burden of Disease

Estimates of the UAE Burden of Disease from Climate Change

The model results show that climate change currently has minimal effects on human health relative to the other modeled priority areas. There were approximately 410 additional health-care facility visits and 3 additional fatalities from cardiovascular disease in the UAE in 2008 due to the added risks of climate change, assuming the "business-as-usual" scenario. As expected, estimates of mortality due to climate change increase over time as the climate warms, with larger effects on health (see Figs. 7.5 and 7.6).

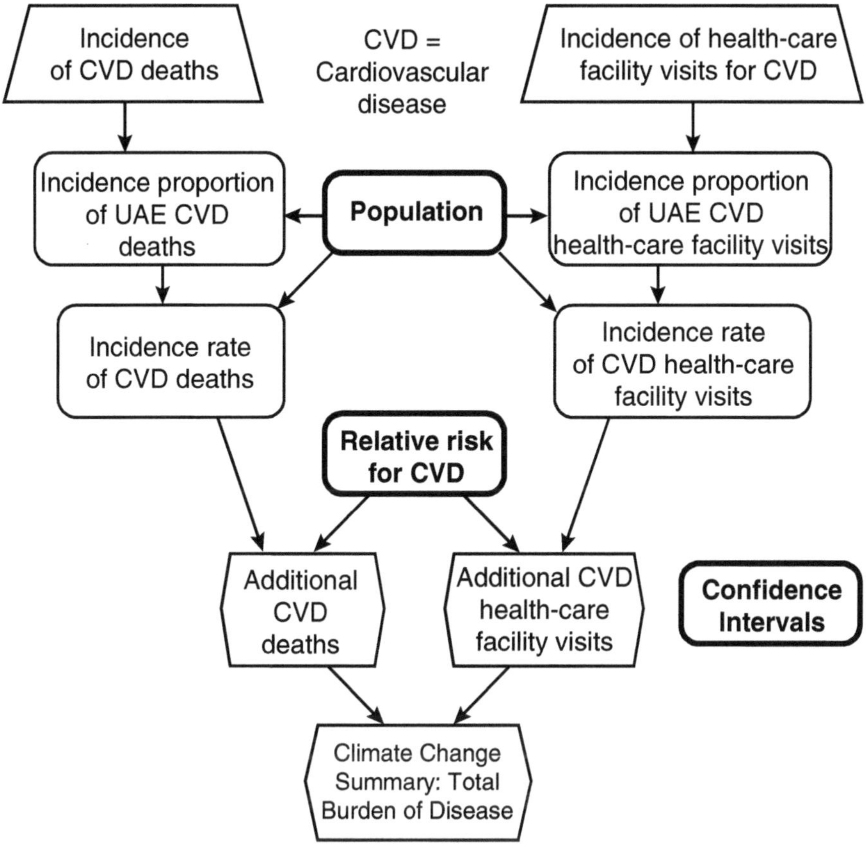

Fig. 7.4 General structure of model calculation of cardiovascular disease effects from climate change

Our model predicts that without mitigation measures, the environmental burden of disease due to climate change will be approximately 27 extra fatalities from cardiovascular disease by 2030, compared with 16 fatalities for either of the s750 or s550 scenarios. The model results are shown in 95% confidence interval outcomes and expressed by citizenship and total population for three time projections and three climate change scenarios, shown in Tables 7.7 and 7.8.

Comparison of Results with Regional and Global Studies

The overall results for 2030 show that the unmitigated climate change scenario yields the highest level of fatalities, 26.6, and the stabilizations at 550 and 750 ppm both yield approximately 16 fatalities. These UAE-specific estimates differ from effects extrapolated from previous global and regional estimates.

Table 7.5 Input variables in the ***Climate Change*** module

Input	Description	Input values	Source
Incidence of deaths related to cardiovascular disease	Number of reported deaths from cardiovascular disease (ICD-9 codes 393–459); scaled to represent entire UAE from available Abu Dhabi emirate data	Female noncitizens: 311.3 Male noncitizens: 1,207 Female citizens: 326.7 Male citizens: 465.7	HAAD (2009)
Incidence of health-care facility visits related to cardiovascular disease	Number of reported health-care facility visits due to cardiovascular disease (ICD-9 codes 393–459); scaled to represent entire UAE from available Abu Dhabi emirate data	Female noncitizens: 96,770 Male noncitizens: 196,724 Female citizens: 7,046 Male citizens: 7,128	HAAD (2009)
Total population	Population by gender and nationality. (estimates for 2020 and 2030 determined with exponential growth function)	*2008 estimates* *Female noncitizens: 978,000* Male noncitizens: 2,646,000 Female citizens: 426,000 Male citizens: 438,000 Equation for future predictions $Pop = e^{(0.0536t+14.145)}$ t = number of years since 1985	Ministry of Economy, Central Statistics Department midyear estimates
Relative risk for cardiovascular disease	Cardiovascular disease mortality relative-risk estimates for the EMR-B subregion	Triangular distribution indexed with time and climate change scenario (Table 7.4)	McMichael et al. (2004)

According to the WHO, the EMR-B subregion had approximately 5.65 excess deaths per million people due to climate change in 2000 (Campbell-Lendrum and Woodruff 2007, Table A.2). Extrapolating these regional estimates to the 2008 UAE population indicates that 25.4 fatalities were predicted to occur from climate change in the UAE alone in 2008, compared with the model's lower predictions of just 3.1 deaths. The numbers extrapolated from the regional estimates are significantly higher than the model's estimates, most likely due to the fact that the regional estimates were preliminary, rough estimates. We have updated our model, taking into consideration country-specific characteristics of the UAE. Upon further examination

Table 7.6 Calculated variables in the ***Climate Change*** module

Name of node	Description	Equation
Incidence proportion of deaths due to cardiovascular disease	The proportion of people by gender and citizenship who died of cardiovascular disease in 2008	$\frac{D_{cv}}{P_{2008}}$ D_{cv} = Incidence of deaths related to cardiovascular disease P_{2008} = 2008 population
Incidence proportion of health-care facility visits related to cardiovascular disease	The proportion of people by gender and citizenship who visited a health-care facility in 2008 for cardiovascular disease	$\frac{H_{cv}}{P_{2008}}$ H_{cv} = Incidence of health-care visits related to cardiovascular disease P_{2008} = 2008 population
Incidence rate of deaths due to cardiovascular disease	The number of annual cardiovascular disease deaths for the three time scenarios (2008, 2020, 2030)	$DP_{cv} \times P$ DP_{cv} = Incidence proportion of deaths due to cardiovascular disease P = population
Incidence rate of health-care facility visits related to cardiovascular disease	The number of annual health-care facility visits related to cardiovascular disease for the three time scenarios (2008, 2020, 2030)	$HP_{cv} \times P$ HP_{cv} = Incidence proportion of health-care visits related to cardiovascular disease P = population
Additional deaths due to cardiovascular disease	The total number of deaths related to cardiovascular disease due to the added risk of climate change, expressed for three time periods, three climate change scenarios, and by gender and citizenship	$D_{cv,t}(RR-1)$ $D_{cv,t}$ = Incidence rate of deaths due to cardiovascular disease for year t RR = relative risk
Additional health-care facility visits related to cardiovascular disease	The total number of health-care facility visits related to cardiovascular disease in one specific year due to the added risk of climate change, expressed for three time periods, three climate change scenarios, and by gender and citizenship	$H_{cv,t}(RR-1)$ $H_{cv,t}$ = Incidence of health-care facility visits related to cardiovascular disease for year t RR = relative risk

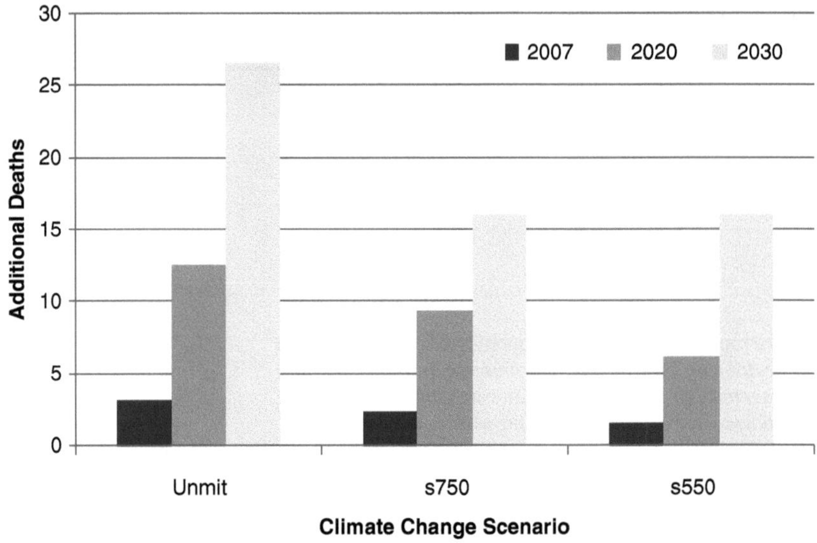

Fig. 7.5 Predicted deaths from cardiovascular disease due to climate change over three time periods and climate change scenarios

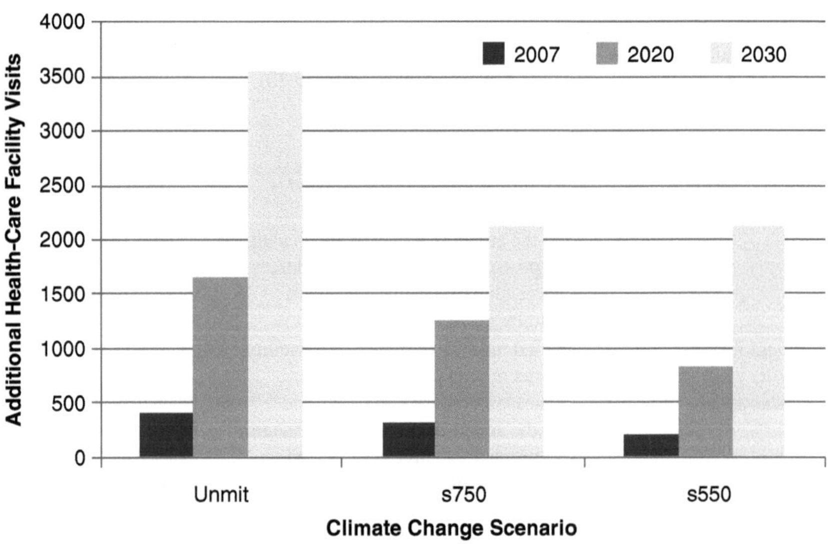

Fig. 7.6 Predicted health-care facility visits related to cardiovascular disease due to the added risk of climate change, shown over three time periods and three climate change scenarios

Table 7.7 Predicted cardiovascular disease fatalities due to climate change, by citizenship and totals over three time periods

	Mean (95% confidence interval)		
	Citizen	Noncitizen	Total
Climate change scenario: Unmitigated, business as usual (Unmit)			
2008	1.06 (0.22, 2.07)	2.02 (0.42, 3.96)	3.08 (0.63, 6.02)
2020	4.27 (0.98, 7.23)	8.18 (1.89, 13.8)	12.5 (2.87, 21.1)
2030	9.13 (1.96, 17.0)	17.5 (3.76, 32.5)	26.6 (5.72, 49.4)
Climate change scenario: Stabilization at 750 ppm by 2210 (s750)			
2008	0.79 (0.18, 1.41)	1.52 (0.34, 2.69)	2.31 (0.52, 4.10)
2020	3.20 (0.72, 5.69)	6.14 (1.34, 10.9)	9.34 (2.11, 16.6)
2030	5.48 (1.23, 9.77)	10.5 (2.35, 18.7)	16.0 (3.58, 28.5)
Climate change scenario: Stabilization at 550 ppm by 2170 (s550)			
2008	0.53 (0.13, 0.78)	1.01 (0.24, 1.50)	1.54 (0.37, 2.28)
2020	2.14 (0.44, 4.19)	4.09 (0.84, 8.02)	6.23 (1.28, 12.2)
2030	5.48 (1.23, 9.78)	10.5 (2.35, 18.7)	16.0 (3.58, 28.5)

Table 7.8 Predicted health-care facility visits for cardiovascular disease due to climate change, by citizenship and totals over three time periods

	Mean (95% confidence interval)		
	Citizen	Noncitizen	Total
Climate change scenario: Unmitigated, business as usual (Unmit)			
2008	19 (4, 37)	391 (80, 765)	410 (84, 802)
2020	76 (18, 129)	1,582 (364; 2,677)	1,659 (382; 2,806)
2030	163 (35, 303)	3,382 (726; 6,278)	3,545 (761; 6,581)
Climate change scenario: Stabilization at 750 ppm by 2210 (s750)			
2008	14 (3, 25)	293 (66, 520)	307 (69, 546)
2020	57 (13, 102)	1,187 (268; 2,107)	1,244 (281; 2,209)
2030	98 (22, 175)	2,029 (454; 3,617)	2,128 (476; 3,791)
Climate change scenario: Stabilization at 550 ppm by 2170 (s550)			
2008	9 (2, 14)	196 (46, 290)	205 (49, 304)
2020	38 (8, 75)	791 (163; 1,550)	829 (171; 1,625)
2030	98 (22, 175)	2,029 (455; 3,623)	2,127 (477; 3,798)

we chose to exclude malnutrition, flooding, malaria, and diarrhea for reasons previously stated. The WHO's worldwide model estimates included these health effects and thus had a significantly higher fatality prediction than the current model. The risk assessment conducted by RAND Corp. for the risk ranking exercise described in Chap. 2 also estimated a number of deaths, with a range of 0–50 deaths in 2030 and a best estimate of 20 deaths (see Appendix A). These numbers are close to our model's estimate of 26.6 deaths in 2030.

In the WHO's global burden of disease estimates, the UAE and the Middle Eastern region have low effects from climate change relative to other countries, with an estimated 14 DALYs per 100,000 people. The most vulnerable subregions are AFR-E (Africa), EMR-D (Eastern Mediterranean), SEAR-D (Southeast Asia), and AFR-D (Africa), with an estimated 358, 213, 207, and 201 DALYs per 100,000 people, respectively (WHO 2002). Thus, while the UAE will inevitably be affected by climate change, the country will experience fairly mild human health effects in the short term relative to countries worldwide.

Uncertainty Analysis

This model is an initial attempt to model climate change effects in the UAE and should be improved and updated over time in order to expand its predictive capacity. Various assumptions made to construct the model increase the uncertainty within the model and should be replaced with UAE-specific empirical data when available. The limited number of indicators in the model has constrained it, resulting in an underestimation of health effects, and should be expanded to include more health effects, even if their impacts are relatively minor. Furthermore, the model also excludes nonhealth effects that are extremely important when considering climate change effects, and these should be added as methodology develops to do so accurately.

Model Assumptions

The model has a number of assumptions, which cause a higher level of uncertainty. As McMichael et al. (2004) discuss, the relative-risk estimate for cardiovascular disease was developed under the assumption that short-term heat-related mortality could be applied to long-term climate change. Also, our assumption that the relative risks of the EMR-B subregion apply to the UAE invites significant uncertainty given the unique age and gender distribution in the country as well as the wide wealth disparity between citizens and noncitizens. For example, the relative risk of cardiovascular disease from heat exposure may be overestimated because of a young population distribution and high per capita income in the UAE. There are fewer elderly people, who are highly vulnerable to cardiovascular disease, which most likely reduces incidence of cardiovascular disease. In addition, the country's wealth means that air conditioning is very common, potentially decreasing the number of cardiovascular incidences due to heat stroke. In the future, developing a relative-risk estimate specifically for the UAE would greatly improve the accuracy of the model. In addition, the relative-risk estimate used in the model represents temperature-related *mortality*; we have assumed that this relationship can also be applied to temperature-related *morbidity*, which could result in an underestimation. The relative risk for morbidity is likely significantly higher because most heat-related health effects, such as heat stroke or heat stress, do not result in death.

Limited Scope of Model

The overall output results of the model are most likely an underestimation of health impacts because the model only evaluates the effect of climate change on cardiovascular disease and no other health effects, which is evident when comparing the model's estimate with previous studies. Agricultural pathogen and pest effects, possible destruction of health infrastructure due to extreme weather events and rising sea level, respiratory illnesses, and possible health effects from large population displacement due to an increase in sea level are all health consequences omitted from this model. We cannot model these effects because there is not yet a proven acceptable methodology to quantify them. In addition, many of these climate change effects, such as health consequences from coastal population displacement, are highly dependent on how people adapt to climate change, and thus are extremely difficult to accurately predict. Moreover, while we assume the effects of malaria, malnutrition, flooding deaths, and diarrhea—all excluded from the model—to be small, they may not be zero. Thus, the model's estimated health outcomes are likely underestimated since the model is only calculating the effect on cardiovascular disease and is unable to quantify the many other health effects likely to occur.

Nonhealth Effects of Climate Change

In addition, nonhealth-related climate change effects were purposely excluded from this analysis but will have an important impact on the UAE. Nonhealth effects such as coastal infrastructure destruction, the effect of temperature and precipitation change on dust storms, destruction of coastal ecosystems, and decrease in biodiversity will adversely affect the UAE's GDP and will likely have a more prominent impact in the UAE than health-related effects. Higher mean temperature may increase the occurrences of inability to work during extreme temperature events. Coastal tourism infrastructure vulnerable to rising sea level could greatly decrease GDP if destroyed in the future. The potential infrastructure destruction and environmental degradation from these nonhealth effects could eventually negatively affect national health.

The increased risk of coastal destruction is one of the more important climate change risks the UAE faces. Eight of the nine major cities in the UAE are on the coast, making the UAE extremely vulnerable to sea-level rise. This large coastal population could possibly face dislocation in the future if adaptation measures are not taken to prevent coastal damage. Current trends of development in the UAE are exacerbating this problem and increasing vulnerability to potential destruction from sea-level rise. The coastline continues to inch forward as developers add fill to build new structures on the beach and create artificial islands for added coastline that is only a few meters above current sea level. Commercial buildings and industrial facilities along the coast, including desalination and oil and gas manufacturing and refining facilities, will also be vulnerable to the threat of sea-level rise.

Table 7.9 Sensitivity analysis for cardiovascular disease incidence, population, and excess relative risk

	−10%	Mean	+10%
Cardiovascular disease deaths in 2030			
Unmit	24.0	26.6	29.3
s750	14.4	16.0	17.6
s550	14.4	16.0	17.6
Cardiovascular disease health-care facility visits in 2030			
Unmit	3,191	3,545	3,900
s750	1,915	2,128	2,340
s550	1,915	2,127	2,340

Results in table occur when any of the variables increase or decrease 10%, holding all other variables constant.

Sensitivity Analysis

The model estimates are highly sensitive to various input predictions. The value of the relative risk used in the model plays an important role in determining the final predicted health effect. If the relative risk for the unmitigated scenario in 2030 is increased from the current EMR-B subregional estimate to the relative risk of Southeast Asia's SEAR-B subregion (triangular distribution [1, 1.004, 1.009]), the resulting mortality value would increase to 34.6 deaths, a 30% increase in the number of fatalities compared with the original EMR-B relative-risk estimates for the 2030 unmitigated scenario. The baseline rates of cardiovascular disease mortality and morbidity highly influence the overall outcome. If the baseline levels were to decrease over time, the UAE would see a substantial decrease in predicted rates of cardiovascular disease. Linked closely with the baseline rate, population growth is the main reason for an increase in cardiovascular disease deaths and morbidity over time. If the population for 2008, 2020, and 2030 were to remain constant at 4.488 million people, it is estimated that only an estimated 7.89 related excess deaths would occur in 2030, compared with the original 26.6 excess deaths predicted. Regardless of the input variables, risks of climate change will continue to increase as climate change continues beyond 2030, even under the most aggressive abatement scenario.

The sensitivity of the model's variables can also be examined by changing the value of a variable and then observing the change in the final model outputs of deaths and health-care facility visits related to cardiovascular disease. Three variables were examined in our model: cardiovascular disease incidence, population, and excess relative risk. Ordinarily, all the variables in the model would remain the same, except for one variable that is slightly changed to test the sensitivity of that variable. However, the climate change model multiplies these variables together; the associative property of multiplication makes the choice of which variable to change unimportant, as the resulting change in outcome is the same among all three variables. Table 7.9 shows the change in deaths and health-care facility visits related to cardiovascular disease in the year 2030, when any of the three inputs examined are increased or decreased by 10%. In other words, changing any of the three variables mentioned above would result in approximately the same change in the final outcome, making none of the variables more influential than any other.

Information Needed to Improve Future Burden of Disease Predictions

Develop UAE-Specific Relative-Risk Estimates

Developing a relative-risk estimate for cardiovascular disease specific to the UAE, and not the EMR-B subregion, is an essential first step toward improving the predictive power of the model. To create cardiovascular disease relative-risk estimates for the UAE, climate change scenarios and time periods must be selected and data must be acquired on the future climate given each scenario and time. Second, the temperature-mortality relationship must be quantified. This step "consists of conducting a time-series regression of variations in (usually daily) mortality rates against variations in temperature, with controlling for confounding factors such as air pollution and secular and seasonal trends" (Campbell-Lendrum and Woodruff 2007). Long-term records of daily temperature and daily mortality from cardiovascular disease are required. Given this newly constructed temperature-mortality relationship, a low and high temperature threshold should be chosen to "yield a heat/cold *coefficient* (i.e., the percentage increase in mortality for each degree Celsius increase in mean temperature above a threshold level)" (Campbell-Lendrum and Woodruff 2007). Third, future temperatures predicted with each climate change scenario are used to decipher how many more days per year the temperature will be above the threshold and then to derive the additional fatalities from cardiovascular disease. This process is explained in detail by Campbell-Lendrum and Woodruff (2007) in a comprehensive guide for how to compute the national burden of disease due to climate change.

Develop Relative-Risk Estimates Specific to Citizen and Noncitizen Populations

Ideally, in the future the UAE would develop two separate relative-risk estimates: one for citizens and another for noncitizens. The wealth distribution in the UAE is unequal between citizens and noncitizens. Therefore, when making certain assumptions considering the wealth of the country, the less affluent noncitizens might have underestimated outcomes. For example, less affluent people might have less access to air conditioning and also might tend to work more outside in hot temperatures, making them especially vulnerable to temperature-related illnesses. For example, outside day laborers who work in hot temperatures are perhaps the most at risk for heat-related illness from climate change. It is notable that the average age of

Table 7.10 Average age at death from cardiovascular disease, by gender and citizenship

	Female noncitizens	Male noncitizens	Female citizens	Male citizens
Age	69.02	53.81	74.93	71.15

HAAD (2009)

cardiovascular-disease-related mortality for male noncitizens was significantly lower than for all other groups examined in this study (see Table 7.10). Diet, medical history, working conditions, and more are all factors that could help explain this disparity. The current model outcomes include separate results for citizens and non-citizens in an effort to document the differences in risk. Upon improving the model, the relative-risk estimates for each subpopulation could be better differentiated.

Include More Health Effects

The model should also be expanded to include other health effects, such as future population displacement, in which people might become temporarily homeless and health-care facilities might be destroyed. The method for modeling these health effects is still in its infancy and in some cases unexplored. As modeling will undoubtedly improve over time, the UAE model should be updated to include more parameters as they become available. In addition, the model only estimates the relative risk of climate change due to gradual increases in average temperature, excluding, due to data availability, risks from increased extreme events and, due to lack of scientific consensus, risks from abrupt climate change possibilities. The impact of these potential catastrophic global risks should be studied in the future if more compelling evidence develops to support this possibility.

In general, the process of modeling climate change effects on human health is still in its infancy. Many underlying complexities complicate predicting climate change as well as its effects on human health, which are highly dependent on future human choices such as adaptation. Since climate change impacts will increase in magnitude over time as GHG concentrations increase, the most severe effects will occur over a long time frame. Human health effects are just a few of many probable consequences of climate change. Acting now to ease climate change as much as possible is essential.

Conclusions

Two important components of a successful strategy for dealing with climate change are mitigation and adaptation. Mitigation involves reducing the potential adverse impacts of future climate change, whereas adaptation involves developing preventative measures for an effective hazard response system (Keim 2008). Mitigation measures will therefore help to stabilize global atmospheric GHG concentrations to limit the extent of future impacts, whereas adaptation measures will prepare the UAE and other nations for the inevitable adverse effects of climate change. As the climate change problem is caused by global GHG emissions, the UAE cannot directly control the impacts it will experience. However, its mitigation efforts can help reduce the overall problem as well as set an example for other countries to follow in creating their own mitigation plans and enacting global policies. It is essential that the UAE begin now to actively anticipate these consequences and adapt to them in order to reduce the impacts and costs of climate change. The UAE has already taken some essential first steps and developed a Climate Change Policy for Abu Dhabi, which includes many of the recommendations in this report, but it has yet to implement this policy and still lacks a national policy to address climate change.

Mitigation

Updated Greenhouse Gas Inventory

In order to develop a successful mitigation strategy, the UAE must first assess its GHG emissions among various sources; this is extremely important for effective decision-making. Any comprehensive national climate change action, whether domestic or international, will require an accurate emissions inventory. A GHG registry and software to assist in tracking GHG emissions are therefore essential components of a successful climate change strategy in the UAE. Building institutional capacity and inter-organizational cooperation will be important to facilitate current and future emissions inventories. Once an appropriate system is in place to conduct regular inventories, the UAE will benefit from updating its emissions data annually in order to track the progress and success of various policy measures. In addition, increasing climate change data specific to the UAE will improve the accuracy of its emissions estimates. An emissions inventory is necessary to develop emission reduction targets, whether on the industry, emirate, or national level. As mentioned previously, Abu Dhabi is already updating its emissions inventory.

Active Participation in International Negotiations

The UAE can play an important role in international negotiations on climate change, which are now focusing on binding emissions reductions for when the Kyoto Protocol expires in 2012. In particular, as a prosperous oil-producing nation, the UAE can promote new energy technologies and play a leading role in reducing emissions since its per capita GHG emissions are currently among the highest in the world (Baumert et al. 2005). In order to serve in this leadership role, an up-to-date, accurate inventory of the UAE's GHG emissions is imperative. This inventory will allow the UAE to determine effective and realistic targets for its own GHG emissions, as well as to be an effective participant in the post-Kyoto negotiations and to help shape international GHG emissions targets. A GHG inventory is a prerequisite for participation in any "cap-and-trade" program or international mitigation agreement.

Reduction Targets for National GHG Emissions

Currently, the UAE is treated as a non-Annex I nation under the UNFCCC, a designation corresponding to countries with a developing country status in 1992. However, the UAE was removed from the World Bank's list of developing nations in 1995 and is now considered a "high income" country (World Bank 2009). The UAE is currently receiving money for three projects under the Clean Development Mechanism, through which Annex I countries fund GHG reductions in non-Annex I countries and receive credit for the subsequent reductions in GHG emissions. Unlike other developed countries, the UAE has not committed to binding reduction targets, but a number of proposed targets are currently under consideration.

Reduction in Emissions from Energy Use

In the UAE, energy use accounted for 95% of all GHG emissions in 1994 (UAE Ministry of Energy 2006). Many energy-related mitigation measures can be taken to reduce current and future emissions. GHG emissions from energy production, and consumption can be reduced through demand-side management strategies, such as energy efficiency and conservation, as well as through supply-side reduction of the carbon intensity of energy sources. Many of these actions are recommended in the Abu Dhabi Climate Change Policy (EAD 2009) and may be included in the forthcoming Abu Dhabi Energy Policy.

Transportation

The UAE could implement minimum fuel economy standards for imported vehicles in an attempt to reduce emissions from the personal vehicle fleet. Increasing the provision of alternative public and non-motorized transportation would also likely reduce personal transportation emissions. The first metro system in the region is already operating in Dubai, and the success of this program may pave the way for expansion of public transit in other UAE cities. Additional transportation measures, such as discouraging driving by charging drivers tolls or fees for using high-volume roadways during times of overcongestion, can play an important role in decreasing transportation demand. Dubai already has implemented some of these measures, including a 24-h toll collection system.

Buildings

Significant mitigation potential exists in the buildings of the UAE. Worldwide, mitigation activities in the buildings sector are deemed to be one of the most cost-effective options for reducing GHG emissions from energy (Urge-Vorsatz and Novikova 2008). The UAE government already is working to improve energy efficiency through mandated energy efficiency standards in building design and materials, appliances, and air conditioners in homes and in larger commercial, office, and multifamily residential buildings.

Industry

The industrial sector consumed 58.4% of the primary energy in the UAE in 1998, with a significant portion used in desalination facilities (Kazim 2007). The government can play a crucial role in encouraging industrial efficiency and conservation. The UAE has only recently begun to implement energy conservation and efficiency measures and should continue these efforts, especially in the industrial sector and desalination facilities. A recent agreement between the Emirates Foundation and Exxon Mobil will provide grants for various energy efficiency projects across the emirates and represents an important first step for energy efficiency in the UAE (Hassan 2009).

Energy Conservation and Public Awareness

Energy conservation programs to raise public awareness as well as energy pricing to encourage conservation are highly recommended for the UAE to reduce per capita energy consumption.

Energy Supply

The UAE could help increase the market share of clean energy sources and develop and implement carbon capture and sequestration (CCS) technology. The UAE can directly reduce GHG emissions by increased substitution of renewable energy sources for natural gas or oil combustion. A renewable portfolio standard is one option for encouraging the production of renewable energy. A renewable portfolio standard is a policy that requires a certain percentage of total energy to be produced from renewable sources such as wind, solar photovoltaic, solar thermal, energy efficiency, or geothermal, among others. The Abu Dhabi Climate Change Policy already proposes a renewable portfolio standard of 10% by 2030, which will be useful in achieving this goal (EAD 2009). The Abu Dhabi Government has committed to a target of 7% of installed energy capacity from renewable sources by 2030. The UAE has abundant solar resources, which may therefore play an important role in increasing its renewable energy supply (Islam et al. 2009). The UAE has already shown leadership in some forward-looking and high profile actions, such as the construction of Masdar City,[2] to reduce emissions. The UAE should continue to implement renewable energy projects as well as support further research and development of these technologies, as this will lead to future growth in the UAE's energy sector beyond its petroleum and natural gas reserves.

By increasing research and implementation of CCS technology, the UAE can reduce emissions from fossil fuel combustion. Globally, only a few example CCS projects currently exist, but this technology has the potential to grow rapidly in the future. In the UAE, CCS can take the form of using CO_2 for enhanced oil recovery, where CO_2 is injected into current oil deposits in order to increase the extraction of oil from existing wells. New power plants can be designed and built to allow for future CCS, which will be more cost-effective than retrofitting existing power plants.

Adaptation

In addition to the mitigation measures mentioned above, adaptation to future climate change risks will be an important component of the UAE's strategy. The UAE will likely face multifaceted risks due to climate change; however, the focus of this report has been on addressing human health risks. An expanded public health infrastructure is one key priority for adapting effectively to the impacts of climate change. Actions should be taken to improve the understanding of how climate change will affect human health, including improved monitoring of the health of all populations in the UAE, education of health care professionals for improved awareness of the threats of climate-related health impacts, and identification of priority actions for health protection, disease prevention, and health-care infrastructure to address UAE-specific climate-related health impacts. An early warning system for extreme heat waves, for instance, is one highly recommended measure to decrease heat-related illnesses associated with extreme heat events (Kovats and Hajat 2008). The UAE already has in place an extensive and successful system for dealing with malaria, with added benefits for dengue fever, and it should continue this program to address potential future risks (Emirates News Agency 2006).

---

[2]The Masdar Initiative is a multibillion-dollar investment in renewable and alternative energy sources and other sustainability initiatives in the UAE. This initiative includes an innovative plan for the world's first carbon-neutral and zero-waste city, called Masdar City, located in the emirate of Abu Dhabi. Masdar City will be able to support a population of 40,000 residents and 50,000 commuters sustainably and will integrate sustainable building design, renewable wind and solar energy, "smart growth" city design, wastewater management, and an elaborate public transportation system to minimize its GHG emissions (Raouf 2008). An important goal of this project is to establish Masdar City as a regional science and technology research hub dedicated to alternative energy solutions.

Declines in water availability may accelerate due to global warming. The UAE is projected to experience a possible 2% decrease in annual precipitation by 2030 due to climate change, requiring greater reliance on desalination facilities that already provide more than 95% of the domestic water supply (EAD 2006). According to Kazim (2007), reverse osmosis or multistage flash desalination processes in the UAE use about 5 kW h/m^3 of desalinated water, energy that is supplied primarily by the combustion of natural gas. As the demand for water increases, therefore, more energy will be required to power desalination plants, exacerbating GHG emissions from this sector. Water conservation will help reduce the effects of increasing water scarcity due to climate change and rising demand—the most pressing threat for water supply and a strategic priority for the emirate of Abu Dhabi—through measures such as improving efficiency of water use, increasing public awareness, and implementing tiered price scales. Water conservation will not only help prepare the nation to adapt to a decrease in water supply but will also mitigate climate change by using less energy for desalination plant operations. Abu Dhabi initiated a water conservation campaign in 2010 and has actively promoted it throughout the emirate.

Finally, nonhealth-related adaptation measures also can reduce the economic and social impacts of climate change. One of the most important nonhealth adaptations will be preparing the coastlines of the UAE for future increases in sea level. Coastal adaptation strategies will include building farther away from the coast and planning now for the consequences of rising sea level in the emirates' extensive seaside urban areas and infrastructure. Zoning ordinances to minimize the built infrastructure in these high-risk coastal flooding areas are imperative. Modifying the built environment to reduce the heat island effect will allow urban areas in the UAE to adapt to increased average temperatures in the future. Sustainable soil management practices that reduce GHG emissions from soil may have positive benefits for agriculture.

Increasing the quality and availability of climate change data specific to the UAE and developing an up-to-date GHG inventory will improve the ability of the UAE to prepare adaptation strategies and adequately plan for the impacts of climate change. Measures should also be taken to improve the scientific understanding of climate change in the UAE and how to adapt effective policies to local conditions and needs.

References

Ballester, D.F., D. Corella, S. Perez-Hoyos, M. Saez, and A. Hervas. 1997. Mortality as a function of temperature: A study in Valencia, Spain, 1991–1993. *International Journal of Epidemiology* 26: 551–561.

Baumert, K.A., T. Herzog, and J. Pershing. 2005. *Navigating the numbers: Greenhouse gas data and international climate policy*. World Resources Institute, 21–24. http://pdf.wri.org/navigating_numbers_chapter4.pdf.

Beljaev, A.E. 2002. Determinants of malaria in the Middle East and North Africa. In *The contextual determinants of Malaria*, ed. E.A. Casman and H. Dowlatabadi, 137–166. Washington, DC: Resources for the Future Press.

Campbell-Lendrum, D.H., and R. Woodruff. 2007. Climate change: Quantifying the health impact at national and local levels. *World Health Organization environmental burden of disease series*, No. 14. http://whqlibdoc.who.int/publications/2007/9789241595674_eng.pdf

Carbon Dioxide Information Analysis Center (CDIAC). *2009. Preliminary 2007–08 global and national estimates by extrapolation.* Oak Ridge National Laboratory. http://cdiac.ornl.gov/trends/emis/meth_reg.html

Central Intelligence Agency. 2009. *The world factbook: United Arab Emirates*. https://www.cia.gov/library/publications/the-world-factbook/geos/ae.html

Confalonieri, U., B. Menne, R. Akhtar, K.L. Ebi, M. Hauengue, R.S. Kovats, B. Revich, and A. Woodward. 2007. Human health. In *Climate change 2007: Impacts, adaptation and vulnerability. Contribution of working group II to the fourth assessment report of the intergovernmental panel on climate change*, ed. M.L. Parry, O.F. Canziani, J.P. Palutikof, P.J. van der Linden, and C.E. Hanson, 391–431. Cambridge: Cambridge University Press.

D'Souza, R., N. Becker, G. Hall, and K. Moodie. 2004. Does ambient temperature affect food-borne disease? *Epidemiology* 15: 86–92.

Easterling, W.E., P.K. Aggarwal, P. Batima, K.M. Brander, L. Erda, S.M. Howden, A. Kirilenko, et al. 2007. Food, fibre and forest products. In *Climate change 2007: Impacts, adaptation and vulnerability. Contribution of working group II to the fourth assessment report of the intergovernmental panel on climate change*, ed. M.L. Parry, O.F. Canziani, J.P. Palutikof, P.J. van der Linden, and C.E. Hanson, 273–313. Cambridge, UK: Cambridge University Press.

Emirates News Agency. 2006. *Anti-malaria drive keeps dengue at bay*. Inter Press Service News Agency, May 13. http://www.ipsnotizie.it/wam_en/news.php?idnews=3099

Environment Agency–Abu Dhabi (EAD). 2006. Water resources of Abu Dhabi Emirate, United Arab Emirates.

Environment Agency–Abu Dhabi (EAD). 2009. Abu Dhabi climate change policy.

Ezzati, M., A.D. Lopez, A. Rodgers, S. Vander Hoorn, C.J.L. Murray, and Comparative Risk Assessment Collaborating Group. 2002. Selected major risk factors and global and regional burden of disease. *Lancet* 360: 1347–1350.

Fencl, A., and R. Klein. 2008. *Part II: Climate change impacts, vulnerability and adaptation: Coastal zones in the UAE*. Prepared for the Environment Agency–Abu Dhabi by the Stockholm Environment Institute, Stockholm, September.

Gouvenia, N., S. Hajat, and B. Armstrong. 2003. Socio-economic differentials in the temperature-mortality relationship in Sao Paulo, Brazil. *International Journal of Epidemiology* 32: 390–397.

Hassan, H. 2009. Emirates Foundation and Exxon sign agreement. *The National*, May 15. http://www.thenational.ae/article/20090516/NATIONAL/705159862/1010/NEWS

Health Authority–Abu Dhabi (HAAD). 2009. *2008 health data for Abu Dhabi emirate*. Abu Dhabi: Health Authority–Abu Dhabi (HAAD).

Intergovernmental Panel on Climate Change (IPCC). 1998. In *The regional impacts of climate change: An assessment of vulnerability*, ed. R.T. Watson, M.C. Zinyowera, and R.H. Moss. Cambridge: Cambridge University Press.

Intergovernmental Panel on Climate Change (IPCC). 2007a. Climate change 2007: Synthesis report, summary for policymakers. http://www.ipcc.ch/pdf/assessment-report/ar4/syr/ar4_syr_spm.pdf

Intergovernmental Panel on Climate Change (IPCC). 2007b. Summary for policymakers. In *Climate change 2007: The physical science basis. Contribution of working group I to the fourth assessment report of the intergovernmental panel on climate change*, ed. M.L. Parry, O.F. Canziani, J.P. Palutikof, P.J. van der Linden, and C.E. Hanson, 7–22. Cambridge: Cambridge University Press.

Intergovernmental Panel on Climate Change (IPCC). 2008. HadCM2 GCM model information. IPCC Data Distribution Center. http://www.ipcc-data.org/is92/hadcm2_info.html

Islam, M.D., I. Kubo, M. Ohadi, and A.A. Alili. 2009. Measurement of solar energy radiation in Abu Dhabi, UAE. *Applied Energy* 86: 511–515.

Kalkstein, L.S., and J.S. Greene. 1997. An evaluation of climate/mortality relationship in large US cities and the possible impacts of a climate change. *Environmental Health Perspectives* 105: 83–93.

Kalkstein, L.S., and K.E. Smoyer. 1993. The impact of climate change on human health: Some international implication. *Experientia* 49: 969–979.

Kazim, A.M. 2007. Assessments of primary energy consumption and its environmental consequences in the United Arab Emirates. *Renewable and Sustainable Energy Reviews* 1: 426–446.

Keim, M.E. 2008. Building human resilience: The role of public health preparedness and response as an adaptation to climate change. *American Journal of Preventative Medicine* 35(5): 508–516.

Kinney, P.L. 2008. Climate change, air quality, and human health. *American Journal of Preventative Medicine* 35(5): 459–467.

Kovats, R.S., and S. Hajat. 2008. Heat stress and public health: A critical review. *Annual Review of Public Health* 29: 41–55.

Kovats, R.S., D. Campbell-Lendrum, C. Reid, and P. Martens. 2000. *Climate and vector-borne disease: An assessment of the role of climate in changing disease patterns*. Maastricht: International Centre for Integrative Studies, Maastricht University.

Levine, M., D. Ürge-Vorsatz, K. Blok, L. Geng, D. Harvey, S. Lang, G. Levermore, et al. 2007. Residential and commercial buildings. In *Climate change 2007: Mitigation. Contribution of working group III to the fourth assessment report of the intergovernmental panel on climate change*, ed. B. Metz, O.R. Davidson, P.R. Bosch, R. Dave, and L.A. Meyer, 387–446. Cambridge, UK: Cambridge University Press.

Luber, G., and M. McGeehin. 2008. Climate change and extreme heat events. *American Journal of Preventative Medicine* 35(5): 429–435.

McGeehin, M.A., and M. Mirabelli. 2001. The potential impacts of climate variability and change on temperature-related morbidity and mortality in the United States. *Environmental Health Perspectives* 109: 185–189.

McMichael, A.J., D. Campbell-Lendrum, S. Kovats, S. Edwards, P. Wilkinson, T. Wilson, R. Nicholls, et al. 2004. Global climate change. In *Comparative quantification of health risks: Global and regional burden of disease attributable to selected major risk factors*, vol. 1, ed. M. Ezzati, A.D. Lopez, A. Rodgers, and C.J.L. Murray, 1543–1649. Geneva: World Health Organization.

McMichael, A.J., R.E. Woodruff, and S. Hales. 2006. Climate change and human health: Present and future risks. *Lancet* 367: 859–869.

Meleigy, M. 2007. The quest to be free of malaria. *Bulletin of the World Health Organization* 85(7): 501–568. http://www.who.int/bulletin/volumes/85/7/07-020707/en/

Miraglia, M., H.J.P. Marvin, G.A. Kleter, P. Battilani, C. Brera, E. Coni, F. Cubadda, et al. 2009. Climate change and food safety: An emerging issue with special focus on Europe. *Food and Chemical Toxicology* 47: 1009–1021.

Mouchet, J., and S. Manguin. 1999. Global warming and malaria expansion. *Annales de la Société Entomologique de France* 35: 549–555.

Radhi, H. 2009. Evaluating the potential impact of global warming on the UAE residential buildings—A contribution to reduce the CO_2 emissions. *Building and Environment* 32(6): 2451–2462.

Raouf, M.A. 2008. Climate change threats, opportunities, and the GCC countries. *The Middle East Institute Policy Brief No. 12*, April.

Reiter, P. 2001. Climate change and mosquito-borne disease. *Environmental Health Perspectives* 109: 141–161.

Rogers, D.J., and S.E. Randolph. 2000. The global spread of malaria in a future, warmer world. *Science* 289(5485): 1763–1766.

Trenberth, K.E., P.D. Jones, P. Ambenje, R. Bojariu, D. Easterling, A. Klein Tank, D. Parker, et al. 2007. Observations: Surface and atmospheric climate change. In *Climate change 2007:*

The physical science basis. Contribution of working group I to the fourth assessment report of the intergovernmental panel on climate change, ed. S. Solomon, D. Qin, M. Manning, Z. Chen, M. Marquis, K.B. Averyt, M. Tignor, and H.L. Miller, 235–336. Cambridge, U.K.: Cambridge University Press.

U.S. Environmental Protection Agency (EPA). 2009. Assessment of the impacts of global change on regional U.S. air quality: A synthesis of climate change impacts on ground-level ozone (An interim report of the U.S. EPA Global Change Research Program). EPA/600/R-07/094 F. http://cfpub.epa.gov/ncea/cfm/recordisplay.cfm?deid=203459

UAE Ministry of Economy. 2007. UAE in numbers. http://www.economy.ae/English/EconomicAndStatisticReports/StatisticReports/Pages/UAEinNumbers.aspx

UAE Ministry of Energy. 2006. Initial national communication to the United Nations Framework Convention on Climate Change. http://unfccc.int/resource/docs/natc/arenc1.pdf

UAE Ministry of Health. 2003. *Integrated vector management for malaria and other vector-borne diseases in the UAE*. Central Malaria Control Department. Workshop in Khartoum, Sudan, January 21–23. http://www.emro.who.int/RBM/CountryReports/UAE/Vector%20managment%20for%20Malaria%20and%20other%20Vector1.doc

UAE Ministry of Health. 2007. *Annual report 2006*. Preventative Medicine Sector, November.

Urge-Vorsatz, D., and A. Novikova. 2008. Potentials and costs of carbon dioxide mitigation in the world's buildings. *Energy Policy* 36: 642–661.

Weekly Epidemiological Record. 2007. United Arab Emirates declared malaria-free. January 26, 4: 30–32. http://www.emro.who.int/rbm/PDF/wer8204.pdf

World Bank. 2009. World Bank list of economies. http://go.worldbank.org/D7SN0B8YU0

World Health Organization (WHO). 2002. Global estimates of burden of disease caused by the environment and occupational risks. http://www.who.int/quantifying_ehimpacts/global/globclimate/en/index.html

World Health Organization (WHO). 2009. Mortality and burden of disease estimates for WHO Member State in 2004, Table 4. http://www.who.int/healthinfo/global_burden_disease/estimates_2000_2002/en/index.html

World Resources Institute. 2009. Climate analysis indicators tool. Version 6.0.

Zhang, Y., P. Bi, and J.E. Hiller. 2007. Climate change and disability-adjusted life years. *Journal of Environmental Health* 70(3): 32–36.

Chapter 8
Burden of Disease from Drinking Water Contamination

Abstract The Gulf Coast countries, including the United Arab Emirates (UAE), have the lowest supplies of fresh water per capita in the world. The groundwater extraction rate has become unsustainable, and desalinated water has become the main source of drinking water, either through piping (tap water) or as bottled water in the UAE. The product water from desalination is generally of high quality but may contain some contaminants, including microbial contaminants, chemicals that may cause acute illness, chemicals that may cause cancer, and radiological contaminants. Chemical disinfectants destroy microbes and prevent their growth in water, but they also produce unwanted chemical by-products that could affect health, including by causing cancer. Water traveling from the desalination plant can be recontaminated within the distribution system via infiltration, corrosion, and bacteria associated with biofilms. Past intermittent service and concerns about having sufficient water in the event a major desalination plant is taken offline due to a technical issue, oil spill, or hostile act have led to the widespread use of rooftop and in-ground storage tanks. These vented tanks are often outdoors and are subject to high temperatures, intense sunlight, animal activity, and windborne contaminants. It is unclear how well these tanks are cleaned and maintained. Our model focuses on the health effects of microbial contamination and disinfection by-products. Drinking water quality data at the point of use for the UAE could not be found, which made it necessary to rely on data from the published literature on samples from Kuwait. Comparisons of water quality data from water treatment plants and distribution networks in Kuwait and Abu Dhabi suggest that Kuwaiti data on water quality at the consumer's tap are a reasonable proxy for Abu Dhabi tap water quality. Data collected by the Abu Dhabi Distribution Company and provided for this study for comparison to Kuwaiti data included 471 samples from 79 stations at endpoints in the distribution system sampled throughout 2008. Based on this assessment, the burden of disease attributable to drinking water in the UAE appears to be small, with a mean estimate of 12 deaths from all causes. This study estimates 340 health-care facility visits each year due to drinking-water-related cancer and 46,000 visits due to gastroenteritis.

J. MacDonald Gibson et al., *Environmental Burden of Disease Assessment*, Environmental Science and Technology Library 24, DOI 10.1007/978-94-007-5925-1_8,

Keywords Drinking water contamination • Environmental burden of disease • Relative risk • Attributable fraction • Premature deaths and health-care facility visits • United Arab Emirates • Brominated disinfection by-products • Thermal and reverse-osmosis desalination • Desalinated water remineralization • Contaminant concentrations • Abu Dhabi emirate

Overview: Nature and Sources of Drinking Water Contamination

Access to clean drinking water is essential for health. The Arabian Gulf Coast countries, including the United Arab Emirates (UAE), have the lowest supplies of fresh water per capita in the world (World Bank 2005). In the past, most drinking water in the UAE came from groundwater and a few surface water sources, both natural and constructed. Recently, however, the groundwater extraction rate has become unsustainable (Brook 2005; Ali Murad et al. 2007; Canada Agri-Food 2008), and desalinated water has become the main source of drinking water, either through piping (tap water) or as bottled water (UAE Ministry of Environment and Water 2004; Brook 2005; Ali Murad et al. 2007). Desalinated water meets 95% of the domestic water use demand in Abu Dhabi emirate (Environment Agency–Abu Dhabi (EAD) 2005), which is likely representative of Dubai as well. Information concerning the production of desalinated and groundwater for the other emirates shows that desalinated water is, by far, the prevalent source of potable water, although groundwater still serves some rural areas for domestic water use (UAE Ministry of Energy 2006).

The UAE has at least 35 desalination plants with a total production of 700 million cubic meters per year (Ali Murad et al. 2007). About 96% of desalinated water is produced by evaporation methods, including multistage-flash and multieffect distillation processes, which use waste heat from electrical production. The remaining amount is produced by reverse osmosis (Brook 2005). Recent technological advances in energy recovery have made reverse osmosis a less expensive option, and it is likely that production by this method will be responsible for a greater fraction of overall desalinated water production in the future.

Drinking water contaminants include microbial contaminants, chemicals that may cause acute illness, chemicals that may cause cancer, and radiological contaminants. The UAE has established water quality guidelines for drinking water that are comparable to international guidelines. A World Health Organization (WHO) estimate indicates that risks associated with water, sanitation, and hygiene in the UAE are comparable to those in other industrialized nations. Despite this, uncertainties remain about the quality of piped water at the point of use, the quality of bottled water, the size of the population still receiving its drinking water from sources other than desalination and bottled water, and the extent to which water quality guidelines are enforced.

Contaminants can change concentrations in many ways along the process of water production and delivery (Fig. 8.1). Estimating contaminant concentrations in drinking

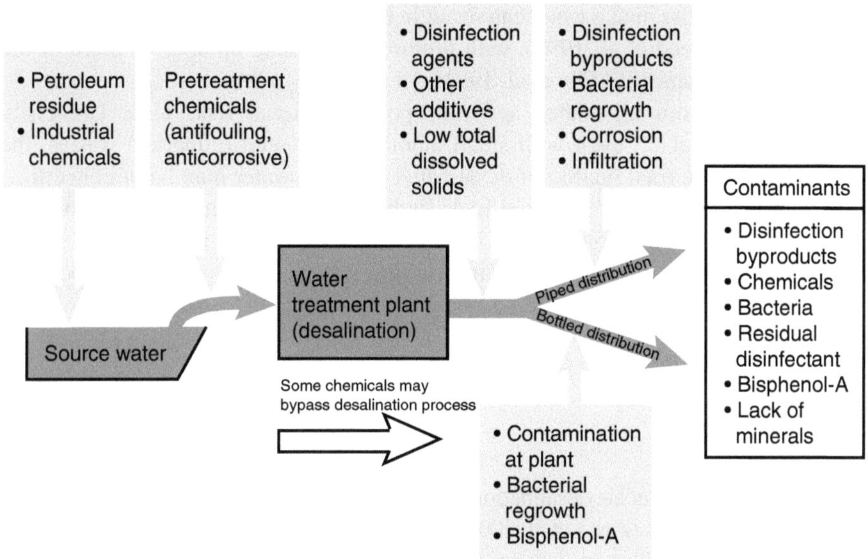

Fig. 8.1 Possible sources of contaminants in drinking water

water involves accounting for these changes through models of the relevant biological, chemical, and physical processes. UAE-specific data are not always available to make accurate estimates of the parameters in the models and to validate the analysis. The use of regional or international data introduces uncertainty into the calculations of contaminant levels.

The product water from desalination is generally of high quality but may contain some contaminants. A larger threat is posed by contaminants that enter drinking water throughout the distribution system. This section discusses the most likely possible sources of contaminant introduction along the path from the water source to the point of use (see Fig. 8.1).

Source Water

Desalination plants in the UAE primarily draw their feed water from the Arabian Gulf and the Gulf of Oman. The surrounding area contains more than 67% of the world's oil reserves. Oil-related activities, the high rate of development, and the concentration of the population along the coast contribute to the pollution of these waters (Elshorbagy 2005; Elshorbagy and Elkaheem 2008; Saeed et al. 1999). Chlorination is used at the intake of desalination plants as a biocide to prevent fouling of the intake water. The availability of organic matter and the presence of bromide ions in the intake water can result in high levels of brominated disinfection by-products (Saeed et al. 1999), with bromoform making up 95% or more of the total trihalomethanes (El Din et al. 1991; Ali and Riley 1989).

Chemicals from oil-related activities pose additional risk. UAE University (2008) found that "...even with small amounts of treated refineries' wastes, the impacts upon the final quality of desalinated drinking water may be of concern.... The removal capacity of potential hydrocarbons by the thermal desalination processes, the most common in the UAE, is either limited or unidentified." Some hydrocarbons pose a potential risk in drinking water because of their association with cancer (WHO 2008).

Desalination

Although reverse-osmosis desalination processes are highly effective at removing disinfection by-products (Agus et al. 2009), significant levels have been found in the distillate of thermal distillation systems, like those that produce most of the potable water in the UAE (Elshorbagy and Abdulkarim 2006; Ali and Riley 1989, 1990). Research also suggests that hydrocarbon precursors to disinfection by-product formation evaporate in thermal desalination plants and condense in the early stages of multistage flash distillation units, resulting in disinfection by-product formation upon chlorination of the finished water (Elshorbagy and Abdulkarim 2006; Saeed et al. 1999).

Reverse-osmosis systems can allow some ions and dissolved organic molecules to pass through the membrane into the permeate water. Boron is one constituent of concern that can be present in the permeate water at levels above WHO guidelines. Bromide can also pass into the permeate water. This process is facilitated by a reaction between bromide in sea water and chlorination used for antifouling. Greater removal of these constituents can be achieved with a second reverse-osmosis stage, or these constituents can be reduced to more acceptable levels by blending with thermally desalinated water (National Research Council 2008).

Treatment After Desalination

Prior to distribution, thermally desalinated water is treated to restore some mineral content and to provide a disinfection residual. In the UAE, remineralization is accomplished by blending the desalinated water with crushed limestone, replacing an earlier practice of mixing the treated water with seawater. Concerns about the formation of bromate and re-contamination of the treated water caused the UAE to abandon its previous practice of blending desalinated water with a small amount of untreated seawater (Saeed et al. 1999; Ali and Riley 1990; Elshorbagy and Abdulkarim 2006).

Chlorination is usually needed to guard against microbial contamination and growth in the distribution system. In 2006 some desalination plants in the UAE had unacceptably high levels of bromate because the chlorine used to provide a disinfection residual in the finished water was derived from seawater (Krane 2005). This problem was resolved by switching to alternate sources of chlorine. The added chlorine reacts with the disinfection by-product precursors that passed through the desalination plant or were added during remineralization to form disinfection by-products.

Piped Distribution System

Water traveling from the desalination plant can be recontaminated within the distribution system via infiltration, corrosion, and bacteria associated with biofilms. Infiltration can occur when interruptions in water service or transient low-pressure episodes allow water surrounding the pipe to enter the distribution system through breaches in the network. This infiltrate can contain contaminants from the surface, contaminants present in the soil, or contaminants from nearby wastewater pipe.

Distribution network breaches are unavoidable; even well-maintained systems have physical leakage rates of approximately 10%. Leakage is a result of the failure of pipes or joints due to corrosion or physical stress. It is not generally considered a problem for water quality since it primarily results in a loss of clean water flowing from the system into the environment surrounding the pipe (National Research Council 2006). Preserving water quality, however, depends on constant pressure in the system to ensure that water always flows out of the distribution system. Loss of supply pressure, transient pressure events from sudden valve closings, and pumps withdrawing water directly from the distribution system can all reverse the flow and cause infiltration.

Physical leakage in the system controlled by the Abu Dhabi Distribution Company is reported as being in the range of 16–18% (EAD 2009), and anecdotal reports of leakage rates throughout the UAE range as high as 40%. Great uncertainty may be associated with these estimates. Leakage rates are difficult to estimate because it is difficult to separate leakage from other sources of unaccounted-for water such as unmetered service, inaccurate meters, and unregistered users.

Anecdotal information also points to intermittent service in at least some parts of the UAE, although no authoritative information has been provided. Intermittent service has historically been a result of inadequate desalinated water supply in the face of high rates of population growth and declining groundwater supplies. Abu Dhabi generally has a high rate of water supply service with the possible exception of some areas near Al Ain where water is transported from the Fujairah desalination plant. Another plant is being built in Fujairah to address this problem. Water supply reliability information for the other six emirates was not available for the analysis in this book.

Past intermittent service and concerns about having sufficient water in the event a major desalination plant is taken offline due to a technical issue, oil spill, or hostile act have led to the widespread use of rooftop and in-ground storage tanks. These vented tanks are often outdoors and are subject to high temperatures, intense sunlight, animal activity, and windborne contaminants. Some of these tanks may not be adequately cleaned and maintained, and many people do not trust the water from their taps because of uncertainty about the water quality in these storage tanks.

Summer temperatures can be hot enough that the growth of *Legionella pneumophila* is a concern. *Legionella* can grow in water between 30 and 50°C, temperatures that are typical of storage tanks in the summer. Because the tanks are vented, the disinfection residual can volatilize and leave the water less protected against microbes in the tank and in the building's interior plumbing. Dead spots inside the tank where water does not routinely circulate and noncontinuous use of a residence increase these concerns.

Ali and Riley (1990) showed that water stored in rooftop storage units in Kuwait had lower-than-expected levels of disinfection by-products. This was attributed to volatilization enhanced by a higher surface-area-to-volume ratio, smaller overall capacity, and photodegradation from sunlight exposure. Underground storage and larger municipal storage facilities had higher disinfection by-product levels.

Ziadat (2005) found higher levels of heavy metals in water from rooftop tanks in Jordan. The presence of these metals was likely due to corrosion of the metal storage tanks.

Bottled Water

A recent study of bottled-water quality in the UAE found that most bottled water is almost free of trace ions and minor constituents; however, this study also pointed out the high variability of water sources and the large number of companies producing and distributing drinking water (Rizk 2009). The author tested 17 brands of bottled water and 14 samples of desalinated water and found total dissolved solid levels from 109 to 230 mg/L and 94 to 169 mg/L, respectively. The Abu Dhabi Food Control Authority (ADFCA) tested hundreds of bottled-water samples for total dissolved solids and bromate and found the majority of the samples to have sufficient minerals and bromate levels below WHO guidelines (ADFCA 2007–2008). Samples in violations of the bromate standards primarily occurred in 2006 and 2007 and were likely the result of using seawater-derived chlorine after desalination. The lack of bromate violations in the 2008 data reflects the fact that this method of chlorination is no longer practiced.

Microbial contamination is another potential risk posed by bottled water, as with other water sources. One study conducted on 80 commercial bottled-water samples from four different companies in the UAE showed that 75% of 20-L bottles were contaminated with ten different species of bacteria. This study did not calculate the health risk of this contamination, nor did it determine the source of the contamination (Nsanze et al. 1999). These bacteria are likely representative of heterotrophic bacteria, a type of bacteria commonly found in drinking water systems, which are measured using heterotrophic plate counts. A workgroup convened by the WHO concluded that heterotrophic bacteria were not associated with adverse health effects (Bartram et al. 2003).

Al Mudhaf et al. (2009) analyzed 113 samples from 71 brands of bottled water available in Kuwait and produced in several countries, including the UAE, for volatile organic compounds (VOCs) and semivolatiles, including pesticides. At least one VOC was found in 93% of the bottled-water samples, but these were primarily trihalomethanes, compounds associated with disinfection by-products. Styrene was found as a major pollutant in brands produced in the UAE and Saudi Arabia and packaged in 200 and 250 mL polystyrene containers. As storage time increased, so did the levels of styrene, toluene, ethyl benzene, and xylenes, which indicates that those VOCs were migrating from the container. Changes in storage temperature had no detectable effect. No phthalates or pesticides were found in the 25 bottled-water brands that were analyzed for extractable semivolatiles.

Key Health Effects of Drinking Water Contamination

In general, drinking water hazards include microbial contamination, which can lead to diarrheal and other diseases; compounds that may cause acute toxicity; compounds that may cause cancer; and radiological contamination. Drinking water may also be a source of essential minerals that may be removed as a result of some water production processes.

Effects of Microbial Contamination

Particularly in developing nations, poor water quality and sanitation take a heavy toll on public health, particularly on the health of children. This is due mostly to microbial contamination, which can be common in drinking water and is the focus of many water quality guidelines and standards (Barrell et al. 2000). Lack of safe drinking water contributes to a variety of intestinal infections that can cause malnutrition and anemia in children (Stephenson et al. 2000). Chronic diarrheal disease can also exacerbate malnutrition. Early childhood malnutrition, anemia, and associated diarrheal disease can cause permanent adverse effects on brain development (Pollitt 1995) and cognitive ability (Dillingham and Guerrant 2004).

Microbial contamination in the drinking water of developed countries usually is the result of contaminated wells from which water is used without treatment or the result of recontamination after treatment during water distribution. Water in distribution systems is subject to bacterial recontamination from infiltration and the presence of biofilms that can provide protection for bacteria from chlorine disinfection and may be responsible for bacterial regrowth (LeChevallier et al. 1988).

The bacteria responsible for Legionnaire's disease, *Legionella pneumophila*, grow well in very warm water (30–50 °C) and are frequently recovered from residential hot-water tanks. The bacteria are embedded in a biofilm that lines the inner walls of the plumbing system's pipes and containers. Infection occurs via inhalation of contaminated water droplets, not by drinking the water. Community-acquired *Legionella pneumophila* is becoming very common (Pedro-Botet et al. 2002), although most cases occur in immunocompromised individuals.

Effects of Disinfection Agents and Disinfection By-products

To prevent microbial contamination of drinking water, water intended for drinking is usually disinfected through a variety of methods, including chlorination, bromination, and ozone treatment. Free chlorine is the most commonly used disinfectant, with a target residual concentration in the range of 0.2–1 mg/L. Although high levels of chlorination could cause respiratory irritation, exposure to free chlorine at levels typically found in water treatment causes no specific adverse health effects. The WHO guideline value of 5 mg/L is well above the taste and odor threshold for most consumers. Chloramines (a mixture of monochloramine, dichloramine, and trichloramine formed when ammonia is present in chlorinated water) are also commonly used disinfectants.

Chemical disinfectants are intended to react with organic matter and thereby destroy microbes and prevent their growth in water. These reactions also produce unwanted organic or inorganic by-products that could affect health, including by causing cancer (Morris et al. 1992; Cantor 1997; Villanueva et al. 2003). The first recognized disinfection by-products were trihalomethanes, which are produced when free chlorine reacts with natural organic matter. When bromide is present in the source water, brominated by-products may be formed along with chlorinated ones. Although trichloromethane (chloroform) is the most prevalent by-product in drinking water produced from fresh water sources, brominated by-products (dibromochloromethane, bromodichloromethane, and tribromomethane (bromoform)) are considered a more serious cancer risk (Richardson et al. 2003; Plewa et al. 2002; Ali and Riley 1990). Other suspected chlorination by-products include haloacetic acids, halogenated ketones, and haloacetonitriles. WHO has set guideline values for 14 by-products (WHO 2006; UNICEF 2008).

Ozone treatment of bromide-containing water can result in the production of bromate, which is thought to increase the risk of renal cancer (Weinberg et al. 2003; Kurokawa et al. 1990). Bromate also can be created during the production of chemicals used for the chlorination of drinking water, especially when such chemicals are produced by electrolyzing seawater (Belluati et al. 2007). Water treated with either desalination or ozonation processes requires the addition of chemical disinfectant to maintain water quality during distribution. When marine water is the source for desalination, it is convenient and economical to produce the disinfectant on-site from seawater, but this can result in unacceptably high levels of bromate, although there is some debate about what concentration of bromate is needed to trigger health effects (Krane 2005).

Effects of Chemical Contamination

WHO lists guideline values for nearly 200 chemicals, ranging from naturally occurring arsenic and fluoride to synthetic chemicals found only in industrial settings (World Health Organization 2006; UNICEF 2008). Exposures to high levels of arsenic during early childhood can have significant neurobiological effects (Tsai et al. 2003; Wasserman et al. 2004; UNICEF 2008). High levels of manganese in water can also cause neurological effects (Wasserman et al. 2006). Industrial chemicals and pesticides have been linked to cancers, birth defects, nervous disorders, and respiratory problems. These can enter the water supply from unlined dump sites, illegal dump sites, and leaking storage facilities.

Boron is found in marine water at much higher concentrations than typically found in fresh surface water used as a drinking water source. Although little data exist on direct human exposure to boron, animal models indicate that at high levels it may have adverse effects on fetal development and on reproductive systems (Institute of Medicine 2000). The current WHO guideline for boron is 0.5 mg/L, but this may increase to 1 mg/L. This is still below the no-observed-adverse-effect level seen in animal models (Price et al. 1997), but this level makes an allowance for other sources of boron intake.

Effects of Demineralization

Along with removing contaminants, the desalination process often removes minerals that may be important for health. A 2003 meeting organized by WHO (2005) addressed the question of whether mineral composition of drinking water represents a significant part of overall dietary intake. The meeting concluded that magnesium and calcium were most likely significant contributors to good health in some populations and that high concentrations of these minerals in water may possibly reduce the incidence of ischemic cardiovascular disease. Other concerns, such as the effect of reduced calcium on bone growth, were not addressed due to a lack of data.

Demineralized water is corrosive and can promote the leaching of metals from distribution systems, which could introduce lead, copper, and other undesirable contaminants into drinking water. It is routine, however, to reintroduce minerals to control corrosion and to enhance taste. As mentioned previously, in the UAE crushed limestone is mixed with desalinated water to restore mineral content.

Desalinated water is often the source for bottled water, and this, too, is often remineralized prior to bottling. Although most bottled-water labels reflect a level and profile of minerals that are consistent with "natural" groundwater, it is unclear whether the labels are consistently accurate. The adverse health effects from drinking demineralized water were seen primarily in places where minerals were not reintroduced. Given the common practice of remineralizing desalinated water for piped and bottled distribution, such health effects should be less likely to occur in the UAE.

Effects of Chemicals from Plastic Bottles

The storage and distribution of drinking water in plastic bottles may also pose a health risk. For example, bisphenol-A (BPA) in polycarbonate plastic bottles has been shown to migrate into the contained water, and this contaminant can pose health risks, such as developmental problems among infants and children. Although it banned baby bottles made with plastics containing BPA in 2008, Health Canada has declared that the risk of BPA is low (Aglukkaq 2008). Polystyrene containers can leach styrene, a chemical that can compromise liver, kidney, and blood cell functions. The WHO guideline for styrene is no more than 20 μg/L, while the guideline in the United States is no more than 100 μg/L. Polyethylene terephthalate (PET) is another common plastic used for bottling water, and there has been some public and scientific concern that this plastic could contribute phthalates, thought to be estrogenic, to the drinking water (Wagner and Oehlmann 2009). Several studies on chemicals in bottled water did not find a significant release of phthalates from PET containers (Al Mudhaf et al. 2009; Signorile et al. 2007).

Although not all plastic bottles contain BPA, styrene, or phthalates, there is concern that other chemical constituents of plastic can leach from bottles into drinking water (Heath et al. 2007). This leaching may be enhanced by higher temperatures and direct sunlight exposure during long periods of storage.

Effects of Radiological Contamination

Drinking water also can be contaminated by radioactivity, including, for example, radon (Hopke et al. 2000). The contribution of drinking water to overall radioactive exposure is very small (typically less than 5%) and is principally due to the presence of naturally occurring elements in the uranium and thorium decay series (UNICEF 2008). Most of the cancer risk from radon in water arises from the transfer of radon into indoor air and subsequent exposure via inhalation (U.S. Environmental Protection Agency EPA 2006). Groundwater typically contains more radioactivity, such as from radon (National Research Council 1999), than does surface water. Radon is effectively removed during desalination.

Method for Estimating the Burden of Disease from Drinking Water Contamination

Contaminants Included in Model

As discussed previously, many contaminants may be present in drinking water, and several are suspected to affect drinking water in the UAE specifically. The model developed for the project described in this book focuses on contaminants of primary concern as determined by prior evidence of their presence in UAE drinking water and their potential to cause adverse health effects. Specifically, the model considers the health effects of microbial contamination and disinfection by-products.

Chemical contamination, whether anthropogenic or of natural origin, is usually associated with ground and surface water and so is not likely to be present in the thermally desalinated water that is the primary source of drinking water in the UAE. Chemical contamination from marine-source water should be largely removed during the desalination process. Boron is a concern in reverse-osmosis desalination, but these plants provide only a small fraction of the drinking water and the reverse-osmosis-product water is usually blended with thermally desalinated water to reduce the boron concentration and increase mineral content. Radiological contamination is associated with groundwater and is therefore not considered a major health risk associated with drinking water in the UAE.

The effects of drinking demineralized water are not modeled because they are clearly present only when the population is both drinking water with a low mineral content and consuming an inadequate diet (Donohue et al. 2005). Furthermore, most desalinated water in the UAE is remineralized prior to bottling or entering the distribution system, and most people in the UAE have sufficient intake of minerals from dietary sources to make the overall effect of demineralization small.

Modeling Method

The estimate for the burden of disease associated with contaminants in drinking water requires knowledge of the levels of contaminants in drinking water, the amount of water consumed, and the relationship between contaminants and diseases of concern. Information on these factors is uncertain under the best of circumstances, but the lack of site-specific drinking water data in the UAE increases the uncertainty associated with estimating the burden of disease. Models can be used to reduce this uncertainty, but they rely on data for calibration and verification and are a poor substitute for a comprehensive dataset on water quality.

The burden of disease is calculated by using exposure data for each contaminant to estimate the fraction of the population at various exposure levels and the associated relative risk for the associated diseases. Chapter 3 provides detailed information on how to calculate the attributable fraction using the exposure and relative-risk calculations, as seen in Eq. 3.6. The result can then be used to estimate the attributable disease burden using Eq. 3.7 (Prüss-Üstün et al. 2003).

Contaminant Concentrations

The burden of disease calculation relies primarily on measured data for estimating drinking water contaminants. A more complex distribution system model, while possible, would require a large investment of resources in order to obtain distribution system information and treatment data and in order to create and validate the model. Given that the goal of this assessment was to compare drinking water risks with risks of pollutants from other exposure routes, constructing a complex drinking water model was beyond the scope of the project.

Drinking water quality data at the point of use for the UAE could not be found, which made it necessary to rely on data from the published literature on samples from Kuwait. Household drinking water in Kuwait, where water also is obtained through the desalination of seawater, was comprehensively tested (Al Mudhaf et al. 2009). More than 600 drinking water samples were tested for VOCs, including the primary disinfection by-products. Mean and variance information were given for each constituent. These data were compared to water quality data in the UAE, where samples were taken at a desalination plant and in the distribution system. Data collected by the Abu Dhabi Distribution Company and provided for this study included 471 samples from 79 stations at endpoints in the distribution system sampled throughout 2008. Data on disinfection by-products at nine sampling locations in part of Abu Dhabi's distribution system sampled throughout a 24-h period were presented by Elshorbagy et al. (2000). Elshorbagy and Abdulkarim (2006) reported on water quality throughout the desalination process, including the distillate before and after post-treatment additives. Ali and Riley (1989) published contaminant levels in the finished water of desalination plants in Kuwait and provided information on the distribution of individual disinfection by-products.

Levels of disinfection by-products were estimated from the reported values from the Kuwait drinking water study (Al Mudhaf et al. 2009). Samples from a UAE desalination plant and distribution system were used to qualitatively determine that it was reasonable to use Kuwaiti data given the lack of drinking water data from the UAE; however, no adjustments were made to the Kuwaiti data based on the UAE information since the Kuwaiti data were for samples at the point of use while the UAE data were not. Figure 8.2 shows a comparison of these data.

Bottled-water consumption is high in the UAE, and, assuming bottled water has negligible disinfection by-products, it is important to estimate the percentage of the population drinking bottled versus tap water in order to get an accurate description of disinfection by-product exposure. Lacking better data, we assumed, based on personal communications, that the fraction of UAE citizens who use bottled water for drinking falls between 70 and 100% while the use of bottled water for drinking among noncitizens falls between 20 and 60%.

UAE data on bacteria concentrations in drinking water delivered through a distribution system at the point of use have not been identified. Unpublished data from the Abu Dhabi Distribution Company show no positive bacteria results in 2,800 samples taken at 78 locations in the distribution system. It is unclear whether this is typical of all distribution systems, some of which have regular episodes of low pressure that increase the chance of infiltration. These data also do not include possible changes to water quality when it is stored in in-ground and rooftop storage tanks and distributed through internal plumbing to buildings.

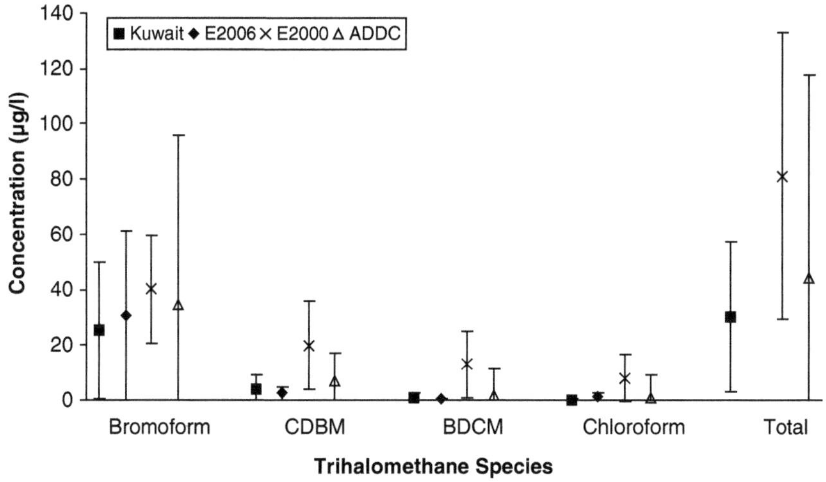

Identifier	Source, Description of Data Used	What is Plotted
Kuwait	Al Mudhaf, Alsharifi, and Abu Shady (2009): Summary data of 623 samples collected from private residences and government buildings at 99 locations in 69 neighborhoods of the six Governorates of the State of Kuwait.	The reported mean and 95th percentile values based on the reported standard deviation.
E2006	Elshorbagy and Abdulkarim (2006): Summary data of samples at the influent to drinking water storage and two outlets of drinking water storage from a Ruwais desalination plant sampled seven times in 2003.	The midpoint of the range and the max and min values. Total trihalomethanes were not reported.
E2000	Elshorbagy, Abu Qdais, and Elsheamy (2000): Six samples at each of seven sampling locations in Abu Dhabi's distribution system during a 24 hour period.	Mean and 95th percentile interval calculated from the standard deviation of the 42 reported values.
ADDC	Abu Dhabi Distribution Company (2008, unpublished): 471 samples from 79 stations at endpoints in the distribution system sampled throughout 2008	Mean and 95th percentile interval calculated from the 471 samples.

Fig. 8.2 Plot of ranges of disinfection by-product levels for Kuwaiti and UAE data

Health Effect Prediction

Meta-analyses of studies of the effects of chlorinated drinking water and disinfection by-products have identified bladder and rectal cancers as the primary causes of concern. Of the 12 types of cancer reviewed by Morris et al. (1992), only these two had relative-risk values that were significantly above 1 at the $p=0.05$ level. Villanueva et al. (2003) also performed a meta-analysis using the odds ratio to assess the association between chlorinated drinking water and bladder cancer and confirmed the significant relationship found by Morris. Bladder and rectal cancer are the primary outcomes of interest for burden of disease measurements associated with disinfection by-products. Colon cancer was included in our analysis as well

Table 8.1 Relative risk associated with disinfection by-products

Site	Category	Group	Relative-risk estimate	95% confidence interval (CI)
Bladder	All studies		1.21	(1.09, 1.34)
	Gender	Female	1.17	(1.03, 1.34)
		Male	1.24	(0.97, 1.57)
		Combined	1.22	(1.08, 1.37)
Rectum	All studies		1.38	(1.01, 1.87)
	Gender	Female	1.10	(0.90, 1.36)
		Male	1.24	(0.86, 1.79)
		Combined	1.42	(0.99, 2.03)
Colon	All studies		1.11	(0.91, 1.37)
	Gender	Female	1.19	(0.93, 1.53)
		Male	1.09	(0.81, 1.48)
		Combined	1.02	(0.86, 1.22)

Morris et al. (1992)

because it is often included in statistics associated with rectal cancer and has an estimated relative risk above 1. Table 8.1 shows the relative risks of these cancers associated with drinking chlorinated water. These relative risks are calculated from many studies—some of which did not differentiate between men and women. Data from all studies were used in the calculation of the "All Studies" entries, but those without a male/female distinction were not used for the remaining three categories in each cancer site.

Villanueva et al. (2004) presented relative risks for bladder cancer from a pooled analysis of six case-control studies where the relative risk was presented for both men and women and for four levels of trihalomethane exposure in addition to the reference group. The relative risks are only significant for men, with the results for women having values near 1 and with confidence intervals broadly including the null result. Table 8.2 shows the data from the pooled analysis that was used for disease-burden estimation.

If data were available, disease burdens from microbial contamination could be estimated using dose response curves and infectious dose information. Bartram and others (2003) presented a table (reproduced and modified here as Table 8.3) of the infectious dose and typical frequency of occurrence of several organisms found in drinking water. Disease estimates rely on knowing how the number of organisms present in drinking water varies and, accounting for variations in water consumption, calculating the frequency at which individuals are exposed in excess of the infectious dose.

Research on human and animal subjects supports dose-response models for microbial disease; the probability of getting a disease is calculated from the dose (i.e., number of organisms) received during an exposure event. WHO (2001) compiled dose-response models for several organisms, and these models follow one of two forms. The simpler model is an exponential probability model (Eq. 8.1) where

Table 8.2 Association of exposure to average trihalomethanes with bladder cancer, by gender

	Men		Women		Both genders
Average trihalomethanes (µg/L)	Cases/ Controls	Odds ratio (95% CI)	Cases/ Controls	Odds ratio (95% CI)	Odds ratio (95% CI)
0	328/605	1.00	94/221	1.00	1.00
>0	1,798/2,909	1.32 (1.10, 1.59)	509/1,415	0.85 (0.60, 1.19)	1.18 (1.00, 1.39)
0–1	711/1,365	1.00	189/506	1.00	1.00
>1	1,415/2,149	1.24 (1.09, 1.41)	414/1,130	0.95 (0.76, 1.20)	1.18 (1.06, 1.32)
0–1	711/1,365	1.00	189/506	1.00	1.00
>1–5	366/574	1.10 (0.92, 1.31)	96/231	0.99 (0.72, 1.36)	1.08 (0.93, 1.26)
>5–25	314/499	1.26 (1.05, 1.51)	97/309	0.86 (0.63, 1.18)	1.15 (0.98, 1.35)
>25–50	399/647	1.25 (1.04, 1.50)	128/356	1.04 (0.76, 1.43)	1.22 (1.04, 1.42)
>50	336/429	1.44 (1.20, 1.73)	93/234	0.93 (0.67, 1.28)	1.31 (1.12, 1.54)
p value		<0.001		0.753	<0.001

Villaneuva et al. (2004)

Table 8.3 Infectious doses and frequency of isolation in drinking water

Bacteria	Infectious dose (number)	Frequency of isolation in drinking water (%)
Pseudomonas aeruginosa	108–109	<1–24
Aeromonas hydrophila	>1,010	1–27
Mycobacterium avium complex	104–107	<1–50
Xanthomonas maltophilia	106–109	<1–2
Moraxella spp.	Unknown	10–80
Legionella pneumophila	105	3–33
Acinetobacter spp.	106–108	5–38

Modified from Bartram et al. (2003), originally from Rusin et al. (1997)

the probability of disease (P_I) is a function of the dose (d). The rate constant (k) determines the shape of the relationship.

$$P_I = 1 - e^{-kd} \tag{8.1}$$

The second form is called a beta Poisson model and is shown mathematically in Eq. 8.2. Two parameters, α and N_{50}, determine the relationship between the dose and the probability of disease. Values for the parameters for these equations for several organisms are given in Table 8.4.

$$P_I = 1 - \left[1 + \frac{d}{N_{50}}\left(2^{1/\alpha} - 1\right)\right]^{-\alpha} \tag{8.2}$$

Table 8.4 Table of best fit dose-response parameters (human)

	Exponential	Beta Poisson	
Organism	k	N_{50}	α
Poliovirus I (Minor)	109.87		
Rotavirus		6.17	0.2531
Hepatitis A virus[a]	1.8229		
Adenovirus 4	2.397		
Echovirus 12	78.3		
Coxsackie[b]	69.1		
Salmonella[c]		23,600	0.3126
Salmonella typhosa		3.60×10^{6}	0.1086
Shigella[d]		1,120	0.2100
Escherichia coli[e]		8.60×10^{7}	0.1778
Campylobacter jejuni		896	0.145
Vibrio cholera		243	0.25
Entamoeba coli		341	0.1008
Cryptosporidium parvum	238		
Giardia lamblia	50.23		

Modified from WHO (2001)
[a]Dose in grams of feces (of excreting infected individuals)
[b]B4 and A21 strains pooled
[c]Multiple (non-typhoid) pathogenic strains (*S. pullorum* excluded)
[d]Flexnerii and dysenteriae pooled
[e]Nonenterohemorrhagic strains (except O111)

When population exposure information is known or can be estimated, the probability of disease can be integrated over the population to generate an expected number of disease cases.

Because data on microbial exposure were not available for this study, we estimated the burden of disease with a method used by WHO that begins with a general survey of the drinking water and wastewater infrastructure (Fewtrell et al. 2007; Prüss-Üstün et al. 2008). The information on infrastructure is used to estimate the fraction of the population in each of seven levels of water supply and sanitation infrastructure quality. The relative risk of disease associated with each of these levels has been estimated, and the relative risk and exposure information are combined to estimate the attributable fraction of disease. Estimates of improved water supply and sanitation infrastructure are provided in a related report (WHO and UNICEF 2006). Although this method was recently used to estimate the burden of disease in the UAE, it is repeated in this work using updated and more comprehensive local health data.

Baseline Health Outcome Data

Mortality and the number of health-care facility visits are the health outcome endpoints of interest in this work. Data for these outcomes for Abu Dhabi emirate for 2008 were provided by the Health Authority–Abu Dhabi (HAAD). The data include information on nationality and gender and were cleaned, coded, and validated by

Table 8.5 Baseline number of cancer deaths and health-care facility visits

		Deaths		Health-care facility visits	
Cancer		Citizen	Noncitizen	Citizen	Noncitizen
Bladder	Males	3.5	10	6	830
	Females	0	9	0	94
Colon	Males	17	16	32	1,600
	Females	22	25	28	580
Rectal	Males	7	7.5	13	500
	Females	3	12	28	98

Table 8.6 Baseline number of gastroenteritis deaths and health-care facility visits

	Deaths	Health-care facility visits
Citizen	0	29,800
Noncitizen	6	51,300

HAAD. The total number of deaths in Abu Dhabi emirate in 2008 was 2,949, but 286 records were excluded due to missing documentation of year of death. Estimates of baseline mortality incidence for the remaining six emirates were obtained by applying the rate in Abu Dhabi emirate to the population of the remaining emirates.

For cancer, health-care facility visit data are based on 2008 Abu Dhabi insurance records from HAAD covering 73% of Abu Dhabi emirate's population. The data are subdivided based on nationality and gender. The records include every visit to any medical facility (i.e., hospital, clinic, center, etc.). The total number of Abu Dhabi emirate health-care facility visits was estimated by linearly scaling the results to represent 100% of the population. Estimates for the remaining six emirates were obtained by applying the rate of visits in Abu Dhabi emirate to each remaining emirate. Table 8.5 shows the baseline health endpoint data along with the range of the fraction drinking chlorinated water used in the model. Health-care facility visit information for gastroenteritis was obtained from the same source but included data for the first half of 2009. These data were doubled to represent a full year and then treated similarly to the 2008 data to generate nationwide numbers. Table 8.6 shows the gastroenteritis baseline data used in this analysis.

Estimated Burden of Disease

Disinfection By-products

Table 8.7 presents the burden of disease for bladder, colon, and rectal cancer calculated based on the association determined by Morris et al. (1992), where the relative risks (Table 8.1) apply to those drinking chlorinated water, compared with the reference

Table 8.7 Attributable burden of cancers associated with drinking chlorinated water, using relative risks from Morris (mean and 95% confidence interval)

	Attributable fraction (%)		Attributable deaths		Attributable health-care facility visits	
	Citizen	Noncitizen	Citizen	Noncitizen	Citizen	Noncitizen
Bladder cancer						
Males	3.4 (0.1, 8.8)	13 (4.1, 22)	0.12 (0, 0.31)	1.27 (0.42, 2.2)	0.22 (0.01, 0.55)	100 (34, 180)
Females	2.5 (0.1, 5.8)	9.2 (4.0, 15)	0	0.85 (0.37, 1.41)	0	8.6 (3.8, 14)
Colon cancer						
Males	1.6 (0.0, 5.5)	6.2 (0.4, 15)	0.28 (0, 0.95)	0.99 (0.05, 2.4)	0.52 (0, 1.8)	95 (4.9, 240)
Females	2.8 (0.1, 7.5)	10 (2.5, 19)	0.61 (0.023, 1.7)	2.6 (0.63, 4.8)	0.79 (0.029, 2.1)	60 (15, 110)
Rectal cancer						
Males	3.5 (0.1, 9.9)	13 (2.5, 25)	0.24 (0, 0.66)	0.95 (0.20, 1.8)	0.45 (0.016, 1.3)	63 (13, 120)
Females	1.6 (0.0, 4.8)	6.0 (0.7, 13)	0.054 (0, 0.16)	0.74 (0.091, 1.6)	0.45 (0.013, 1.4)	5.9 (0.72, 13)

group of those drinking unchlorinated water. These values are very sensitive to the fraction of the population assumed to be drinking chlorinated water.

The meta-analysis of Villanueva et al. (2003) allows better exposure resolution than that from Morris et al. (1992) in that relative risks for various levels of trihalomethanes are estimated. However, the Villanueva study only includes bladder cancer and concludes there is no significant risk increase for women. Assuming the trihalomethane concentrations from the Kuwaiti study (Al Mudhaf et al. 2009) follow a normal curve with a mean and standard deviation as given in that study, one can calculate the attributable fraction of bladder cancer disease for men (Table 8.8). The burden of disease calculated using this method is very similar to that for bladder cancer in men calculated using the relative risks from Morris et al. (1992) (Table 8.7). Because the meta-analysis of Villenueva showed no significant relative risk for women, the attributable fraction for women is zero. To incorporate the results of both attributable fraction calculations, the bladder cancer estimates for both men and women are averaged in Table 8.9 and the subsequent discussion.

Bacterial Contamination

Table 8.10 presents calculations for the burden of diarrheal disease due to water, sanitation, and hygiene, following WHO methods (Fewtrell et al. 2007). The mean values for the fraction of the population at each scenario were obtained from the

Table 8.8 Attributable burden of male bladder cancer associated with drinking chlorinated water, using relative risks from Villanueva (mean and 95% confidence interval)

THM[a] (μg/L) exposure range	Fraction of UAE males in exposure range		Relative risk for exposure range	Attributable fraction (%)		Attributable deaths		Attributable health-care facility visits	
	Citizen	NC[b]		Citizen	NC	Citizen	NC	Citizen	NC
0–1	0.853	0.409	1	3.7	13.4	0.13	1.4	0.23	110
>1–5	0.002	0.010	1.10 (0.92, 1.31)	(0.18, 8.4)	(6.3, 21)	(0.0061, 0.30)	(0.62, 2.2)	(0.01, 0.54)	(51, 180)
>5–25	0.048	0.193	1.26 (1.05, 1.51)						
>25–50	0.086	0.345	1.25 (1.04, 1.50)						
>50	0.011	0.043	1.44 (1.20, 1.73)						

[a]Trihalomethane
[b]Noncitizen

Table 8.9 Attributable burden of cancers associated with drinking chlorinated water, using averaged Morris-Villanueva methods (mean and 95% confidence interval)

	Attributable deaths		Attributable health-care facility visits	
	Citizen	Noncitizen	Citizen	Noncitizen
Bladder cancer				
Males	0.13 (0.01, 0.28)	1.3 (0.7, 2.0)	0.23 (0.01, 0.51)	110 (57, 160)
Females	0 (0, 0.01)	0.43 (0.19, 0.70)	0 (0, 0)	4.3 (1.9, 7.1)
Colon cancer				
Males	0.28 (0, 0.95)	0.99 (0.05, 2.4)	0.52 (0, 1.8)	95 (4.9, 240)
Females	0.61 (0.023, 1.7)	2.6 (0.63, 4.8)	0.79 (0.029, 2.1)	60 (15, 110)
Rectal cancer				
Males	0.24 (0, 0.66)	0.95 (0.20, 1.8)	0.45 (0.016, 1.3)	63 (13, 120)
Females	0.054 (0, 0.16)	0.74 (0.091, 1.6)	0.45 (0.013, 1.4)	5.9 (0.72, 13)

WHO study, while the ranges around those estimates were generated to have the mean value as the central point and to include the extreme case of everyone in the population included in water delivery scenario II.

Table 8.10 Burden of diarrheal disease due to water, sanitation, and hygiene (mean and 95% confidence interval)

Water delivery scenario	Fraction of population	Relative risk	Attributable fraction (%)	Attributable deaths	Attributable health-care facility visits
II: Access to improved drinking water and sanitation	0.96–1	2.5	0.57 (0.19, 0.75)	3.8 (1.2, 5.0)	46,100 (14,700; 60,900)
Vb: Improved water supply but not improved sanitation	0–0.04	8.7			

This study estimates 1.2–5 deaths and 14,700–60,900 health-care facility visits each year to be associated with bacteria in drinking water. The low number of deaths is consistent with the large percentage of the population with improved water supply, but the number of health-care facility visits is surprisingly high. Gastrointestinal disease rates are difficult to estimate because people often feel they do not need to visit a doctor, and therefore cases go unreported, so the number of reported cases represents only a small fraction of the true number of cases. As a result, this large number of visits may indicate a much higher number of associated cases.

The WHO method (Fewtrell et al. 2007) assigns exposure classification scenario II, with a relative risk of 2.5, to developed countries and does not assign the ideal scenario I, with a relative risk of 1.0, to any country. Because of this, the lowest attainable attributable fraction is 60%. In the analysis, the relative risk was given a range of 1–4. The lower 95th percentile bound of the attributable fraction estimate in the analysis is about 20%, corresponding to 14,700 health-care facility visits. This lower bound would be a better burden of disease estimate if the relative risk were believed to be much closer to 1 than 2.5.

Comparison with Preliminary Estimate

In its preliminary analysis (in Appendix A) the RAND Corporation estimated 0–147 deaths per year from drinking water contaminants with a best estimate of "nonzero but low." The analysis here agrees with that assessment with a range of 7–18 deaths from all causes and a mean estimate of 12. The estimates for illness or injury are not directly comparable because the RAND health endpoint is different than the one used here (cases of illness versus the number of health-care facility visits). RAND estimated that "zero" long-term illnesses and "nonzero but low" short-term illnes-sess result from drinking water contamination—results that are more consistent with a number lower than the 340 health-care facility visits due to cancer and 46,000 visits due to gastroenteritis estimated in this analysis. The prediction for cancer may not be as inconsistent as it seems because a single case of cancer could generate many health-care facility visits.

The large number of cases of gastroenteritis cannot be easily reconciled with the RAND estimate, however, and this may indicate that a more comprehensive analysis is warranted, based on actual measured levels of bacteria in drinking water and accurate estimates of water consumption patterns. Also, as mentioned previously, the method used assumes developed countries have, at best, an exposure scenario with a relative risk of 2.5. A relative risk closer to 1 could reduce the estimate of the number of attributable health-care facility visits by as much as two-thirds.

Sources of Uncertainty

Summing the lower and upper bounds of the 95% confidence intervals in Table 8.9 provides an estimate for the number of cancer deaths due to drinking chlorinated water of 2–17 per year, or about 8% of the deaths from these cancers reported in the UAE in 2008. The low numbers are in agreement with the assumptions that the water supply complies with modern contaminant standards. However, much uncertainty is associated with the inputs of the model, specifically the percentage of people drinking bottled water and the disinfection by-product concentrations. The estimate of the number of cancer-related health-care facility visits attributable to drinking chlorinated water is 90–660 visits. This reflects the much larger number of visits compared with deaths from cancer (340 versus 8) since there may be many health-care facility visits related to one case of cancer. It is also unclear whether the health-care facility visits include cases of screening and diagnostic procedures that do not result in a cancer diagnosis.

The number of people drinking bottled water is a key parameter as it determines the percentage of people exposed to higher levels of disinfection by-products. Tap-water consumption rates were set based on anecdotal information and could be estimated more accurately using bottled-water marketing data, bottled-water production data, and surveys. The citizen tap-water use rate is low enough that the true estimate could be several times that value.

Levels of disinfection by-products were obtained from Kuwaiti data. Data on drinking water quality at the point of delivery to the home are not available for the UAE, but there is reason to believe such data would be similar to the Kuwaiti data. Data obtained at the point of use, such as at the tap in the home, would be most useful. Internal plumbing conditions, intermediate storage, and usage patterns can affect water quality. Future modeling efforts would benefit if such data were to become available for the UAE.

The analysis of the effects of disinfection by-products assumes a long-term exposure to the levels used in the calculations. The UAE has grown dramatically over the past few decades, with a corresponding change from groundwater consumption to desalinated water consumption and, hence, a likely change in the quality of water being consumed. It is unlikely that many of the residents are drinking the same type of water they did a decade ago because either they lived elsewhere or the water source has changed. The attributable fraction estimates are more appropriately interpreted as the fraction of disease attributable to disinfection by-products should current conditions persist indefinitely.

Perhaps the largest uncertainty relates to the incidence and death statistics associated with the diseases covered. The majority of the residents are noncitizens who will not likely spend most of their lives in the UAE. Even if they are exposed to carcinogens while in the UAE, the resulting cancer cases will likely occur in another country. In this way, the effects of the exposure are not reflected in the country's mortality and morbidity statistics. Those who are getting cancer now are from a subgroup of the entire population who have lived in the UAE for an extended time and are old enough to begin getting these diseases in relatively large numbers. The true incidence rate should be adjusted to reflect this smaller baseline population.

Sensitivity Analysis

The sensitivity of model inputs is evaluated by varying their values and reporting the effect on the health outcomes. Inputs of interest are those for which uncertainty is not well quantified or those that can be affected by policy changes. Knowing which inputs have the greatest proportional effect on the health outcomes helps identify activities and policies that might be worthwhile to reduce model uncertainty or reduce the overall burden of disease.

In this work, inputs were increased by 10% and the percentage change in the health outcomes was recorded. (When inputs were reduced by 10%, similar results were obtained and have been omitted for simplicity.) Since both the number of health-care facility visits and the number of deaths attributable to drinking water contaminants are derived from the same attributable fraction calculation, the percentage change is the same regardless of which health outcome is chosen for the sensitivity analysis. When an input's ideal value is zero (e.g., a contaminant level), it is simply multiplied by 1.1 for the sensitivity runs. When the ideal value is 1 (e.g., a relative risk), the 10% change is in the difference of the value from 1. Table 8.11 reports the results of the sensitivity analysis.

The percentage of people drinking tap water (instead of bottled water) is important because it is almost directly related to the number of health outcomes. It is also a parameter for which there are no good estimates. Surprisingly, the levels of trihalomethanes as described by the mean and the standard deviation had very little effect. This is partially due to the large number of people assumed to be drinking bottled water, who are thus not affected by trihalomethane levels, and also the insensitivity of the relative-risk estimates to trihalomethane levels. Changing relative-risk values has the expected large corresponding effect on health outcomes. For microbial contamination there is only a small effect from exposure scenario classification

Table 8.11 Percent change in health-care outcomes when model parameters increase 10%

Parameter increased by 10%	Change in outcomes (%)
Cancer associated with trihalomethane levels	
Percentage of people drinking tap water	9.0
Mean trihalomethane level in drinking water	0.5
Standard deviation of trihalomethane levels in drinking water	0.1
Relative risk of cancers associated with trihalomethane level	9.0
Gastroenteritis associated with microbial contamination	
Fraction of the population not at Scenario II	0.3
Relative risk of gastroenteritis associated with microbial contamination	3.8

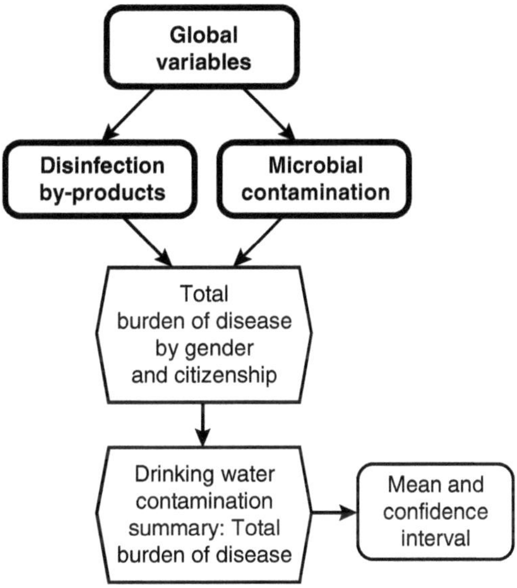

Fig. 8.3 Top-level diagram of ***Drinking Water*** module

and a larger effect from the relative-risk estimates. As previously discussed, the relative risk associated with the lowest exposure scenario may be higher than warranted and this confirms the importance of better understanding that risk.

Model Description

Top Layer

The top layer of the burden of disease model shows just the overall structure (Fig. 8.3). No inputs or calculations are performed (Table 8.12).

Table 8.12 Description of top-level influence diagram nodes

Node	Type	Description	Source
Global variables	Module	Contains data common to models for all subject areas, including population and baseline health data	N/A
Disinfection by-products	Module	Model for diseases associated with disinfection by-products	See Table 8.13
Microbial contamination	Module	Model for gastroenteritis cases associated with drinking water	See Table 8.17
Burden of disease by disease, gender, and citizenship	Objective	Burden of diseases associated with drinking water categorized by disease, gender, and citizenship	Calculated by model
Burden of disease by gender and citizenship	Objective	Burden of diseases associated with drinking water categorized by gender, and citizenship	Calculated by model
Overall burden of disease	Objective	Burden of disease summed over all diseases, genders, and citizenship	Calculated by model
Health endpoint	Index	Definition of endpoints included in the model	Determined by data availabilty from HAAD and Ministry of Health
Uniform distribution parameters	Index	Labels for uniform distribution parameters	N/A
Disease	Index	List of diseases included in this analysis	Diseases associated with drinking water in literature

Disinfection By-products Module

This module calculates the burden of disease for diseases associated with disinfection by-products in drinking water. Exposure information and relative-risk estimates (using two methods) for particular exposure levels are combined to calculate the fraction of diseases attributable to the exposure. This is applied to the baseline disease rates to estimate the burden of disease, and an average burden is calculated to arrive at a single result based on the two different relative-risk methods (Fig. 8.4 and Table 8.13).

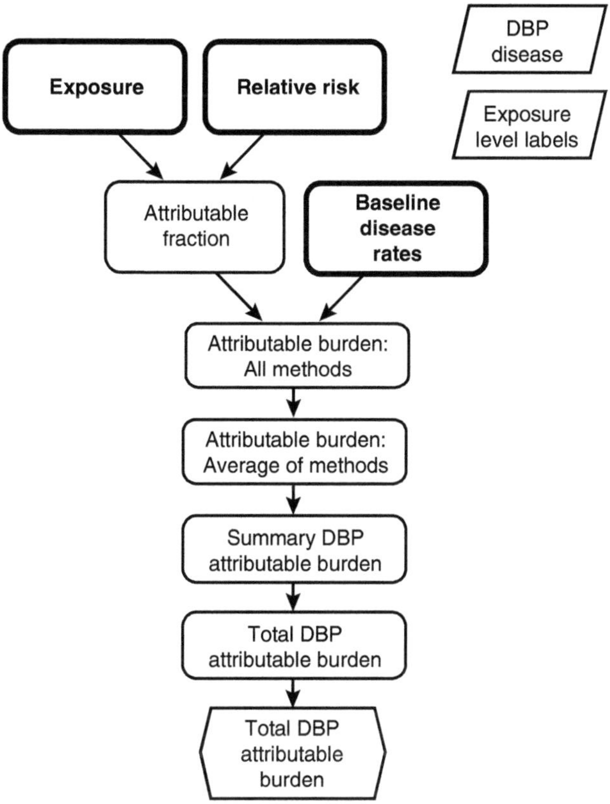

Fig. 8.4 ***Disinfection By-products*** module

Exposure Submodule

This submodule calculates the fraction of the population in each of five exposure categories ranging from no exposure to greater than 50 μg/L. Summary statistics from a drinking water sampling study in Kuwait are used to generate estimates of the mean and variation of disinfection by-products in the UAE. From this, the population that drinks chlorinated water is divided into exposure categories. The fraction of the population assumed not to be drinking chlorinated water is then added to come up with an overall exposure distribution (Fig. 8.5 and Table 8.14).

Relative Risk Submodule

This submodule generates a probabilistic relative-risk estimate for the diseases of interest at each exposure level based on literature estimates (Fig. 8.6 and Table 8.15).

Table 8.13 Description of ***Disinfection By-products*** module influence diagram nodes

Node	Type	Description	Source
Exposure	Module	Calculates the population exposure to trihalomethanes	See Table 8.14
Relative risks	Module	Contains the relative risk for cancers associated with chlorinated drinking water	See Table 8.15
Baseline disease rate	Module	Assembles estimates of mortality and health-care facility visits for cancers associated with chlorinated drinking water	See Table 8.16
Attributable fraction	Variable	Calculates the fraction of cancers attributable to chlorinated drinking water	Prüss-Üstün et al. (2003)
Attributable burden: All methods	Variable	Burden of disease associated with chlorinated drinking water categorized by disease, gender, citizenship, and methods used to estimate relative risks	Calculated by model
Attributable burden: Average of methods	Objective	Average burden of disease over all methods used to estimate relative risk	Calculated by model
DBP disease	Index	List of diseases associated with chlorinated drinking water	Villanueva et al. (2003), Morris et al. (1992)
Exposure level labels	Index	List of discrete levels of exposure corresponding to levels used in epidemiologic studies	Villanueva et al. (2003)

Baseline Disease Rates Submodule

This submodule assembles baseline disease rates into a data structure that is needed for the model (Fig. 8.7 and Table 8.16).

Microbial Contamination Module

This module calculates the burden of gastroenteritis from microbial contamination in drinking water. Exposure information and relative-risk estimates for particular exposure levels are combined to calculate the fraction of disease attributable to the exposure. This is applied to the baseline disease rates to estimate the burden of disease (Fig. 8.8 and Table 8.17).

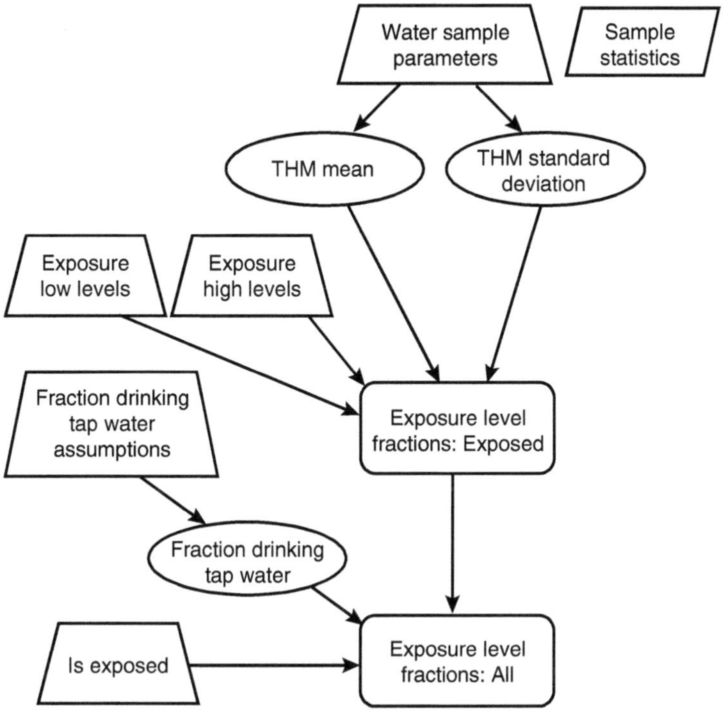

Fig. 8.5 ***Exposure*** submodule

Information Needed to Improve Future Burden of Disease Predictions

Most of the information gathered for this project came from Abu Dhabi, and the calculations reflect the assumption that the other six emirates are similar to Abu Dhabi. Given the different population, economic activity, and governance of the other six emirates, it would be of great benefit to have data concerning water conveyance infrastructure, usage patterns, water sources, and water quality from individual emirates.

No bacterial data applicable to the UAE have been found for this analysis. Measured bacteria levels from a comprehensive sampling program would allow for better estimates of the burden of gastrointestinal disease from drinking water. Because rooftop and other local storage methods might serve as entry points for contamination or places where bacteria can thrive, testing water that has traversed these systems and understanding how well these systems are maintained is important. Barring direct bacterial evidence, estimates might be strengthened with information about infrastructure, water residence times in distribution systems, and estimates of distribution system integrity.

Table 8.14 Description of ***Exposure*** submodule influence diagram nodes

Node	Type	Description	Source
Water sample parameters	Constant	Statistical summary information from Kuwait study of trihalomethane concentrations in drinking water	Al Mudhaf et al. (2009)
Trihalomethane mean	Chance	Mean concentration of trihalomethanes in drinking water	Al Mudhaf et al. (2009), variation estimated as standard error
Trihalomethane standard deviation	Chance	Standard deviation of the concentration of trihalomethanes in drinking water	Al Mudhaf et al. (2009), variation estimated from Chi-square distribution
Exposure low levels	Constant	Numerical values associated with the low end of the exposure ranges used to assign relative risks	Villanueva et al. (2003)
Exposure high levels	Constant	Numerical values associated with high end of the exposure ranges used to assign relative risks	Villanueva et al. (2003)
Exposure level fractions: exposed	Variable	Fraction of the population that drinks chlorinated water in each exposure category	Calculated by model
Fraction drinking tap water assumptions	Constant	Minimum and maximum percentages of citizens and noncitizens drinking chlorinated water	Anecdotal information
Fraction drinking tap water	Chance	Fraction of people assumed to be drinking tap water	Calculated by model
Is exposed	Constant	An array corresponding to exposure categories	N/A
Exposure level fractions: All	Variable	Fraction of the entire population in each of the exposure categories	Calculated by model
Sample statistics	Index	Statistical parameters available in the ***Water Sample Parameters*** node	N/A

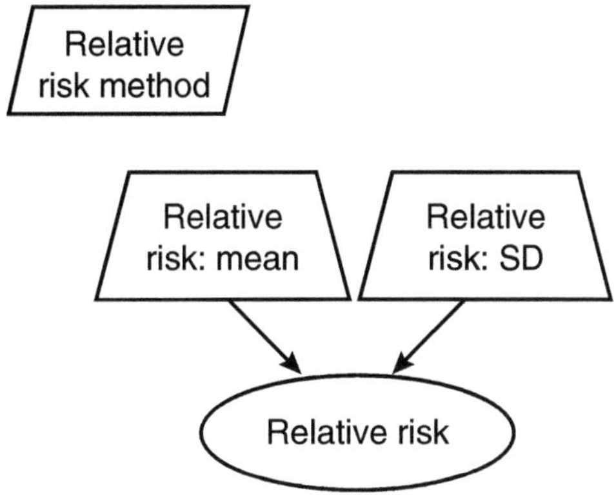

Fig. 8.6 ***Relative Risk*** submodule

Table 8.15 Description of ***Relative Risk*** submodule influence diagram nodes

Node	Type	Description	Source
Relative risk: Mean	Constant	Mean relative risk for cancer given a particular exposure level	Villanueva et al. (2003), Morris et al. (1992)
Relative risk: SD	Constant	Standard deviation of the relative risk for cancer given a particular exposure level	Villanueva et al. (2003), Morris et al. (1992)
Relative risk	Chance	Relative risk generated from the mean and standard deviation information	Calculated by model
Relative risk method	Index	Labels for two different approaches to applying relative risks	N/A

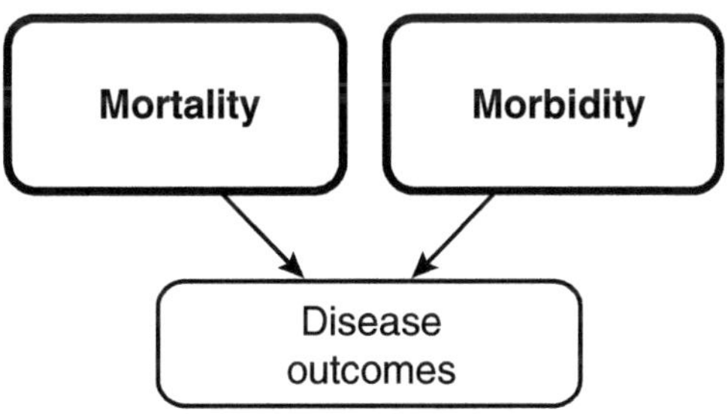

Fig. 8.7 ***Baseline Disease Rates*** submodule

Table 8.16 Description of ***Baseline Disease Rates*** submodule influence diagram nodes

Node	Type	Description	Source
Mortality	Module	Baseline mortality data from global module	N/A
Morbidity	Module	Baseline health-care facility visit data from global module	N/A
Disease outcomes	Variable	Combined mortality and morbidity data	Calculated by model

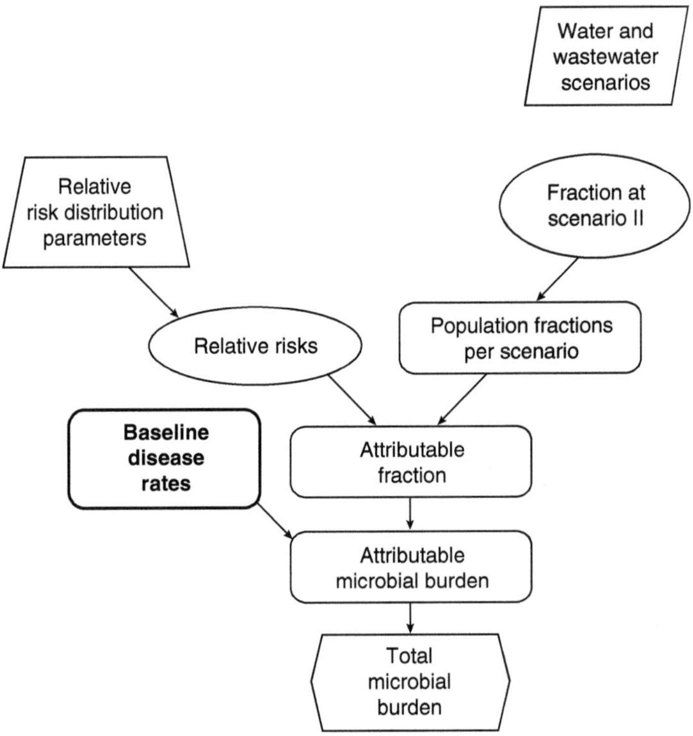

Fig. 8.8 ***Microbial Contamination*** module

Conclusions

The burden of disease attributable to drinking water appears to be small. The following recommendations are aimed at maintaining a high level of drinking water protection where it already exists and to extend these good practices nationally. Disinfection by-products are a result of chemical disinfection for the purpose of preventing bacterial contamination, and Ashbolt (2004) demonstrated an overall

Table 8.17 Description of ***Microbial Contamination*** module influence diagram nodes

Node	Type	Description	Source
Water and wastewater scenarios	Index	Categories used by WHO to classify drinking and wastewater infrastructure	Fewtrell et al. (2007), Prüss-Üstün et al. (2008)
Relative-risk distribution parameters	Constant	High and low values for a uniform distribution describing the relative risks associated with each exposure scenario	Midrange values from Fewtrell et al. (2007), Prüss-Üstün et al. (2008); high and low estimates assumed
Relative risks	Chance	Relative risks associated with each exposure scenario drawn from a uniform distribution	Calculated by model
Fraction at scenario II	Chance	Fraction of the population at exposure scenario II drawn from a triangular distribution	Midrange values from Fewtrell et al. (2007); high and low estimates assumed
Population fractions per scenario	Variable	Fraction of the population at each of the two scenarios applicable to the UAE	Calculated by model
Attributable fraction	Variable	Fraction of cases of gastroenteritis attributable to microbial contamination	Calculated by model
Baseline disease rates	Module	Mortality and morbidity data combined into one data structure	See Table 8.6
Attributable microbial burden	Variable	Burden of gastroenteritis associated with microbial contamination in drinking water	Calculated by model

health benefit from chemical disinfection. The goal, therefore, is to reduce the amount of chemicals needed to maintain the appropriate disinfection residual in the distribution system by reducing contaminant loading due to infiltration and intermediate storage.

Primary recommendations to prevent drinking water-related illnesses in the UAE are to increase water quality monitoring, prevent contamination through infiltration into the piped distribution system, and prevent water degradation in local storage tanks. Some emirates have already implemented many of the recommendations, and their expertise will facilitate implementing these practices nationally.

Routine monitoring efforts that statistically sample water throughout the distribution system and at the point of use can help assess water quality against drinking water quality standards that incorporate international standards and local needs. Uniform national standards would facilitate this effort, and adherence to nationally recognized standards of sample tracking and testing will assure the validity of the results. The results of this testing should be available not only to the managers of the water production and distribution systems but also to the water users.

Ensuring the quality of drinking water as it travels through the piped distribution systems of the various municipalities is a challenge. Rapid population growth has stressed existing supply and distribution systems and has resulted in the expansion of distribution systems with a consequential increase in water residence times. Contamination can happen through physical breaks that occur in every distribution system, the same breaks that are responsible for water loss through leakage. Water supply challenges have resulted in intermittent water supply in some areas, creating low-pressure events that allow infiltration of contaminated water through these leakage sites.

Addressing infiltration requires improving the integrity of the distribution system and reducing the number of low-pressure episodes. Distribution companies should document current leakage rates and develop programs that allow them to monitor the integrity of their systems while reducing the leakage rates to an acceptable level. National standards that ensure the proper maintenance and oversight of distribution systems are also recommended.

Preventing infiltration also requires constant pressure in the distribution system so that unavoidable breaks in the system result in water flowing out of the system, not into the system. Some systems in the UAE experience intermittent water supply and therefore have periods of low pressure. Goals for reducing the number of low-pressure events should be established. In addition, loss-reduction, capacity-building, and demand-management strategies should be identified through analysis to bring systems within the established goals. Monitoring should be implemented to measure progress towards these goals.

Bottled water is the prevalent source of drinking water for those who can afford that option. Tap water is seen as unhealthy due to the historic use of highly-saline groundwater and the perceived degradation of water quality during storage in rooftop and underground storage tanks. Although standards exist in some emirates as to how new storage tanks should be built (Regulation and Supervision Bureau 2009), the existing infrastructure may not meet these standards. Furthermore, water quality may be degraded because of the high ambient temperatures to which these tanks are exposed, airborne contamination through vents, and lack of mixing within the tanks.

References

Abu Dhabi Food Control Authority (ADFCA). 2007–2008. *Bottled water statistics*. Abu Dhabi: Laboratories Department, Chemistry and Radiation Section, Inorganic Chemistry Unit, Water Analysis Lab.

Aglukkaq, L. 2008. Minister's remarks on bisphenol A. Speech given by Canadian Minister of Health on April 18. http://www.hc-sc.c.ca/ahc-asc/minist/speeches-discours/008_04_18_e.html

Agus, E., N. Voutchkov, and D.L. Sedlak. 2009. Disinfection by-products and their potential impact on the quality of water produced by desalination systems: A literature review. *Desalination* 237(1–3): 214–237.

Al Mudhaf, A.F., F.A. Alsharifi, and A.-I. Abu Shady. 2009. A survey of organic contaminants in household and bottled drinking waters in Kuwait. *Science of the Total Environment* 407: 1658–1668.

Ali Murad, A., H. Al Nuaimi, and M. Al Hammadi. 2007. Comprehensive assessment of water resources in the United Arab Emirates (UAE). *Water Resource Management* 21: 1449–1463.

Ali, M.Y., and J.P. Riley. 1989. The production of brominated methanes in desalination plants in Kuwait. *Water Research* 23(9): 1099–1106.

Ali, M.Y., and J.P. Riley. 1990. Distribution of halomethanes in potable waters of Kuwait. *Water Research* 24(4): 533–538.

Ashbolt, N.J. 2004. Risk analysis of drinking water microbial contamination versus disinfection by-products (DBPs). *Toxicology* 198: 255–262.

Barrell, R.A., P.R. Hunter, and G. Nichols. 2000. Microbiological standards for water and their relationship to health risk. *Communicable Disease and Public Health* 3: 8–13.

Bartram, J., J. Cotruvo, M. Exner, C. Fricker, and A. Glasmacher. 2003. *Heterotrophic plate counts and drinking-water safety: The significance of HPCs for water quality and human health.* London: IWA Publishing.

Belluati, M., E. Danesia, G. Petruccia, and M. Rosellini. 2007. Chlorine dioxide disinfection technology to avoid bromate formation in desalinated seawater in potable waterworks. *Desalination* 203(1–3): 312–318.

Brook, M. 2005. *Water resources of Abu Dhabi emirate UAE.* Abu Dhabi: Water Resources Department, Environment Agency–Abu Dhabi.

Canada Agri-Food. 2008. *Food in the United Arab Emirates: Potable water.* Dubai, UAE, Agri-Food Program Section, Consulate of Canada. http://www.ats.agr.gc.ca/africa/4396_e.htm

Cantor, K.P. 1997. Drinking water and cancer. *Cancer Causes and Control* 8: 292–308.

Dillingham, R., and R.L. Guerrant. 2004. Childhood stunting: Measuring and stemming the staggering costs of inadequate water and sanitation. *Lancet* 363: 94–95.

Donohue, J.M., C.O. Abernathy, P. Lassovszky, and G. Hallberg. 2005. The contribution of drinking water to total daily dietary intakes of selected trace mineral nutrients in the United States. In *Nutrients in drinking water*, 75–91. Geneva: World Health Organization.

El Din, A.M.S., R.A. Arian, and A.A. Hammoud. 1991. A contribution to the problem of trihalomethane formation from the Arabian Gulf water. *Desalination* 85(1): 13–32.

Elshorbagy, W. 2005. *Overview of marine pollution in the Arabian Gulf with emphasis on pollutant transport modeling.* ArabianCoast 2005 Keynote Address. Al Ain: United Arab Emirates University.

Elshorbagy, W., and M. Abdulkarim. 2006. Chlorination by-products in drinking water produced from thermal desalination in United Arab Emirates. *Environmental Monitoring and Assessment* 123(1–3): 313–331.

Elshorbagy, W., and A. Elkaheem. 2008. Risk assessment maps of oil spill for major desalination plants in the United Arab Emirates. *Desalination* 228: 200–216.

Elshorbagy, W., H. Abu Qdais, and M.K. Elsheamy. 2000. Simulation of THM species in water distribution systems. *Water Resources* 34(13): 3431–3439.

Environment Agency–Abu Dhabi (EAD). 2009. *Abu Dhabi water resources master plan.* Abu Dhabi: Environment Agency–Abu Dhabi (EAD). Draft, January.

Environment Agency–Abu Dhabi (EAD). 2005. *State of the environment: Pollution of groundwater.* Abu Dhabi: Environment Agency–Abu Dhabi (EAD).

Fewtrell, L., A. Prüss-Üstün, R. Bos, F. Gore, and J. Bartram. 2007. *Water, sanitation, and hygiene: Quantifying the health impact at national and local levels in countries with incomplete water supply and sanitation coverage.* Environmental Burden of Disease Series, No. 15. Geneva: World Health Organization.

Heath, L., S. Smith, and J. Fitzgerald. 2007. Bottled water: Some health considerations. *Public Health Bulletin South Australia, Water and Public Health* 4(2): 13–17. http://www.health.sa.gov.au/pehs/publications/0707-PHB-water-vol4-no2.pdf

Hopke, P.K., T.B. Borak, J. Doull, J.E. Cleaver, K.F. Eckerman, L.C.S. Gundersen, N.H. Harley, et al. 2000. Health risks due to radon in drinking water. *Environmental Science and Technology* 34(6): 921–926.

Institute of Medicine. 2000. *Dietary reference intakes for vitamin A, vitamin K, arsenic, boron, chromium, copper, iodine, iron, manganese, molybdenum, nickel, silicon, vanadium, and zinc.* Washington, DC: National Academies Press.

Krane, J. 2005. *UAE working to rid drinking water of chemical linked to cancer.* The Associated Press, December 31. http://www.accessmylibrary.com/coms2/summary_0286-12201558_ITM

Kurokawa, Y., A. Maekawa, M. Takahashi, and Y. Hayashi. 1990. Toxicity and carcinogenicity of potassium bromate—A new renal carcinogen. *Environmental Health Perspectives* 87: 307–335.

LeChevallier, M.W., C.D. Cawthon, and R.G. Lee. 1988. Factors promoting survival of bacteria in chlorinated water supplies. *Applied and Environmental Microbiology* 54(3): 649–654.

Morris, R.D., A. Audet, I.F. Angelillo, T.C. Chalmers, and F. Mosteller. 1992. Chlorination, chlorination by-products, and cancer: A meta-analysis. *American Journal of Public Health* 82(7): 955–963.

National Research Council. 1999. *Risk assessment of radon in drinking water.* Committee on Risk Assessment of Exposure to Radon in Drinking Water. Board on Radiation Effects Research, Commission on Life Sciences, National Research Council. Washington, DC: National Academy Press. http://www.nap.edu/openbook.php?isbn=0309062926

National Research Council. 2006. *Drinking water distribution systems.* Washington, DC: National Academy Press.

National Research Council. 2008. *Desalination: A national perspective.* Washington, DC: National Academy Press.

Nsanze, H., Z. Babarinde, and H. Al Kohaly. 1999. Microbiological quality of bottled drinking water in the UAE and the effect of storage at different temperatures. *Environment International* 25(1): 53–57.

Pedro-Botet, M.L., J.E. Stout, and V.L. Yu. 2002. Legionnaire's disease contracted from patient homes: The coming of the third plague? *European Journal of Clinical Microbiology and Infectious Diseases* 21: 699–705.

Plewa, M.J., Y. Kargalioglu, D. Vankerk, R.A. Minear, and E.D. Wagner. 2002. Mammalian cell cytotoxicity and genotoxicity analysis of drinking water disinfection by-products. *Environmental and Molecular Mutagenesis* 40: 134.

Pollitt, E. 1995. The relationship between undernutrition and behavioral development in children. *The Journal of Nutrition* 125: 2212.

Price, C.J., P.L. Strong, F.J. Murray, and M.M. Goldberg. 1997. Blood boron concentrations in pregnant rats fed boric acid throughout gestation. *Reproductive Toxicology* 11(6): 833–842.

Prüss-Üstün, A., C. Mathers, C. Corvalán, and A. Woodward. 2003. *Introduction and methods: Assessing the environmental burden of disease at national and local levels.* Environmental Burden of Disease Series, No. 1. Geneva: World Health Organization.

Prüss-Üstün, A., R. Bos, F. Gore, and J. Bartram. 2008. *Safer water, better health: Costs, benefits and sustainability of interventions to protect and promote health.* Geneva: World Health Organization.

Regulation and Supervision Bureau. 2009. *Guide to water supply regulations.* Abu Dhabi: Regulation and Supervision Bureau for Water, Wastewater, and Electricity Sector in the Emirate of Abu Dhabi.

Richardson, S.D., A.D. Thruston Jr., C. Rav-Acha, L. Groisman, I. Polilevsky, O. Juraev, V. Glezer, A.B. McKague, M.J. Plewa, and E.D. Wagner. 2003. Tribromopyrrole, brominated acids, and other disinfection by-products produced by disinfection of drinking water rich in bromide. *Environmental Science and Technology* 37: 3782–3793.

Rizk, Z.S. 2009. Inorganic chemicals in domestic water of the United Arab Emirates. *Environmental Geochemical Health* 31: 27–45.

Rusin, P.A., J.B. Rose, C.N. Haas, and C.P. Gerba. 1997. Risk assessment of opportunistic bacterial pathogens in drinking water. *Reviews of Environmental Contamination and Toxicology* 152: 57–83.

Saeed, T., H. Khordagui, and H. Al Hashash. 1999. Contribution of power/desalination plants to the levels of halogenated volatile liquid hydrocarbons in the coastal areas of Kuwait. *Desalination* 121: 49–63.

Signorile, G., A. Neve, F. Lugoli, M.C. Piccinni, R. Arina, and R. Di Marino. 2007. Evaluation of toxic chemical parameters and ecotoxicity levels in bottled mineral waters. *Journal of Preventative Medicine and Hygiene* 48: 10–16.

Stephenson, L.S., M.C. Latham, and E.A. Ottesen. 2000. Malnutrition and parasitic helminth infections. *Parasitology* 121: S23–S38.

Tsai, S.-Y., H.-Y. Chou, H.-W. The, C.-M. Chen, and C.J. Chen. 2003. The effects of chronic arsenic exposure from drinking water on the neurobehavioral development in adolescence. *Neurotoxicology* 24(4–5): 747–753.

U.S. Environmental Protection Agency (EPA). 2006. Initial EPA perspectives on NAS report: Risk assessment of radon in drinking water. http://www.epa.gov/ogwdw/radon/remove/nasdw.html

UAE Ministry of Energy. 2006. Initial national communication to the United Nations Framework Convention on Climate Change. http://unfccc.int/resource/docs/natc/arenc1.pdf

UAE Ministry of Environment and Water. 2004. Annual Statistical Report of 2004. Abu Dhabi.

UAE University. 2008. Impact assessment of refinery wastewater on desalinated water: Background. http://www.engg.uaeu.ac.ae/Jccp_b/

UNICEF. 2008. *Handbook on water quality*. New York: United Nations Children's Fund. http://www.unicef.org/wes/files/WQ_Handbook_final_signed_16_April_2008.pdf

Villanueva, C.M., F. Fernández, N. Malats, J.O. Grimalt, and M. Kogevinas. 2003. Meta-analysis of studies on individual consumption of chlorinated drinking water and bladder cancer. *Journal of Epidemiology and Community Health* 57: 166–173.

Villanueva, C.M., K.P. Cantor, S. Cordier, J.J.K. Jaakkola, W.D. King, C.F. Lynch, S. Porru, and M. Kogevinas. 2004. Disinfection by-products and bladder cancer: A pooled analysis. *Epidemiology* 15(3): 357–367.

Wagner, M., and J. Oehlmann. 2009. Endocrine disruptors in bottled mineral water: Total estrogenic burden and migration from plastic bottles. *Environmental Science and Pollution Research* 16: 278–286.

Wasserman, G.A., X. Liu, F. Parvez, H. Ahsan, P. Factor-Litvak, A. van Geen, V. Slavovich, et al. 2004. Water arsenic exposure and children's intellectual function in Araihazar, Bangladesh. *Environmental Health Perspectives* 112: 1329–1333.

Wasserman, G.A., X. Liu, F. Parvez, H. Ahsan, D. Levy, P. Factor-Litvak, J. Kline, et al. 2006. Water manganese exposure and children's intellectual function in Araihazar, Bangladesh. *Environmental Health Perspectives* 114(1): 124–129.

Weinberg, H.S., C.A. Delcomyn, and V. Unnam. 2003. Bromate in chlorinated drinking waters: Occurrence and implications for future regulation. *Environmental Science and Technology* 37(14): 3104–3110.

World Bank. 2005. *A water sector assessment report on the countries of the Cooperation Council of the Arab States of the Gulf.* Water, Environment, Social and Rural Development Department, Middle East and North Africa Region, 32539-MNA. Washington, DC: World Bank.

World Health Organization. 2001. Risk assessment. In *Water quality—Guidelines, standards and health: Assessment of risk and risk management for water-related infectious disease*, Ch. 8. London: IWA Publishing.

World Health Organization. 2005. *Nutrients in drinking water*. Water, Sanitation and Health. http://www.who.int/water_sanitation_health/dwq/nutrientsindw.pdf

World Health Organization. 2006. *Guidelines for drinking-water quality*, 3rd ed. Geneva: World Health Organization. http://www.who.int/water_sanitation_health/dwq/gdwq3rev/en/index.html

World Health Organization. 2008. *Petroleum products in drinking-water: Background document for development of WHO guidelines for drinking-water quality*. WHO/SDE/WSH/05.08/123. Geneva: World Health Organization.

World Health Organization and UNICEF. 2006. *Meeting the MDG drinking water and sanitation target: The urban and rural challenge of the decade*. Geneva: World Health Organization. http://www.who.int/water_sanitation_health/monitoring/jmp2006/en/index.html

Ziadat, A.H. 2005. Impact of storage tanks on drinking water quality in Al-Karak province—Jordan. *Journal of Applied Sciences* 5(4): 634–638.

Chapter 9
Burden of Disease from Coastal Water Pollution

Abstract Contamination of coastal water with pathogenic microorganisms is generally considered a relatively lesser environmental concern in developed nations such as the United Arab Emirates (UAE) that require treatment of municipal wastewater prior to discharge. In the Arabian Gulf in the 1990s, measurements of coliform bacteria (which indicate possible contamination with fecal matter and hence disease-causing organisms) generally supported this assumption, with reports of low average coliform concentrations along the coast. However, more recent data provided by the Environment Agency–Abu Dhabi indicate that concentrations of enterococci, which are the most important indicators of fecal contamination in coastal waters, at the city's public beaches occasionally exceed World Health Organization recommended water quality standards. Additional observations of high coliform concentrations in nearby Abu Dhabi shipping channels and news reports of high *E. coli* concentrations along the Dubai coastline suggest that episodic events involving high concentrations of microorganisms of fecal origin may need to be addressed to preserve coastal water quality. Microorganisms in coastal water can enter the body through ingestion, inhalation, or transport through existing skin lesions or open wounds. Gastroenteritis (including diarrhea, vomiting, and associated symptoms) is the most common type of illness resulting from such exposures in developed countries. The analysis in this chapter focuses on risks due to pathogenic microorganisms of fecal origin because these organisms pose the most important public health risk from short-term exposure to coastal water as occurs during swimming and other coastal recreational activities. Further, at present, industrial effluents appear to be sufficiently regulated with no recorded violations, and a public notification system exists to warn beachgoers of "catastrophic" single incidents such as local oil spills or red tides. Our model estimates that swimming in UAE coastal waters could contribute to 1,300 medical visits for gastroenteritis per year among UAE residents. If the potential total number of illnesses (reported through medical visits and unreported) is considered, the estimate rises to 64,000. In addition to the cases among residents, another 24,000 cases could be expected among tourists. This suggests that virtually immediate reductions in health risks related to recreational waters could be achieved through increased

J. MacDonald Gibson et al., *Environmental Burden of Disease Assessment*, Environmental Science and Technology Library 24, DOI 10.1007/978-94-007-5925-1_9,

monitoring of fecal indicator bacteria concentrations at local beaches and closure of areas in violation of international recommended standards until adequate reductions in microbial concentrations are achieved. Ideally, to provide consistent public health safeguards throughout the nation, a standardized coastal water monitoring program should be instituted at the federal level by the Ministry of Environment and Water and implemented by relevant emirate-level agencies.

Keywords Coastal water pollution • Environmental burden of disease • Relative risk • Attributable fraction • Premature deaths and health-care facility visits • United Arab Emirates • Harmful algal blooms ("red tides") • Pathogenic microorganisms of fecal origin • Coliform bacteria • Fecal streptococci • Gastrointestinal illness • Gastroenteritis • No observed adverse effects level

Overview: Nature and Sources of Coastal Water Pollution

Coastal recreational activities such as swimming, fishing, and sailing are popular in the United Arab Emirates, just as they are in many other countries with ample coastlines and favorable climates. Tourism is becoming a major industry, attracting millions of foreign visitors annually, particularly to Dubai and Abu Dhabi, where many visitors engage in coastal recreational activities (Dubai Department of Tourism and Commerce Marketing 2007; Shekhar 2009). Several of the northern emirates are aiming to increase tourism through promotion of swimming and snorkeling at their beaches as well. The preservation of marine water quality and the attractiveness of local beaches to foreign visitors are therefore essential to maintaining this source of national revenue.

Several potential sources of UAE coastal water contamination exist (Fig. 9.1). Perhaps the most significant concern is evidence that municipal wastewater treatment plants are overburdened, particularly in cities undergoing unprecedented growth. Periodic high levels of coliform bacteria attributed to insufficient wastewater treatment were observed in Dubai Creek even 15 years ago when growth was relatively modest (Abu Hilal et al. 1994). Chronic overloading of wastewater treatment plants would be expected to reduce effluent water quality, resulting in increased risks that pathogenic organisms will be released into coastal water. Heavy reliance on the transport of sewage by tanker trucks, particularly in Dubai, may also result in illegal direct discharges of sewage into coastal tributaries or the Arabian Gulf itself, with the potential to seriously compromise the water quality of nearby local beaches (Setrakian 2009).

In addition to elevated concentrations of fecal microorganisms, untreated or poorly treated sewage discharges result in high concentrations of nutrients (e.g., nitrogen, phosphorous) in coastal waters, which can encourage the growth of harmful algal blooms (Glibert 2007; Anderson et al. 2002). Commonly referred to as "red tides," these blooms host algal species that can produce toxins with respiratory or neurological effects on fish and/or humans (Graneli and Turner 2006; Henrickson et al. 2001). Although the obvious appearance of a red tide generally deters recreational activity, consumption of fish exposed to toxins, inhalation of water droplets containing toxins, or the intake of contaminated water by desalination plants can pose a public

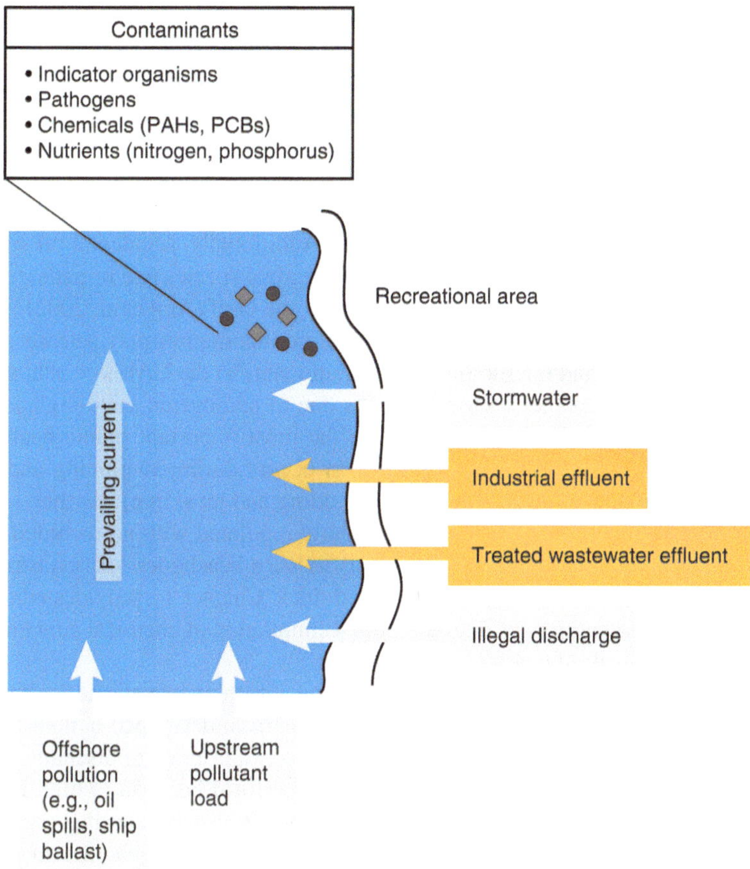

Fig. 9.1 Major sources of contamination of UAE coastal waters. *White boxes* indicate nonpoint sources, while *yellow boxes* indicate point sources, which can be controlled by effluent requirements

health threat (Baker and McGillicuddy 2006). Red tides have recently forced the closure of desalination plants in Sharjah (Sambidge 2008) and beaches in Dubai (Menon 2009). Though bloom organisms clearly pose a serious ecological issue, with continued monitoring and public notice of red tide events, human illness resulting from recreational exposure to these organisms is likely very small. Harmful algal blooms are discussed further in Chap. 11.

Wastewater streams from industrial processes also are important potential contamination sources for coastal waters. As a primary producer of the world's oil supply, the UAE has numerous oil refineries, particularly along the Abu Dhabi coast. A recent study investigating wastewater discharges at a petroleum refinery at Al Ruwais, United Arab Emirates, observed high biochemical and chemical oxygen demand levels, as well as high concentrations of polyaromatic hydrocarbons (PAHs) and occasional detection of polycholorinated biphenyls (PCBs) and dioxins (Al Zarooni and Elshorbagy 2006). Results support the implementation of secondary treatment strategies and/or more stringent enforcement to ensure effluent adherence to recommended UAE national standards.

The unique hydrology of the Arabian Gulf makes all types of contaminant loadings to coastal areas of particular concern. Because the Arabian Gulf is linked to the Gulf of Oman by the narrow Straits of Hormuz, dilution or flushing of the system is relatively slow (7–8 years), potentially allowing pollutants to accumulate and concentrate. Low average depths and high average temperatures result in conditions that are ideal for microbial and algal growth. Additionally, accidental oil spills or ballast discharge potentially involving nonnative algal species and/or pathogens are a concern because of heavy shipping throughout the Gulf (Anil et al. 2002).

Although the multiple potential sources of contamination may release a wide variety of chemical and microbiological contaminants to the UAE's coastal waters, the analysis in this chapter focuses on risks due to pathogenic microorganisms of fecal origin, because these organisms pose the most important public health risk from short-term exposure to coastal water as occurs during swimming and other coastal recreational activities (such as water skiing and jet skiing). Further, at present, industrial effluents appear to be sufficiently regulated with no recorded violations, and a public notification system exists to warn beachgoers of "catastrophic" single incidents such as local oil spills or red tides. Chapter 11 provides additional information on potential risks due to the bioaccumulation of chemical contaminants in seafood as a result of coastal contamination.

Contamination of coastal water with pathogenic microorganisms is generally considered a relatively lesser environmental concern in developed nations such as the UAE that require treatment of municipal wastewater prior to discharge. In the Arabian Gulf in the 1990s, measurements of coliform bacteria (which indicate possible contamination with fecal matter and hence disease-causing organisms) generally supported this assumption, with reports of low average coliform concentrations along the coast (Banat et al. 1998, 1993; Abu Hilal et al. 1994). However, more recent data provided by the Environment Agency–Abu Dhabi (EAD) indicate that concentrations of enterococci (fecal streptococci), which are the most important indicators of fecal contamination in coastal waters, at the city's public beaches occasionally exceed World Health Organization (WHO) recommended water quality standards (EAD 2007). Additional observations of high coliform concentrations in nearby Abu Dhabi shipping channels (EAD 2007) and news reports of high *E. coli* concentrations along the Dubai coastline (Setrakian 2009; *Telegraph* 2009) suggest that episodic events involving high concentrations of microorganisms of fecal origin may need to be addressed to preserve coastal water quality.

Key Health Effects of Coastal Water Pollution

Microorganisms in coastal water can enter the body through ingestion, inhalation, or transport through existing skin lesions or open wounds. Gastroenteritis (including diarrhea, vomiting, and associated symptoms) is the most common type of illness

Table 9.1 Selected human pathogens transmissible through marine water

Pathogen	Type	Illness
Escherichia coli, *Salmonella* spp.	Bacteria	Gastroenteritis
Rotavirus, Norwalk virus	Virus	Gastroenteritis
Cryptosporidium, *Giardia*	Protozoan	Gastroenteritis
Entamoeba histolytica	Amoeba	Amebiasis (severe gastroenteritis)
Vibrio cholera	Bacteria	Cholera (severe gastroenteritis)
Staphylococci spp., *Streptococci* spp.	Bacteria	Skin and ear infections
Vibrio spp.	Bacteria	Necrotizing wound infections
Naegleria fowleri	Amoeba	Meningoencephalitis
Hepatitis	Virus	Hepatitis
Poliovirus	Virus	Poliomyelitis
Harmful algal bloom organisms (e.g., *Gymnodinium*, *Pfiesteria*)	Zoo-/phytoplankton	Various human illnesses, including respiratory and neurological symptoms

resulting from such exposures in developed countries (Craun et al. 2005). While other types of illness (e.g., respiratory, ear, skin infections) have been documented (Henrickson et al. 2001; Corbett et al. 1993), the relationship between exposure to contaminated beach water and health risks is at present unclear, perhaps due to a much lower incidence rate, the quantification of which would require studies of larger populations than have yet been carried out (Fleming et al. 2006; Turbow et al. 2003; Wade et al. 2003; Prüss 1998). Table 9.1 provides a selection of potential waterborne pathogens and the types of illnesses they can cause.

As Table 9.1 illustrates, a variety of pathogenic organisms may be transmitted via recreational use of contaminated coastal waters. Because many of these pathogens are difficult, time-consuming, or prohibitively expensive to detect, recreational water quality is generally assessed through the monitoring of indicator organisms. Indicator organisms (e.g., coliforms, *E. coli*, enterococci) are typically bacteria native to the digestive tract of warm-blooded animals that are assumed to be reasonable surrogates for the presence, fate, and transport of pathogens (Savichtcheva and Okabe 2006). Enterococci bacteria, historically referred to as "fecal streptococci," are widely regarded as the most useful indicator of health risk for swimmers in marine waters (Boehm et al. 2009). Recommended standards for marine recreational areas in the United States (U.S. EPA 1986) and Europe (Kay et al. 2004) are based on enterococci levels, because increasing enterococci concentrations are more strongly correlated with an increased risk of illness in swimmers than are fecal coliform bacteria or *E. coli* (Wade et al. 2003, 2006; Turbow et al. 2003; Prüss 1998).

For several reasons, considerable debate currently surrounds the strategy of using indicator organisms as surrogates for pathogens. First, pathogens have been recovered from marine waters that comply with indicator standards (Polo et al. 1998; Morinigo et al. 1990). Second, fecal indicators by definition do not account for native pathogenic species (e.g., *Vibrio* spp.) that are not fecal in origin (Fleming et al. 2006). Third, the majority of attempts to link swimmer illness with indicator organism exposure have focused on populations swimming in coastal waters contaminated by municipal wastewater effluents (Boehm et al. 2009), but in developed countries that require wastewater treatment, stormwater discharges to recreational waters are generally of greater concern, and the applicability of current standards to these waters is unclear (Colford et al. 2007; Kay et al. 1999; Calderon et al. 1991). This criticism is likely of less concern to a desert nation such as the UAE, where rainfall is minimal and wastewater discharges to the coast are more likely the primary source of contamination. Fourth, although many studies have documented an increased probability of contracting gastrointestinal illnesses as a result of exposure to beach water with elevated levels of indicator organisms (Boehm et al. 2009; Wade et al. 2003; Prüss 1998; Kay et al. 1994; Corbett et al. 1993; Cabelli et al. 1982), the relationships between indicator exposure and human health impacts are uncertain and may vary significantly by location, possibly as a result of regional differences in immunity or endemic pathogens (Boehm et al. 2009; Colford et al. 2007; Fleisher 1991). Also of concern are uncertainties in interpreting the results of epidemiologic studies designed to establish mathematical relationships (known as dose-response relationships) between exposure to a specific concentration of indicator organisms and the probability of becoming ill. These studies depend on surveys of beachgoers who volunteer to have their health status tracked in the days after their beach visit. These volunteers are separated into exposure categories depending on the amount of time they report spending in the water. Potential differences in prior health status that may contribute to whether these survey volunteers choose to swim can make the establishment of a true dose-response relationship difficult (Prüss 1998). The widely varying analytical techniques and sampling strategies used to quantify microbial concentrations also hinder comparisons of indicator organism exposure (Wade et al. 2006; Haugland et al. 2005; Prüss 1998).

Despite these limitations, no superior surrogate for measuring fecal contamination of coastal waters has been identified. As a result, water quality standards based on indicator organisms retain broad acceptance in regulatory programs and remain the primary means of public health protection for coastal recreational areas in both the United States and Europe (Kay et al. 2004, U.S. EPA). Hence, this analysis predicts the burden of disease attributable to recreational exposure to coastal water in the UAE based on concentrations of enterococci in previous water quality surveys of UAE beaches.

Method for Estimating the Burden of Disease from Coastal Water Pollution

Model Overview

The method for estimating the number of gastroenteritis cases attributable to recreational exposure to coastal water follows the WHO attributable fraction approach described in Chap. 3. As Chap. 3 explains, this approach requires as input information about

1. the fraction of the population exposed (in this case, the fraction of the UAE population that immerses their face or entire head in UAE coastal waters and the frequency of such events)—represented as P_i;
2. the magnitude of the exposure (here, the concentration of enterococci in beach water), expressed as $C_{ent,i}$;
3. the relative risk of contracting an illness (in this case, gastroenteritis) as a function of the magnitude of the exposure (the concentration of enterococci), denoted as $RR(C_{ent,i})$; and
4. the total baseline incidence rate of the illness (gastroenteritis) in the study population during the time period of interest, denoted D_{total}.

These information sources are combined using Eqs. 3.6 and 3.7 to yield

$$D_{attrib} = \frac{\sum_i P_i \times RR(C_{ent,i}) - 1}{\sum_i P_i \times RR(C_{ent,i})} \times D_{total} \tag{9.1}$$

where D_{attrib} is the number of gastrointestinal illness cases attributable to exposure to coastal water pollution. The following sections explain the information sources and assumptions used to develop these input data.

Outbreaks of illness related to exposure to contaminated marine waters have the potential to negatively affect the tourism industry and, hence, the UAE economy. Consequently, in contrast with the other chapters in this report, burden of disease calculations in this chapter include illnesses afflicting both residents (citizens and expatriates) and foreign tourists. We provide separate estimates for tourists and for UAE residents, and these estimates are based on separate (but similar) models. Figure 9.2 illustrates these models, and Table 9.2 defines all the model variables and the data sources used to characterize them.

Exposed Population

Resident populations of all seven emirates were obtained from the UAE Ministry of Economy and are the same as population estimates used throughout this report. Monthly tourist numbers were obtained from Dubai's Department of Tourism and Commerce Marketing website quarterly reports for the year 2006, with the total number of tourists assumed to be the sum of "hotel" and "hotel-apartment" guests for each month. This likely represents a large under estimation because it does not account for the large number of family members of expatriates living in the UAE who visit their relatives periodically. The annual number of tourists visiting Abu Dhabi has been estimated at 1.5 million for 2008 (Shekhar 2009). This value was distributed throughout the 12 months of the year based on the distribution of tourism numbers for Dubai. A similar technique was used to distribute the 125,000 non-Emirati visitors to Fujairah each year (Carvalho 2009). No tourism data were available for

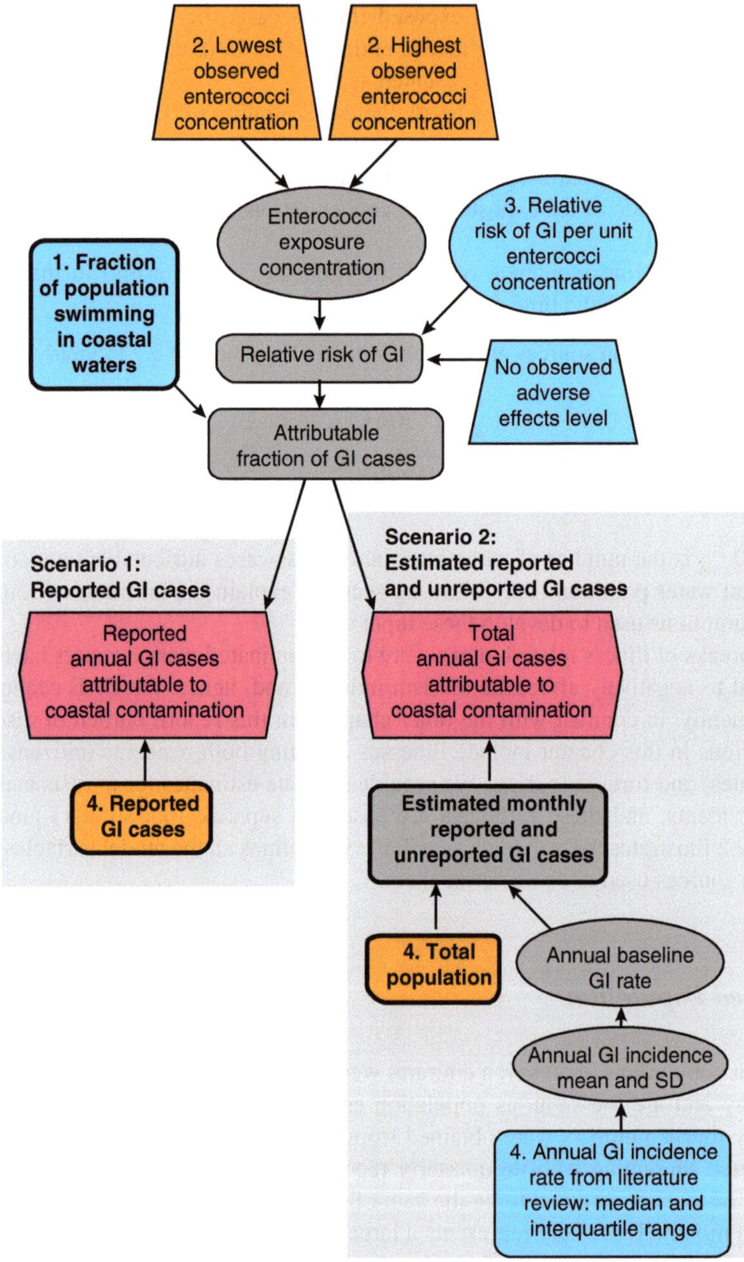

Fig. 9.2 Influence diagrams for model constructed to estimate the burden of disease attributable to recreational exposure to coastal waters for UAE residents (*above*) and UAE tourists (*next page*). Numbers correspond to the category of information to which the variable or module pertains: *1* exposed population, *2* magnitude of exposure; *3* relative risk of illness per unit exposure, *4* baseline illness

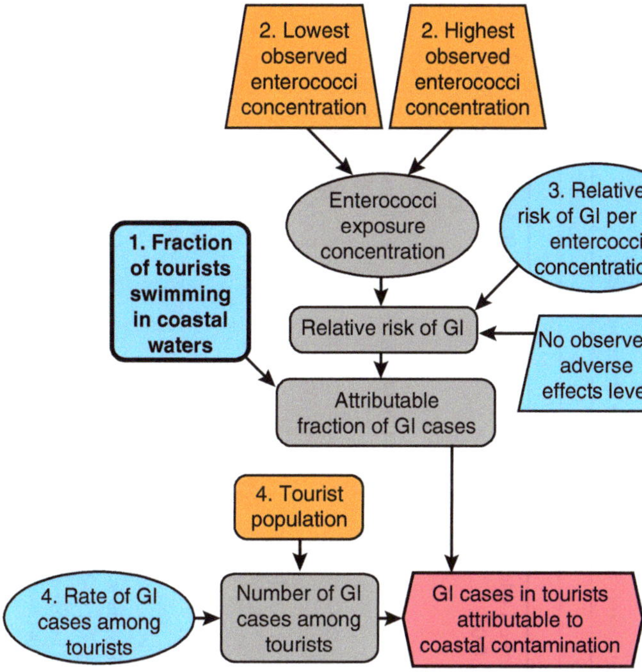

Fig. 9.2 (continued) incidence rate. Nodes without numbers indicate intermediate computations. Nodes outlined in bold represent modules containing further layers of computations. The color scheme corresponds to source of data: *orange*, observational data from the UAE; *blue*, data estimated from scientific literature; *grey*, intermediate computations using data in parent nodes; *pink*, model output

the emirates of Sharjah, Ajman, Ras Al Khaimah, or Umm Al Quwain; however, this information can be easily added to the model when available.

In accordance with epidemiologic studies investigating the health effects of microbial contamination of beach water, we define exposure as the recreational immersion of the face or entire head in the water (Wiedenmann et al. 2006; Wade et al. 2003). We use different estimates of swimming behavior for residents and for tourists.

For residents, in order to account for limitations in available information on population behavior, we used two different approaches to estimate the percentage of the population that swims, and we estimated the resulting disease burden separately for each approach. (As will be discussed in the results section, the choice of approach made little difference in the burden of disease estimate.)

Our first approach relied on the results of a questionnaire about physical activity behaviors (including frequency of swimming) included in the previous UAE Health and Lifestyle Survey (Badrinath et al. 2002). Badrinath and others report estimates of swimming frequency by gender and age group for UAE citizens. We estimated based on a study of UAE drownings by Barss and others that 67% of the time citizens swim, they swim in marine water (rather than in swimming pools). Since the Health

Table 9.2 Parameter inputs for estimating burden of disease attributable to recreational exposure to coastal water contamination

Variable	Description	Definition	Units	Source
UAE residents				
Fraction of population swimming in coastal waters (S)	Fraction of population swimming in coastal water each month	Scenario 1: see Table 9.3 Scenario 2: 0.047	None (ratio)	Badrinath et al. (2002), Barss et al. (2009), Australian Sports Commission (2011)
Lowest observed enterococci concentration (C_{low})	2006 observed concentrations at Al Raha Beach	Monthly values	CFU[a]/100 mL	EAD monitoring (see Table C.1, Appendix C)
Highest observed enterococci concentration (C_{high})	2006 observed concentrations at the Abu Dhabi public beach	Monthly values	CFU/100 mL	EAD monitoring (Table C.1)
Enterococci exposure concentration (C)	Exposure to fecal contamination	Uniform (C_{low}, C_{high})	CFU/100 mL	Calculated
Relative risk (RR)	Risk of gastroenteritis per log(*C*)	Triangular (1.00, 1.34, 1.75)	None (ratio)	Wade et al. (2003)
No observed adverse effects level (NOAEL)	Concentration below which relative risk of illness = 1	25	CFU/100 mL	Wiedenmann et al. (2006)
Relative risk of GI [RR(C)]	Relative risk of GI for the given exposure concentration	If *C* > NOAEL, then (*RR*)*log(*C*); else, 1	None (ratio)	Calculated
Attributable fraction of GI cases	Fraction of GI cases attributable to swimming in contaminated coastal water	[*S***RR*(*C*) + (1 − *S*) − 1]/[*S***RR*(*C*) + (1 − *S*)]	None (ratio)	Eq. 3.6
Reported GI cases	GI cases per month by nationality, age, gender	Table 9.4	# patient visits to healthcare facility	Scaled to entire UAE population from HAAD (2009)

Annual GI incidence rate from literature review (medR, IQRR)	Estimated median (med) and interquartile range IQR) of incidence of acute GI illness per person per year	By age: <5, med=1.8, IQR=1.1–3.4; 5–14, med=0.56, IQR=0.34–1.1; 15–54, med=0.37, IQR=0.20–0.75; >55, med=0.79, IQR=0.50–1.1	#/person-year	Walker and Black (2010), Bern (2004)
Annual incidence mean and standard deviation (medR, gsdR)	Geometric mean and geometric standard deviation of *R*	Estimated from *medR* and *IQRR*	#/person-year	Calculated
Annual baseline GI rate (R)	Incidence of acute GI illness per person per year	Logormal (*medR*, g*sdR*)	#/person-year	Calculated
Total population (P)	UAE population by emirate, age group, citizenship, gender	Tables 1.1, 9.6	# of people	UAE Ministry of Economy (2008)
Estimated monthly reported and unreported GI cases	Number of GI cases per month by emirate, age group, citizenship	*R*P**(days in month)/365	# of cases	Calculated
Tourists				
Fraction of tourists swimming in coastal waters	Fraction of tourists who swim in UAE coastal water	0.164	None (ratio)	Hsieh et al. (1992)
Tourist population (T)	Monthly tourist visits (by emirate)	Monthly values	Number of tourists	Dubai Department of Tourism and Commerce Marketing (2007), Shekhar (2009), Carvalho (2009)
Rate of GI cases among tourists (RT)	Percentage of tourists who contract acute GI	Lognormal (mean=0.053, sd=0.00875)	None (ratio)	Riddle et al. (2006)
Number of GI cases among tourists	Number of GI cases among tourists	*RT*T*	# cases	Calculated

[a]Colony-forming units

Table 9.3 Percentage of UAE resident population swimming each month, by age

	Females				Males			
Emirate	<5	5–14	15–54	>55	<5	5–14	15–54	>55
Citizens (%)	0.9	0.9	0.0	0.0	3.8	3.8	1.4	1.4
Noncitizens (%)	6.2	6.2	6.2	6.2	6.2	6.2	6.2	6.2

and Lifestyle Survey did not include noncitizens, we derived separate estimates of swimming behavior for noncitizens from the annual Participation in Sport, Recreation and Exercise Survey from Australia (Australian Sports Commission 2011), using the results from New South Wales as a proxy for non-Emiratis in the UAE. In this case, swimming participation was reported as frequency per year, and we included all those who reported swimming at least once a month. We assumed that, as for Emiratis, 67% of swims occurred in coastal waters. We were not able to stratify by age group and gender for the non-Emirati population because the reporting format for swimming frequency in the Australia study did not allow for such stratification. Table 9.3 shows the resulting estimates of the percentage of the population that swims each month by citizenship, gender, and age group.

Our second approach estimated swimming frequency from the UAE drowning rate reported by Barss and others (2009), who estimated that 0.5 drownings per 100,000 people occur each year in the UAE. Of these, 67% occur in coastal waters. In Australia, the incidence of drowning is 1.3 per 100,000, and 18.2% of the population reports swimming (in any type of venue) per month (Mitchell et al. 2010). We assumed that the ratio of drowning rate to swimmers matches that in Australia. With this assumption, on average, $18.2\% \times (0.5/1.3) = 7.0\%$ of the UAE population swims each 4 weeks. Since 67% of UAE drownings occur in UAE coastal waters, we assumed that $0.67 \times 7\% = 4.7\%$ of people swim in coastal waters each month, on average. In this case, we did not differentiate between citizens and noncitizens or by gender and age.

For tourists, the average length of stay in Dubai is less than 3 days, and we assume tourist stays are similar in Abu Dhabi and Fujairah. We assumed each visitor who chooses to swim spends a single day at the beaches, i.e., each swimmer equals a single potential exposure. We could not find data on the proportion of tourists who visit and swim at beaches. As a proxy, we used a survey by Hsieh et al. (1992) of international pleasure travelers to Hong Kong. Given that both Hong Kong and the UAE are coastal destinations with warm climates, it is not unreasonable to assume that international tourist behavior would be similar in both locales. According to Hsieh, O'Leary, and Morrison, 16.4% of Hong Kong tourists engage in sunbathing/beach activities. As a conservative estimate, we therefore assume that 16.4% of visitors to the UAE are exposed to coastal water.

Due to the limitations of available UAE data on frequency of swimming in coastal waters by UAE residents and tourists, studies similar to that presented by Dwight et al. (2007) for the southern California beaches in the United States are strongly recommended for UAE beaches to improve future burden of disease estimates.

Magnitude of Exposure

For this study, the EAD provided single monthly measurements of enterococci concentrations at two Abu Dhabi beaches—Al Raha beach and the public beach—for 2006 and 2007 (see Table C.1, Appendix C). No further information on enterococci concentrations was available from previous local studies, and hence we had to rely on observations from these two beaches alone.

Concentrations observed at Al Raha beach were consistently lower than those at the public beach. Hence, lacking any better information, we assumed that the enterococci contamination at Abu Dhabi beaches follows a uniform distribution with Al Raha concentrations for each month as the lower limit and public beach concentrations as the upper limit. Under this assumption, model simulation runs randomly choose an enterococci concentration from between the observed values at these two beaches in each month to determine the exposure level for that month. Estimates of the burden of disease were calculated using the 2006 data set because no violations were recorded for any month during 2007. However, monthly measures of water quality generally underestimate actual health risk (Leecaster and Weisberg 2001), and urbanization in the UAE is increasing the volume of sewage effluent released to coastal waters (Saunders et al. 2007). Therefore, while modeling results using only the 2006 data set might be considered upper limits, they are likely more indicative of future conditions.

Reports suggest that fecal contamination of Dubai's coastal waters may be more widespread, with more frequent violations of the recommended bathing standards in beach areas. A recent investigation of the effects of organic contamination on benthic macroinvertebrates in Dubai Creek (Saunders et al. 2007) reported generally degraded aquatic ecology at many points in the creek, particularly below the Al Aweer sewage outfall. Although this study did not include water quality testing for indicator organisms, the presence of high biochemical oxygen demand and nutrient levels strongly suggests fecal contamination, which would corroborate anecdotal reports (Setrakian 2009; *Telegraph* 2009). As no bacteriological water quality data were available for Dubai's coastal waters, monthly enterococci concentrations were assumed to be twice those observed for the beaches at Abu Dhabi to reflect the likelihood of significant fecal contamination.

No coastal water quality data are currently available for the five remaining emirates. In the absence of region-specific data, concentrations of enterococci were assumed to be equivalent to the values observed in Abu Dhabi.

Due to the extremely crude nature of these exposure estimates (all that was possible within the bounds of this study), the results in this chapter should be regarded as preliminary estimates of the potential magnitude of risk due to coastal water pollution. Indeed, extremely high spatial and temporal variability in enterococci concentrations would be expected at UAE beaches due not only to tidal action and variation in sewage discharge rates over time but also to the complex geomorphology of the UAE coastline. The coastline contains many small channels and islands and is constantly being altered by infill and dredging. Hence, detailed studies to characterize the spatial and temporal variation in enterococci concentrations at UAE beaches are strongly recommended. Such studies already are under way in Abu Dhabi, and in the future the results could be used to update the burden of disease analyses presented in this report.

Relative Risk

A recent systematic meta-analysis of the many previous epidemiologic studies evaluating the risk of gastroenteritis associated with recreational water exposure estimated that the relative risk of gastrointestinal illness for every 10-fold (i.e., log-10) increase in enterococci concentration per 100 mL of water is 1.34 (95% confidence interval = 1.00, 1.75) (Wade et al. 2003). Represented in mathematical terms, $RR(C_{ent,i})$ in Eq. 9.1 becomes:

$$RR(C_{ent,i}) = 1.34 log(C_{ent,i}) \tag{9.2}$$

where C_{ent} is the enterococci concentration (CFU/100 mL). Wade's calculation of this average value involved the review of more than 900 candidate studies with the ultimate synthesis of data from the 17 studies deemed the most rigorous. As the most thorough meta-analysis of the literature available, combining observations of health risk from studies conducted in the United States, Asia, the South Pacific, Europe, Africa, and the Middle East, the relative risk proposed by Wade et al. (2003) appears the most appropriate for calculations of disease burden among swimmers in the UAE.

Current bacteriological water quality standards for bathing areas were established to minimize but not eliminate public health risk. Maximum allowable concentrations do therefore imply that some level of risk exists (e.g., eight illnesses per 1,000 swimmers in the U.S. EPA 1986 criteria). While an excellent synthesis of available data, the 2003 study by Wade et al. does not provide a zero risk level. In order to ensure that application of Eq. 9.1 to low concentrations does not result in inappropriate protective estimates of risk (e.g., RR < 1), a no observed adverse effect level (NOAEL) of 25 enterococci/100 mL was adopted in accordance with the results of a recent, large, randomized control study in Germany (Wiedenmann et al. 2006). For the purposes of this study, adverse health effects resulting from exposure to waters with enterococci concentrations below this level are assumed to be zero (i.e., relative risk = 1.00).

Baseline Gastroenteritis Rate

Data on health-care facility visits by Abu Dhabi residents were provided by the Health Authority–Abu Dhabi (HAAD). The data include information on citizenship

Table 9.4 Baseline number of medical visits for gastroenteritis among UAE residents

Emirate	<5	5–14	15–54	>55	<5	5–14	15–54	>55
	Female citizens (by age)				*Male citizens (by age)*			
Abu Dhabi	2,426	1,300	2,304	143	2,728	1,527	2,380	187
Dubai	890	477	845	53	990	554	863	68
Sharjah	890	477	845	53	1,056	591	922	72
Ajman	256	137	243	15	294	165	257	20
Umm Al Quwain	110	59	104	6	107	60	93	7
Ras Al Khaimah	585	314	556	35	655	367	572	45
Fujairah	390	209	370	23	428	240	373	29
	Female noncitizens (by age)				*Male noncitizens (by age)*			
Abu Dhabi	1,985	762	2,056	120	1,887	764	7,425	236
Dubai	1,732	665	1,794	105	2,762	1,119	10,869	345
Sharjah	1,453	558	1,504	88	1,312	531	5,161	164
Ajman	360	138	373	22	309	125	1,214	39
Umm Al Quwain	70	27	72	4	57	23	226	7
Ras Al Khaimah	226	87	234	14	225	91	885	28
Fujairah	129	50	134	8	141	57	555	18
Totals	11,503	5,258	11,434	689	11,503	5,258	11,434	689

and gender and were cleaned, coded, and validated by HAAD. We used these data to estimate the number of annual healthcare facility visits for gastroenteritis in the other emirates. In developing these estimates, we applied different rates for different ethnic backgrounds and genders. Table 9.4 shows the results.

Gastroenteritis is generally extremely underreported, as those afflicted rarely seek medical attention (Palmer et al. 1997). For this reason, the reported number of healthcare facility visits underestimates the true number of cases. Hence, for UAE residents (not including tourists), we estimated not only the number of reported healthcare facility visits but also the potential total number of cases, both reported and unreported. Estimates for those ages 5 and over are derived from a meta-analysis of gastroenteritis incidence by world region prepared by Walker and Black (2010), based on a review of studies published between 1980 and 2008. We used incidence rates for European countries, rather than for the Eastern Mediterranean Region, to reflect the UAE's comparatively high development status relative to most other countries in the region. For children under age 5, we used the rate for established market economies reported in Bern (2004). Note that the Walker article provides median and inter-quartile values for incidence rates; we assumed incidence rates are lognormally distributed within each age group and used the median and inter-quartile values to estimate distribution parameters. Table 9.2 shows the assumed median and inter-quartile values for each age group. In the absence of better data, we assumed these parameters do not vary by month—an assumption that should be corrected to account for seasonal variation as more health data become available.

For tourists, the estimated rate of gastroenteritis is based on a meta-analysis by Riddle et al. (2006) of the incidence of GI illness among long-term travelers. We use the incidence reported for the Middle East and North Africa region: 5.3% with a

95% CI of 3.6–7.1%. Riddle et al. do not report incidence rates for long-term travelers in Europe or other highly developed regions, but the mean rate for the Middle East and North Africa is lower than for all other regions (sub-Saharan Africa, Latin America and the Caribbean, and Southeast Asia).

Estimated Burden of Disease

Model Estimates of Burden of Disease

A run of 10,000 Monte Carlo simulations (which led to stable results) estimated that swimming in UAE coastal waters could contribute to 1,300 (95% CI: 800, 1,900) medical visits for gastroenteritis per year among UAE residents. If the potential total number of illnesses (reported through medical visits and unreported) is considered, the estimate rises to 64,000 (95% CI: 15,000, 200,000). These estimates use swimming frequencies estimated from the UAE Health and Lifestyle Survey (for citizens) and from the Australia Sport, Recreation, and Exercise Survey in New South Wales for noncitizens. Employing swimming frequency estimates based on the frequency of drowning results in very similar estimates (1,300 reported cases and 59,000 total cases), and therefore the results presented here use only the swimming frequency estimates developed from the UAE Health and Lifestyle Survey. In addition to the cases among residents, another 24,000 cases (95% CI: 12,000, 39,000) could be expected among tourists. Figure 9.3 shows the resulting estimates for UAE residents, comparing the number of reported medical visits with the total estimated number of cases for each age group. Figure 9.4 compares the estimated total number of cases (using the literature-derived baseline case estimates) by emirate and also shows estimates of the number of cases among tourists. Figure 9.5 shows the total estimated number of cases among residents and tourists by month.

While the potential total number of illnesses as estimated using literature-derived information on gastroenteritis rates is high, it is worth noting that the most common types of gastroenteritis are self-limiting for immunocompetent individuals (i.e., likely swimmers), last only a few days, and do not require medical treatment.

As another measure of the impact of coastal water pollution on the burden of disease in the UAE, we used Eq. 3.1 (see Chap. 3) to convert the number of gastrointestinal illness cases to the unit of disability-adjusted life years (DALYs), a common metric for comparing health effects with differing levels of severity. The "years of life lost due to premature death" (YLL) component of the DALY equation was assumed to be zero for this calculation (i.e., no deaths occur) because gastroenteritis, while unpleasant, is generally not life threatening among immunocompetent people and older (noninfant) children, and individuals healthy enough to be able to swim generally fall into one of those categories. The "years lived with disability" component (YLD) can be calculated with Eq. 3.3. The disability weight for diarrheal disease is given as 0.105 by WHO (2004). Although gastrointestinal diseases

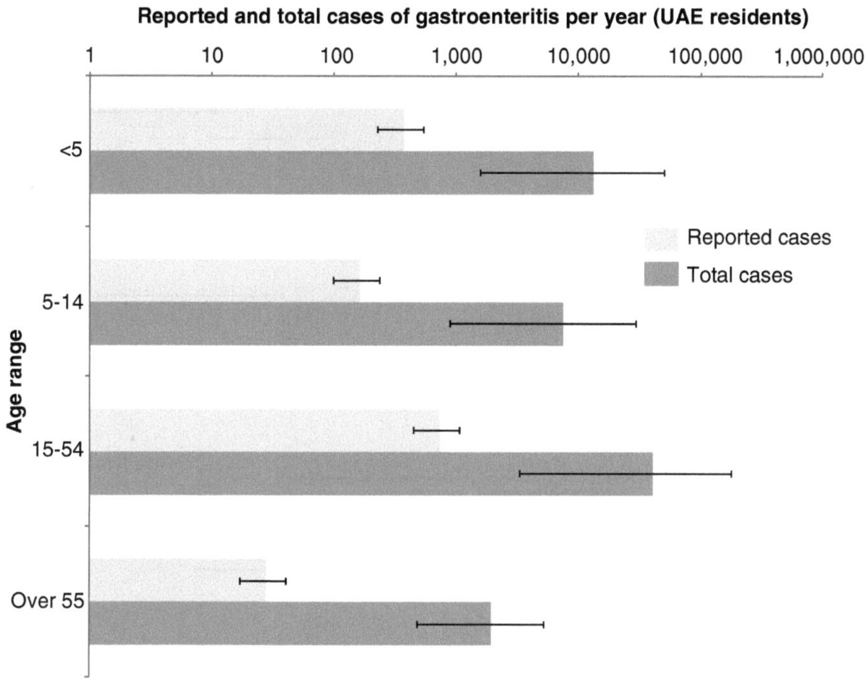

Fig. 9.3 Number of reported medical visits and total estimated number of cases per year of gastroenteritis attributable to exposure to coastal water pollution among UAE residents. Error bars show 95% confidence intervals

acquired through exposure to recreational waters vary in etiological agent, viral illnesses are assumed to be the most common (Boehm et al. 2009; Colford et al. 2007; Craun et al. 2005). Typical cases of viral gastroenteritis last 1–10 days (Centers for Disease Control and Prevention 2009). For this example calculation, the average length of illness was assumed to be 5 days (0.014 years).

Applying the above parameters and Eq. 3.1, 88,000 cases of gastroenteritis (the total for residents and tourists) is equivalent to 130 DALYs. This low value relative to the large number of cases reflects the short duration and less serious nature of the illness.

Sensitivity Analysis

A lack of nation-specific data meant that several model inputs had to be estimated or approximated based on scant coastal water quality observations, UAE physical activity surveys that did not ask specifically about participants' use of coastal waters and that included only UAE citizens, incomplete tourist activity data, and relative risk information from international literature. To investigate the possible effects of

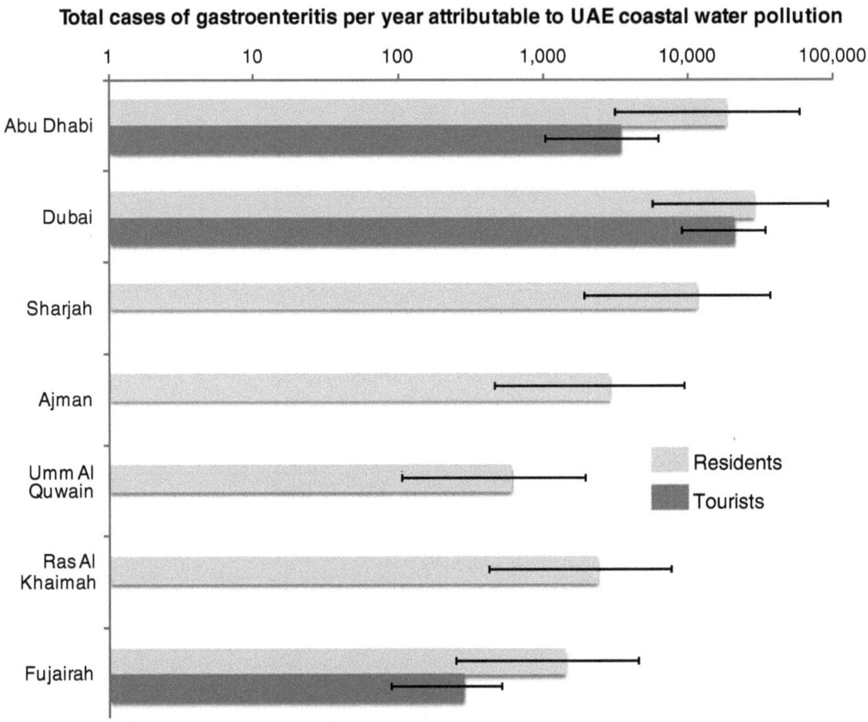

Fig. 9.4 Total estimated number of cases per year of gastroenteritis attributable to exposure to coastal water pollution among UAE residents and tourists, by emirate. Error bars show 95% confidence intervals

these assumptions on the final estimate of the burden of disease, the key model input variables were varied one at a time by ±10% in value while all other values were held constant. Figure 9.6 shows the results of this sensitivity analysis. The figure shows how the median estimated number of reported gastroenteritis cases (1,445) among UAE residents changes as each input value is increased or decreased by 10%. As shown, the estimates are most sensitive to changes in the assumed increase in relative risk per 10-fold (log) increase in enterococci concentration. The estimates also are relatively sensitive to the annual baseline rate of reported gastroenteritis cases, percentage of the population swimming, and highest observed enterococci levels at Abu Dhabi beaches.

Figure 9.7 shows the factor by which the burden of disease estimate changes (e.g., doubles, triples, etc.) as each of these variables changes by factors ranging from 0 to 10. As shown, the estimated burden of disease *decreases* most with decreases in the maximum observed enterococci concentration at beaches—an input that was highly uncertain in this analysis as observational data were available only from two Abu Dhabi beaches and only from single monthly samples at each of these beaches. High spatial and temporal variability in these concentrations is expected due to tidal changes, the many inlets and islands along the coast that

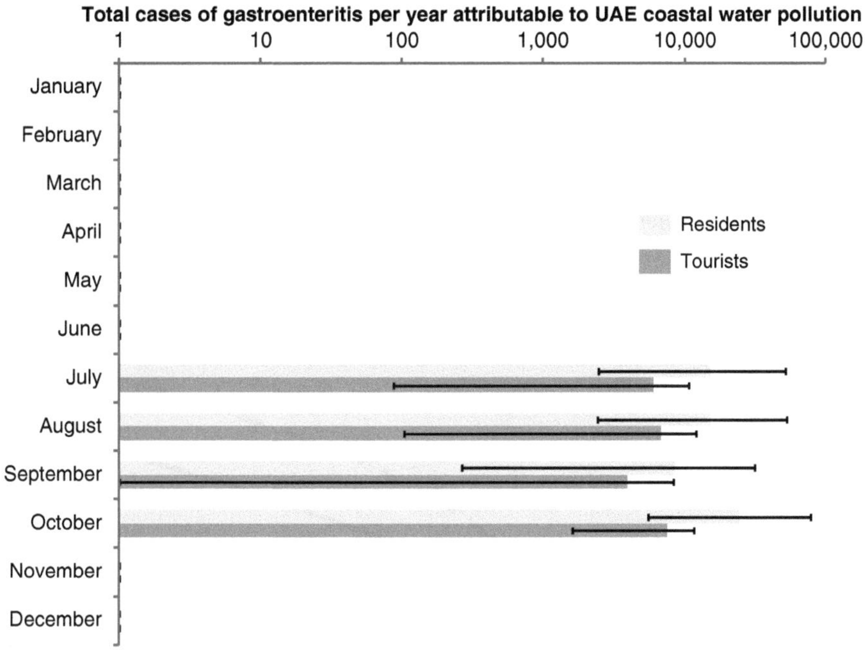

Fig. 9.5 Total estimated number of cases per year of gastroenteritis attributable to exposure to coastal water pollution among UAE residents and tourists, by month. Error bars show 95% confidence intervals

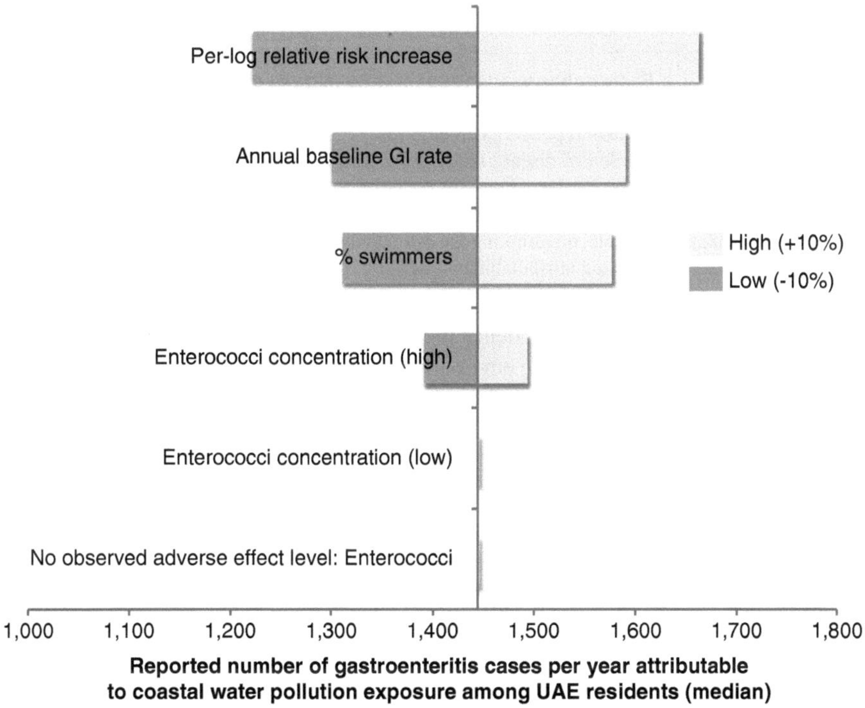

Fig. 9.6 Median estimate of reported gastroenteritis cases attributable to coastal water contamination when burden of disease model input variables are varied by ±10%

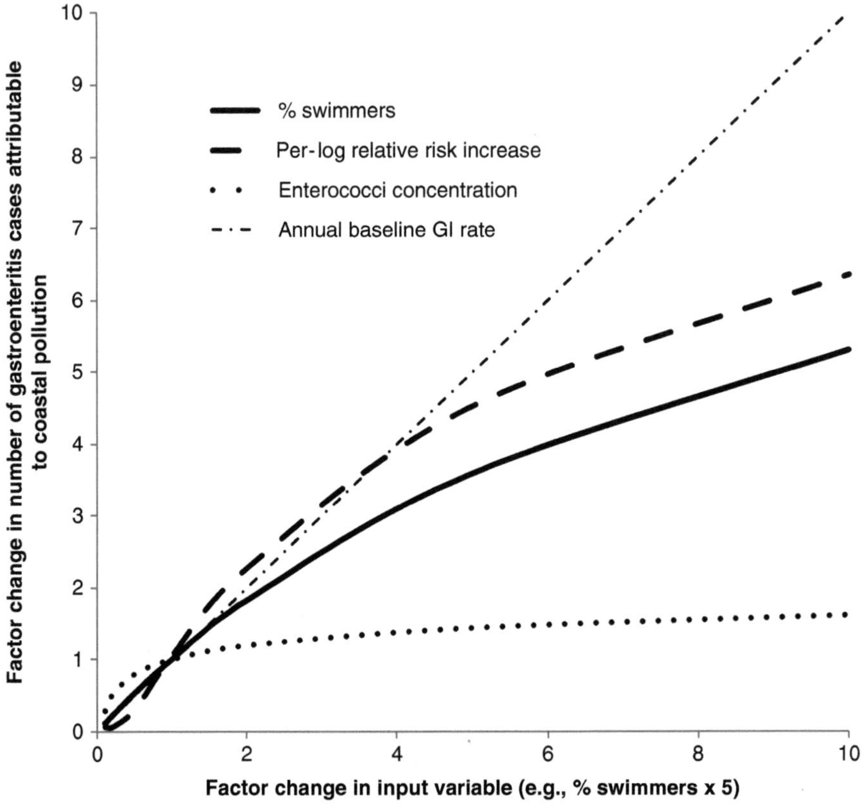

Fig. 9.7 Change in estimate of reported gastroenteritis cases attributable to coastal water contamination with changes in burden of disease model input variables by factors ranging from 0 to 10. The vertical axis depicts the burden of disease estimate as computed with baseline values of all model parameters divided by the estimate when the indicated model parameter is changed by the indicated amount. For example, quadrupling the estimated percentage of each population subgroup that swims triples the estimated attributable number of gastroenteritis cases

change the hydraulic regime and hence the fate of pathogens, and the variable nature of pathogen releases to the marine environment. It is very likely that actual enterococci concentrations could vary by several orders of magnitude throughout the year, rather than only by a factor of 10. This would produce substantial changes in estimates of cases of gastroenteritis among swimmers—beyond those shown in Fig. 9.7. Hence, obtaining better estimates of the pathogen concentrations at UAE beaches and how these vary in space and time is extremely important for establishing a reasonable estimate of the possible number of cases attributable to coastal water pollution.

Figure 9.7 shows that the burden of disease estimate *increases* most with increases in the assumed baseline rate of gastroenteritis, per-log increase in the relative risk of gastroenteritis upon exposure, and percentage of the population swimming. As is expected based on the form of the equations used to calculate the burden of disease, the estimate increases linearly with the estimated number of baseline gastroenteritis cases and nonlinearly with the other variables. Hence, improved data on these other variables also is very important to develop a more credible burden of disease estimate. In particular, it is worth noting that the estimate depends nearly as much on the assumed fraction of the population swimming as it does on epidemiologic estimates of the relative risk of enterococci exposure. Collecting data on the number of people using the UAE's beaches, the demographics of beachgoers, and the temporal variation in their beach use patterns would be a relatively straightforward task, in comparison to carrying out an epidemiologic study to improve the relative risk estimates, although the latter also would be worthwhile.

Comparison with Previous Estimates

In the absence of UAE-specific coastal water quality data RAND Corporation's initial estimate of burden of disease from swimming (see Appendix A) was "nonzero but probably low." The results in this chapter are consistent with this preliminary, qualitative assessment. Even though our best estimate of the total number of gastroenteritis cases attributable to coastal water pollution among UAE residents was high (64,000), when standardized for severity of illness, it is equivalent to fewer than 100 DALYs per year among UAE residents.

The initial RAND estimate also included an estimate of 0–33 deaths due to drowning; however, as this is not considered a health outcome related to environmental exposure, drowning was not considered in this analysis.

Information Needed to Improve Future Burden of Disease Predictions

Relative-risk values were obtained from the most recent and relevant studies in the international scientific literature. Nonetheless, it is important to recognize that these numbers reflect an analysis of data collected outside of the Arabian Gulf region. Though this information is likely generally applicable to the UAE, some evidence exists that susceptibilities can vary geographically (Fleisher 1991), potentially altering relative risks for the local population. Epidemiologic studies that track the health of UAE beach users in the days following their beach visit would provide a stronger basis for developing coastal water burden of disease estimates.

Estimates of the number of swimmers and the baseline rate of gastrointestinal illness in the UAE population also contribute substantially to the uncertainty in these results. For this analysis, we developed estimates of beach use from previous health and lifestyle studies in the UAE and elsewhere, and we estimated tourist numbers based on monthly tourist information from Dubai and annual tourist information from Abu Dhabi and Sharjah. Better characterization of the population using the UAE's beaches and of their behaviors while at the beach would substantially improve the burden of disease estimates. Furthermore, baseline gastroenteritis prevalence rates are notoriously difficult to determine as those affected generally recover relatively quickly and do not require formal medical treatment (Palmer et al. 1997); therefore, hospital or clinic-scale data are generally less appropriate than are health surveys for estimating baseline illness rates as medical visit data reflect only a very small fraction of the affected population. Additional baseline health surveys are required in conjunction with beach visitor studies to determine UAE-specific baseline rates of gastrointestinal illness, beach use patterns, and the proportion of beach visitors in different demographic groups who swim (defined as immersing their face or entire head in water) during their beach outings.

Ideally, because water quality varies considerably in space and time—particularly along the irregular UAE coastline, with its many and changing patterns of inlets and islands—the model presented here would be redeveloped at the beach scale, rather than at the emirate scale, with numbers of visitors and concentration data determined separately for each recreational area (Dwight et al. 2007). A 2001 study by Leecaster and Weisberg of a comprehensive data set of California beach water quality determined that random single monthly samples likely fail to detect 95% of water quality violations. Therefore, at the Abu Dhabi beaches for which data were available, it is very likely that there were at least some days during months with single observations below the NOAEL when enterococci concentrations were elevated and posed some risk to swimmers. Similarly, as the majority of water quality violations in coastal recreational areas are of relatively short duration (Leecaster and Weisberg 2001), it is unlikely that the high concentrations observed during the months of July to October 2006 actually persisted for the entire month. Uncertainty modeling as presented here can be used to give a crude indication of the potential magnitude, from a public health perspective, of coastal water contamination with fecal pathogens, but it is a poor substitute for actual environmental data in describing exposure.

Conclusions

Virtually immediate reductions in health risks related to recreational waters could be achieved through increased monitoring of fecal indicator bacteria concentrations at local beaches and closure of areas in violation of international recommended standards until adequate reductions in microbial concentrations are achieved. Although the cities of Abu Dhabi and Dubai are planning expansion of their coastal water quality monitoring programs, these plans differ in targeted indicator organisms, analytical methods, and proposed sampling frequencies. Similar programs exist in the northern emirates, but no specific information was available for this report.

Ideally, to provide consistent public health safeguards throughout the nation, a standardized coastal water monitoring program should be instituted at the federal level by the Ministry of Environment and Water and implemented by relevant emirate-level agencies. Beaches should be monitored frequently for enterococci bacteria (fecal streptococci) at a minimum, which have been identified as the most useful fecal indicator organism for regulation of marine water quality by WHO (Boehm et al. 2009). Although the necessary frequency of sample collection may differ due to the specific popularity of a given location and its proximity to potential pollution sources, current monthly sampling programs are likely insufficient to protect swimmer health (Leecaster and Weisberg 2001).

Consistent guidelines for beach closures or postings due to periodic high enterococci concentrations also should be established specifying concentration levels necessary to trigger closure, reductions in concentration required to permit reopening, and procedures for public notification. Creation of a national database as a repository for required beach monitoring data is highly recommended to allow for continued examination of water quality trends over time and to encourage collaboration among all emirates to preserve the shared waters along the coast. Adherence to these recommendations should enable additional beaches throughout the UAE to join Abu Dhabi's Corniche Beach in being recognized by the international Blue Flag Programme (http://www.blueflag.org), which certifies beaches as eco-conscious and swimmer friendly. Joining is currently a major goal of Dubai's tourism industry.

National standards for coastal water quality monitoring and beach closures will prevent exposure to waterborne pathogens but will not reduce actual coastal contamination. To achieve such a reduction throughout the nation, consistent national discharge standards for point source effluents, including mandatory daily source monitoring for pathogen indicators and future potential contaminants of concern (e.g., dioxins, PCBs, etc.), should be established to ensure consistently safe recreational areas. Although some effluent discharge recommendations currently exist, there remains considerable confusion at the emirate level as to which government entities are responsible for monitoring and enforcement. Without enforcement, including the levying of meaningful fines or penalties in response to violations, improvements in water quality are unlikely to be achieved. Clarification of the roles and responsibilities of each government entity for ensuring adherence to national discharge standards at the emirate level is critical.

Planned expansions of major municipal wastewater treatment plants, which are particular point sources of concern, will undoubtedly reduce the overloading of current facilities and should continue in order to minimize use of tanker trucks and illegal dumping. Increased sewage treatment will likely result in improved effluent quality that will not only reduce pathogen concentrations in coastal areas but will also reduce nutrient loadings to the Gulf, which are probably at least partially responsible for increasing numbers of hazardous algal blooms (Anderson et al. 2002; Glibert 2007).

References

Abu Hilal, A.H., A.B. Adam, I.M. Banat, and E.S. Hassan. 1994. Sanitary conditions in three creeks in Dubai, Sharjah, and Ajman emirates on the Arabian Gulf (UAE). *Environmental Monitoring and Assessment* 32: 21–36.

Al Zarooni, M., and W. Elshorbagy. 2006. Characterization and assessment of Al Ruwais refinery wastewater. *Journal of Hazardous Materials* A136: 398–405.

Anderson, D.M., P.M. Glibert, and J.M. Burkholder. 2002. Harmful algal blooms and eutrophication: Nutrient sources, composition, and consequences. *Estuaries* 25(4b): 704–726.

Anil, A.C., K. Venkat, S.S. Sawant, M. Dileepkumar, V.K. Dhargalkar, N. Ramaiah, S.N. Harkantra, and Z.A. Ansari. 2002. Marine bioinvasion: Concern for ecology and shipping. *Current Science* 83(3): 214–218.

Australian Sports Commission. 2011. *Participation in exercise, recreation, and sport: 2010 annual report*. Sydney: Government of Australia, Standing Committee on Recreation and Sport.

Badrinath, P., Q.A. Al-Shboul, T. Zoubeidi, A.S. Gargoum, and O.E. El-Rufaie (eds.). 2002. *Measuring the health of the nation: United Arab Emirates health and lifestyle survey 2000*. Al Ain: UAE University, Faculty of Medicine and Health Sciences and College of Business and Economics.

Baker, L.C., and D.J. McGillicuddy. 2006. Harmful algal blooms: At the interface between coastal oceanography and human health. *Oceanography* 19(2): 94–106.

Banat, I.M., E.S. Hassan, A.H. Abu Hilal, and A.B. Adam. 1993. Microbial and nutrient pollution assessment of coastal and creek waters of northern UAE. *Environment International* 19(6): 569–578.

Banat, I.M., E.S. Hassan, M.S. El Shahawi, and A.H. Abu Hilal. 1998. Post Gulf War assessment of nutrients, heavy metal ions, hydrocarbons, and bacterial pollution levels in the United Arab Emirates coastal waters. *Environment International* 24(1/2): 109–116.

Barss, P., A.M. Subait, M. Hassan Al Ali, and M. Grivna. 2009. Drowning in a high-income developing country in the Middle East: Newspapers as an essential resource for injury surveillance. *Journal of Science and Medicine in Sport* 12: 164–170.

Bern, C. 2004. Diarrhoeal diseases. In *The global epidemiology of infectious diseases*, ed. C.J.L. Murray, A.D. Lopez, and C.D. Mathers. Geneva: World Health Organization. Chapter I.

Boehm, A.B., N.J. Ashbolt, J.M. Colford, L.E. Dunbar, L.E. Fleming, M.A. Gold, J.A. Hansel, et al. 2009. A sea change ahead for recreational water quality criteria. *Journal of Water and Health* 7(1): 9–20.

Cabelli, V.J., A.P. Dufour, L.J. McCabe, and M.A. Levin. 1982. Swimming-associated gastroenteritis and water quality. *American Journal of Epidemiology* 115(4): 606–616.

Calderon, R.L., E.W. Mood, and A.P. Dufour. 1991. Health effects of swimmers and nonpoint sources of contaminated water. *International Journal of Environmental Health Research* 1: 21–31.

Carvalho, S. 2009. Dh2b Fujairah resort lures tourists. *Gulf News*, May 3. http://archive.gulfnews.com/indepth/atm/more_stories/10037279.html

Centers for Disease Control and Prevention. 2009. Viral gastroenteritis. http://www.cdc.gov/ncidod/dvrd/revb/gastro/faq.htm

Colford, J.M., T.J. Wade, K.C. Schiff, C.C. Wright, J.F. Griffith, S.K. Sandhu, S. Burns, M. Sobsey, G. Lovelace, and S. Weisberg. 2007. Water quality indicators and the risk of illness at beaches with nonpoint sources of fecal contamination. *Epidemiology* 18(1): 27–35.

Corbett, S.J., G.L. Rubin, G.K. Curry, and D.G. Kleinbaum. 1993. The health effects of swimming at Sydney beaches. *American Journal of Public Health* 83(12): 1701–1706.

Craun, G.F., R.L. Calderon, and M.F. Craun. 2005. Outbreaks associated with recreational water in the United States. *International Journal of Environmental Health Research* 15(4): 243–262.

Dubai Department of Tourism and Commerce Marketing. 2007. Hotel Statistics Summary. http://www.dubaitourism.ae/EServices/HotelStatistics/tabid/167/language/en-US/Default.aspx

Dwight, R.H., M.V. Brinks, G. SharavanaKumar, and J.C. Semenza. 2007. Beach attendance and bathing rates for Southern California beaches. *Ocean and Coastal Management* 50(10): 847–858.

Environment Agency–Abu Dhabi (EAD). 2007. *Report on water quality of Abu Dhabi coastal waters*. Abu Dhabi: Environment Agency–Abu Dhabi (EAD).

Fleisher, J.M. 1991. A re-analysis of data supporting U.S. federal bacteriological water quality criteria governing marine recreational waters. *Research Journal of the Water Control Pollution Federation* 63(3): 259–265.

Fleming, L., K. Broad, A. Clement, E. Dewailly, S. Elmir, A. Knap, S.A. Pomponi, S. Smith, H. Solo-Gabriele, and P. Walsh. 2006. Oceans and human health: Emerging public health risks in the marine environment. *Marine Pollution Bulletin* 53: 545–560.

Glibert, P.M. 2007. Eutrophication and harmful algal blooms: A complex issue, examples from the Arabian Seas, including Kuwait Bay, and an introduction to the Global Ecology and Oceanography of Harmful Algal Blooms (GEOHAB) Programme. *International Journal of Oceans and Oceanography* 2(1): 157–169.

Graneli, E., and J.T. Turner. 2006. An introduction to harmful algae. In *Ecology of harmful algae*, ed. E. Graneli and J.T. Turner, 3–21. Berlin: Springer.

Haugland, R.A., S.C. Siefring, L.J. Wymer, K.P. Brenner, and A.P. Dufour. 2005. Comparison of *Enterococcus* measurements in freshwater at two recreational beaches by quantitative polymerase chain reaction and membrane filter culture analysis. *Water Research* 39: 559–568.

Health Authority–Abu Dhabi (HAAD). 2009. *2008 health data for Abu Dhabi emirate*. Abu Dhabi: Health Authority–Abu Dhabi (HAAD).

Henrickson, S.E., T. Wong, P. Allen, T. Ford, and P.R. Epstein. 2001. Marine swimming-related illness: Implications for monitoring and environmental policy. *Environmental Health Perspectives* 109(7): 645–650.

Hsieh, S., J.T. O'Leary, and A.M. Morrison. 1992. Segmenting the international travel market by activity. *Tourism Management* 13(3): 209–223, June.

Kay, D., J.M. Fleisher, R.L. Salmon, F. Jones, M.D. Wyer, A.F. Godfree, Z. Zelenauch-Jacquotte, and R. Shore. 1994. Predicting likelihood of gastroenteritis from sea bathing: Results from randomized exposure. *Lancet* 334: 905–909.

Kay, D., M.D. Wyer, J. Crowther, and J. Fewtrell. 1999. Faecal indicator impacts on recreational waters: Budget studies and diffuse source modelling. *Journal of Applied Microbiology (Symposium Supplement)* 85: 70S–82S.

Kay, D., J. Bartram, A. Prüss, N. Ashbolt, M.D. Wyer, J.M. Fleisher, L. Fewtrell, A. Rogers, and G. Rees. 2004. Derivation of numerical values for the World Health Organization guidelines for recreational waters. *Water Research* 38: 1296–1304.

Leecaster, M.K., and S.B. Weisberg. 2001. Effect of sampling frequency on shoreline microbiology assessments. *Marine Pollution Bulletin* 42(11): 1150–1154.

Menon, P. 2009. Red tide closes two Dubai beaches. *The National*, April 7. http://www.thenational.ae/article/2009/0407/NATIONAL/138119746/1186

Mitchell, R.J., A.M. Williamson, and J. Olivier. 2010. Estimates of drowning morbidity and mortality adjusted for exposure to risk. *Injury Prevention* 16: 261–266.

Morinigo, M.A., R. Cornax, M.A. Munoz, P. Romero, and J.J. Borrego. 1990. Relationships betwen *Salmonella* spp. and indicator microorganisms in polluted natural waters. *Water Research* 24(1): 117–120.

Palmer, S., H. Houston, B. Lervy, D. Riberio, and P. Thomas. 1997. Problems in the diagnosis of foodborne infection in general practice. *Epidemiology and Infection* 117: 479–484.

Polo, F., M.J. Figueras, I. Inza, J. Sala, J.M. Fleisher, and J. Guarro. 1998. Relationship between presence of *Salmonella* and indicators of faecal pollution in aquatic habitats. *FEMS Microbiology Letters* 160: 253–256.

Prüss, A. 1998. Review of epidemiological studies on health effects from exposure to recreational water. *International Journal of Epidemiology* 27: 1–9.

Riddle, M.S., J.W. Sanders, S.D. Putnam, and D.R. Tribble. 2006. Incidence, etiology, and impact of diarrhea among long-term travelers (U.S. military and similar populations): A systematic review. *American Journal of Tropical Medicine and Hygiene* 74(5): 891–900.

Sambidge, A. 2008. "Red tide" forces desalination plant closure. *ArabianBusiness.com*, November 17. http://www.arabianbusiness.com/538468-red-tide-forces-desalination-plant-closure

Saunders, J.E., K.M. Al Zahed, and D.M. Paterson. 2007. The impact of organic pollution on the macrobenthic fauna of Dubai Creek (UAE). *Marine Pollution Bulletin* 54: 1715–1723.

Savichtcheva, O., and S. Okabe. 2006. Alternative indicators of fecal pollution: Relations with pathogens and conventional indicators, current methodologies for direct pathogen monitoring and future application perspectives. *Water Research* 40: 2463–2476.

Setrakian, L. 2009. Filthy rich: Dubai choking on sewage. *ABC News*, February 1. http://abcnews.go.com/International/Story?id=6781673&page=1

Shekhar, S. 2009. Abu Dhabi aims for significant rise in tourism spend. *Emirates Business 24-7*, May 7.

Telegraph. 2009. Dubai's polluted beaches closed to public. January 29. http://www.telegraph.co.uk/travel/travelnews/4380051/Dubais-polluted-beaches-closed-to-public.html

Turbow, D.J., N.D. Osgood, and S.C. Jiang. 2003. Evaluation of recreational health risk in coastal waters based on enterococcus densities and bathing patterns. *Environmental Health Perspectives* 111(4): 598–603.

U.S. Environmental Protection Agency (EPA). 1986. *Ambient water quality for bacteria–1986*. Office of Water: Washington, DC. http://www.epa.gov/waterscience/beaches/files/1986crit.pdf

UAE Ministry of Economy. 2008. United Arab Emirates: Population and Vital Statistics 2008. Abu Dhabi: UAE Ministry of Economy Central Statistics Department.

Wade, T.J., N. Pai, J.N.S. Eisenberg, and J.M. Colford. 2003. Do U.S. Environmental Protection Agency water quality guidelines for recreational waters prevent gastrointestinal illness? A systematic review and meta-analysis. *Environmental Health Perspectives* 111(8): 1102–1109.

Wade, T.J., R.L. Calderon, E. Sams, M. Beach, K.P. Brenner, A.H. Williams, and A.P. Dufour. 2006. Rapidly measured indicators of recreational water quality are predictive of swimming-associated gastrointestinal illness. *Environmental Health Perspectives* 114(1): 24–28.

Walker, C.L., and R.E. Black. 2010. Diarrhoea morbidity and mortality in older children, adolescents, and adults. *Epidemiology and Infection* 138: 1215–1226.

Wiedenmann, A., P. Kruger, K. Dietz, J.M. Lopez-Pila, R. Szewzyk, and K. Botzenhart. 2006. A randomized controlled trial assessing infectious disease risks from bathing in fresh recreational waters in relation to the concentration of *Escherichia coli*, intestinal enterococci, *Clostridium perfringens*, and somatic coliphages. *d* 114(2): 228–236.

World Health Organization (WHO). 2004. Global burden of disease 2004 update: Disability weights for diseases and conditions. http://www.who.int/healthinfo/global_burden_disease/GBD2004_DisabilityWeights.pdf

Chapter 10
Burden of Disease from Soil and Groundwater Contamination

Abstract Soil and groundwater contamination due to waste disposal may pose an increasing public health threat in the United Arab Emirates (UAE) if measures are not taken to improve waste management practices and prevent exposure to wastes disposed of improperly in the past. The UAE currently has one of the highest rates of solid waste generation per capita of any country in the world. In addition, waste disposal in the UAE historically has been inadequately controlled, with wastes of a wide variety disposed of in open, unlined dump sites in the desert. Chemicals can leach from uncontrolled waste disposal sites and contaminate soil and the underlying groundwater. The soil in much of the UAE is silty and sandy with low cation exchange capacity. This soil type is highly permeable, and thus contaminants that leach from waste sites have the potential to migrate rapidly and contaminate large areas. At the time this project was carried out, no data were available on the nature and amounts of hazardous chemicals found in soil and groundwater in the UAE from waste disposal sites, but the types of chemicals present due to releases from waste disposal sites are likely to be similar to those found in groundwater contaminated from past waste disposal practices in other developed countries. These chemicals are associated with a range of effects, from cancer to neurological and reproductive effects to suppression of the immune system. Current information is not sufficient to assess the burden of disease due to soil and groundwater contamination from waste sites in the UAE. At present, this disease burden is likely to be small because of the small size of the potentially exposed population. However, given the plans to invest in developing the Western Region, it would be prudent for the UAE to begin to collect the information needed to assess risks from these sites to the current population and to future residents. Our primary recommendation is a two-part process that first would provide approximate estimates of the potential burden of disease from individual waste disposal sites and then develop detailed risk assessments for sites showing a significant health risk potential. The information needed for the first part of this process should be relatively easy to obtain, with the primary effort required for additional visual inspections of a selected number of waste sites. Once these basic site inspections and preliminary risk assessments are

J. MacDonald Gibson et al., *Environmental Burden of Disease Assessment*, Environmental Science and Technology Library 24, DOI 10.1007/978-94-007-5925-1_10,

completed, the UAE government will have greatly improved estimates of which waste sites may be cause for concern, allowing it to focus on sites that have significant risk potential.

Keywords Soil and groundwater contamination • Environmental burden of disease • Relative risk • Attributable fraction • United Arab Emirates • Dose-response assessment • Uncontrolled waste disposal sites • Screening-level risk assessment • Quantitative risk assessment

Overview: Nature and Sources of Soil and Groundwater Contamination

Groundwater and soil can become contaminated from a number of human activities. Common sources of groundwater and soil contamination include agricultural application of pesticides and fertilizers, leaking septic tanks, leaking underground storage tanks at petrol stations and industrial sites, abandoned and uncontrolled hazardous and solid waste disposal sites, underground injection wells, routine use of solvents and other industrial chemicals for activities such as cleaning metal parts, municipal landfills, runoff from roadways and urban areas, and air pollutants that have settled on the land surface.

Humans can become exposed to pollutants that reach soil and groundwater from any of these sources through a number of pathways. These include drinking contaminated groundwater, inhaling vapors from contaminants either at the site of contaminant release or in buildings into which these vapors have migrated, directly contacting and then unintentionally ingesting contaminated soil, and eating food that has been grown on contaminated soil or irrigated with contaminated groundwater.

Although all of the contamination sources mentioned above may be important, this chapter focuses on soil and groundwater contamination from the disposal of solid and hazardous waste.

Soil and groundwater contamination due to waste disposal may pose an increasing public health threat in the United Arab Emirates if measures are not taken to improve waste management practices and prevent exposure to wastes disposed of improperly in the past. The UAE's rate of waste production is growing rapidly. Indeed, the UAE has one of the highest rates of solid waste generation per capita of any country in the world. As an example, average waste production in Abu Dhabi is 2.3 kg/day per person, whereas waste production rates in countries belonging to the Organisation for Economic Co-Operation and Development are estimated at 1.5 kg/day (Environment Agency–Abu Dhabi 2007).

In addition, waste disposal in the UAE historically has been inadequately controlled, with wastes of a wide variety disposed of in open, unlined dump sites in the desert. Although these waste disposal areas currently are sparsely inhabited, plans are under way for extensive developments that will substantially increase the number of people at risk of exposure. As this development proceeds, it will be important for the UAE to plan for potential risks that may arise due to past waste disposal practices.

Although the scope of this chapter is limited, future assessments of the state of the environment in the UAE should address all potential sources of groundwater and soil contamination and associated health risks. Agricultural pesticides and fertilizers, leaking underground storage tanks, and industrial use of solvents are likely to be important potential sources of groundwater contamination that should be analyzed.

Solid and Hazardous Waste Disposal Practices in the UAE

Until recently, wastes in the UAE were disposed of in open dumps that lacked systems to prevent leaching of the wastes into soil and groundwater (Al Ashram 2005). Most municipalities gathered waste and roughly sorted it into organic matter, which was composted, and inorganic matter, which was taken to a dump. However, even the major dumps (such as Al Dhafra, which handled most of the waste from Abu Dhabi City) lacked controls such as liners to prevent leaching of contaminants into groundwater. A recent review of waste management practices in Abu Dhabi emirate noted, "Unlined landfills without gas collection systems receive commingled, untreated wastes that may be leaching into the limited groundwater supply. There are no facilities for the treatment or disposal of hazardous wastes from nonoil industries; moreover, there is a lack of reliable waste generation and characterization data and an ineffective system for recording and tracking hazardous wastes" (Environment Agency–Abu Dhabi 2009). Further, although the major cities exercise some control over waste disposal, the sparsely inhabited Western Region of Abu Dhabi has "no recycling facilities, transfer stations or household compost plants, and the waste collected is transported directly to various dumping sites" (Al Ashram 2005). As a result, the Western Region is littered with uncontrolled dump sites.

The UAE has recognized the potential risks of its past waste disposal practices and over the past several years has begun taking steps to vastly improve its solid and hazardous waste management systems. In 2001, the federal Handling of Hazardous Materials, Hazardous Wastes, and Medical Wastes bylaw to Federal Law 24 for the first time established the authority of the government to begin to control hazardous wastes (Environment Agency–Abu Dhabi 2009). Three other subsequent government policies (Law 21 in 2005, Decree 18 in 2007, and Decree 20 in 2008) have further strengthened the government mandate to control waste disposal and to track hazardous wastes (Environment Agency–Abu Dhabi 2009). A 2007 government decree established the Center of Waste Management–Abu Dhabi (CWM) "to coordinate waste management throughout Abu Dhabi emirate" (Environment Agency–Abu Dhabi EAD 2009). The CWM began operations in February 2008 and soon after issued an ambitious waste management strategy aimed at elevating waste management practices in Abu Dhabi emirate to the best international standards (CWM 2008).

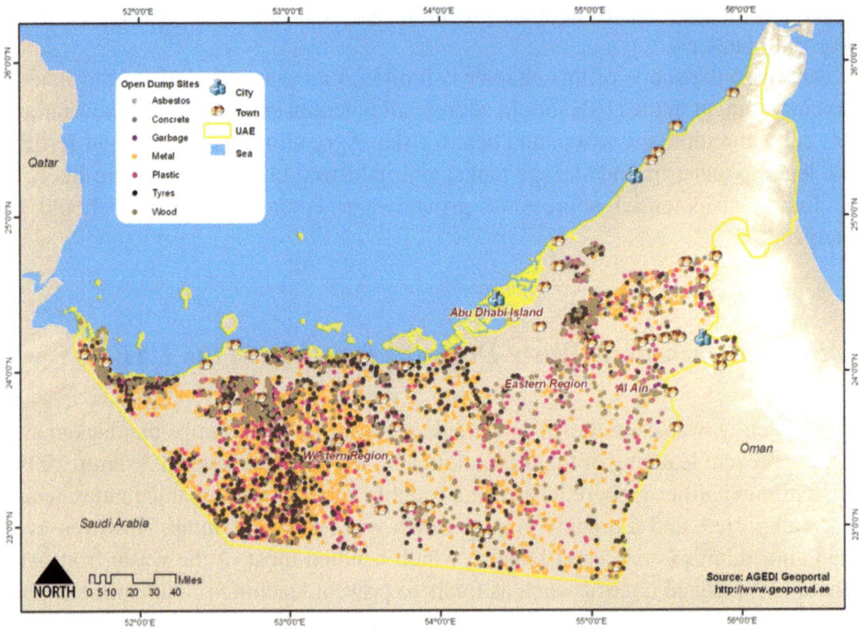

Fig. 10.1 Locations of uncontrolled waste disposal sites in Abu Dhabi emirate. As shown, these sites are concentrated in Abu Dhabi's Western Region (map created by Chidsanuphong Chart-asa, University of North Carolina, Chapel Hill)

Locations of Uncontrolled Waste Disposal Sites in the UAE

Despite the ambitious plans now under way to improve the UAE's handling of solid and hazardous wastes, the UAE will increasingly face the potential for risks stemming from a legacy of uncontrolled waste disposal. To begin to understand the potential magnitude of risks posed by these legacy waste disposal sites, we obtained from the Environment Agency–Abu Dhabi (EAD) data on locations of historic dump sites. These data are from 2007 aerial surveys of waste locations across Abu Dhabi emirate. The location of each waste dumping point was recorded with a global positioning system, and the total quantity of waste was estimated. The waste locations were classified into seven groups based on waste class: asbestos, concrete, garbage, metal, plastic, tires, and wood (EAD 2008). Figure 10.1 shows the locations of the waste sites found in this aerial survey. Figure 10.2 shows the distribution of wastes by type (garbage, plastic, etc.). In total, the aerial survey found 18,640 dump sites containing nearly 8 million kg of wastes.

As Fig. 10.1 illustrates, the dump sites are concentrated in the Western Region (recently renamed Al Gharbia) and, to a lesser extent, the sparsely inhabited portions of the Eastern Region of Abu Dhabi. Currently, the Western Region contains a small population, estimated at 138,000, representing only about 8% of the total population of Abu Dhabi emirate and about 2% of the total UAE population (Abu Dhabi Tourism Authority 2010). Chemicals leaching from these legacy dump sites thus pose a risk

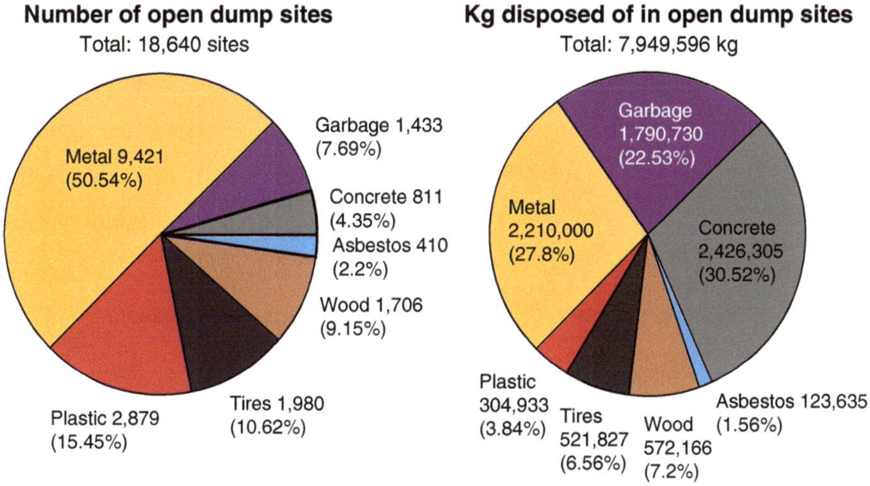

Fig. 10.2 Contents of waste sites shown in Fig. 10.1, by number and weight

for a relatively small population group, especially in comparison to outdoor air pollution, which has a much broader reach. However, plans have been announced for a $27 billion initiative to develop the Western Region. These plans include substantial expansions to the existing settlements in the region. New residential areas, government centers, schools, parks and sports centers, and industrial facilities all are planned (Urban Planning Council 2009). Thus, the size of the potential population exposed to these legacy risks will increase substantially in the coming years.

Risks to Soil and Groundwater from Uncontrolled Waste Disposal

Chemicals can leach from uncontrolled waste disposal sites such as many of those shown in Fig. 10.1 and contaminate soil and the underlying groundwater. The soil in much of the UAE is silty and sandy with low cation exchange capacity (EAD 2006). This soil type is highly permeable, and thus contaminants that leach from waste sites have the potential to migrate rapidly and contaminate large areas.

At the time this project was carried out, no data were available on the nature and amounts of hazardous chemicals found in soil and groundwater in the UAE from waste disposal sites. In 2005, EAD undertook an extensive program to sample water wells for pesticides and fertilizers. They found positive results for pesticides in only 3 of 228 samples collected from wells throughout the emirate, but the positive results from these samples were not confirmed in duplicate samples (Al Ashram 2005). Nitrates (associated with agricultural fertilizers) were found in all samples and were present at levels above WHO guidelines in 80% of the samples. However, this sampling effort did not include the kinds of industrial and commercial chemicals that could be expected at waste disposal sites. Thus, at the time this project was carried out, no data were available on the extent to which the waste sites shown in Fig. 10.1 may be affecting groundwater and soil quality.

Table 10.1 Chemicals most frequently detected at U.S. hazardous waste sites

Rank	Compound	Common sources
1	Trichloroethylene	Dry cleaning; metal degreasing
2	Lead	Gasoline (prior to banning lead in gasoline); mining; construction materials (pipes); manufacturing
3	Tetrachloroethylene	Dry cleaning; metal degreasing
4	Benzene	Gasoline; manufacturing
5	Toluene	Gasoline; manufacturing
6	Chromium	Metal plating
7	Methylene chloride	Degreasing; solvents; paint removal
8	Zinc	Manufacturing; mining
9	1,1,1-Trichloroethane	Metal and plastic cleaning
10	Arsenic	Mining; manufacturing
11	Chloroform	Solvents
12	1,1-Dichloroethane	Degreasing; solvents
13	1,2-Dichloroethene-trans	Transformation product of 1,1,1-trichloroethane
14	Cadmium	Mining; plating
15	Manganese	Manufacturing; mining; occurs in nature as oxide
16	Copper	Manufacturing; mining
17	1,1-Dichloroethane	Manufacturing
18	Vinyl chloride	Plastic and record manufacturing
19	Barium	Manufacturing; energy production
20	1,2-Dichloroethane	Metal degreasing; paint removal
21	Ethylbenzene	Styrene and asphalt manufacturing; gasoline
22	Nickel	Manufacturing; mining
23	Di(2-ethylhexyl)phthalate	Plastics manufacturing
24	Xylenes	Solvents; gasoline
25	Phenol	Wood treating; medicines

National Research Council (1994)

Types of chemicals present due to releases from waste disposal sites are likely to be similar to those found in groundwater contaminated from past waste disposal practices in other developed countries, including the United States. Table 10.1 shows the 25 most commonly detected chemicals in groundwater at U.S. hazardous waste sites. Due to their common use in industrial and commercial applications in developed economies, it is likely that these same kinds of chemicals may be found in soil and groundwater at some of the uncontrolled waste disposal sites in the UAE. The third column of Table 10.1 shows common sources of these contaminants. Many of the sources shown (such as metal degreasing, dry cleaning, construction, and manufacturing of a variety of products) are likely to be found in the UAE.

Key Health Effects of Soil and Groundwater Contamination

A variety of adverse health effects may be associated with chemicals commonly found at waste disposal sites. The chemicals listed in Table 10.1 are associated with a range of effects, from cancer to neurological and reproductive effects to suppression of the immune system. Whether any of the kinds of adverse health outcomes potentially influenced by these chemicals may occur in a population exposed to contamination via waste disposal sites is highly uncertain for a number of reasons.

A major cause of uncertainty in predicting the health effects of waste disposal sites results from uncertainty in knowledge of the health effects of specific contaminants. Most of the information on potential health effects comes from toxicological studies in animals (Persad and Cooper 2008). For example, the Integrated Risk Information System (IRIS) is perhaps the most extensive available database on human health effects associated with chemicals in the environment. Health effects are derived from human studies for only 44 of the 545 chemicals in the database, with the rest of the health effects estimates based on animal studies (Persad and Cooper 2008). Even for those 44 contaminants, the data generally are derived from high levels of exposure in occupational settings rather than from the lower exposure levels that typically result from environmental contamination. The extent to which the effects observed in high doses in workers also will be observed in populations exposed to much lower doses through environmental media is not known.

Another cause of uncertainty is that waste disposal sites contain complex mixtures of chemicals. Mixtures of chemicals may cause health effects beyond those due to any single chemical acting alone (Carpenter et al. 2002). Different contaminants may affect the same kinds of human cells, causing effects that may either increase or decrease risks compared with those that would be expected based on one contaminant alone. As Carpenter, Arcaro, and Spink note, "In the extreme case, there may be synergistic effects, in which case the effects of two substances together are greater than the sum of either effect alone."

Over the past few decades, a number of epidemiologic studies have sought to estimate the health effects in populations potentially exposed to chemicals from waste disposal sites. Vrijheid (2000) of the London School of Hygiene and Tropical Medicine conducted an exhaustive review to locate all such studies in the United States and Europe between 1980 and 1998. She found 41 studies that had assessed health risks due to specific waste disposal sites (most of them in the United States). Of these 41 studies, 33 found positive associations between potential exposure to chemicals from waste disposal sites and at least one health effect. In general, the health effects could be grouped into three categories: adverse pregnancy outcomes such as low birth weight and birth defects; cancers; and self-reported health outcomes such as headaches, respiratory symptoms, sleepiness, and psychological stress. A few of the studies evaluated other health outcomes (such as height of exposed children, chromosomal aberrations, liver function, and immunologic function). Figure 10.3 shows the total number of studies that assessed these different groups of health outcomes and the percentage with positive findings.

As shown in Fig. 10.3, more than 90% of the studies that analyzed self-reported health outcomes found elevated rates of such outcomes in populations exposed to waste disposal sites. These increased rates may be in part due to fears associated with knowing about the potential for exposure to chemicals from a waste site. Thus, exposed populations may be more alert to potential health symptoms. However, Vrijheid suggests, "From a public health point of view, the findings of high symptom

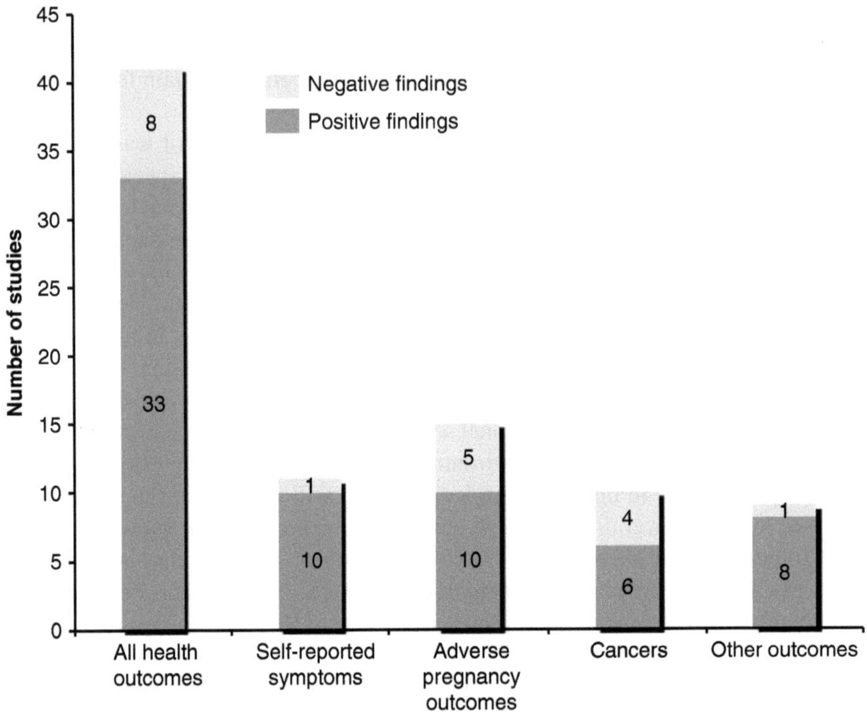

Fig. 10.3 Number of studies assessing health risks due to specific waste disposal sites, mostly in the United States (Vrijheid 2000). "Positive findings" indicates positive associations between potential exposure to chemicals from waste disposal sites and at least one health effect

reporting, whether or not due to differential self-reporting, may indicate the impact that stress and concerns related to landfill can have on ill health and/or perceived ill health."

Figure 10.3 shows that adverse pregnancy outcomes were found in two-thirds of the studies that tracked such outcomes. Among these studies, the most consistent finding was an increase in low birth weight. According to Vrijheid, "These were generally well-designed studies and low birth weight is thought to be a sensitive marker of effects of chemical exposures.... Fetuses, infants, and children are generally thought to be more vulnerable and therefore experience toxic effects at lower doses than the adult population."

The findings on cancer risks due to exposure to waste sites are less conclusive. Of ten studies that assessed cancer risks, six found positive associations between these risks and waste site exposure. However, one might expect a 20% chance of six such positive findings based on chance alone. That is, if a waste site were as likely not to cause cancer as it were to cause cancer, then among ten waste sites we would expect a 20% chance of six positive findings for cancer. According to Vrijheid, many of the studies that assessed cancer endpoints were weak because they did not include sufficient information on confounding factors (i.e., other factors, such as smoking, that could lead to elevated cancer risks). Further, the long latency period between exposure to chemicals and the onset of cancer poses difficulties for accurately assessing cancer risks due to exposure to chemicals from waste sites.

While the weight of evidence from studies such as the 41 summarized in Fig. 10.3 increasingly indicates that adverse health effects may be associated with exposure to chemicals from waste sites, a great deal of uncertainty on the nature of these effects remains. Further, individual waste sites may vary considerably in their contents and thus in the nature of chemicals that could be released into the environment. For example, as shown in Fig. 10.1, many of the waste sites in the Western Region appear to contain only wood or concrete, which may not pose health risks at all (unless the wood is treated with preservatives that could leach into the soil and groundwater), while others contain garbage that could contain a variety of chemicals. Thus, information from studies such as those summarized by Vrijheid and Fig. 10.3 cannot, in general, be extrapolated to other waste sites.

Method for Estimating the Environmental Burden of Disease in the Future

At this time, information is not sufficient to assess the burden of disease due to soil and groundwater contamination from waste sites in the UAE. At present, this disease burden is likely to be small because of the small size of the potentially exposed population. However, given the plans to invest in developing the Western Region, it would be prudent for the UAE to begin to collect the information needed to assess risks from these sites to the current population and to future residents.

As Fig. 10.1 illustrates, the number of uncontrolled waste disposal sites and landfills in the UAE is extremely large. Clearly, simultaneously investigating and assessing risks from all of these sites—the uncontrolled sites as well as municipal landfills—is not feasible. Thus, we recommend a two-part process that first narrows the list of sites where detailed risk assessment would be undertaken and then proceeds with detailed, quantitative risk assessments of these high-priority sites.

Part 1: Conduct Screening-Level Risk Assessments

As a first step, we recommend screening all potential waste disposal sites and then developing preliminary risk estimates for a subset of these sites, using the attributable fraction approach described in Chap. 3. This screening-level assessment would proceed as described in the next sections.

Step A: Identify Sites for Visual Inspection

The first step would be to sort the database of waste disposal sites upon which Fig. 10.1 is based, plus all of the known municipal landfills, by waste type and mass. Priority would be given to sites containing large masses of substances that may pose immediate groundwater and soil contamination risks. Municipal landfills and other garbage sites would be given high priority. Sites containing metals and plastics would be given priority for inspection as well, in case they contain drums or other containers of hazardous liquids that could leak into the soil and groundwater. Asbestos disposal sites also would be included. Sites containing only wood or concrete could be postponed for future inspection. Studies have indicated that health risks due to exposure to chemicals from waste tires is likely to be minimal, even at playgrounds that use tire crumb as a play surface (Birkholz et al. 2003); thus, sites containing only tires also could be deferred for future inspection.

Step B: Visually Inspect Sites

In this step, ground crews would visually inspect each site identified as a priority in Step A. The crews would record in a database whether a site appears to contain any wastes (e.g., cleaners and solvents, petroleum products, pesticides, paints, batteries) with potential to contaminate the soil and groundwater beneath or adjacent to the waste site. Sites not containing such wastes would be removed from further consideration.

Step C: Assess the Potential Burden of Disease

Steps A and B would result in a list of sites for which screening-level risk assessments would be carried out. These screening-level risk assessments would use the attributable fraction approach described in Chap. 3 to estimate the numbers of adverse pregnancy outcomes and cancers that could be attributable to chemicals from each site.

For this analysis, the total population within a vicinity of approximately 2 km from each waste site would be identified. (This distance is recommended because previous studies of adverse health effects of waste sites typically have focused on populations within 2 km or less of the site.) Health statistics for these populations then would be sought from the Health Authority–Abu Dhabi (HAAD), which has plans to collect high-resolution data on the place of residence of medical patients. The key health statistics needed would be

- the total number of cases of leukemia and of bladder, lung, and stomach cancer in the potentially exposed population (i.e., among those residing within 2 km of each site) and
- the total number of adverse pregnancy outcomes (low birth weight, pre-term birth, and birth defects) in the potentially exposed population.

A thorough literature review would be conducted to identify relative risks assessed for previous waste sites for these groups of health end points, building on the information contained in Vrijheid (2000) and updating it with recent studies. The attributable number of cancers and adverse reproductive outcomes could then be calculated for each priority waste site using:

$$Attributable\,Health\,Outcomes = \frac{RR-1}{RR} \times I_0 \tag{10.1}$$

The *RR* value would be a random variable characterized from a meta-analysis of epidemiologic studies (using a procedure similar to that used to characterize relative risks from indoor air exposures as described in Chap. 5). The baseline rate of disease, I_0, for each site would be obtained from HAAD health statistics for the population living within 2 km of a waste site.

Step D: Prioritize Sites for Further Quantitative Risk Assessment

Sites from Step C with the largest attributable burden of disease would be identified for detailed risk assessment (and, optimally, for rehabilitation) as described in Part 2.

Part 2: Conduct Detailed Risk Assessments for Priority Sites

The attributable fraction approach in Part 1 will indicate how many adverse health outcomes might be associated with any one waste site. However, this approach relies on epidemiologic data from other waste sites, and, because each site is unique, conducting a detailed risk assessment is necessary to more accurately predict the potential risks to exposed populations. Such detailed risk assessments are costly and time-consuming, however. The screening approach of Part 1 will help reduce the number of sites for which detailed risk assessments are needed. The screening study may reveal that some sites pose no or very minimal risks, and therefore a detailed assessment is not justified.

Detailed guidelines from a number of sources provide the specifics of how to go about assessing health risks from waste disposal sites (U.S. Environmental Protection Agency 1989; Agency for Toxic Substances and Disease Registry 2005). The following sections describe the steps of this detailed risk assessment process in general terms (National Research Council 1983).

Step A: Hazard Identification

In this step, the waste disposal site is carefully investigated to determine the types of hazardous substances that may be present. This step can begin with visual inspection. Based on the results of this inspection, samples of soil and groundwater in the vicinity of the site can be collected to check for chemicals that may be associated with wastes disposed of at the site. High-priority chemicals for testing would be identified based on the types of waste at the site and associated knowledge of the chemicals that may leach from such wastes.

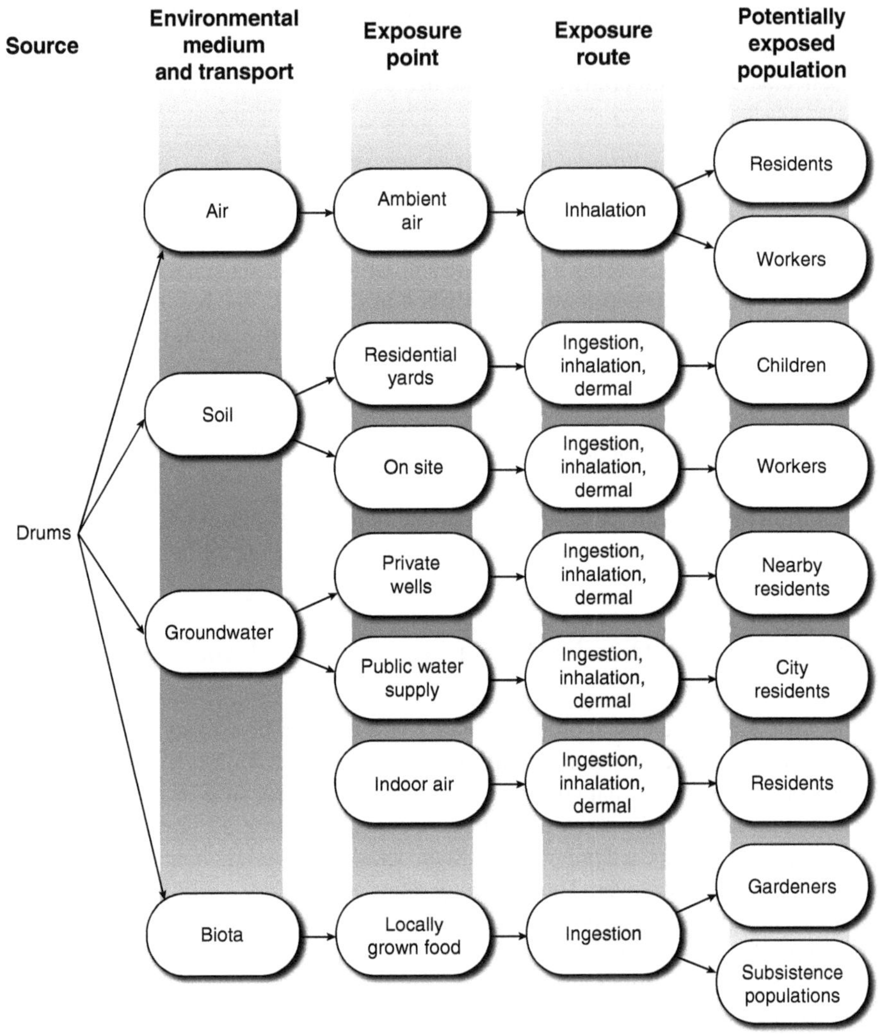

Fig. 10.4 Potential routes of exposure to chemicals originating from waste disposal sites

Step B: Exposure Assessment

This step involves systematically identifying (1) the routes by which people may be exposed to the contaminants, (2) the number of people potentially exposed via each route, and (3) the dose of contaminant potentially received by each of these routes.

Figure 10.4 shows the potential routes of exposure to chemicals from waste sites. During the exposure assessment, the relative importance of these different exposure routes is identified, and some routes may be eliminated as concerns. An often-overlooked exposure route, the importance of which is increasingly recognized, is inhalation of vapors inside buildings that overlie plumes of contaminated groundwater (Johnson and Ettinger 1991; Johnson 2005). In one study of multiple pathways, exposure to tetrachloroethylene—one of the most common contaminants at hazardous waste sites (Table 10.1)—from soil-gas accounted for 75% of the total exposure—more than ingestion via drinking water, inhalation of vapor while showering, and inhalation of vapors from outdoor air (Hodgson et al. 1992).

Ideally, the doses of contaminants to which people are exposed can be directly measured from sampling of water, soil, air, and/or food at the points of exposure. When this is not possible, these concentrations can be estimated through environmental modeling. A number of texts and guidance documents provide details on how to estimate contaminant concentrations in groundwater, soil, air, and food based on concentrations found in the vicinity of the waste disposal site (U.S. Environmental Protection Agency 1989; Anderson and Woessner 1991; Johnson and Ettinger 1991; Johnson 2005; Fetter 2008).

Step C: Dose-Response Assessment

This step involves calculating the probability that an exposed member of the population will contract an adverse health outcome. Multiple chemicals likely will need to be assessed for each site; some of these chemicals may cause multiple adverse effects that also will need to be assessed individually. Detailed guidance on conducting the dose-response assessment is available from many sources (U.S. Environmental Protection Agency 1989; Agency for Toxic Substances and Disease Registry 2005). In general, the probability of an adverse health outcome is estimated from a function that resembles:

$$P(\textit{health outcome}) = \textit{risk factor} \times \textit{exposure concentration} \qquad (10.2)$$

The exposure concentration is determined from the exposure assessment step (through either environmental sampling at the points of exposure or through modeling). Risk factors for a wide variety of chemicals commonly found at waste sites are available in the IRIS database previously mentioned, available at http://www.epa.gov/iris/.

Step D: Risk Characterization

The last step of the risk assessment process is to characterize risks to the exposed population, based on information from the previous steps. Risks may be characterized in many ways, including

- the number of adverse health outcomes of specific types expected to occur in the exposed population (i.e., in communities surrounding the waste site),
- the number of adverse health outcomes in sensitive populations (such as pregnant women or children),

- the probability that an "average" member of the exposed community will develop an adverse outcome, and/or
- the probability that the most sensitive or most highly exposed member of the community will develop an adverse health outcome.

Based on this information, decisions can be made about whether the waste site needs to be cleaned up and whether steps need to be taken to protect current populations from further exposure.

Protecting Future Populations

While in the immediate future the UAE should give highest priority to waste sites posing risks to the current population, carrying out similar assessments focused on future populations is essential given the major expansion planned for the Western Region and the large number of uncontrolled waste sites in that region. Through such advance risk assessments, the UAE can identify steps to protect future populations from the kinds of consequences that have occurred elsewhere around the world when former waste sites have been redeveloped without adequate protections in place. In North America, perhaps the best-known example of such a site was Love Canal (see Box 10.1). Failure to adequately plan for the protection of residents of a community that grew up around this former waste disposal site led to the eventual evacuation of entire neighborhoods, with the U.S. government purchasing houses from affected homeowners. In the United States, such legacy waste sites have led to tragic consequences and hardships for communities that could have been avoided with advance risk assessment and appropriate planning to address identified risks before construction of the new communities. The UAE is in a position to avoid such events through planning, remediation of sites that may pose risks to future developments, and land-use controls where remediation cannot provide the necessary level of protection.

Box 10.1 Love Canal: An Example Consequence of Failing to Track and Clean Up Former Waste Disposal Sites

Say the words "Love Canal" to any American born in the last 80 years and a host of images will come to mind: families displaced from their homes, children with disabilities, a quiet community ruined by substances in their own back yard. Love Canal is the name of a middle class neighborhood in Niagara Falls, New York. About 900 families were living there in 1978 when a local mother named Lois Gibbs started investigating the health problems she saw in area school children, including her son. After months of researching, Gibbs discovered that the Love Canal was the receptacle of decades of industrial waste from Hooker Chemical, a company in Niagara Falls. Some estimate that the area was built on more than 21,000 tons of chemical waste. Worse, the

(continued)

Box 10.1 (continued)

local school, the 99th Street School, had knowingly been built on the waste site in 1954, one year after Hooker Chemical stopped its dumping practices. In addition, chemicals from this waste disposal site had migrated via groundwater and soil into the yards and gardens in Love Canal, where children played for years. In her book *Love Canal: My Story*, Gibbs writes of dogs that came inside the house after playing in the backyard with burns on their noses and children whose hands would itch and burn after touching the grass surrounding their homes (Gibbs 1982). Leakage from the site was detected in loc.l creeks, sewers, soil, and the indoor air of homes. A media storm brewed over the community.

Though everyone agreed that something was wrong at Love Canal, there was no precedent in American environmental history. The community engaged in a years-long battle with Hooker Chemical, which repeatedly failed to acknowledge that its chemicals caused the health crisis at Love Canal. Scientists visiting the site detected dioxin in the water and benzene (a well-studied carcinogen) in the soil but could not conclusively agree on how the chemicals got there. In 1979 the U.S. Environmental Protection Agency (EPA) found that 33% of Love Canal residents had chromosomal damage, compared with 1% of people in normal population studies. This discovery prompted President Jimmy Carter to declare Love Canal a health emergency. One thing was certain: The story was an internationally talked-about nightmare, an extreme example of what can happen when the government, industrial, and civil sectors do not track hazardous waste and address problems due to the legacy of past waste disposal practices.

By the early 1980s, the federal government relocated the families of Love Canal and demolished homes. The 99th Street School was also demolished. Love Canal opened a Pandora's box for the EPA and local governments throughout the United States, which started to question the safety of waste sites. The Love Canal mess inspired the government to pass the Comprehensive Environmental Response, Compensation, and Liability Act, commonly known as the Superfund Act. In the years since Love Canal, Superfund has spent billions of dollars cleaning up hundreds of hazardous waste sites found throughout the United States.

Information Needed to Estimate the Environmental Burden of Disease from Soil and Groundwater Contamination

Carrying out the two-part process to prioritize waste sites and assess associated risks will require information beyond what is currently available in the UAE. The amount of information needed for each of the two parts is different, with much more data (and as a result time and resources) needed for the detailed quantitative risk estimates (Part 2) than for Part 1. Key information needs for the screening-level risk assessments (Part 1) include:

- **Geographic coordinates of sites identified as containing wastes of potential concern**. These sites would need to be identified by a team of inspectors who would visit each location containing a significant mass of garbage, plastic, metal, and/or asbestos. The inspectors would identify which sites appear to contain materials that could release contaminants into the soil and groundwater (e.g., containers of solvents, petroleum products, cleaning products, pesticides, and other chemicals; batteries; and solid materials with significant potential to leach chemicals).
- **Population living within 2 km of each site (ideally stratified by age)**. As described in Chap. 4, our research team mapped the population distribution across the UAE at a fine spatial scale, in order to estimate exposure to ambient air pollution, using the LandScan™ Global Population Database from the U.S. Oak Ridge National Laboratory. This fine-scale population map is based on predictions from satellite data showing nighttime lighting, number of roads, land cover and other information. Ideally, census data specific to the areas in the vicinity of each priority waste disposal site could be obtained to improve estimates of the current population near each site.
- **Observed health outcomes in at-risk populations**. While we previously obtained data from HAAD on the numbers of specific health outcomes observed in 73% of the Abu Dhabi population (see Chap. 3), information on the places of residence of those experiencing the health outcomes was not complete. Ideally, for the screening-level risk estimates, approximate locations of the places of residence of each person experiencing a health endpoint of potential concern (adverse pregnancy outcomes and certain cancers) would be made available.

Information needed to characterize risks for the sites selected for detailed assessments (Part 2) is much more detailed. The kinds of information required will vary depending on the approach—direct sampling at the points of exposure or modeling—used to characterize exposure. The resources referenced in the earlier section on Part 2 provide details on information needs. Whether exposure is estimated through direct sampling or through modeling, the information necessary for such detailed assessments is quite substantial compared with what is needed for the recommended screening-level assessment. Hence, the screening-level assessments are important in order to reduce the number of sites requiring detailed assessments to a tractable number.

Conclusions

The UAE currently faces a legacy of improper waste disposal practices, as illustrated by the vast number of uncontrolled waste disposal sites in Fig. 10.1. Wastes improperly disposed of on land in the past may cause the contamination of soil and groundwater as chemicals from these wastes leach into the ground and then migrate to the water beneath. Currently, the population potentially exposed to these wastes appears to be small, primarily concentrated in the Western Region and in less-inhabited areas of the Eastern Region of Abu Dhabi. Thus, the current risks attributable to groundwater and soil contamination from improper waste disposal likely will appear small when compared with other risks, such those attributable to outdoor air pollution, to which a large fraction of the population is exposed. Indeed, participants in the risk-ranking exercise described in Chap. 2 ranked soil and groundwater contamination risks as low in part due to the small population potentially exposed.

Despite the relatively small size of the current population at risk of exposure to chemicals from waste disposal sites, steps should be taken to assess risks to these at-risk communities so that appropriate protections can be put in place where necessary. Further, steps are needed to assess and prevent potential risks to future populations that will be drawn to areas proposed for future, large-scale developments. These risk assessments should include not just the illegal waste disposal sites, but also all existing municipal landfills, because these have historically not included the engineering controls or waste segregation procedures needed to protect soil and groundwater from contamination (Al Ashram 2005).

The primary recommendation is to carry out a two-part process that first would provide approximate estimates of the potential burden of disease from individual waste disposal sites and then develop detailed risk assessments for sites showing a significant health risk potential. The information needed for the first part of this process should be relatively easy to obtain, with the primary effort required for additional visual inspections of a selected number of waste sites. Once these basic site inspections and preliminary risk assessments are completed, the UAE government will have greatly improved estimates of which waste sites may be cause for concern. The UAE can then focus on sites that have significant risk potential.

Through such staged risk assessments, plus remediation of sites posing risks and improved waste management practices into the future, the UAE can prevent public health problems that otherwise might arise due to waste disposal.

References

Abu Dhabi Tourism Authority. 2010. Population. http://www.visitabudhabi.ae/en/uae.facts.and.figures/population.aspx

Agency for Toxic Substances and Disease Registry. 2005. *Public health assessment guidance manual (2005 update)*. Atlanta: Agency for Toxic Substances and Disease Registry. http://www.atsdr.cdc.gov/HAC/PHAManual/toc.html

Al Ashram, O. 2005. *Wastes and pollution sources in Abu Dhabi emirate*. Abu Dhabi: Environment Agency–Abu Dhabi, Environment Protection Department.

Anderson, M.P., and W.W. Woessner. 1991. *Applied groundwater modeling: Simulation of flow and advective transport*. London: Academic.

Birkholz, D.A., K.L. Belton, and T.L. Guidotti. 2003. Toxicological evaluation for the hazard assessment of tire crumb for use in public playgrounds. *Journal of the Air and Waste Management Association* 53: 903–907.

Carpenter, D.O., K.A. Arcaro, and D.C. Spink. 2002. Understanding the human health effects of chemical mixtures. *Environmental Health Perspectives* 110(Suppl. 1): 25–42.

Center of Waste Management—Abu Dhabi (CWM). 2008. *Abu Dhabi waste management strategy: A cleaner Abu Dhabi*. Abu Dhabi: CWM.

Environment Agency–Abu Dhabi (EAD). 2006. *Water resources of Abu Dhabi Emirate*. Abu Dhabi: EAD, Abu Dhabi Global Environmental Data Initiative.

Environment Agency–Abu Dhabi (EAD). 2007. *State of the environment Abu Dhabi.* Abu Dhabi: EAD. http://www.agedi.ae/SOEAssessment/default.aspx

Environment Agency–Abu Dhabi (EAD). 2008. *Waste and pollution sources of Abu Dhabi Emirate, United Arab Emirates.* Abu Dhabi Global Environmental Data Initiative. Abu Dhabi: Environment Agency–Abu Dhabi.

Environment Agency–Abu Dhabi (EAD). 2009. *Draft gap analysis report on Abu Dhabi environment-related policies*. Abu Dhabi: EAD Policy Sector.

Fetter, C.W. 2008. *Contaminant hydrogeology*, 2nd ed. Long Grove: Waveland Press.

Gibbs, L. 1982. *Love Canal: My story*. Albany: State University of New York Press.

Hodgson, A.T., K. Garbesi, R.G. Sextro, and J.M. Daisey. 1992. Soil-gas contamination and entry of volatile organic compounds into a house near a landfill. *Journal of the Air and Waste Management Association* 42(3): 277–283.

Johnson, P.C. 2005. Identification of application-specific critical inputs for the 1991 Johnson Ettinger vapor intrusion model. *Groundwater Monitoring and Remediation* 25(1): 63–78.

Johnson, P.C., and R.A. Ettinger. 1991. Heuristic model for predicting the intrusion rate of contaminant vapors into buildings. *Environmental Science and Technology* 25(8): 1445–1452.

National Research Council. 1983. *Risk assessment in the federal government: Managing the process*. Washington, DC: National Academies Press.

National Research Council. 1994. *Alternatives for groundwater cleanup*. Washington, DC: National Academies Press.

Persad, A.S., and G.S. Cooper. 2008. Use of epidemiologic data in Integrated Risk Information System (IRIS) assessments. *Toxicology and Applied Pharmacology* 223: 137–147.

U.S. Environmental Protection Agency (EPA). 1989. *Risk assessment guidance for superfund volume I: Human health evaluation manual (Part A)*. Washington, DC: EPA, Office of Emergency and Remedial Response.

Urban Planning Council. 2009. *Plan Al Gharbia 2030*. Abu Dhabi: Urban Planning Council. www.algharbia2030.com

Vrijheid, M. 2000. Health effects of residence near hazardous waste landfill sites: A review of epidemiologic literature. *Environmental Health Perspectives* 108(1): 101–112.

Chapter 11
Burden of Disease from Produce and Seafood Contamination

Abstract Eating fruits and vegetables is beneficial to human health but exposes people to risk if the produce contains hazardous contaminants. Two potential contaminants are human pathogens (e.g., *Salmonella, E. coli*) and agricultural pesticides (e.g., organophosphates, carbamates), both of which can be reduced with proper food handling and preparation. Foodborne pathogens can cause and/or contribute to an array of human illnesses, including acute gastroenteritis as well as more complex chronic conditions such as organ failure, arthritis, and heart disease. Agricultural pesticide exposure can result in dizziness, nausea, abdominal cramps, diarrhea, tremors, anxiety, confusion, neurological disorders, developmental/reproductive disorders, and death. Because large percentages of fruit, vegetables, grains, and legumes consumed in the United Arab Emirates are produced abroad, pesticide use and other farm management practices in countries exporting to the UAE will affect contamination levels of food consumed in the UAE. Domestically harvested seafood has historically been a primary staple of the Emirati diet. More than 90% of citizens eat fish during at least one meal every week. Consumption of fish provides numerous documented health benefits, including a reduction in risk of chronic heart disease; however, fish can also serve as a vector for pathogenic microorganisms (e.g., *Vibrio* spp.), heavy metals (e.g., mercury) and other toxins (e.g., dioxin). Estimates of illness resulting from seafood consumption focus on exposure to mercury. Although numerous metals can result in adverse health effects if consumed in seafood, mercury is generally regarded as of greatest concern. Chronic mercury poisoning results in a host of neurological and psychological symptoms, including tremors, motor/cognitive dysfunction, and memory loss. Exposure in utero can result in serious lifetime illness, including mental retardation, sensory loss, developmental delay, cerebral palsy, and seizures. In lieu of estimating foodborne mortality and morbidity cases, our modeling approach directly calculates the probability of exceeding international guidelines for exposure to specific hazardous chemicals in fruit, vegetables, and seafood in the UAE. For fruits and vegetables, the model estimates the number of daily incidents in which UAE residents are exposed to a particular type of pesticide residue above a prespecified benchmark dose, due to eating

J. MacDonald Gibson et al., *Environmental Burden of Disease Assessment*, Environmental Science and Technology Library 24, DOI 10.1007/978-94-007-5925-1_11,

a particular type of fruit or vegetable. For seafood, the model estimates the number of daily incidents in which UAE residents are exposed to mercury levels above the reference dose maintained by the U.S. Environmental Protection Agency due to eating fish. Results of daily cases in which a UAE resident may be at risk of overexposure to methylmercury from eating seafood and exceeding the reference dose suggest 2,927 women and 11,882 men—with the gender imbalance an artifact of the male-dominated expatriate workforce—could be at risk for health effects. Of all pesticides and crops, chlorpyrifos on tomato has the highest mean ratio (0.26) of average estimated pesticide exposure (0.000078 mg/kg) to its chronic population adjusted dose (cPAD) value (0.0003 mg/kg), making tomatoes the most suitable candidate for a worst-case hypothetical scenario. Considering an atypical but theoretical UAE resident eating 100% tomatoes, and assuming no reduction in pesticide due to washing, peeling, and/or cooking, the model estimates this person has 20.6% (chlorpyrifos) and 1.0% (vinclozolin) chances of exceeding cPAD values each day. Overall, this model estimates 631,074 worst-case daily incidents (cPAD exceedance) contributing to potential chronic illness. Although these probabilities may seem high, daily cPAD incidents are assumed contributory toward potential cases of annual chronic illness; the model assumes (worst-case) no reduction in pesticide due to washing, peeling, and/or cooking for all incidents; and, only very limited human epidemiologic studies exist to objectively link chronic pesticide exposure with adverse health effects—a major reason for the safety factors already built into the cPAD and other benchmarks.

Keywords Produce and seafood contamination • Environmental burden of disease • United Arab Emirates • Marine water quality monitoring • Agricultural pathogens • Food safety standards • Harmful algal blooms ("red tides") • Organophosphate and carbamate pesticides • Acetylcholinesterase inhibition • Acceptable daily intake • Acute population adjusted dose • Chronic population adjusted dose • Organic methylmercury • Reference dose

Overview: Nature and Sources of Produce and Seafood Contamination

Factors Affecting Produce and Seafood Contamination

Progress toward the United Arab Emirates' goal of national self-sufficiency in agricultural production has been slowed in recent decades by four factors: globalization of food supply markets, higher food demand due to rapid UAE population growth, changes in residents' food preferences due to the country's economic growth, and the inherent difficulties in transforming arid desert land into arable crop fields. Recently reported figures (Khan and Salama 2008) suggest that about 85% of the UAE's food supplies are now imported, at an estimated annual cost of AED 11 billion (US$3 billion).

Further complicating matters is a rise in salinity of crop irrigation water and soil, stemming from dwindling groundwater supplies, saltwater intrusion, lack of proper drainage and evapotranspiration of irrigation water, estimated to reach full depletion within the next 50 years at current extraction levels (Environment Agency–Abu Dhabi EAD 2006). Water trends have already spurred the development of wastewater treatment plants for irrigation of city-maintained decorative plants and shade trees. The UAE does not yet irrigate agricultural crops with treated wastewater, although feasibility studies for this process—including analysis of potential pathogenic transmission—are under way (UAE Ministry of Environment and Water 2006).

Looking past food availability, agricultural land use, and irrigation issues, remaining food concerns typically include risk of contamination by environmental pollutants, especially those introduced into the production environment by humans. Food consumed in the UAE, whether grown domestically or abroad, is susceptible to—and therefore tested for—a variety of hazardous pathogens and toxic compounds. Each of the UAE's seven emirates maintains a food-control authority, department, or municipality section responsible for regulating, testing, and enforcing food safety standards as the product moves from the domestic farm (or port of entry) to the marketplace. In the case of domestic crops, the Ministry of Environment and Water (MOEW) works alongside the emirate-level food control authorities and new efforts such as the Abu Dhabi Farmers Services Centre (ADFSC) to promote agricultural practices geared toward reducing the risk of food contamination.

Eating fruits and vegetables is beneficial to human health but exposes people to risk if the produce contains hazardous contaminants. Two potential contaminants are human pathogens (e.g., *Salmonella, E. coli*) and agricultural pesticides (e.g., organophosphates, carbamates), both of which can be reduced with proper food handling and preparation. Because large percentages of fruit, vegetables, grains, and legumes consumed in the UAE are produced abroad, pesticide use and other farm management practices in countries exporting to the UAE will affect contamination levels of food consumed in the UAE.

Because the UAE is a coastal desert nation with only a relatively recent substantial quantity of land-based agriculture, domestically harvested seafood has historically been a primary staple of the Emirati diet. A recent national health and lifestyle survey indicates that seafood consumption remains high. More than 90% of citizens eat fish during at least one meal every week (Badrinath et al. 2002). Consumption of fish provides numerous documented health benefits, including a reduction in risk of chronic heart disease (Mozaffarian and Rimm 2006); however, fish can also serve as a vector for pathogenic microorganisms (e.g., *Vibrio* spp.), heavy metals (e.g., mercury) and other toxins (e.g., dioxin) (Fleming et al. 2006). While pathogenic microorganisms can be minimized by proper food-handling procedures and cooking, toxins and heavy metals cannot. Prevention of exposure to heavy metals and toxins in seafood depends on limiting consumption (particularly by sensitive groups such as pregnant women or small children) and/or reduction of pollutant levels in the coastal environment to prevent seafood contamination.

The UAE has instituted several relevant pieces of legislation to continue to preserve coastal water and seafood quality, including Federal Law 24 of 1999 which aims to protect the marine environment from pollution (UAE Federal Government 1999). Marine water quality monitoring generally is the responsibility of relevant emirate-level agencies, although federal level guidelines can be set by MOEW.

This chapter describes potential sources of UAE food contamination, explains the probabilistic modeling process of the burden of disease due to contaminated food, and discusses results of the model within a context of existing UAE legislation concerning agricultural production and monitoring. For the purposes of this chapter, "food" is primarily considered to be fruit, vegetables, and seafood. Consumed meats (e.g., chicken, mutton, beef) are beyond the scope of this discussion.

Food Contaminants Affecting UAE Residents

Major contamination pathways for crops and seafood are illustrated in Figs. 11.1 and 11.2, respectively. Discussion of these pathways, resultant potential health effects, and current UAE-specific conditions are detailed below.

Agricultural Pesticides

Proper application of pesticides—including insecticides, fungicides, herbicides and other compounds used for the control of pests—is a major contributor to higher and more uniform agricultural crop yields. Misuse or misapplication of pesticides, however, is believed to significantly affect human health.

When pesticides are restricted, farmers must generally substitute other means of pest control. In addition, pesticide-use restriction or prohibition can increase the cost of particular foods, potentially leading consumers to substitute less healthy foods from other sources. Pesticide regulation can therefore have an effect on the overall health risk consumers actually experience and should be considered in policy decisions (Gray and Hammitt 2000; Ragsdale 2000). Conversely, pesticide regulation is often useful in protecting public health if particular formulations or usage patterns are determined to be causing harm.

More than 400 pesticide formulations are used in the UAE (Environment Agency–Abu Dhabi 2007b). Pesticide monitoring activities are conducted by the food control authorities of individual emirates (e.g., Abu Dhabi Food Control Authority). The Codex Alimentarius Commission—a joint initiative of the World Health Organization (WHO) and the Food and Agricultural Organization of the United Nations—establishes guidelines for maximum allowable pesticide residue levels (MRLs) for specific crop/pesticide combinations. Although the food control departments of each UAE municipality generally follow these MRL guidelines (or those similarly adopted by the European Food Safety Authority or the U.S. Environmental Protection Agency), these levels are not set by UAE federal law and are subject to adjustment by each individual emirate.

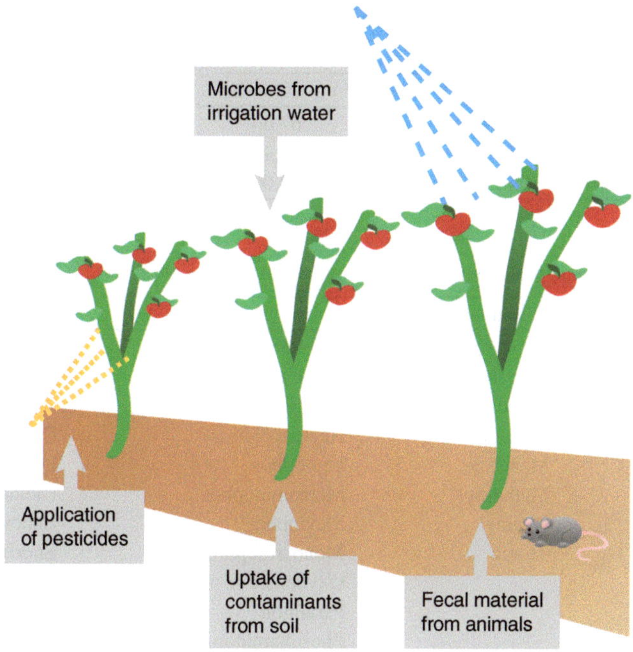

Fig. 11.1 Potential sources of UAE crop contamination

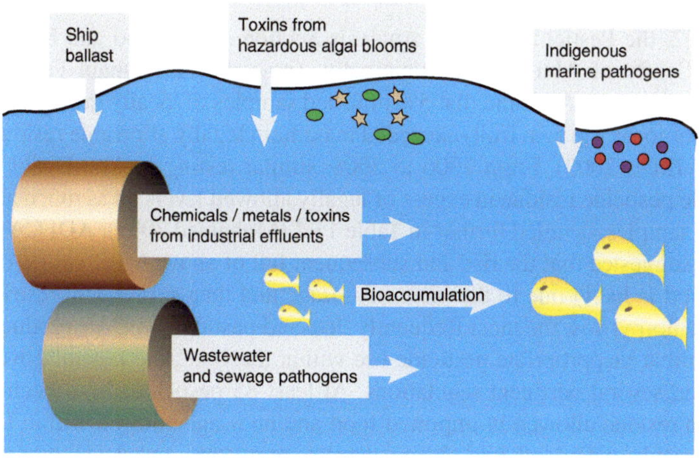

Fig. 11.2 Potential sources of UAE seafood contamination

Table 11.1 2006–2008 pesticide residue testing results for Abu Dhabi emirate

	Crops grown in UAE			Imported crops		
	2006 (%)	2007 (%)	2008 (%)	2006 (%)	2007 (%)	2008 (%)
Zero residue detected	79.70	73.30	67.60	86.60	82.19	64.30
Residue below MRL	19.70	22.33	32.40	12.80	13.69	32.30
Residue above MRL	0.57	4.36	0.00	0.58	4.10	3.50

Table 11.2 Synthetic pesticides banned in the UAE, by chemical family

Triazines	Carbamates	Organophosphates	Organochlorines
Atrazine	Aldicarb	Dichlorvos	Aldrin
Cyanazine	Carbaryl	Disulfoton	Chlordane
Simazine	Carbofuran	Fenthion	Chlordecone
	Mancozeb	Heptenophos	DDT
	Maneb	Leptophos	Dicofol
	Methomyl	Methamidophos	Dieldrin
	Oxamyl	Mevinphos	Endosulfan
	Thiram	Monocrotophos	Endrin
		Oxydemeton–methyl	Heptachlor
		Parathion	Hexachlorobenzene
		Phosphamidon	Kelevan
		Tetrachlorvinphos	Methoxychlor
			Mirex
			Pentachlorophenol
			Strobane

Al Ashram (2005)

In 2002, the Pesticide Residue Analysis Section at the Food and Environment Control Centre of Abu Dhabi Municipality (now the Abu Dhabi Food Control Authority, or ADFCA) found that 5% of tested samples of locally grown vegetables (and 0% of locally grown fruit) exceeded maximum legally allowable residue levels (*Khaleej Times* 2002). From 2006 to 2008, similar testing in Abu Dhabi showed detectable pesticide residue in excess of legally allowed levels on as much as 4.36% of tested samples, detailed further in Table 11.1 (Mohamed 2009). ADFCA personnel have suggested that the rise and subsequent fall of above-MRL values over this time period is likely due to the increased usage and then subsequent restriction in 2007 of chlorpyrifos, the most frequently detected pesticide in UAE produce.

To combat inappropriate pesticide use within the UAE, the federal government has passed several pertinent regulations. At least 85 pesticides have been banned (i.e., zero residue allowed in imported food and no usage allowed within UAE) so far as a result of Federal Law 24 concerning protection and development of the environment (UAE Federal Government 1999) and the subsequent regulations generated by the former Ministry of Agriculture and Fisheries (UAE Federal Government 2004). Additionally, the manufacture and formulation of any pesticide is prohibited in the UAE, and a federal registry controls which particular pesticides can be imported or used legally in the country (Al Ashram 2005). Table 11.2 lists

Table 11.3 Estimated annual U.S. illnesses and deaths from known foodborne pathogens

	Illnesses	% of all foodborne illnesses	Deaths	% of all foodborne deaths
Norwalk-like viruses	9,200,000	66.59	124	6.85
Campylobacter	1,963,141	14.21	99	5.47
Salmonella	1,341,873	9.71	553	30.57
Staphylococcus	185,060	1.34	2	0.11
Escherichia coli O157:H7	62,458	0.45	52	2.87
Listeria monocytogenes	2,493	0.02	499	27.58
Cholera (*Vibrio cholerae*)	49	< 0.01	0	0.00

Mead et al. (1999)

the banned pesticides in the four categories of pesticides considered in this chapter.

Due to their widespread farming use, potential for human health effects, detected presence in domestic and imported food in the UAE, and availability of data, this chapter focuses on organophosphates, carbamates, and pyrethroids (excepting the banned chemicals listed in Table 11.2).

Agricultural Pathogens

Approximately 30% of all worldwide emerging infectious diseases from 1940 to 2004 were caused by pathogens commonly transmitted through food (Jones et al. 2008; Kuchenmüller et al. 2009). Singularly large, attention-raising outbreaks of foodborne illness—especially those associated with specific restaurants or food processing plants—are more likely to be publicly reported and followed up with research efforts (Batz et al. 2005). But recorded outbreak events are an insufficient indicator of illness attributable to foodborne pathogens. These data should therefore be supplemented with expert advice when considering food safety policy issues, since rare or difficult-to-identify pathogens are thought to be underrepresented in such reports (Hoffmann et al. 2008).

Regional varieties in agricultural production methods, food safety standards, and consumer handling of food suggest that one country's foodborne disease statistics may not be accurately scalable to another nation. Table 11.3 is restricted to selected pathogen-specific cases in the United States and is shown here as one regional example, but U.S. trends may not necessarily reflect UAE conditions.

Heavy Metals in Seafood

Numerous metals and metalloids are discharged into the world's oceans via industrial and municipal effluents. Aquatic plants and animals living in waters with high metal concentrations will incorporate these metals into their tissues, which can pose a health risk to humans if consumed. Shellfish from waters of high metal content pose a unique potential threat, as these animals can accumulate and concentrate metals from the environment via natural filter feeding. Larger predatory fish are also of concern as some metals, including mercury and cadmium, can bioaccumulate up the food chain (Fleming et al. 2006).

Prevention of metal toxicity in humans as a result of exposure to contaminated seafood depends upon monitoring fish and shellfish tissues as well as monitoring harvesting waters and underlying sediments. Concentrations of cadmium, manganese, and nickel in fish in the Arabian Gulf region are reportedly below U.S. Environmental Protection Agency (EPA) or WHO levels of concern (Kosanovic et al. 2007; Al Yousuf et al. 2000; El Shahawi and Al Yousuf 1998; Ahmad and Al Ghais 1996; Al Ghais 1995). Concentrations of zinc in the tissue of fish caught in the Gulf are generally low (Kosanovic et al. 2007; Ahmad and Al Ghais 1996; Al Ghais 1995), although skin concentrations can be high (Al Yousuf et al. 2000). The most recent study of mercury concentrations in domestically harvested UAE fish indicates levels below those of concern (Kosanovic et al. 2007; Agah et al. 2006), but observations from the previous decade have been higher (Ahmad and Al Ghais 1996; Al Ghais 1995).

Recent monitoring data from the Environment Agency–Abu Dhabi (EAD) indicate that concentrations of metals—cadmium, chromium, copper, mercury, manganese, nickel, lead and zinc—in open water off the coast of Abu Dhabi city are generally low or undetectable; however, water and sediment concentrations in coastal channels that receive urban industrial and wastewater effluents are higher than those in open water (EAD 2007a). Monitoring data from other emirate-level environmental agencies are not yet available.

Pathogens in Seafood

Shellfish harvested from microbially contaminated waters are of particular concern because these animals concentrate pathogens from overlying waters through filter feeding (Fleming et al. 2006; Stewart et al. 2008). Although a large number of pathogens can be inactivated through proper preparation procedures, shellfish are very frequently eaten raw.

Prevention of pathogen contamination in the marine environment is generally achieved through monitoring fish or shellfish harvesting areas for fecal indicator organisms (e.g., coliforms, *E. coli*, enterococci) (Fleming et al. 2006; Stewart et al. 2008). These microorganisms are native to the intestinal tract of warm-blooded animals, and their presence is therefore considered indicative of fecal pollution and elevated health risk. Although the use of indicator organisms in water quality monitoring programs is ubiquitous throughout the world, significant criticisms of the concept exist. Pathogens, including *Salmonella* and *Cryptosporidium*, have been recovered from marine waters where indicators were absent or below regulatory levels (Polo et al. 1998; Ferguson et al. 1996; Morinigo et al. 1990). Conversely, evidence also shows that in some cases indicator bacteria may persist and regrow in the environment, potentially signaling a health threat unnecessarily (Stewart et al.

2008; Lee et al. 2006). Perhaps most critically, some of the pathogens most commonly associated with seafood-related illness, including *Vibrio cholerae* and *V. vulnificus,* are native to the marine environment and are therefore not associated with fecal contamination or its indicators (Fleming et al. 2006).

Despite the serious concerns surrounding the indicator concept, the use of indicator organisms is still recommended by the U.S. Food and Drug Administration (FDA) for monitoring seafood harvesting areas. The current recommended maximum concentration of fecal coliforms in shellfish harvesting areas is 14 CFU (colony forming units) per 100 mL (U.S. FDA 2007), significantly lower than standards recommended for recreational use by the EPA in order to account for concentration through filter feeding. Additional microbial water quality data from the UAE would be beneficial for further comparison with these standards. In Abu Dhabi emirate, monitoring data for coastal areas near Abu Dhabi city is mostly focused on beaches and recreational areas. Monitoring of fish harvesting areas focuses mostly on algal blooms rather than on microbial pathogens with the potential to infect humans through seafood consumption. It is unclear how much recreational or large-scale commercial fishing occurs near the city. But high concentrations of fecal indicator bacteria in coastal channels suggest that nontrivial discharges of fecal material to the marine environment are occurring (EAD 2007a). Reports from Dubai of overloaded wastewater treatment plants (Setrakian 2009) suggest that sewage effluents are likely contributing human pathogens to the Arabian Gulf off the coast of that emirate as well.

Harmful Algal Blooms ("Red Tides")

Heavy anthropogenic inputs of nitrogen and phosphorus to coastal waters can result in the sudden proliferation of native phytoplankton, some of which may produce toxins. This phenomenon is more commonly referred to as a "red tide" because of the reddish pigment of some plankton (Anderson et al. 2002; Fleming et al. 2006; Glibert et al. 2005). Many of the toxins produced by these algal blooms bioaccumulate as they move up the food chain from smaller organisms to larger predators, and they cannot be deactivated by any method of food preservation or cooking (Baker and McGillicuddy 2006).

Harmful algal blooms have emerged as a serious threat to coastal nations around the globe, particularly those undergoing significant urbanization and development. Increases in population generally result in greater discharges of nutrients to the marine environment in the form of domestic sewage and agricultural fertilizers, resulting in rapid coastal eutrophication and ideal conditions for algal blooms (Anderson et al. 2002; Glibert et al. 2005). Perhaps not surprisingly, as coastal populations have undergone unprecedented growth and cities have rapidly developed around the Arabian Gulf, harmful algal blooms in the Gulf have become more frequent and of greater concern (Glibert 2007), leading in some cases to massive fish kills, such as the death of more than 2,500 metric tons of wild mullet in Kuwait Bay in 1999 (Glibert et al. 2002).

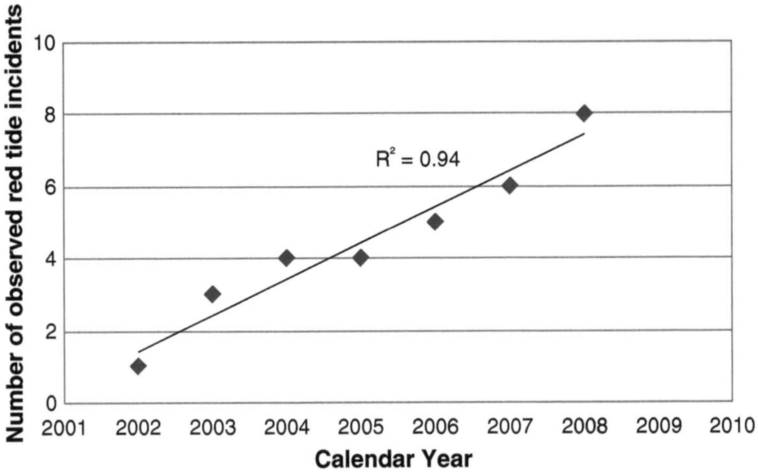

Fig. 11.3 Red tide incidents observed by EAD in UAE coastal waters

Harmful algal blooms pose a threat not only to the larger Gulf region but specifically to UAE coastal waters as well. Since the Kuwait Bay fish kill, the number of red tide incidents observed annually off the coast of the UAE by EAD has increased steadily (Fig. 11.3). Most recently, red tides have forced temporary closures of the Sharjah desalination plant (Sambidge 2008) and Dubai beaches (Menon 2009). As in the larger Gulf region, the increasing frequency and intensity of harmful algal blooms in Emirati waters is likely due to the unprecedented scale of development in coastal cities and accompanying increases in marine pollution. Overloaded wastewater treatment plants, particularly in Dubai (Setrakian 2009) but also in Abu Dhabi, are likely contributing high loadings of nutrients to coastal areas in addition to microorganisms. In Abu Dhabi, nitrogen and phosphorus levels in coastal channels receiving wastewater and other municipal and industrial effluents are two to three orders of magnitude greater than those observed in open water (EAD 2007a). These high nutrient levels are likely at least partly responsible for greater proliferation of harmful algal blooms. Although there have been no recorded outbreaks of UAE fish poisoning related to red tides, studies from other parts of the world do indicate that coastal development can be correlated with increased incidence of paralytic shellfish poisoning (Glibert et al. 2005), emphasizing the need to address prevention of harmful algal blooms in order to safeguard public health.

Despite rising global concern over the increasing incidence of harmful algal blooms, the processes responsible for initiating blooms, particularly those with toxic constituent species, are at present very poorly understood (Glibert et al. 2005; Graneli and Turner 2006). The U.S.-based nonprofit National Research Council, along with many global organizations, has targeted red tide modeling as a major research need for the preservation of the marine environment (Fleming et al. 2006;

National Research Council 1999). In the absence of any known quantitative relationships between contaminant inputs and resultant blooms or recorded outbreaks in the UAE, the burden of disease resulting from these blooms is extremely uncertain and cannot be easily modeled.

Other Potential Seafood Contaminants

A myriad of additional potential marine water contaminants are often present in municipal and industrial discharges. Some of these contaminants—dioxins, polychlorinated biphenyls (PCBs) and polyaromatic hydrocarbons (PAHs)—can have deleterious human health effects if concentrated in seafood (Fleming et al. 2006). Limited data on concentrations of these contaminants exist for UAE waters, with correspondingly little data on seafood concentrations in the UAE. A recent examination of petroleum refinery discharges at an Abu Dhabi plant detected dioxin and PCB concentrations above regulatory values in effluents (Al Zarooni and Elshorbagy 2006); however, initial monitoring of ambient dioxin levels in waters off the coast of Abu Dhabi city found concentrations generally below 0.4 ppm (EAD 2007a). In the absence of UAE-specific information, it is difficult to assess the risk specifically posed by these contaminants. Recently instituted emirate-level monitoring programs should therefore continue and be expanded to include the monitoring of actual seafood levels of these toxins.

Key Health Effects of Produce and Seafood Contamination

Agricultural Pesticides and Pathogens

Inorganic compounds such as sulfur and arsenic have been used for centuries to control pests, but synthetic chemical compounds gained widespread use as pesticides in developed countries during the mid- to late-twentieth century. Environmental and human health concerns have resulted in the reevaluation and occasional restriction of particular compounds. Current common pesticide formulations vary in their methods of toxicity to agricultural pests and humans. Human health effects of some organochlorine pesticides (e.g., DDT) include acute problems with blood clotting, severe confusion, and seizures, as well as possible long-term brain development issues. Many organochlorine compounds are being phased out in favor of other families of synthetic chemicals—pyrethrins, pyrethroids, organophosphates, carbamates, triazines—thought to be less harmful (McKinlay et al. 2008; U.S. EPA 1997).

Although some debate has ensued as to whether organophosphate pesticides share a common route of toxicity in humans, the EPA is currently acting on the basis that they do (Ragsdale 2000). This primary mechanism involves the inhibition of the acetylcholinesterase enzyme in the brain and peripheral nervous systems. Acetylcholinesterase is responsible for recycling acetylcholine, a neurotransmitter compound necessary for proper activation of muscle cells and regulation of neurons in the brain. N-methyl carbamate pesticides share this primary route of toxicity, though to a lesser extent. Symptoms of organophosphate and carbamate exposure can include abdominal cramps, nausea, diarrhea, dizziness, tremors, anxiety, and confusion (Steenland 1996; U.S. EPA 1999). Exposure has also been previously linked to serious, long-term neurological, developmental, and reproductive disorders.

Pyrethroid pesticides are chemically similar to naturally occurring pyrethrins, extracted from dried chrysanthemum flowers. They are often combined with at least one nonpesticidal synergistic compound to help reduce their rate of degradation in an outdoor environment. Unlike organophosphates and carbamates, pyrethroids do not inhibit acetylcholinesterase; instead, they induce insect paralysis by affecting nerve cell function. Toxicity to humans from oral ingestion is believed to be reduced due to limited absorption rates, rapid breakdown by liver enzymes, and timely excretion by the kidney (U.S. EPA 1999). Symptoms of exposure to pyrethroid compounds include seizures, dizziness, headache, fatigue, vomiting, and diarrhea.

The American Institute for Cancer Research has noted that the risk of potential carcinogenic effects from pesticide exposure is greatly outweighed by known health benefits—including anticancer properties and prevention of cardiovascular disease—gained from a diet rich in fruit and vegetables (Ragsdale 2000). Significant overexposure to organophosphate, carbamate, or pyrethroid compounds, however, can result in an increase in the severity of symptoms or even death.

Foodborne pathogens can cause and/or contribute to an array of human illnesses. Although acute gastroenteritis (also known as "food poisoning") is the most common illness, more complex chronic conditions such as organ failure, arthritis, and heart disease may be partially attributable to foodborne pathogens as well. Properly cooking produce contaminated with pathogens can reduce the risk of food-to-human microbial pathogen transmission. But past outbreaks of illness have historically demonstrated that peeling and washing alone may not eliminate pathogen hazard in raw food products.

Heavy Metals and Pathogens in Seafood

Toxic effects of metal consumption depend upon the dose ingested, itself a function of the amount of contaminated seafood consumed and the concentrations of metals present. Potential manifestations of human illness include neurological effects following mercury, lead, or manganese exposure; cancers resulting from cadmium exposure; and cardiovascular disease from arsenic exposure (U.S. EPA 2009).

Seafood can also serve as a vector for pathogens (e.g., *Vibrio* spp., *Salmonella* spp.) if harvest waters are contaminated (Fleming et al. 2006; Stewart et al. 2008). These microorganisms can cause serious human illness, including severe gastroenteritis, if not removed or inactivated prior to human consumption. Improper storage of seafood at insufficient refrigeration temperatures prior to consumption will elevate this threat, as microbes can proliferate and produce toxins during storage times, thus becoming more difficult to eliminate during cooking. The purchase and consumption of nonrefrigerated fish has been recently linked to an outbreak of severe gastroenteritis in Sharjah, which resulted in the death of a small child (Kakande and Kwong 2009) and has highlighted the need to enforce proper food handling practices throughout the emirates.

Several common algal bloom organisms produce extremely powerful toxins that cause gastroenteritis, neurological damage, or respiratory distress if ingested or inhaled by animals (Baker and McGillicuddy 2006; Fleming et al. 2006).

Method for Estimating the Burden of Disease from Produce and Seafood Contamination

Targeted Exposures and Health Effects

Estimates of the burden of disease resulting from fruit and vegetable contamination focus on consumption of pesticide residues as a primary exposure. For organophosphate and carbamate pesticides, acetylcholinesterase inhibition can manifest as dizziness, nausea, abdominal cramps, diarrhea, tremors, anxiety, confusion, neurological disorders, developmental/reproductive disorders, and death. For pyrethroids and pesticides not falling into any above categories, health effects can include seizures, dizziness, headache, fatigue, vomiting, diarrhea, pulmonary edema, and death.

While the transmission mechanisms of bacterial, parasitic, and viral pathogens from food to humans are well understood, exposure rates and subsequent burdens of disease are difficult to estimate in a probabilistic fashion for the UAE without additional region-specific data combining food testing results with known disease outbreaks. Although WHO has initiated efforts (WHO 2006) toward compiling estimates of the global disease burden of foodborne illness—much of which focuses on collecting information on pathogen-related illness—this work was unavailable at the time the study described in this book was carried out. Without region-specific estimates of pathogen instances on food from UAE studies, the scope of this report forgoes further quantitative modeling analysis of agricultural pathogens. However, recommendations for initiating best public health practices to monitor and prevent pathogenic outbreaks are detailed in the "Conclusions" section of this chapter.

Estimates of illness resulting from seafood consumption focus on exposure to mercury. Although numerous metals can result in adverse health effects if consumed in seafood, mercury is generally regarded as of greatest concern. Chronic mercury poisoning results in a host of neurological and psychological symptoms, including tremors, motor/cognitive dysfunction, and memory loss (WHO 2007; Jarup 2003; Eto 1997). Exposure in utero can result in serious lifetime illness,

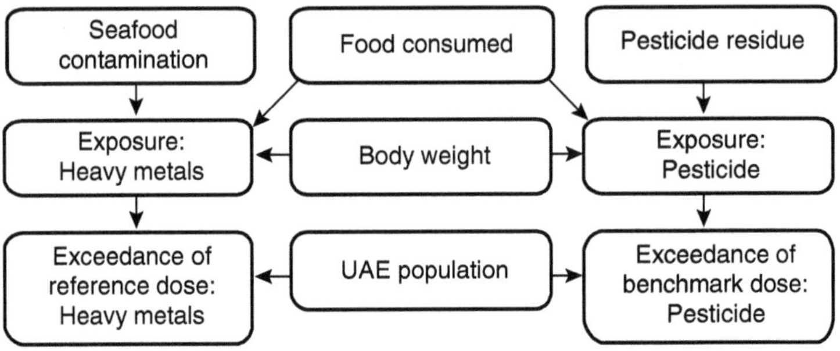

Fig. 11.4 Top-level influence diagram of the ***Foodborne Contamination*** module

including mental retardation, sensory loss, developmental delay, cerebral palsy, and seizures (WHO 2007; Mozaffarian and Rimm 2006; Jarup 2003).

General Approach

In lieu of estimating foodborne mortality and morbidity cases, this modeling approach directly calculates the probability of exceeding international guidelines for exposure to specific hazardous chemicals in fruit, vegetables, and seafood in the UAE. This is currently the standard approach used in assessing the noncancer risks associated with the foodborne chemicals considered in this analysis (see, for example, Cardoso et al. 2010). Therefore, concepts of relative risk and attributable fractions of population are not needed in the food contamination model, unlike the risk area models discussed in other chapters of this report.

For fruits and vegetables, the model estimates the number of daily incidents in which UAE residents are exposed to a particular type of pesticide residue above a prespecified benchmark dose, due to eating a particular type of fruit or vegetable. These benchmark doses include the acceptable daily intake (ADI), the acute population adjusted dose (aPAD) and the chronic population adjusted dose (cPAD). ADI values are derived from WHO; aPAD and cPAD values are taken from EPA.

For seafood, only mercury (as organic methylmercury) was specifically considered, although the ability to include other heavy metals and/or chemical contaminants as future data and time become available is included in the present model. The model estimates the number of daily incidents in which UAE residents are exposed to mercury levels above the reference dose (RfD) maintained by EPA due to eating fish.

The population was delineated by gender, age, and body weight whenever data existed to support this type of separate calculation. Figure 11.4 illustrates the top-level influence diagram of the process model for estimating exposure to hazardous chemicals in contaminated fruits, vegetables, and seafood. Model outputs and specifics of each individual module are discussed separately in the following sections.

Food Consumption

In order to estimate current UAE fruit, vegetable, and fish consumption patterns, UAE-specific survey results—including daily frequencies, not actual amounts of food—were combined with UAE nutrition data for daily portion sizes. An age-indexed modifying coefficient was used to account for age-related differences in amounts of food typically eaten. Finally, the totals of fruits and vegetables eaten were split into itemized lists of specific kinds of fruits and vegetables (e.g., date, banana, tomato, lettuce) to more accurately depict a UAE resident's typical daily fare. The food consumption model influence diagram (Fig. 11.5) shows how these components conceptually connect to one another.

Daily food frequency—the proportion of UAE residents (1.0 = 100%) eating a specific kind of food (fruit or vegetable) each day—was determined by a UAE health and lifestyle study (Badrinath et al. 2002) for adults and by a UAE student survey (Al Matroushi 2005) for children. This frequency was assumed to follow a normal statistical distribution; mean and standard deviation values are given for fruits and vegetables in Table 11.4. Daily frequency of consumed fish is deterministic, with values (Musaiger and Abuirmeileh 1998) also given in Table 11.4.

Sizes of daily servings of fruits, vegetables, and fish were determined by UAE food frequency questionnaire results (Dehghan et al. 2005), lognormally modeled and detailed in Table 11.5.

An attempt to account for age-related differences in food eating habits is abstracted by the ***Age Effect on Diet*** node, an age-indexed table of proportions (1.0 = 100%) shown in Table 11.6. Values were obtained by considering 50th percentile body weights (National Center for Health Statistics 2000) of children and comparing them with those of adults.

For seafood, the total amount of consumed fish is represented by a single node:

$$\text{Daily fish eaten} = A_{diet} N_{fish} E_{fish} (\text{kg} / (\text{day} \bullet \text{person})) \tag{11.1}$$

Where:

A_{diet} = Age effect on diet, Table 11.6
N_{fish} = Serving of fish (kg/(day·person)), Table 11.5
E_{fish} = Fish eaten, Table 11.4

For fruits and vegetables, two model nodes store temporary, nonitemized amounts of consumed food:

$$\text{Daily fruit eaten} = A_{diet} N_{fruit} E_{fruit} (\text{kg} / (\text{day} \bullet \text{person})) \tag{11.2}$$

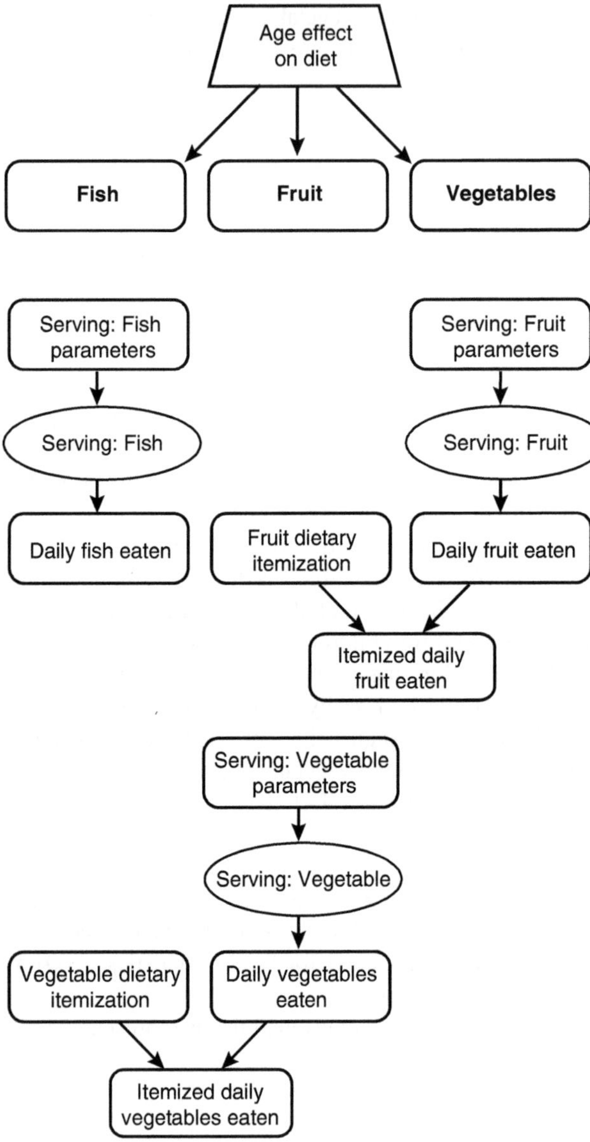

Fig. 11.5 Modeling food consumption habits of UAE residents

$$\text{Daily vegetables eaten} = A_{diet} N_{veg} E_{veg}\ (\text{kg}/(\text{day} \bullet \text{person})) \tag{11.3}$$

Where:

A_{diet} = Age effect on diet, Table 11.6
N_{fruit} = Serving of fruit (kg/(day·person)), Table 11.5
N_{veg} = Serving of vegtables (kg/(day·person)), Table 11.5

Table 11.4 Daily frequencies of fruit, vegetable, and fish consumption

	Mean (standard deviation)					
	Fruit		Vegetable		Fish	
Age	Female	Male	Female	Male	Female	Male
2–15	0.442 (0.017)	0.508 (0.013)	0.532 (0.014)	0.589 (0.012)	0.100	0.090
16–29	0.490 (0.021)	0.490 (0.021)	0.466 (0.021)	0.466 (0.021)	0.100	0.090
30–44	0.424 (0.025)	0.424 (0.025)	0.366 (0.020)	0.366 (0.020)	0.100	0.090
45–59	0.374 (0.036)	0.374 (0.036)	0.356 (0.036)	0.356 (0.036)	0.180	0.170
60–74	0.480 (0.044)	0.480 (0.044)	0.410 (0.044)	0.410 (0.044)	0.180	0.170
75+	0.385 (0.077)	0.385 (0.077)	0.308 (0.073)	0.308 (0.073)	0.180	0.170

Al Matroushi (2005), Badrinath et al. (2002), and Musaiger and Abuirmeileh (1998)

Table 11.5 Serving sizes of food used in the model, in kg/(day·person)

	Mean (standard deviation)	
	Female	Male
Fruit	0.300 (0.170)	0.480 (0.200)
Vegetable	0.300 (0.150)	0.370 (0.160)
Fish	0.054 (0.223)	0.073 (0.029)

Dehghan et al. (2005)

Table 11.6 Proportional effect of age on food consumption as used in the model

Age	Female	Male
2	0.207	0.179
3	0.236	0.203
4	0.271	0.230
5	0.308	0.261
6	0.348	0.293
7	0.391	0.327
8	0.440	0.363
9	0.498	0.404
10	0.565	0.452
11	0.639	0.508
12	0.715	0.573
13	0.787	0.646
14	0.848	0.722
15	0.894	0.797
16+	1.000	1.000

National Center for Health Statistics (2000)

E_{fruit} = Fruit eaten, Table 11.4
E_{veg} = Vegetables eaten, Table 11.4

To allow the model to consider variations in dietary structure—since usage of pesticides is registered separately for different types of produce—fruits and vegetables feature two more model nodes: ***Fruit Dietary Itemization*** is a table of proportion values, indexed by fruit; ***Vegetable Dietary Itemization*** is a table of proportion values, indexed by vegetable.

Table 11.7 Proportional dietary considerations used by the model

UAE dietary proportion (1.0 = 100%)			
Fruits		Vegetables	
Apple	0.114	Cabbage	0.091
Apricot	0.007	Carrot	0.019
Banana	0.012	Cauliflower	0.046
Date	0.102	Celery	0.091
Grape	0.066	Coriander	0.012
Guava	0.007	Cucumber	0.091
Kiwi	0.007	Eggplant	0.046
Lemon	0.048	Fennel	0.012
Mandarin	0.133	Green bean	0.091
Mango	0.007	Green onion	0.064
Melon	0.007	Ginger	0.019
Orange	0.133	Hot pepper	0.003
Pear	0.007	Lettuce	0.091
Peach	0.007	Marrow (zucchini)	0.046
Rambutan	0.007	Okra	0.046
Tomato	0.325	Parsley	0.012
Watermelon	0.007	Pumpkin	0.046
		Radish	0.019
		Riglah (arugula)	0.046
		Sweet pepper	0.003
		Watercress	0.046
		White onion	0.064

FAO (2003)

Both are adapted from UAE-specific food consumption estimates from the FAOSTAT database of the Food and Agriculture Organization of the United Nations (FAO 2003) and shown in Table 11.7. Several assumptions were made: FAOSTAT ***pimento*** mapped to equal distributions of UAE ***hot pepper*** and. ***sweet pepper***; FAOSTAT ***starchy roots*** mapped to equal distributions of UAE ***carrot, ginger***, and ***radish***; FAOSTAT ***spices*** mapped to equal distributions of ***parsley, coriander***, and ***fennel***; staple crops ***cabbage, celery, cucumber, green bean***, and ***lettuce*** each were assumed to have portions twice as large as remaining nonstaple crops not already mapped above.

The addition of this itemized produce list resulted in the last two nodes of this module:

$$\text{Itemized daily fruit eaten} = Z_{fruit} D_{fruit} (\text{kg} / (\text{day} \bullet \text{person})) \tag{11.4}$$

$$\text{Itemized daily vegetables eaten} = Z_{veg} D_{veg} \; (\text{kg} / (\text{day} \bullet \text{person})) \tag{11.5}$$

Where:

Z_{fruit} = Fruit dietary itemization, Table 11.7

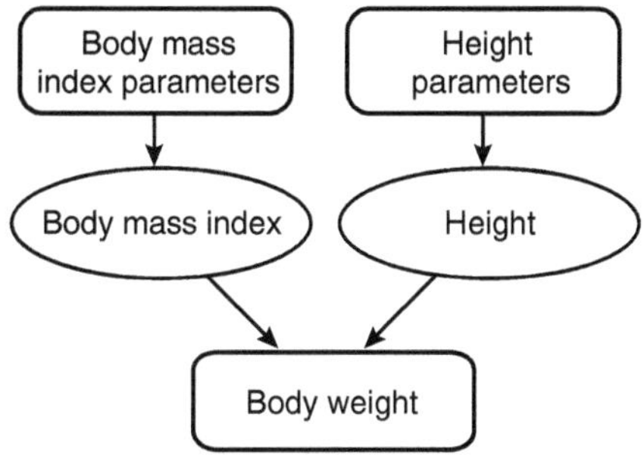

Fig. 11.6 Modeling body weight of UAE residents

Z_{veg} = Vegetable dietary itemization, Table 11.7
D_{fruit} = Daily fruit eaten, Eq. 11.2
D_{veg} = Daily vegetables eaten, Eq. 11.2

Body Weight

Because health effects due to foodborne illness depend on the concentration of a contaminant within the human body, illness is also considered to be dependent on human body weight. Reference doses, acceptable daily intakes, and similar benchmarks are typically issued in units of mass of a substance per mass of a person for a specific duration of time (e.g., mg/(kg·day)). Care must be taken not to confuse these units with contaminant levels, which are measured in mass of a substance (pesticide, heavy metal) per overall mass of vector (fruit, vegetable, fish). For modeling purposes, Fig. 11.6 shows the simple influence diagram connecting the relevant nodes.

Many studies are interested in the body mass index (BMI) values for a population; subsequently, relevant survey data from the UAE report BMI—a person's body weight divided by the square of their height—instead of actual body weight. The foodborne illness model, however, needs to consider weight values, and so it calculates this value using UAE-specific BMI results combined with worldwide average height data, delineated by gender and age.

The model's ***Body Mass Index*** node is a reference for selecting a random variable to represent a UAE resident's BMI value. Shown in Table 11.8, it matches the BMI ranges presented by a UAE-wide survey (Badrinath et al. 2002). Accordingly, worldwide ***Height*** data as represented in the model is considered to be normally distributed, using the statistical parameters shown in Table 11.9.

The last node in the influence diagram, ***Body Weight***, solves the BMI equation for weight:

$$\text{Weight} = BMI(H^2)\ (\text{kg}) \tag{11.6}$$

Table 11.8 Random variable selection for body mass index used in the model

	Distribution of BMI random variable
Normal	Uniform (min: 20, max: 25)
Overweight	Uniform (min: 25, max: 30)
Obese	Uniform (min: 30, max: 30)

Badrinath et al. (2002)

Table 11.9 Height data used in the model (cm)

	Mean (standard deviation)	
Age	Female	Male
2	86.42 (3.23)	87.12 (3.06)
3	95.05 (3.81)	96.08 (3.71)
4	102.73 (4.31)	103.33 (4.19)
5	109.42 (4.76)	109.96 (4.63)
6	115.12 (5.12)	115.95 (4.93)
7	120.81 (5.47)	121.73 (5.29)
8	126.56 (5.80)	127.27 (5.65)
9	132.49 (6.11)	132.57 (6.01)
10	138.63 (6.40)	137.78 (6.37)
11	144.99 (6.65)	143.11 (6.73)
12	151.23 (6.84)	149.08 (7.09)
13	156.37 (6.94)	156.04 (7.43)
14	159.79 (6.94)	163.18 (7.69)
15	161.67 (6.88)	168.96 (7.80)
16+	163.15 (6.54)	176.54 (7.30)

WHO (2009a, b)

Where:

BMI = Body mass index (kg/m^2), Table 11.8
H = Height (m), Table 11.9

In the ***UAE Population*** module, discussed later in this chapter, UAE-specific health survey data on BMI ranges are combined with the results from this section to form overall body weight estimates for the entire UAE population. See the section titled "UAE Population" for further details.

Pesticide Residue

During testing for traces of pesticide on domestic and imported food, UAE agencies are generally concerned with results in exceedance of maximum residue levels (MRLs) established and maintained by the Codex Alimentarius Commission, the European Commission (EC), European Food Safety Authority (EFSA), and the EPA. To match these efforts, this model similarly used these sources as guidance for deciding theoretical maximum residue values (Tables 11.10 and 11.11), which

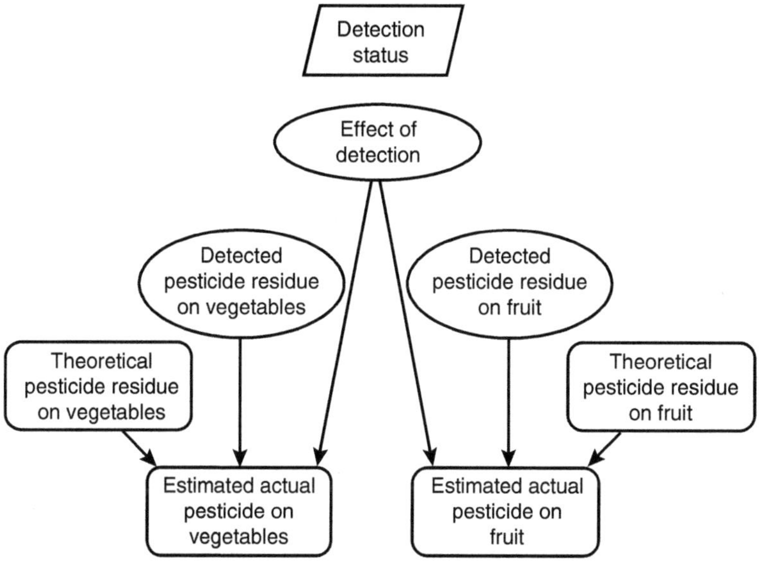

Fig. 11.7 Modeling pesticide residue on food in the UAE

vary by crop type and by pesticide. Codex data were used whenever applicable as a primary reference; EFSA/EPA data were used to fill in gaps for crop/pesticide combinations not included in Codex tables. Previously conducted UAE pesticide residue study results were then used to further refine expected residue values used by the model. Types of crops and pesticides considered were restricted only to those also appearing in UAE pesticide residue studies. Figure 11.7 shows the influence diagram for modeling pesticide residue.

Locally representative residue data from previous Abu Dhabi testing (ADFCA 2006) were combined with the MRLs (Tables 11.10 and 11.11) to further refine the estimated pesticide residue in the model. This modification uses these guidelines:

1. If no pesticide was detected during previous testing: model also assumes zero pesticide as a best estimate.
2. If previous tests detected a pesticide below its MRL: modeled pesticide is MRL × Uniform(0,1).[1]
3. If previous tests detected a pesticide above its MRL: modeled pesticide is MRL × 110%.

Tables 11.12 and 11.13 correspond to the ***Detected Pesticide Residue on Vegetables*** and ***Detected Pesticide Residue on Fruit*** nodes of the model. The remaining nodes in this section, ***Effect of Detection, Estimated Actual Pesticide on Vegetables*** and ***Estimated Actual Pesticide on Fruit***, correspond to the equations explained above for the two residue modeling scenarios.

[1]Uniform distribution, min = 0, max = 1.

Table 11.10 Maximum theoretical vegetable pesticide residue levels used in the model, in mg/kg

	Acephate	Bifenthrin	Bromopropylate	Chlorpyrifos	Cyhalothrin	Dimethoate
Arugula	0.02	N/A	0.05	0.05	1	N/A
Cabbage	0.02	N/A	0.05	1	0.2	1
Carrot	0.02	0.05	0.05	0.1	N/A	0.02
Cauliflower	2	N/A	0.05	0.05	0.1	2
Celery	0.02	0.05	0.05	N/A	0.3	0.5
Coriander	0.02	N/A	0.05	0.05	0.3	0.02
Cucumber	0.02	0.1	0.5	0.05	0.1	0.02
Eggplant	0.02	N/A	N/A	N/A	N/A	N/A
Fennel	0.02	0.05	0.05	0.05	0.3	0.02
Green bean	5	0.5	1	0.05	0.2	N/A
Green onion	0.02	0.02	0.05	N/A	0.05	2
Ginger	0.02	N/A	N/A	N/A	N/A	N/A
Hot pepper	50	N/A	N/A	20	N/A	1
Lettuce	0.02	2	0.05	0.05	1	2
Okra	0.02	N/A	N/A	N/A	N/A	N/A
Parsley	0.02	0.05	0.05	0.05	1	0.02
Pumpkin	0.02	0.1	0.05	N/A	0.05	0.02
Radish	0.02	0.05	0.05	0.2	0.1	0.02
Sweet pepper	0.02	0.2	N/A	2	0.1	1
Watercress	0.02	0.05	0.05	0.05	0.02	0.02
White onion	0.02	0.05	0.05	0.2	1	N/A
Zucchini	0.02	0.1	0.5	0.05	0.05	0.02

Codex Alimentarius Commission (2009), European Food Safety Authority (2009), European Commission (1990) and U.S. Department of Agriculture (USDA) (2009)

Malathion	Methamidophos	Methidathion	Pirimicarb	Procymidone	Quintozene	Vinclozolin
0.02	0.01	0.02	N/A	0.02	N/A	0.05
0.02	0.5	0.1	N/A	2	0.1	1
0.02	0.01	0.02	N/A	0.02	0.02	0.05
0.5	0.5	0.02	N/A	0.02	0.02	1
1	0.01	0.02	N/A	0.02	0.02	0.05
0.02	0.01	0.02	N/A	0.02	0.02	0.05
0.2	1	0.05	1	2	0.02	1
0.5	N/A	0.02	N/A	N/A	N/A	N/A
0.02	0.01	0.02	N/A	0.02	0.02	0.05
1	1	0.1	N/A	2	0.02	2
5	0.01	0.1	0.1	0.2	0.02	1
N/A	N/A	0.02	0.05	N/A	N/A	N/A
0.5	2	0.02	20	50	0.1	30
8	0.2	0.02	5	5	0.02	5
0.02	N/A	0.02	N/A	N/A	N/A	N/A
0.02	0.01	0.02	N/A	0.02	0.02	0.05
0.02	0.01	0.02	1	1	0.02	1
0.02	0.01	0.02	N/A	0.02	N/A	0.05
0.5	1	0.02	N/A	5	0.05	3
0.02	0.01	0.02	N/A	0.02	0.02	0.05
1	0.01	0.02	0.05	0.2	0.02	1
0.02	0.01	0.02	1	1	0.02	1

Table 11.11 Maximum theoretical fruit pesticide residue levels used in the model, in mg/kg

	Acephate	Bifenthrin	Bromopropylate	Chlorpyrifos	Cyhalothrin	Dimethoate	Fenitrothion
Apple	0.02	0.3	2	0.5	0.1	0.02	0.5
Avocado	0.02	N/A	0.05	N/A	0.02	N/A	N/A
Apricot	0.02	0.2	0.05	N/A	0.2	1	N/A
Banana	0.02	3	0.05	3	0.02	N/A	N/A
Date	0.02	N/A	0.05	N/A	0.02	N/A	N/A
Grape	0.02	0.05	2	0.5	0.2	0.02	N/A
Guava	0.02	N/A	0.05	N/A	N/A	N/A	N/A
Kiwi	0.02	N/A	0.05	2	0.02	N/A	N/A
Lemon	0.02	0.1	2	0.2	0.2	5	N/A
Mandarin	0.02	0.1	2	2	0.2	5	N/A
Mango	0.02	N/A	0.05	N/A	0.02	1	N/A
Melon	0.02	0.05	0.5	0.05	0.05	0.02	N/A
Orange	0.02	0.1	2	1	0.1	5	N/A
Pear	0.02	0.5	2	0.5	0.1	1	N/A
Peach	0.02	0.2	0.05	0.5	0.2	1	N/A
Rambutan	0.02	N/A	N/A	0	N/A	N/A	N/A
Tomato	0.02	0.2	1	0.5	0.1	N/A	N/A
Watermelon	0.02	0.05	0.05	0.05	0.05	0.02	N/A

Codex Alimentarius Commission (2009), European Food Safety Authority (2009), European Commission (1990) and USDA (2009)

Malathion	Methamidophos	Methidathion	Pirimicarb	Procymidone	Quintozene	Vinclozolin
2	0.01	0.5	N/A	N/A	0.02	1
0.02	0.01	N/A	N/A	N/A	0.02	0.05
0.02	0.01	0.02	N/A	N/A	0.02	2
N/A	0.01	N/A	N/A	N/A	0.02	0.05
0.02	0.01	N/A	N/A	N/A	0.02	0.05
8	0.01	1	1	5	0.02	5
0.02	0.01	N/A	N/A	N/A	0.02	0.05
N/A	0.01	N/A	N/A	5	0.02	10
7	0.2	2	3	0.02	0.02	0.05
7	0.2	5	3	0.02	0.02	0.05
0.02	0.01	N/A	N/A	N/A	0.02	0.05
0.02	0.01	0.02	0.2	1	0.02	1
7	0.2	2	3	0.02	0.02	0.05
0.5	0.01	1	N/A	1	0.02	1
6	0.01	0.2	N/A	2	0.02	5
N/A	0.01	N/A	N/A	N/A	0.02	0.05
3	0.5	0.1	0.5	5	0.02	3
0.02	0.01	0.02	1	1	0.02	1

Table 11.12 Abu Dhabi vegetable pesticide residue testing results

	UAE food samples with detected pesticide (%)		
	None	Below MRL	Above MRL
Cabbage	100		
Carrot	100		
Cauliflower	91.7	Chlorpyrifos: 8.3	
Celery	87.5	Chlorpyrifos: 12.5 Vinclozolin: 12.5	
Coriander	100		
Cucumber	96.9	Bifenthrin: 3.1 Chlorpyrifos: 3.1 Vinclozolin: 3.1	
Eggplant	87.0	Chlorpyrifos: 13.0 Dimethoate: 4.4	
Fennel	66.7	Acephate: 33.3 Chlorpyrifos: 33.3	
Green bean	100		
Green onion	100		
Ginger	80.0	Chlorpyrifos: 20.0	
Hot pepper	85.7	Chlorpyrifos: 14.3 Ethion: 7.1	
Lettuce	96.6	Chlorpyrifos: 3.4 Pirimicarb: 3.4	
Marrow (zucchini)	92.3	Chlorpyrifos: 7.7	
Okra	94.4	Chlorpyrifos: 5.6 Malathion: 5.6	
Parsley	73.4	Chlorpyrifos: 20.0	Chlorpyrifos: 6.6
Pumpkin	100		
Radish	100		
Riglah (arugula)	100		
Sweet pepper	88.9	Bromopropylate: 11.1 Chlorpyrifos: 11.1	
Watercress	100		
White onion	100		

ADFCA (2006)

Pesticide Exposure

To estimate rates of human exposure to pesticide via oral ingestion of fruit and vegetables, the model relies on a standard exposure formula for its two relevant nodes, ***Exposure: Pesticide (Fruit)*** and ***Exposure: Pesticide (Vegetables)***:

$$\text{Exposure} = \frac{P_{crop} E_{crop}}{W} (\text{mg} / (\text{kg} \bullet \text{day})) \tag{11.7}$$

Table 11.13 Abu Dhabi fruit pesticide residue testing results

	UAE food samples with detected pesticide (%)		
	None	Below MRL	Above MRL
Apple	68.8	Chlorpyrifos: 31.2	
		Fenitrothion: 6.3	
		Methidathion: 6.3	
Avocado	100		
Apricot	100		
Banana	100		
Date	88.7	Bifenthrin: 1.6	
		Bromopropylate: 11.3	
		Chlorpyrifos: 3.2	
		Ethion: 3.2	
Grape	85.7	Chlorpyrifos: 14.3	
		Procymidone: 7.1	
Guava	100		
Kiwi	100		
Lemon	66.7	Chlorpyrifos: 33.3	Malathion: 11.1
		Methidathion: 11.1	
Mandarin	100		
Mango	100		
Melon	100		
Orange	80.0	Chlorpyrifos: 20.0	
		Dimethoate: 10.0	
Pear	100		
Peach	100		
Rambutan	100		
Tomato	76.0	Chlorpyrifos: 24.0	
		Cyhalothrin: 0.96	
		Phenthoate: 3.9	
		Procymidone: 0.96	
		Quintozene: 0.96	
		Vinclozolin: 0.96	
Watermelon	100		

ADFCA (2006)

Where:

P_{crop} = Pesticide concentration on crop (mg/kg)
E_{crop} = Amount of crop eaten (kg)
W = Body weight (kg)

In this model, the above formula is revised to sum over the itemized list of foods and develop a daily average, as follows:

$$\text{Exposure} = \sum_{crop} \left(\frac{P_{crop} E_{crop}}{W} \right) (\text{mg} / (\text{kg} \bullet \text{day})) \tag{11.8}$$

with resulting units of measurement of mass of pesticide per mass of person per capita per day, or mg/(kg·day).

Seafood Contamination

The mercury content of seafood was estimated using the most recent UAE-specific data identified in the scientific literature (Kosanovic et al. 2007). As a conservative estimate, mean and confidence interval values from fish harvested near Sharjah—exhibiting the highest mercury content—were represented in the model's ***Seafood contamination: Methylmercury*** node as a continuous triangular distribution (min: 0.033, mode: 0.068, max: 0.098), in units of mg/kg. It is important to note that these values were only observed in one fish species (Redspot emperor, *Lethrinus lentjan*) and that both higher (Ahmad and Al Ghais 1996; Al Ghais 1995) and lower muscle concentrations of mercury have been observed (Agah et al. 2006) in the Arabian Gulf; however, these values were either observed some time ago or in areas of the Gulf that are distant from the UAE.

Methylmercury—highly lipophilic and poorly excreted from the human body—is the most toxic form of mercury, although total mercury concentrations are generally reported in fish studies, including the study by Kosanovic et al. (2007). As methylmercury can comprise up to 100% of the total mercury content of fish (Agah et al. 2006), for conservative estimation purposes, the values reported by Kosanovic et al. (2007) were assumed to be equivalent to methylmercury values.

Exposure to Contaminated Seafood

To estimate rates of human exposure to methylmercury via oral ingestion of fish, the model relies on a standard exposure formula for its single relevant node, ***Exposure: Contaminated Seafood***:

$$\text{Exposure} = \frac{M_{fish} E_{fish}}{W} (\text{mg} / (\text{kg} \bullet \text{day})) \tag{11.9}$$

Where:

M_{fish} = Methylmercury concentration in fish (mg/kg)
E_{fish} = Amount of fish eaten (kg)
W = Body weight (kg)

UAE Population

The foodborne illness model calculates exposure rates to various contaminants and then multiplies resulting rates of incidence by the total population of the UAE. The model's ***Demographics*** node uses information collected from the UAE Ministry of Economy (2008) and combined with the ***Body Mass Index Range*** node's source data (Badrinath et al. 2002) to create an overall picture of the UAE population

Table 11.14 UAE population as considered by the model, 2008

	Female			Male			
Age	Normal	Overweight	Obese	Normal	Overweight	Obese	Total
2	20,843	4,916	3,680	22,331	5,267	3,943	60,980
3	19,371	1,560	8,508	20,754	1,672	9,115	60,980
4	18,517	5,034	5,888	19,839	5,394	6,308	60,980
5	17,984	892	9,006	19,544	970	9,787	58,183
6	22,807	1,255	3,792	24,785	1,364	4,121	58,124
7	21,274	4,405	2,203	23,119	4,787	2,394	58,182
8	21,162	3,848	2,872	22,998	4,181	3,121	58,182
9	17,817	3,876	6,190	19,362	4,212	6,727	58,184
10	16,948	6,353	2,109	18,826	7,056	2,343	53,635
11	15,043	7,267	3,100	16,709	8,072	3,443	53,634
12	14,992	6,505	3,913	16,653	7,226	4,347	53,636
13	16,161	6,353	2,897	17,951	7,056	3,218	53,636
14	18,067	2,668	4,675	20,068	2,964	5,193	53,635
15	14,963	2,636	6,151	16,468	2,902	6,770	49,890
16–29	260,770	122,525	79,063	529,532	248,805	160,550	1,401,244
30–44	107,581	144,433	120,237	364,209	488,973	407,057	1,632,490
45–59	32,680	40,558	33,319	112,086	139,104	114,277	472,024
60–74	7,168	6,315	5,481	13,591	11,973	10,391	54,919
75+	4,026	1,905	637	4,502	2,130	712	13,912
Total	**668,174**	**373,304**	**303,721**	**1,303,327**	**954,108**	**763,817**	**4,366,451**

UAE Ministry of Economy (2008) and Badrinath et al. (2002)

averaged over the year 2008, the ***UAE Population*** node (Table 11.14). Because of differences in eating habits of infants and toddlers, UAE residents under 2 years of age are not included in the model's estimates, bringing the total year 2008 population down from 4.488 to 4.366 million.

Pesticide-Related Health Effects

For comparative purposes, three benchmark pesticide exposure levels are considered in this section: acceptable daily intake (ADI), acute dietary population adjusted dose (aPAD), and chronic dietary population adjusted dose (cPAD). This model considers ADI values set by WHO and aPAD/cPAD values from the EPA, available online (U.S. EPA 1997, 2009) and shown in Table 11.15.

Pesticide exposure results averaged over all Monte Carlo runs are used by the model to determine incidence rates of exceeding these three benchmark levels. Finally, incidence rates are multiplied by the total UAE population to obtain final model outputs: the number of cases in which a person eats a particular food contaminated with a particular pesticide and is subsequently exposed to a contaminant concentration above the ADI, aPAD, or cPAD levels. Logically connected conceptual nodes are shown in the influence diagram of Fig. 11.8.

Table 11.15 Pesticide health effect benchmark levels used by the model, in mg/(kg·day)

	ADI	cPAD	aPAD
Acephate	0.005	0.0012	0.005
Bifenthrin	0.015	0.015	0.015
Bromopropylate	0.03	0.2	0.2
Chlorfenvinphos	0.002	0.0007	0.002
Chlorpyrifos	0.010	0.0003	0.005
Dichlorvos	0.004	0.0005	0.008
Dimethoate	0.010	0.0022	0.013
Ethion	0.002	0.0005	0.0005
Fenitrothion	0.005	0.0013	0.13
Cyhalothrin	0.02	0.001	0.005
Malathion	0.02	0.07	0.14
Methamidophos	0.004	0.0001	0.001
Methidathion	0.001	0.0015	0.002
Phenthoate	0.003	0.003	0.003
Pirimicarb	0.02	0.02	0.02
Procymidone	0.1	0.035	0.035
Quintozene	0.007	0.003	0.003
Vinclozolin	0.07	0.0012	0.006

U.S. EPA (1997, 2009)

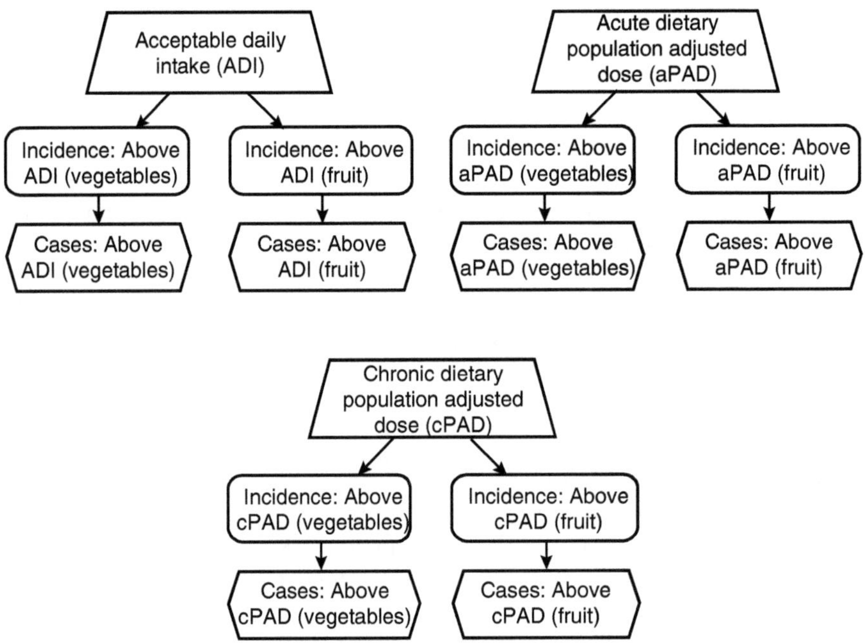

Fig. 11.8 Modeling health effects of pesticides consumed in the UAE

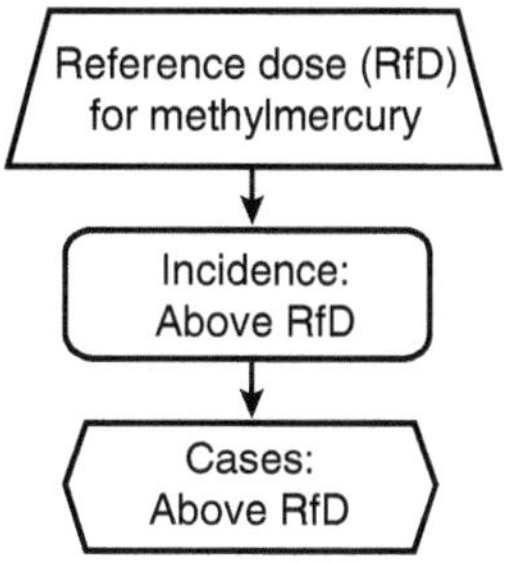

Fig. 11.9 Modeling health effects of methylmercury consumed in the UAE

ADI is considered by WHO to be the maximum amount of a substance that can be eaten daily without significant risk to a person, even over the course of a lifetime. It is similar to the cPAD used in the United States for chronic exposure assessment.

The cPAD and aPAD values likewise are believed to be the best estimates of the maximum amount of a substance that can be eaten daily without significant chronic (cPAD) or immediate (aPAD) health effects. Both values can be adjusted for population-specific factors such as particularly vulnerable groups (e.g., children, expectant mothers) as outlined by federal regulations, including the U.S. Food Quality Protection Act of 1996.

ADI and cPAD/aPAD values do not necessarily match, as they were created by different agencies using different animal-based studies and guidelines. The aPAD values will always be equal to or higher than corresponding cPAD values. Health effects of exceeding any of these values are considered those within the full range of disorders described earlier in this chapter.

Mercury-Related Health Effects

To assess possible human health effects of organic methylmercury exposure, the EPA reference dose (RfD) of 0.0001 mg/(kg·day) is used (U.S. EPA 2002). As with pesticides, methylmercury exposure results are averaged over all Monte Carlo runs and then used by the model to determine incidence rates of exceeding the RfD level. Again, incidence rates are multiplied by the total UAE population to obtain final model outputs: the number of cases in which a person eats fish contaminated with methylmercury and is subsequently exposed to a contaminant concentration above the RfD level. Conceptual nodes to this effect appear in the influence diagram in Fig. 11.9.

Health effects of exceeding the RfD value for methylmercury are considered those within the full range of disorders described earlier in this chapter.

Table 11.16 Probability[a] of UAE seafood consumer exceeding daily methylmercury RfD

	Female			Male		
Age	Normal	Overweight	Obese	Normal	Overweight	Obese
2	1:132	1:476	1:1,111	1:69	1:278	1:435
3	1:179	1:1,000	1:2,000	1:109	1:435	1:625
4	1:204	1:909	1:3,333	1:114	1:476	1:1,000
5	1:196	1:1,250	1:1,667	1:98	1:476	1:769
6	1:172	1:769	1:2,500	1:98	1:500	1:714
7	1:152	1:556	1:1,111	1:105	1:400	1:625
8	1:139	1:588	1:1,000	1:88	1:333	1:714
9	1:108	1:435	1:769	1:72	1:250	1:476
10	1:78	1:303	1:667	1:64	1:263	1:476
11	1:70	1:250	1:455	1:50	1:169	1:286
12	1:63	1:213	1:417	1:40	1:122	1:303
13	1:53	1:200	1:370	1:33	1:112	1:204
14	1:44	1:169	1:303	1:30	1:99	1:179
15	1:38	1:125	1:233	1:27	1:83	1:145
16–29	1:24	1:69	1:120	1:13	1:38	1:64
30–44	1:24	1:71	1:137	1:13	1:38	1:62
45–59	1:23	1:70	1:123	1:13	1:38	1:65
60–74	1:24	1:72	1:128	1:13	1:38	1:62
75+	1:23	1:66	1:123	1:13	1:37	1:58

[a]For instance, 1:54 suggests 1 in 54 seafood consumers in the UAE will exceed the methylmercury RfD on any given day

Estimated Burden of Disease

Exposure to Heavy Metals

Model results suggest some UAE subpopulation groups may be at an elevated risk of health effects due to overexposure to methylmercury from eating seafood, as shown in Tables 11.16 and 11.17. Exposure is inversely related to body weight, as demonstrated by the higher proportion of above-RfD probabilities in the *Normal* BMI columns.

Results of daily cases in which a UAE resident may be at risk of overexposure to methylmercury from eating seafood and exceeding the reference dose suggest 2,927 women and 11,882 men—with the gender imbalance an artifact of the male-dominated expatriate workforce—could be at risk for health effects.

Although probabilities shown in Table 11.16 are much higher than standard regulatory risk targets—typically ranging from 1:10,000 to 1:1,000,000—it should again be noted that this study focuses on how many people exceed the daily RfD, not on how many people are estimated to actually become ill (presumably a much lower probability). Nonetheless, the high reported probabilities of UAE seafood consumers exceeding the daily methylmercury RfD do merit additional policies or regulations to help address this concern and reduce this exposure.

Table 11.17 Estimated daily cases in UAE of people exceeding daily methylmercury RfD

	Female			Male			
Age	Normal	Overweight	Obese	Normal	Overweight	Obese	Total
2	16	1	0	29	2	1	49
3	11	0	0	17	0	1	29
4	9	1	0	16	1	1	28
5	9	0	1	18	0	1	29
6	13	0	0	23	0	1	37
7	14	1	0	20	1	0	36
8	15	1	0	23	1	0	40
9	17	1	1	24	2	1	46
10	22	2	0	26	2	0	52
11	22	3	1	30	4	1	61
12	24	3	1	37	5	1	71
13	30	3	1	49	6	1	90
14	41	2	2	59	3	3	110
15	39	2	3	55	3	4	106
16–29	1,100	178	66	3,632	591	225	5,792
30–44	456	204	88	2,498	1,166	593	5,005
45–59	256	104	49	1,456	620	301	2,786
60–74	55	16	8	174	53	28	334
75+	31	5	1	58	10	2	107
Total	**2,180**	**527**	**222**	**8,244**	**2,470**	**1,165**	**14,808**

The preliminary risk estimates documented in Appendix A and developed by the RAND Corporation for this project show relative agreement with these results. RAND-derived figures estimate 27,000 (67,000 worst-case) potential annual cases of long-term illness—primarily neurological development due to seafood methylmercury overexposure—whereas this model estimates 14,808 daily incidents. It is logical to assume that practically all of these daily incidents are contributory toward chronic illness, not acute illness; therefore, daily incidents can be seen as constituting 14,808 potential annual cases of chronic methylmercury overexposure.

Exposure to Agricultural Pesticides

No vegetable and pesticide combinations have estimated daily exposures above aPAD or ADI levels, and only one combination (chlorpyrifos, hot pepper) exceeds daily cPAD levels, with a resulting exceedance probability of 1:227, or about 8,099 potential daily incidents in the UAE (Tables 11.18 and 11.19). More recent UAE pesticide-residue testing data (newer than 2008) would likely reduce this and other chlorpyrifos-related model results, as this pesticide's use was reportedly limited as a result of previous residue screening.

Table 11.18 Probability[a] of UAE consumer exceeding daily reference levels (pesticide on fruit/vegetable)

	Pesticide on fruits and vegetables	
Average probability of exceeding ADI	Methidathion, lemon	1:1,252
	Dimethoate, orange	1:5,543
Average probability of exceeding aPAD	Vinclozolin, tomato	1:1,167
Average probability of exceeding cPAD	Chlorpyrifos, hot pepper	1:227
	Chlorpyrifos, apple	1:26
	Chlorpyrifos, grape	1:398
	Chlorpyrifos, tomato	1:7
	Chlorpyrifos, orange	1:10
	Dimethoate, orange	1:35
	Vinclozolin, tomato	1:134
	Methidathion, lemon	1:33,490

[a]For instance, 1:100 suggests 1 in 100 fruit/vegetable consumers in UAE will exceed this benchmark on any given day, assuming zero reduction in pesticide residue from washing and/or cooking

Table 11.19 Estimated daily cases in UAE of people exceeding daily reference levels (pesticide on fruit/vegetable)

	Pesticide on fruits and vegetables	
Daily cases exceeding ADI	Methidathion, lemon	1,057
	Dimethoate, orange	310
Daily cases exceeding aPAD	Vinclozolin, tomato	1,684
Daily cases exceeding cPAD	Chlorpyrifos, hot pepper	8,099
	Chlorpyrifos, apple	75,169
	Chlorpyrifos, grape	4,908
	Chlorpyrifos, tomato	270,269
	Chlorpyrifos, orange	201,385
	Dimethoate, orange	56,497
	Vinclozolin, tomato	14,683
	Methidathion, lemon	64

For fruit and pesticide, Table 11.18 indicates one combination (vinclozolin, tomato) may be responsible for a 1:1,167 risk (about 1,684 UAE cases, Table 11.19) of exceeding daily aPAD. Two combinations (methidathion, lemon; dimethoate, orange) resulted in respective 1:1,252 (1,057 cases) and 1:5,543 (310 cases) chances of exceeding ADI levels. Of reduced concern but still of note, Table 11.18 details eight pesticide/crop combinations potentially causing UAE resident exceedances above daily cPAD levels.

Of all pesticides and crops, chlorpyrifos on tomato has the highest mean ratio (0.26) of average estimated pesticide exposure (0.000078 mg/kg) to its cPAD value (0.0003 mg/kg), making tomatoes the most suitable candidate for a worst-case hypothetical scenario. Considering an atypical but theoretical UAE resident eating 100% tomatoes, and assuming no reduction in pesticide due to washing, peeling, and/or cooking, the model estimates this person has 20.6% (chlorpyrifos) and 1.0% (vinclozolin) chances of exceeding cPAD values each day.

In comparison with earlier research conducted for this report (see Appendix A), some variation exists from this chapter's results. RAND-derived figures estimate zero average (89,000 worst-case) potential annual cases of short-term illness—primarily effects of acute pesticide overexposure—whereas this model estimates 631,074 worst-case daily incidents (cPAD exceedance) contributing to potential chronic illness.

Although these probabilities may seem high, it should be remembered that daily cPAD incidents are assumed contributory toward potential cases of annual chronic illness; the model assumes (worst-case) no reduction in pesticide due to washing, peeling, and/or cooking for all ADI, aPAD, and cPAD-related incidents; and, only very limited human epidemiologic studies exist to objectively link chronic pesticide exposure with adverse health effects—a major reason for the safety factors already built into the cPAD and other benchmarks.

Sensitivity Analysis

To determine relative effects of seafood input variables, three were scaled higher (+10%) and lower (−10%) before comparing their corresponding outputs. The procedure was then repeated with respect to an alternative methylmercury health benchmark (FAO and WHO 2003) for comparison with the U.S. reference dose, with results in Table 11.20. A 10% drop in methylmercury contamination or fish serving size results in a 39–40% reduction of daily cases exceeding the U.S. reference dose. Conversely, a 10% rise in either contamination or serving size corresponds to an estimated 59–60% rise in daily RfD exceedance cases. In all these scenarios, very few UAE residents (two baseline; six maximum) are estimated to be in exceedance of FAO/WHO provisional tolerable weekly intake (separately scaled to daily values for children and adults). If fish serving size is dropped to 31% (or methylmercury contamination dropped to 34%) of its baseline value, daily RfD exceedances drop to zero.

As expected, raising or lowering pesticide residue (±10%) affected exceedance probability and cases accordingly (Table 11.21). The ratio of fruits to vegetables appearing in Table 11.21 indicates fruit may have a smaller health safety cushion if pesticide MRL values are not properly heeded by growers.

Information Needed to Improve Future Burden of Disease Predictions

The model estimating health effects due to UAE food contamination as described in this chapter would be improved with additional UAE-specific health data; results of expanded pesticide residue testing studies; data on relative quantities of imported and domestic

Table 11.20 Sensitivity effects of varying seafood parameters

Methylmercury health benchmark considered		Baseline	Methylmercury contamination		Serving size		Daily consumption frequency	
			−10%	+10%	−10%	+10%	−10%	+10%
Avg. probability[a] of exceeding benchmark	U.S. RfD[b]	1:30	1:51	1:19	1:49	1:19	N/A	N/A
	FAO/WHO[c]	1:302,755	1:1,457,726	1:74,349	1:873,362	1:130,890	N/A	N/A
Daily cases exceeding benchmark	U.S. RfD	14,808	8,822	23,537	9,056	23,733	13,795	16,861
	FAO/WHO	2	0	6	0	3	2	2

[a]For instance, 1:100 suggests 1 in 100 seafood consumers in the UAE will exceed the benchmark on any given day
[b]U.S. EPA daily reference dose, all ages: 0.0001 mg/kg body weight
[c]FAO/WHO provisional tolerable weekly intake converted into daily dose: 0.00047 mg/kg (adult), 0.00023 mg/kg (child)

Table 11.21 Sensitivity effects of varying fruit or vegetable pesticide levels

−10%	Baseline	+10%
Average probability of exceeding ADI		
Methidathion, lemon 1:1,770	Methidathion, lemon 1:1,252	Methidathion, lemon 1:732
Dimethoate, orange 1:16,500	Dimethoate, orange 1:5,543	Dimethoate, orange 1:2,490
		Methidathion, apple 1:13,335
Average probability of exceeding aPAD		
Vinclozolin, tomato 1:1,526	Vinclozolin, tomato 1:1,167	Vinclozolin, tomato 1:785
Average probability of exceeding cPAD		
Chlorpyrifos, hot pepper 1:398	Chlorpyrifos, hot pepper 1:227	Chlorpyrifos, hot pepper 1:163
Chlorpyrifos, apple 1:34	Chlorpyrifos, apple 1:26	Chlorpyrifos, apple 1:22
Chlorpyrifos, grape 1:593	Chlorpyrifos, grape 1:398	Chlorpyrifos, grape 1:187
Chlorpyrifos, tomato 1:8	Chlorpyrifos, tomato 1:7	Chlorpyrifos, tomato 1:7
Chlorpyrifos, orange 1:11	Chlorpyrifos, orange 1:10	Chlorpyrifos, orange 1:9
Dimethoate, orange 1:40	Dimethoate, orange 1:35	Dimethoate, orange 1:26
Vinclozolin, tomato 1:140	Vinclozolin, tomato 1:134	Vinclozolin, tomato 1:138
	Methidathion, lemon 1:33,490	Methidathion, lemon 1:10,919
Daily cases exceeding ADI		
Methidathion, lemon 635	Methidathion, lemon 1,057	Methidathion, lemon 2,676
Dimethoate, orange 130	Dimethoate, orange 310	Dimethoate, orange 798
		Methidathion, apple 154
Daily cases exceeding aPAD		
Vinclozolin, tomato 1,291	Vinclozolin, tomato 1,684	Vinclozolin, tomato 2,485
Daily cases exceeding cPAD		
Chlorpyrifos, hot pepper 4,640	Chlorpyrifos, hot pepper 8,099	Chlorpyrifos, hot pepper 11,285
Chlorpyrifos, apple 57,397	Chlorpyrifos, apple 75,169	Chlorpyrifos, apple 88,792
Chlorpyrifos, grape 3,319	Chlorpyrifos, grape 4,908	Chlorpyrifos, grape 10,542
Chlorpyrifos, tomato 251,229	Chlorpyrifos, tomato 270,269	Chlorpyrifos, tomato 274,046
Chlorpyrifos, orange 181,108	Chlorpyrifos, orange 201,385	Chlorpyrifos, orange 215,693
Dimethoate, orange 48,686	Dimethoate, orange 56,497	Dimethoate, orange 74,891
Vinclozolin, tomato 14,090	Vinclozolin, tomato 14,683	Vinclozolin, tomato 14,248
	Methidathion, lemon 64	Methidathion, lemon 185

foods, seafood quality and coastal water quality; and additional scientific understanding of the dose-response relationships linking exposure and health outcomes.

Morbidity data provided by the Health Authority–Abu Dhabi—used in other risk area chapters of this report—do not include cases relevant to the health effects targeted by this study (neurological disorders, pesticide-induced poisoning). The data set includes about 73% of the Abu Dhabi population. It is unclear whether these illnesses are wholly absent in the UAE or if the resolution of available data is insufficiently comprehensive to capture these outcomes. In addition to pesticide and heavy metal poisoning, ideally this model would include estimates of infectious disease for purposes of calculating foodborne illness.

Further details of specific current agricultural practices, including pesticide application practices, would be required to confirm assumptions made during model development. Emirate-specific maximum residue levels (MRL) for all pesticide/crop combinations, assuming they are continuously tested and enforced within each emirate, would further enhance the model's performance. In addition, while numerous studies provide heavy metal concentrations in the flesh of local seafood, these studies target limited numbers of species in limited areas. The relative portion of each species comprising the typical local diet is unknown.

Conclusions

Preservation of the quality of foodstuffs cultivated and harvested in the UAE is essential to expanding agriculture and ensuring public trust of domestic crops. MOEW, the emirate food control authorities, and similar agencies within the UAE must continue to provide leadership on efforts to improve food safety policy and practices, especially via the UAE Codex Committee (operating in conjunction with WHO and the Food and Agriculture Organization of the United Nations). To continue to ensure food safety, the following specific initiatives are recommended:

Reduce environmental discharge of pollutants to prevent the contamination of land crops and seafood harvested in Emirati waters. While it would be impossible to prevent all sources of contamination to crops and seafood, reduction of pollutant loadings to agricultural areas through the institution and enforcement of relevant practices and standards will greatly reduce human exposure. Emirate-wide Farmers' Services Centers and similar UAE agencies must draft and implement Good Agricultural Practices as defined by the FAO. Pesticide residue standards, codes of practice, and auditing and enforcement procedures need to be developed and harmonized across all seven emirates. Inspections of imported food must adopt new standards based on a renewed risk-based focus and should be standardized across all emirates. Similarly, water quality and effluent standards must be harmonized throughout the seven emirates to prevent coastal pollution, which inevitably contaminates fish and shellfish in the Arabian Gulf. Adherence to standards, particularly in the case of wastewater treatment plant effluents, will likely reduce red tide incidents in addition to preventing fish contamination.

Increase monitoring efforts for pollutants of concern in conjunction with enforcement programs. Successful implementation of the recommendation to reduce contaminant discharges to the environment will require regular stringent enforcement. Contaminant levels in both the actual food and the environment in which food is cultivated or harvested (e.g. soil, irrigation water, coastal water) must be monitored regularly by relevant national agencies in order to assure adherence to standards and to alert the public should a contamination event occur. High probabilities of daily methylmercury RfD exceedance due to seafood consumption warrant further study and possible changes to advisory policies.

Reduce human exposure through promotion of proper food handling and storage. Unavoidable pesticide residues and microbial contamination can be substantially reduced or eliminated through proper preparation (e.g., washing, cooking). Proper storage with adequate refrigeration, particularly for seafood, is necessary to prevent outbreaks of microbial illness. Pertinent agencies within the UAE must design and conduct proactive public safety information campaigns on food-related issues to disseminate this information to the general public. Nationwide food labeling should denote organically farmed products. Advisories related to red tide conditions (offshore algal bloom events) and persistent seafood pollutants must be adopted and extended across all seven emirates. Vulnerable groups (e.g. pregnant women) should be instructed to avoid or minimize consumption of seafood species identified as high in heavy metal or toxic chemical content.

Increase the quantity of and improve the availability of data on food-related illnesses. UAE agencies from all seven emirates must continue efforts toward establishing a national medical records surveillance system and ensure that all medical centers report cases of disease suspected as or consistent with foodborne illness. Completion of this task will require substantial interagency communication and collaboration. Therefore, the UAE must include adequate funding for staff to work between agencies to monitor and analyze a national health database.

Improve urban planning to account for environmental impacts. Communication procedures between emirate-wide food control authorities and urban planning offices must be implemented and expanded to assure that environmental impact statements consider food contamination issues (such as effects of additional waste discharges into the coastal environment).

References

Abu Dhabi Food Control Authority (ADFCA). 2006. *Pesticides residue group monitoring program 2006*. Abu Dhabi: Abu Dhabi Food Control Authority (ADFCA).

Agah, H., M. Leermakers, M. Elskens, S.M.R. Fatemi, and W. Baeyens. 2006. Total mercury and methylmercury concentrations in fish from the Persian Gulf and the Caspian Sea. *Water, Air, and Soil Pollution* 181(1–4): 95–105.

Ahmad, S., and S.M. Al Ghais. 1996. Metal contents in the tissues of *Lutjanus fulviflamma* (Smith 1949) and *Epinephelus tauvina* (Forskal 1775) collected from the Arabian Gulf. *Bulletin of Environmental Contamination and Toxicology* 57(6): 957–963.

Al Ashram, O. 2005. *Wastes and pollution sources in Abu Dhabi Emirate*. Abu Dhabi: Environment Agency–Abu Dhabi, Environment Protection Department.

Al Ghais, S.M. 1995. Heavy metal concentrations in the tissue of *Sparus-Sarba* Forskal, 1775 from the United Arab Emirates. *Bulletin of Environmental Contamination and Toxicology* 55(4): 581–587.

Al Matroushi, M.A. 2005. *United Arab Emirates global school-based student health survey 2005.* UAE-GSHS-2005. Centers for Disease Control and Prevention/World Health Organization.

Al Yousuf, M.H., M.S. El Shahawi, and S.M. Al Ghais. 2000. Trace metals in liver, skin, and muscle of *Lethrinus lentjan* fish species in relation to body length and sex. *The Science of the Total Environment* 256(2/3): 87–94.

Al Zarooni, M., and W. Elshorbagy. 2006. Characterization and assessment of Al Ruwais refinery wastewater. *Journal of Hazardous Materials* A136: 398–405.

Anderson, D.M., P.M. Glibert, and J.M. Burkholder. 2002. Harmful algal blooms and eutrophication: Nutrient sources, composition, and consequences. *Estuaries* 25(4b): 704–726.

Badrinath, P., Q.A. Al Shboul, T. Zoubeidi, A.S. Gargoum, R. Ghubash, and O.E. El Rufaie. 2002. *Measuring the health of the nation: United Arab Emirates Health and Lifestyle survey 2000.* Al Ain: UAE University.

Baker, L.C., and D.J. McGillicuddy. 2006. Harmful algal blooms: At the interface between coastal oceanography and human health. *Oceanography* 19(2): 94–106.

Batz, M.B., M.P. Doyle, J.G. Morris Jr., J. Painter, R. Singh, R.V. Tauxe, M.R. Taylor, and D.M. Lo Fo Wong. 2005. Attributing illness to food. *Emerging Infectious Diseases* 11(7): 993–999.

Cardoso, C., N. Bandarra, H. Lourenco, C. Afonso, and M. Nunes. 2010. Methylmercury risks and EPA and DHA benefits associated with seafood consumption in Europe. *Risk Analysis* 30(5): 827–840.

Codex Alimentarius Commission. 2009. Pesticide residues in food: Maximum residue limits, extraneous maximum residue limits. http://www.codexalimentarius.net/mrls/pestdes/jsp/pest_q-e.jsp

Dehghan, M., N. Al Hamad, A. Yusufali, F. Nusrath, S. Yusuf, and A.T. Merchant. 2005. Development of a semi-quantitative food frequency questionnaire for use in United Arab Emirates and Kuwait based on local foods. *Nutrition Journal* 4(18). http://www.nutritionj.com/content/4/1/18

El Shahawi, M.S., and M.H. Al Yousuf. 1998. Heavy metal (Ni, Co, Cr, and Pb) contamination in liver and skin tissues of *Lethrinus lentjan* fish family: *Lethrinidae* (toelost) from the Arabian Gulf. *International Journal of Food Sciences and Nutrition* 49(6): 447–451.

Environment Agency–Abu Dhabi (EAD). 2006. *State of the environment Abu Dhabi: Key findings*. Abu Dhabi: Environment Agency–Abu Dhabi (EAD).

Environment Agency–Abu Dhabi (EAD). 2007a. *Report on water quality of Abu Dhabi coastal waters*. Abu Dhabi: Environment Agency–Abu Dhabi (EAD).

Environment Agency–Abu Dhabi (EAD). 2007b. *State of the environment Abu Dhabi: Pollution of groundwater*. Abu Dhabi: Environment Agency–Abu Dhabi (EAD).

Eto, K. 1997. Pathology of Minamata disease. *Toxicologic Pathology* 25(6): 614–623.

European Commission (EC). 1990. Council directive of 27 November 1990: On the fixing of maximum levels for pesticide residues in and on certain products of plant origin, including fruit and vegetables. 90/642/EEC. Brussels.

European Food Safety Authority (EFSA). 2009. Reasoned opinion of EFSA prepared by the Pesticides Unit on the 2007 annual report on pesticide residues. *EFSA Scientific Report* 305: 1–106.

Ferguson, C.M., B.G. Coote, N.J. Ashbolt, and I.M. Stevenson. 1996. Relationships between indicators, pathogens, and water quality in an estuarine system. *Water Research* 30(9): 2045–2054.

Fleming, L., K. Broad, A. Clement, E. Dewailly, S. Elmir, A. Knap, S.A. Pomponi, S. Smith, H. Solo-Gabriele, and P. Walsh. 2006. Oceans and human health: Emerging public health risks in the marine environment. *Marine Pollution Bulletin* 53: 545–560.

Food and Agriculture Organization of the United Nations (FAO). 2003. *Crops primary equivalent database: United Arab Emirates, food consumption quantity (2003 data).* FAOSTAT. http://faostat.fao.org

Food and Agriculture Organization of the United Nations (FAO) and the World Health Organization (WHO). 2003. *Evaluation of certain food additives and contaminants. Sixty-first report of the joint FAO/WHO expert committee on food additives*, Geneva.

Glibert, P.M. 2007. Eutrophication and harmful algal blooms: A complex issue, examples from the Arabian Seas, including Kuwait Bay, and an introduction to the Global Ecology and Oceanography of Harmful Algal Blooms (GEOHAB) programme. *International Journal of Oceans and Oceanography* 2(1): 157–169.

Glibert, P.M., J.H. Landsberg, J.J. Evans, M.A. Al Sarawi, M. Funaj, M.A. Al Jarallah, A. Haywood, et al. 2002. A fish kill of massive proportion in Kuwait Bay, Arabian Gulf, 2001: The roles of bacterial disease, harmful algae, and eutrophication. *Harmful Algae* 1: 215–231.

Glibert, P.M., D.M. Anderson, P. Gentien, E. Graneli, and K.G. Sellner. 2005. The global, complex phenomena of harmful algal blooms. *Oceanography* 18(2): 136–147.

Graneli, E., and J.T. Turner. 2006. An introduction to harmful algae. In *Ecology of harmful algae*, ed. E. Graneli and J.T. Turner, 3–21. Berlin: Springer.

Gray, G.M., and J.K. Hammitt. 2000. Risk/risk trade-offs in pesticide regulation: An exploratory analysis of the public health effects of a ban on organophosphate and carbamate pesticides. *Risk Analysis* 20(5): 665–680.

Hoffmann, S., P. Fischbeck, A. Krupnick, and M. McWilliams. 2008. Informing risk-mitigation priorities using uncertainty measures derived from heterogeneous expert panels: A demonstration using foodborne pathogens. *Reliability Engineering and System Safety* 93: 687–698.

Jarup, L. 2003. Hazards of heavy metal contamination. *British Medical Bulletin* 68: 167–182.

Jones, K.E., N.G. Patel, M.A. Levy, A. Storeygard, D. Balk, J.L. Gittleman, and P. Daszak. 2008. Global trends in emerging infectious diseases. *Nature* 451(7181): 990–994.

Kakande, Y., and M. Kwong. 2009. Inspectors fail half of Sharjah's restaurants. *The National*, June 6. http://www.thenational.ae/article/20090607/NATIONAL/706069820

Khaleej Times. 2002. Fruits and vegetables safe from pesticides: Study. November 30.

Khan, S., and V. Salama. 2008. UAE may buy Pakistan farms. *The National*, May 5. http://www.thenational.ae/article/20080505/BUSINESS/55399771/1118/rss

Kosanovic, M., M.Y. Hasan, D. Subramanian, A.A.F. Al Ahbabi, O.A.A. Al Kathiri, and E.M.A.A. Aleassa. 2007. Influence of urbanization of the western coast of the United Arab Emirates on trace metal content in muscle and liver of wild Red-spot Emperor (*Lethrinus lentjan*). *Food and Chemical Toxicology* 45(11): 2261–2266.

Kuchenmüller, T., S. Hird, C. Stein, P. Kramarz, A. Nanda, and A.H. Havelaar. 2009. Estimating the global burden of foodborne diseases: A collaborative effort. *Eurosurveillance* 14(18): 1–4.

Lee, C.M., T.Y. Lin, C.-C. Lin, G.A. Kohbodi, A. Bhatt, R. Lee, and J.A. Jay. 2006. Persistence of fecal indicator bacteria in Santa Monica Bay beach sediments. *Water Research* 40: 2593–2602.

McKinlay, R., J.A. Plant, J.N.B. Bell, and N. Voulvoulis. 2008. Calculating human exposure to endocrine disrupting pesticides via agricultural and non-agricultural exposure routes. *Science of the Total Environment* 398: 1–12.

Mead, P.S., L. Slutsker, V. Dietz, L.F. McCaig, L.F. Bresee, J.S. Shapiro, P.M. Griffin, and R.V. Tauxe. 1999. Food-related illness and death in the United States. *Emerging Infectious Diseases* 5(5): 607–625.

Menon, P. 2009. Red tide closes two Dubai beaches. *The National*, April 7. http://www.thenational.ae/article/20090407/NATIONAL/138119746/1186

Mohamed, A. A. 2009. *Abu Dhabi Food Control Authority pesticides monitoring program, 2006–2008*. Presentation, Dubai International Food Safety Conference, February 6, Dubai.

Morinigo, M.A., R. Cornax, M.A. Munoz, P. Romero, and J.J. Borrego. 1990. Relationships between *Salmonella* spp and indicator microorganisms in polluted natural waters. *Water Research* 24(1): 117–120.

Mozaffarian, D., and E.B. Rimm. 2006. Fish intake, contaminants, and human health: Evaluating the risks and benefits. *Journal of the American Medical Association* 296(15): 1885–1899.

Musaiger, A.O., and N.M. Abuirmeileh. 1998. Food consumption patterns of adults in the United Arab Emirates. *Journal of the Royal Society for the Promotion of Health* 118(3): 146–150.

National Center for Health Statistics. 2000. *Data table of weight-for-age charts: Children 2 to 20 years*. Centers for Disease Control and Prevention. http://www.cdc.gov/growthcharts/clinical_charts.htm

National Research Council. 1999. *From monsoons to microbes: Understanding the ocean's role in human health*. Washington, D.C.: The National Academies Press.

Polo, F., M.J. Figueras, I. Inza, J. Sala, J.M. Fleisher, and J. Guarro. 1998. Relationship between presence of *Salmonella* and indicators of faecal pollution in aquatic habitats. *FEMS Microbiology Letters* 160: 253–256.

Ragsdale, N.N. 2000. The impact of the Food Quality Protection Act on the future of plant disease management. *Annual Review of Phytopathology* 38: 577–596.

Sambidge, A. 2008. "Red tide" forces desalination plant closure. *ArabianBusiness.com*, November 17. http://www.arabianbusiness.com/538468-red-tide-forces-desalination-plant-closure

Setrakian, L. 2009. Filthy rich: Dubai choking on sewage. *ABC News*, February 1. http://abcnews.go.com/International/Story?id=6781673&page=1

Steenland, K. 1996. Chronic neurological effects of organophosphate pesticides. *British Medical Journal* 312(7042): 1312–1313.

Stewart, J.R., R.J. Gast, R.S. Fujioka, H.M. Solo-Gabriele, J.S. Meschke, L.A. Amaral-Zettler, E. del Castillo, et al. 2008. The coastal environment and human health: Microbial indicators, pathogens, sentinels, and reservoirs. *Environmental Health* 7(Suppl. 2): S3.

U.S. Department of Agriculture (USDA). 2009. *International maximum residue limit database*. Foreign Agricultural Service. http://www.fas.usda.gov/htp/mrl.asp

U.S. Environmental Protection Agency (EPA). 1997. *Reference dose tracking report*. Office of Pesticide Programs. http://npic.orst.edu/tracking.htm

U.S. Environmental Protection Agency (EPA). 1999. *Recognition and management of pesticide poisonings*, ed. J.R. Reigart, and J.R. Roberts. Office of Prevention, Pesticides and Toxic Substances. http://www.epa.gov/pesticides/safety/healthcare

U.S. Environmental Protection Agency (EPA). 2002. *Integrated risk information system: Methylmercury* (CASRN 22967-92-6). http://www.epa.gov/iris/subst/0073.htm

U.S. Environmental Protection Agency (EPA). 2009. *Pesticide reregistration status*. http://www.epa.gov/opp00001/reregistration/status.htm

U.S. Food and Drug Administration (FDA). 2007. *National Shellfish Sanitation Program*. http://www.fda.gov/Food/Safety/Product-SpecificInformation/Seafood/FederalStatePrograms/NationalShellfishSanitationProgram/default.htm

UAE Federal Government. 1999. Protection and development of the environment. Federal Law 24.

UAE Federal Government. 2004. UAE Ministerial Decree 193, concerning ban on importing and circulation of some harmful pesticides for health and environment.

UAE Ministry of Economy. 2008. *United Arab Emirates: Population and vital statistics 2008*.

UAE Ministry of Environment and Water. 2006. *United Arab Emirates: National report*. Report submitted to International Conference on Agrarian Reform and Rural Development.

World Health Organization. 2006. *WHO initiative to estimate the global burden of foodborne diseases: A summary document*. Geneva: World Health Organization.

World Health Organization. 2007. *Exposure to mercury: A major public health concern*. http://www.who.int/phe/news/Mercury-flyer.pdf

World Health Organization. 2009a. *Child growth standards: Length/height for age*. http://www.who.int/childgrowth/standards/height_for_age/en/index.html

World Health Organization. 2009b. *Growth reference: height for age (5–19 years)*. http://www.who.int/growthref/who2007_height_for_age/en/index.html

Chapter 12
Applying Environmental Burden of Disease Models to Strengthen Public Policy

Abstract The methods described in this book can provide a foundation for the next generation of environment and health strategic plans. Our approach provides an empirically validated means for the kinds of cooperative planning by the various levels of government, nongovernmental organizations and local communities needed in order to reduce human impacts on the environment and environmental impacts on human health. The project documented in this book followed three major steps: (1) developing preliminary environmental burden of disease estimates for 14 risk categories, (2) engaging stakeholders in a systematic process to prioritize these 14 risk categories based on the burden of disease information and other factors, and (3) analyzing in detail the burden of disease for eight key risk categories emerging from the priority-setting exercise. This chapter integrates the environmental burden of disease estimates from Chaps. 4, 5, 6, 7, 8, 9, 10, and 11. It provides a big-picture view of the United Arab Emirates' environmental disease burden across risk categories. It then outlines how the environmental burden of disease model described in these chapters can serve as a foundation for systematically analyzing interventions to improve environmental quality and lessen the associated disease burden. Next, it explains how a process like that in Chap. 2 could provide the foundation for the next generation of environment and health strategic plans, in which stakeholders come together to prioritize environmental interventions from a menu of options. The chapter also explains how ecological impacts of interventions could be incorporated in this priority-setting process. The budget struggles that many nations face as they contend with the continuing global economic crisis underline the need for renewed environment and health strategic planning. The approach outlined in this book paves the way for doing more with less—for increasing the public health gains of environmental interventions without necessarily increasing the economic burden on governments and their citizens.

J. MacDonald Gibson et al., *Environmental Burden of Disease Assessment*,
Environmental Science and Technology Library 24, DOI 10.1007/978-94-007-5925-1_12,

Keywords Environmental burden of disease • United Arab Emirates • Systematic prioritization of public health interventions • Land-use regression modeling • Hydrodynamic modeling • Risk management decision-making • Integrated environment and health modeling system

Introduction

The 1992 United Nations (UN) Conference on International Development in Rio de Janeiro, known as the Earth Summit, sparked a variety of international initiatives to reduce environmental impacts on human health. One outcome was *Agenda 21*, a global action plan to reduce human impacts on the environment and environmental impacts on human health. *Agenda 21* encouraged each of the 178 signatory nations to develop its own systematic plan to reduce environmental risks to health: "Countries ought to develop plans for priority actions [for environment and health protection], … which are based on cooperative planning by the various levels of government, nongovernmental organizations and local communities" (United Nations 1992). Many nations responded by developing national environment and health strategic plans. Yet as Chap. 2 explained, the lack of a systematic process for characterizing the environmental burden of disease across a population proved an obstacle to evidence-based planning. Most plans also lacked an empirically tested, validated basis for engaging stakeholders.

The methods described in this book can provide a foundation for the next generation of environment and health strategic plans. The bulk of the book (Chaps. 3, 4, 5, 6, 7, 8, 9, 10, and 11) provides technical details and a case study (the UAE) illustrating how to quantify the number of preventable illnesses and deaths attributable to environmental degradation—an approach that had not been codified when nations began environmental health strategic planning under *Agenda 21*. In addition, Chap. 2 demonstrates a process—known as the deliberative method for ranking risks—for engaging stakeholders to prioritize environmental risks to health, once the disease burden is quantified. This approach provides an empirically validated means for the kinds of "cooperative planning by the various levels of government, nongovernmental organizations and local communities" that *Agenda 21* promoted but for which, at the time, no systematic process had emerged.

The project documented in this book followed three major steps: (1) developing preliminary environmental burden of disease estimates for 14 risk categories, (2) engaging stakeholders in a systematic process to prioritize these 14 risk categories based on the burden of disease information and other factors, and (3) analyzing in detail the burden of disease for eight key risk categories emerging from the priority-setting exercise. Ideally, this priority-setting process would be repeated with the more comprehensive burden of disease estimates from Chaps. 4, 5, 6, 7, 8, 9, 10, and 11.

In the wake of the 20th anniversary of the original Rio Earth Summit, the time may be right to update previous environment and health strategic plans, not just in the UAE but also around the world. The methods illustrated in this book provide a template for doing so. These methods can help serve the broad goals of the latest Earth Summit, held in June 2012 and known as Rio+20: "to shape how we can reduce poverty, advance social equity and ensure environmental protection on an ever more crowded planet to get to the future we want." In an editorial commenting on preparations for Rio+20, UN Secretary General Ban Ki-moon observed,

> Just as there can be no sustainable growth without development, there can be no sustainable development without protecting the planet. Our collective health, wealth and well-being depends on how we husband the earth's 'natural capital'—the air, rivers and oceans, soils and forests, its full diversity of flora and fauna.... Rio+20 is an opportunity to define a clear path to a better future—a future of integrated solutions to interrelated problems (Ki-moon 2011).

The methods described in this book demonstrate an approach for developing integrated solutions to reduce environmental risks to public health at the national, regional, and local levels.

This chapter integrates the environmental burden of disease estimates from Chaps. 4, 5, 6, 7, 8, 9, 10, and 11. It provides a big-picture view of the UAE's environmental disease burden across risk categories. It then outlines how the environmental burden of disease model described in these chapters can serve as a foundation for systematically analyzing interventions to improve environmental quality and lessen the associated disease burden. Next, it explains how a process like that in Chap. 2 could provide the foundation for the next generation of environment and health strategic plans, in which stakeholders come together to prioritize environmental interventions from a menu of options. The chapter also explains how ecological impacts of interventions could be incorporated in this priority-setting process.

The budget struggles that many nations face as they contend with the continuing global economic crisis underline the need for renewed environment and health strategic planning. The approach outlined in this book paves the way for doing more with less—for increasing the public health gains of environmental interventions without necessarily increasing the economic burden on governments and their citizens. As Ban Ki-moon wrote in his recent editorial, "We all recognize that budgets are stretched thin. For much of the world, fiscal austerity is the new order of the day." For this reason, environment and health strategic planning is more important than ever and holds potential to provide both health benefits and economic relief.

Summary of UAE's Environmental Disease Burden

Figures 12.1 and 12.2 summarize the estimated number of deaths and medical visits attributable to the environmental risks considered in this report for the year 2008. These figures combine the information developed in Chaps. 4, 5, 6, 7, 8, 9, 10, and 11.

From Fig. 12.1, it is clear that outdoor air pollution is the leading environmental cause of premature death, compared to the other risks. From Fig. 12.2, indoor air

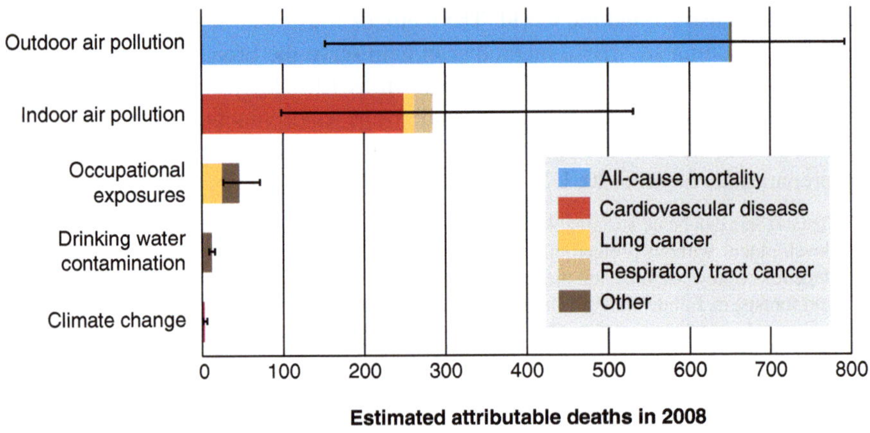

Fig. 12.1 The annual number of deaths potentially attributable to environmental contamination in the UAE, by exposure route and broad disease category

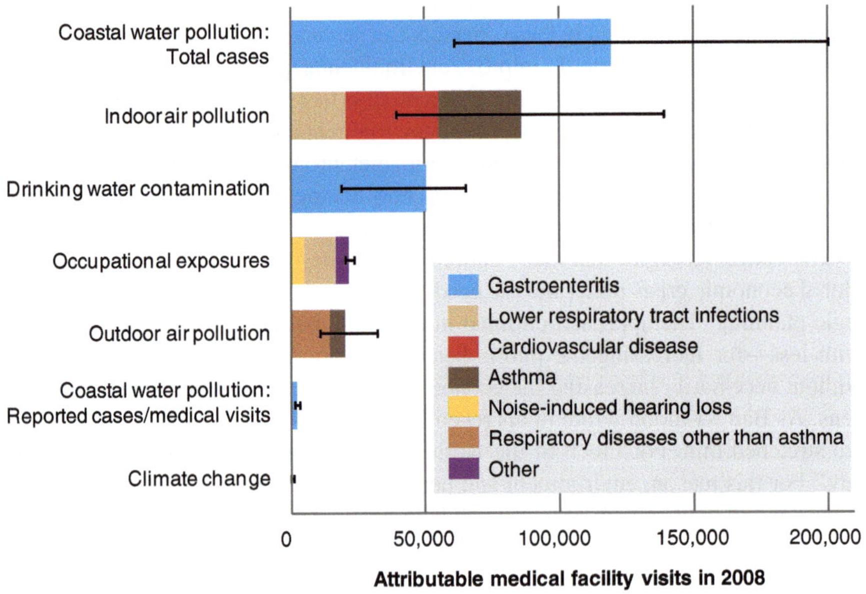

Fig. 12.2 The annual number of visits to health-care facilities potentially attributable to environmental contamination in the UAE, by exposure route and health outcome

pollution is a leading causes of medical visits. If one considers not just medical visits but also the total potential number of illness cases (including those for which no medical assistance is sought), then coastal water contamination rises in importance.

Although fears of cancer tend to dominate many debates about environmental risks to health (Wandersman et al. 1993), Figs. 12.1 and 12.2 illustrate that environmental pollution also is an important driver of cardiovascular disease and other chronic conditions that have become leading causes of death in developed nations and emerging economies, where infectious disease risks have declined due to improved water, sanitation, hygiene, and medical treatments. Many recent analyses have documented that these increased rates of chronic conditions can pose major economic burdens to national governments (Yach et al. 2004). Hence, intervening to reduce environmental pollutant exposures could yield economic benefits by preventing common chronic diseases that escalate government spending on health care.

Quantifying Health Benefits of Environmental Interventions

Understanding environmental impacts on public health under existing conditions is an essential first step in systematically planning how to optimize resources to reduce the environmental burden of disease. Another critical step is identifying potential interventions and the number of excess illnesses they could prevent. The *UAE Environmental Burden of Disease Model*, in combination with other modeling tools, can support such analyses. It can be linked to environmental models to predict how an intervention that decreases pollutant concentrations or the number of people exposed would, in turn, benefit health.

As an example, the sensitivity analysis in Chap. 11 shows that reducing the highest observed microbial pollutant concentration by 10% would reduce the total number of annual gastroenteritis cases attributable to swimming in coastal water by about 6,400. Further models, linking microbial pollution sources in the coastal zone to measured enterococci concentrations, could be used to analyze options for achieving a 10% reduction in maximum enterococci concentrations. Hydrodynamic models that predict the fate and transport of microbes from sewage discharges, urban runoff, ships, and other sources are one option for planning ways to reduce microbiological pollutants in the coastal zone. Such models already have been developed for use in simulating potential oil spills and also for analyzing the effects of discharges from potable water desalination plants (see, for example, Elhakeem et al. 2007). Land-use regression modeling is another, much less computationally intensive approach. A land-use regression model predicts coastal pollutant concentrations based on land-use characteristics such as the number of sewer outfalls in a given region, the percentage of impervious surface (in other words, land coverage with roads, parking lots, buildings, and other surfaces that limit water infiltration into the ground), and population density, along with water quality characteristics (see, for example, Coulliette et al. 2009; Mallin et al. 2000). A land-use regression model could be readily linked to the *UAE Environmental Burden of Disease Model*.

As another example, if the goal were to reduce premature mortality attributable to outdoor air pollution to levels consistent with the nations estimated to have the best air quality, then the environmental burden of disease model could be used to predict outdoor air concentrations necessary to achieve this goal. For example, the World Health Organization (WHO) estimates that Finland has among the lowest number of annual deaths attributable to particulate matter (PM) in outdoor air among developed nations: approximately 0.02 per 1,000 people per year (WHO 2007). The analysis in Chap. 4 suggests that the death rate due to PM in outdoor air in the UAE in 2008 was approximately 0.14 per 1,000 people. (According to the WHO, this rate is lower than in many developed nations, including Japan and the United Kingdom, where WHO estimates about 0.19 deaths per 1,000 people per year are attributable to PM in outdoor air.) As Chap. 4 illustrates, PM levels vary substantially by location (even within the same city). Hence, different interventions that achieve the same average level of pollutant reduction may confer different public health benefits, depending on where the reductions occur in relation to human populations. The environmental burden of disease model can be used to explore these effects. Ideally, the model would be coupled with an outdoor air quality model such as the CMAQ model discussed in Chap. 4 for analyzing how controls on pollutant emissions from specific sources translate into air quality improvements across space and time.

Prioritizing Interventions

Beyond understanding the benefits of improved environmental controls for public health, prioritizing possible interventions requires an understanding of the costs of the alternatives. Vast differences exist in the cost-effectiveness of regulations designed to protect health and safety. As an example, Fig. 12.3 compares the estimated cost per life saved of a variety of health, safety, and environmental regulations implemented in the United States between 1967 and 2005, as compiled by Hahn and Tetlock (2008). As shown, the scatter in these cost-effectiveness estimates tends to increase after 1986. This observation led Hahn and Tetlock to conclude, "There appear to be ample opportunities for refocusing regulations away from those with a high cost per statistical life saved and toward those with a low cost per statistical life saved. The result would be that regulation could either save more lives, or reduce expenditures, or both." The United States, however, has yet to undertake such a comprehensive reanalysis of its environmental and other regulations designed to improve public health. The approach outlined in this chapter could be used as a step toward comparing the costs and effectiveness of alternative interventions.

A frequent criticism of traditional cost-benefit analysis of alternative environment and health protection approaches is that it overlooks attributes for which monetary values are not easily assigned; examples include the distribution of effects across a population and people's level of control over their exposure to risk. Another criticism is that cost-benefit analysis requires the assignment of a monetary value to a human life. The priority-setting approach described in Chap. 2 could be used as the basis for systematic prioritization of interventions, considering not just costs and the potential for population-scale reductions in disease burden but also other attributes characteristically omitted from traditional cost-benefit analyses. As explained in Chap. 2, these attributes include the nature of the health effects (some diseases are more dreaded than others), their distribution across the population (for instance, the extent to which they

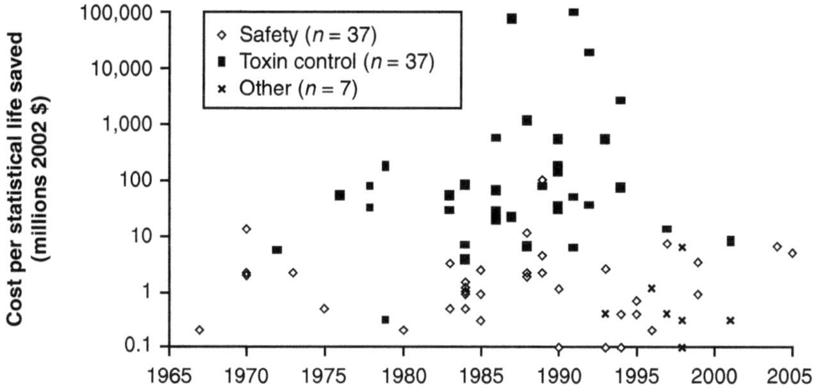

Fig. 12.3 Cost-effectiveness of 79 U.S. regulations enacted between 1967 and 2005 (Hahn and Tetlock 2008). All were aimed at reducing human health and safety risks. Those labeled "safety" were aimed at safety risks; "toxin control" regulations were aimed at toxins associated with cancer; "other" regulations addressed health risks not associated with toxin control or safety

disproportionately affect children and other vulnerable populations), the time between exposure and health effects, the degree of certainty in risk forecasts, and an individual's ability to control exposure. Another advantage of the approach detailed in Chap. 2 is that it does not require monetizing the value of human life or health. This same process could be applied to prioritize interventions. That is, each potential intervention could be characterized according to the attributes described in Chap. 2 (and shown in the risk summary sheets in Appendix A). Then, facilitators could engage stakeholders in focus groups to review and prioritize these interventions, using the analytic-deliberative process of Chap. 2. The results might have greater credence if the priority setting process engaged government representatives elected or appointed to represent the interests of the people. In the United States, for example, members of Congress and/or elected state legislatures could be involved. In Abu Dhabi, the Abu Dhabi Executive Council might nominate delegates to participate in such a process.

Incorporating Ecological Effects into the Priority-Setting Process

The priority setting process employed in this study also could be expanded to include ecological effects among the attributes considered. Willis et al. (2004) demonstrated how the deliberative method for ranking risks can be extended to encompass both human health and ecosystem risks. In essence, this extended process involves characterizing risks according to an expanded attribute list that includes ecological effects such as the spatial area of habitat affected, damage to native species, and changes in landscape appearance. Willis et al. developed such an attribute list and pilot tested this expanded priority setting process using ten hazards in a hypothetical U.S. community. Figure 12.4 shows an example summary of a risk

Road Salt and Road Salt Runoff

Summary:

As in other parts of the country which experience ice and snow in the winter, the DePaul County Highway Department and the Centerville Department of Public Works use salt as a deicing agent on roads in winter months. Salt runoff and spray can adversely affect plants and some other life forms. In some circumstances, sensitive trees near salted roads can be killed. Salt runoff can contaminate groundwater. Salt also contributes to the accelerated deterioration of structural materials such as concrete and steel in structures such as bridges and auto bodies. This deterioration increases the risk of automobile-related injuries

Human Health and Safety Impacts	Low Estimate	**Best Estimate**	High Estimate
Risk of death			
For the average person –			
Chance in a million of death per year	4	**8**	15
Expected number of deaths per year	0.05	**0.1**	0.2
For the person at highest risk –			
Chance in a million of death per year	11	**20**	35
Catastrophic potential, greatest no. of deaths in a single event		**5-20**	
Risk of injury and illness			
Serious injuries and illnesses, number of cases per year	2	**4**	8
Minor injuries and illnesses, number of cases per year	4	**8**	15
Other factors			
Time between exposure and health effects		**immediate**	
Scientific understanding and predictability of health and safety impacts		**high**	
Ability of individual to control exposure to health and safety risks		**medium**	

Environmental Impacts	Low Estimate	**Best Estimate**	High Estimate
Ecological effects			
Habitat affected –			
Acres	3,000	**5,300**	10,000
Square miles	5	**8**	16
Animals killed or displaced, number		**few**	
Effects on variety of native species		**small**	
Ecological significance of affected species and habitat		**medium**	
Effects on natural processes and cycles		**low**	
Catastrophic potential, magnitude of worse-case effects		**low**	
Aesthetic effects			
Changes in landscape appearance		**small negative (–3)**	
Effects on noise, smell, taste, and visibility		**little or no change (0)**	
Other factors			
Time between exposure and health effects		**0–5 years**	
Duration of environmental effects, assuming the current activity or stress does not continue but no other corrective actions are taken		**0–30 years**	
Scientific understanding and predictability of environmental impacts		**somewhat high**	
Negative effects on the environment's capacity to provide goods and services to people		**small**	

Fig. 12.4 Example of including health and ecological impacts in an environmental priority-setting project (Willis et al. 2004)

considered in their project, illustrating the ecological attributes added to the assessment. Similar to the experiences in the UAE described in Chap. 2, participants "reported that they were satisfied with the procedures and results, and indicated their support for using the method to advise real world risk-management decisions" (Willis et al. 2004).

Limitations and Opportunities

A combination of the environmental burden of disease modeling approach described in Chaps. 3, 4, 5, 6, 7, 8, 9, 10, and 11 and the stakeholder engagement process documented in Chap. 2 could provide an improved foundation for future national-, regional-, or local-scale environmental strategies aimed at prioritizing how to invest limited resources in improving environmental conditions. The approach could be further strengthened if several key limitations were addressed.

One limitation is that the *UAE Environmental Burden of Disease Model* as currently constructed does not address risks of exposure to contaminant mixtures. Studies have shown that in some cases, contaminants can act synergistically, so that one contaminant amplifies the risks of exposure to another (Dominici et al. 2010). A commonly cited and well-studied example is the effects of active smoking on risks due to asbestos exposure: smoking greatly increases the probability of experiencing adverse health effects from asbestos, due to the simultaneous exposure to asbestos and pollutants in cigarette smoke (Dominici et al. 2010). The effects of multiple exposures on health risks can be built into the *UAE Environmental Burden of Disease Model* as the toxicologic and epidemiologic information needed to characterize such multipollutant effects becomes available.

Another limitation is the lack of automated links between the *UAE Environmental Burden of Disease Model* and models that estimate the effects of specific interventions on pollutant concentrations—for example, the impacts of increasing requirements for automotive fuel economy on outdoor air pollution. We demonstrated in Chap. 4 that such links are possible: We linked the *UAE Environmental Burden of Disease Model* to output from a model that estimates the effects of changing pollution from traffic and other sources on air quality over a relatively small spatial scale. However, this linkage is not automated. Coupling the models requires first running the air quality model, and then manually transferring the output to the *UAE Environmental Burden of Disease Model*. In the future, options for automating the links between these models could be explored.

A third limitation is the time and personnel required to carry out a detailed environmental burden of disease analysis. Gathering and analyzing the data employed in the assessment presented in this book required a team of 39 and a 2-year timeline. Resources may not be available for such ambitious efforts elsewhere. Furthermore, politicians may not wish to wait 2 years for the results. A long-term solution to this problem is to directly link the *UAE Environmental Burden of Disease Model* to ambient environmental monitors (such as air and water quality monitors) and health tracking systems (such as record-keeping systems for patient encounters

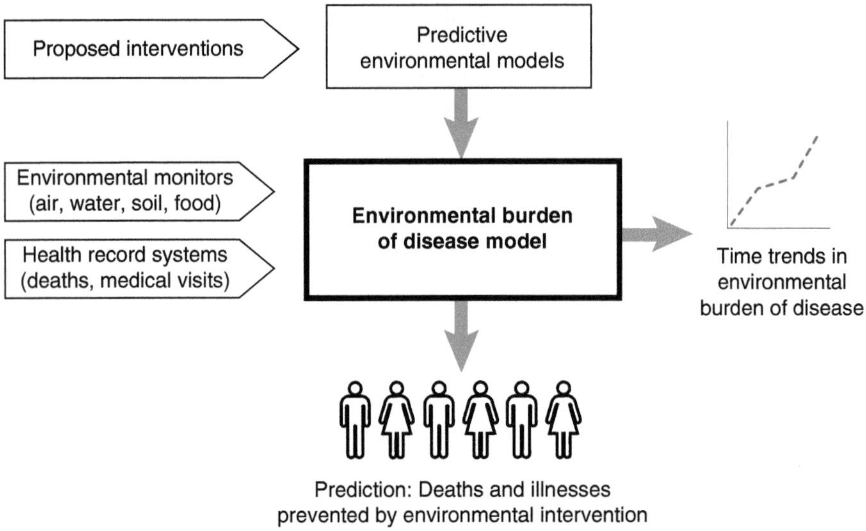

Fig. 12.5 Proposed process for linking the environmental burden of disease model to environmental models and existing data collection systems in order to (1) predict the public health effects of environmental interventions and (2) efficiently track the environmental disease burden over time

at medical facilities). The burden of disease estimates then could be updated annually to reflect the latest observational data. This would provide a mechanism not only for reducing the time and personnel required for future environmental disease burden estimates but also for tracking changes over time.

Figure 12.5 illustrates a flow diagram for such an integrated environment and health modeling system. On the left of the figure is a clipboard representing a list of potential interventions to improve environmental quality and reduce negative health consequences. Moving to the right on the diagram, environmental models can predict how each intervention would alter pollutant concentrations at specific locations and times. Moving downward, these altered pollution patterns could be input to the *UAE Environmental Burden of Disease Model*. The predicted post-intervention disease burden then could be compared to current conditions to establish the number of excess illnesses and deaths (if any) the measure might prevent. The left-hand side of the diagram conceptualizes how data from routine environmental monitoring systems (for example, ambient air quality monitors) and health data collection systems (such as medical patient encounter information collected by health agencies) could be linked—perhaps in real time—to the *UAE Environmental Burden of Disease Model*, generating information about time trends in the environmental disease burden, illustrated on the right-hand side of the diagram.

Progress has been made in developing such integrated modeling systems for individual pollutants and exposure pathways. For example, the U.S. Environmental Protection Agency's Environmental Benefits Mapping and Analysis System is a

Windows-based, publicly available program for estimating health impacts and economic benefits that accrue when United States population exposure to a specific air pollutant decreases (see, for example, Hubbell et al. 2004). However, a multipathway, multipollutant exposure approach such as that employed to assess the UAE's environmental burden of disease would provide a more complete picture for policymakers of options for improving public health through environmental interventions.

Rene Dubos, one of the twentieth century's pre-eminent microbiologists and environmentalists and originator of the phrase, "Think globally, act locally," is frequently quoted as saying, "Man shapes himself through decisions that shape his environment." The methods illustrated in this text can serve as tools for making better-informed decisions about the environment, hence helping to improve the future human condition.

References

Coulliette, A.D., E.S. Money, M.L. Serre, and R.T. Noble. 2009. Space/time analysis of fecal pollution and rainfall in an eastern North Carolina estuary. *Environmental Science & Technology* 43(10): 3728–3735.

Dominici, F., R.D. Peng, C.D. Barr, and M.L. Bell. 2010. Protecting human health from air pollution: Shifting from a single-pollutant to a multipollutant approach. *Epidemiology* 21(2): 187–194. doi:10.1097/EDE.0b013e3181cc86e8.

Elhakeem, A., W. Elshorbagy, and R. Chebbi. 2007. Oil spill simulation in the Arabian (Persian) Gulf with special reference to the UAE coast. *Water, Air, and Soil Pollution* 184(1–4): 243–254.

Hahn, R.W., and P.C. Tetlock. 2008. Has economic analysis improved regulatory decisions? *Journal of Economic Perspectives* 22(1): 67–84.

Hubbell, B.J., A. Hallberg, D.R. McCubbin, and E. Post. 2004. Health-related benefits of attaining the 8-hr ozone standard. *Environmental Health Perspectives* 113(1): 73–82. doi:10.1289/ehp.7186.

Ki-moon, B. 2011. The clock is ticking. *The New York Times*, October 31.

Mallin, M.A., K.E. Williams, E.C. Esham, and R.P. Lowe. 2000. Effect of human development on bacteriological water quality in coastal watersheds. *Ecological Applications* 10(4): 1047–1056.

United Nations. 1992. *Agenda 21: The United Nations programme of action from Rio*. New York: United Nations.

Wandersman, A.H., W.K. Hallman, B. Keenan, L. Lefton, E. Vaughan, L.P. Wandersman, N. Weinstein, et al. 1993. Understanding public concerns about environmental threats: Are people acting irrationally? *American Psychologist* 48(6): 681–686.

World Health Organization. 2007. *Environmental burden of disease: Country profiles*. Geneva: World Health Organization.

Willis, H.H., M.L. DeKay, M.G. Morgan, H.K. Florig, and P.S. Fischbeck. 2004. Ecological risk ranking: Development and evaluation of a method for improving public participation in environmental decision making. *Risk Analysis* 24: 363–378.

Yach, D., C. Hawkes, C.L. Gould, and K.J. Hofman. 2004. The global burden of chronic diseases: Overcoming impediments to prevention and control. *Journal of the American Medical Association* 291(21): 2616–2622. doi:10.1001/jama.291.21.2616.

Appendices

Appendix A: Risk Summary Sheets

Overview

This appendix contains risk summary sheets for 14 environmental health risks in the United Arab Emirates, as shown in Table A.1. The risk estimates shown in these summaries functioned as preliminary estimates for this study. Because the estimates given here were based on early and incomplete information, they are superseded by the results detailed in the preceding chapters.

The 14 risk summary sheets summarize the scope of each risk, describe what is known about the risks from the exposure generally and specifically within the UAE, and provide an overview of what has already been done in the UAE to manage the risk of the exposure. These risk summary sheets were the primary means of educating participants in the risk-ranking workshops (described in detail in Chap. 2) about environmental health risks in the UAE. The first page of each risk summary sheet includes a table with estimates of the environmental health risks in the UAE. The risk summary sheets have been reformatted from their original four-page format to fit this report. Notes on how calculations were performed for each risk are also included.

Definitions of Risk Attributes

Number of Deaths per Year. This is the average number of deaths expected per year among residents in the UAE based on a lifetime of exposure.

Chance in a Million of Death per Year for the Average Resident. This is the average annual lifetime risk of death for a randomly chosen resident in the UAE.

J. MacDonald Gibson et al., *Environmental Burden of Disease Assessment*,
Environmental Science and Technology Library 24, DOI 10.1007/978-94-007-5925-1,

Table A.1 List of risk summary sheets found in Appendix A

The risk to the average resident is simply the estimate of deaths per year in the UAE divided by the population.

Chance in a Million of Death per Year for the Resident at Highest Risk. For some hazards, certain people are known to be more exposed or more susceptible than others. Examples include susceptibility of infants and the elderly to pulmonary effects of outdoor air pollution. For some of the hazards, however, the risk to the most exposed resident is the same as the risk to the average resident because residents who might be more exposed or more susceptible cannot be easily identified.

Greatest Number of Deaths in a Single Episode. Some environmental hazards kill only one person at a time, whereas other hazards can kill a group of people all at once. For instance, people who die from chronic low-level exposure to radon gas will die one at a time, but an industrial accident releasing toxic gases could claim many lives at once. This statistic represents the greatest number of people who could plausibly be killed in a single event involving a given hazard. In estimating this statistic, we have tried to account for the fact that major incidents tend to be much less frequent than smaller incidents.

Illness or Injury. Many environmental hazards present nonfatal risks. These risks vary in both duration and severity. The risk summary tables describe four categories of cases of nonfatal illness or injury per year expected among UAE residents resulting from one year of exposure to a given hazard. These four categories are defined in Table A.2.

Time between Exposure and Health Effect. Some hazards, such as poison, have immediate impacts, whereas hazards such as asbestos have effects years into the future.

Table A.2 Categorization used to describe nonfatal risks

Duration	More severe	Less severe
Long term (>3 months)	Serious chronic conditions, often involving hospitalization. Examples: loss of limb; mental retardation requiring continuous care; blindness; infertility; nonfatal cancer; chronic migraine; disfiguring burns; any condition requiring long-term institutional care; permanent damage to lungs, liver, kidney, or heart resulting in more than 20% loss of organ function	Long-term conditions not requiring hospitalization, except emergency room. Examples: joint damage; loss of finger; mild mental retardation; scars and burns affecting movement; permanent damage to lung, liver, kidney, or heart resulting in less than 20% loss of organ function
Short term (<3 months)	Serious acute conditions requiring hospitalization. Examples: meningitis, pneumonia, severe asthma or allergic attack, compound fracture, severe food poisoning	Conditions that might require medical care, but no hospital admission, and that involve the loss of at least 1 day of work or otherwise restricted activity. Examples: acute infectious disease without hospital stay (e.g., cold, flu, earache), mild food poisoning

Quality of Scientific Understanding. There are two sources of uncertainty in estimating risks for the residents of UAE. One involves how well scientists know the relationship between exposure to a hazard and its resulting health impacts. The other involves how well we can predict exposure of UAE residents to a particular hazard. This statistic characterizes the former. For instance, scientists still do not know whether exposure to electromagnetic fields from power systems causes cancer, but scientists understand very well the physical and biological processes leading to injury from exposure to lead. Three categories are used to rate scientific understanding: high, moderate, and low.

Combined Uncertainty in Deaths, Illness, and Injury. This statistic reflects both uncertain scientific understanding about the risk and uncertainty about the extent of exposure or susceptibility of UAE residents to the particular hazard. The table entry gives the amount of uncertainty in deaths, illness, and injury, expressed qualitatively with respect to other risks assessed in this report. Three categories are used to rate uncertainty: high, moderate, and low.

Ability of Residents to Control Exposure. This statistic characterizes the degree to which people can control their own exposure to a given hazard. Some hazards that UAE residents encounter can be avoided partly or entirely by measures that individuals can take on their own. For instance, people not wishing to incur risks from coastal recreational water can choose not to bathe in the ocean. Three categories are used to rate this controllability: high, moderate, and low.

Outdoor Air Pollution

Summary

Major causes of outdoor air pollution in the UAE include transportation, energy refineries and factories, desalination plants, and seasonal dust and sandstorms. Transportation is a particularly large source of outdoor air pollution in Dubai, which has the largest concentration of vehicles in the country. These activities release several pollutants into the air, including particulate matter, ozone, carbon monoxide, sulfur oxides, nitrogen oxides, heavy metals, and the resultant smog. Breathing these pollutants increases risks for mortality, lung cancer, chronic obstructive pulmonary disease, asthma, and other respiratory illnesses that may develop years or even decades after prolonged exposure.

Risk characteristic	Low estimate	Best estimate	High estimate
Fatalities			
Number of deaths per year	400	1,400	2,600
Chance in a million of death per year for the average resident	90	300	600
Chance in a million of death per year for the resident at highest risk	Not reported	Not reported	Not reported
Greatest number of deaths in a single event	1	Not reported	4,000
Illness or injury			
More serious long-term cases per year	Not reported	Not reported	Not reported
Less serious long-term cases per year	Not reported	Not reported	Not reported
More serious short-term cases per year	Not reported	Not reported	Not reported
Less serious short-term cases per year	Not reported	Not reported	Not reported
Other factors			
Time between exposure and health effects	Immediate (asthma) to 30 years (cancers, lung disease)		
Quality of scientific understanding	Moderate		
Combined uncertainty in death, illness, and injury	High		
Ability of resident to control exposure to hazard	Moderate		

Exposure to outdoor air pollutants depends on proximity to busy roadways and industrial sources like power plants and refineries, length of time spent outdoors, and an individual's level of physical exertion. The very young, the very old, pregnant women, individuals who spend a large amount of time outdoors, and individuals with pre-existing health conditions such as heart or lung disease are the most vulnerable to disease and mortality risks from outdoor air pollution.

What Is Known About the Risk from Outdoor Air Pollution?

Air pollution affects the air quality and health of people living close to pollution sources. Because pollutants can also be transported through the atmosphere over long distances, air pollution may impact health on a broader local or regional scale. The principal pollutants of concern worldwide, and likely also in the UAE, are particulate matter and ground-level ozone, as well as lead and combustion-related air pollutants.

Particulate matter, or PM, consists of solid and liquid particles in air. Fine particles are those that are 2.5 μm or smaller in diameter ($PM_{2.5}$) and coarse particles are particles 10 μm or smaller in diameter (PM_{10}). PM_{10} can reach the upper part of the airway and lungs and cause an increase in respiratory illness and death. $PM_{2.5}$ particles are inhaled more deeply into the lungs and have been linked with poor lung function, aggravation of asthma, respiratory problems, infant mortality, lung cancer, and increased risk of death from cardiovascular and respiratory disease. Sandstorms may provide a source of exposure to particulate matter and may transport dust contaminated with toxic metals, fungi, and bacteria. Exposure to sandstorms has been linked to higher childhood asthma rates and increased hospital admissions related to heart disease.

Ozone at the ground level is the primary component of smog. Short-term effects of ozone include lung inflammation, respiratory symptoms, an increase in hospital admissions for asthma, and an increase in mortality. Smoking worsens these health effects, especially for those with asthma. The mortality risks associated with ozone may be exacerbated in the summer, when ozone levels are higher.

Lead is a toxic heavy metal that is absorbed through inhalation, drinking, and eating. Lead affects kidney function, blood pressure, gastrointestinal symptoms, sexual development, and the brain. Although leaded gasoline was phased out of the UAE in 2003, the health effects of lead may persist for decades because lead-contaminated soil can be spread through windblown dust and sand.

Combustion-related air pollutants such as those generated by the transportation, power, and industrial sectors, including carbon monoxide, nitrogen dioxide, and sulfur dioxide, all have short- and long-term health effects, including respiratory symptoms, worsening symptoms in people with asthma, decreased lung and exercise capacity, and death from cardiovascular and/or respiratory disease. Chronic exposures to nitrogen dioxide and sulfur dioxide can also affect lung development, increase respiratory illnesses, and worsen asthma in children.

What Is the Exposure to Outdoor Air Pollutants in the UAE?

The primary sources of outdoor air pollution in the UAE are transportation, oil and gas production, power plants, and numerous industrial processes. Increases in car ownership and gas consumption, demand for natural gas and oil used for electricity, and desalinated water generation have contributed to rising levels of outdoor air pollution. According to a report by the U.N. Environment Programme, vehicular emissions are the main source of outdoor air pollution in the Middle East. Driven by increased vehicle ownership, the UAE has witnessed an enormous increase in vehicle travel over the past few years. Given the easy mobility of outdoor air pollution, it is also quite possible for the UAE to experience pollution from neighboring countries that blow into the Emirates. Across the region, sulfur dioxide and nitrogen oxide emissions have been steadily increasing in many locations, and these increases are expected to affect outdoor air pollutant levels in the UAE.

$PM_{2.5}$ and PM_{10} are emitted from combustion sources such as diesel-powered engines, power generation, and wood burning. PM_{10} also comes from windblown dust or soil, and construction activities. PM is also formed in the atmosphere from chemical reactions of gases such as sulfur dioxide, nitrogen oxides, ammonia, and volatile organic compounds.

Ozone is formed in the atmosphere as a result of chemical reactions between volatile organic compounds and nitrogen oxides in the presence of sunlight. Therefore, ozone is at its highest concentrations when sunlight is most intense during midday and summer months. Combustion products such as sulfur dioxide, carbon monoxide, and nitrogen dioxide arise from transportation, power, and industrial emissions. Transportation emissions were also a major source of lead exposure during vehicular combustion of leaded gasoline. Exposure to lead is also believed to be related to the resuspension of lead-bearing dust in the air, which is spread by wind, vehicle motion, or human activities.

What Has the UAE Already Done About the Risk from Outdoor Air Pollution?

The first binding regional enforcement regarding air pollution in the region was the Convention on Long-range Transboundary Air Pollution, signed in 1979 and entered into force in 1983. Despite the Convention's lengthy existence, the party countries are still building monitoring and regulatory capacity. Additional UAE initiatives include:

- A zero-flaring strategic objective by the Abu Dhabi National Oil Company to reduce the burning off of waste gas, oil, and hydrocarbons
- A transition to natural gas in power plants and desalination plants to reduce carbon dioxide emissions
- The phase-out of leaded gasoline by January 2003
- Newly developed guidelines that limit the amount of air pollution generated by quarries and associated crushing plants

For Abu Dhabi emirate, the Environment Agency–Abu Dhabi (EAD) has planned, or has already implemented, several initiatives to reduce outdoor air pollution, including:

- Creating an air quality monitoring and management network, including one central, two mobile, and ten fixed air stations covering the emirate, as well as site evaluations to evaluate and implement adherence to the agency's regulations on air quality

- Continuing air quality management by EAD along with the Norwegian Institute for Air Research from 2008 to 2012, which includes implementing noise and air quality management, developing sector-specific emission limits (e.g., for power and transportation sectors) and establishing an online data reserve of outdoor air quality measures from the monitoring network
- Replacing 20% of vehicle fleets with compressed natural gas vehicles by 2012
- Switching the emirate's diesel fuel supply to ultra-low-sulfur diesel by 2015
- Employing environmental impact assessments for air quality management
- Pushing the power sector to rely more on natural gas
- Exploring more stringent controls for the oil and gas sector
- Continuing efforts to reduce emissions from other sources

Notes on Outdoor Air Pollution Risk Calculation

Number of Deaths per Year. We calculated the number of deaths in UAE adults over age 30 and for infants under 1 during 2007 that were attributable to two outdoor air pollutants, $PM_{2.5}$ and ozone, by using the following equation representing a widely accepted function of outdoor air health impacts (Ostro 2004):

$$Mortality = Baseline\ mortality\ rate \times Population \times \left(1 - e^{(-{}^{2}\times{}'C)}\right)$$

Where: β = the concentration-response coefficient for $PM_{2.5}$ or ozone
δC = the difference between theoretical background concentrations and 2007 monitored outdoor levels of PM_{10} or ozone from anthropogenic sources in Abu Dhabi emirate (Whitford 2008)

Since $PM_{2.5}$ levels were not monitored across the UAE, we used the annual average of 10 monitoring stations in Abu Dhabi. For PM_{10} we assumed a $PM_{2.5}$ to PM_{10} ratio of 0.35 for an arid desert region where there is a greater proportion of PM_{10}, based on data from the World Health Organization (Ostro 2004). To calculate δC, the background $PM_{2.5}$ concentration was assumed to be 7.5 μg/m^3 (Pope et al. 2002). For ozone, we used the annual average ozone concentration averaged over 10 monitoring stations in Abu Dhabi emirate. We considered a background concentration of 20 parts per billion (ppb) to be reasonable. β values indicating the percent increase in adult mortality per 1.0 μg/m^3 increase in annual average $PM_{2.5}$ or 1.0 ppb increase in daily average ozone were determined based on U.S. data from Pope et al. (2002) and Bell et al. (2004), respectively, and it was assumed that they applied to the UAE as well. β values for percent increase in infant mortality per 1.0 μg/m^3 increase in annual average $PM_{2.5}$ were obtained from Woodruff et al. (2006). Prior studies that examined mortality in adults older than age 30 and this study based at-risk population on the assumption that 60% of the population in the UAE is over age 30. Baseline adult mortality rate (2.16 deaths/1,000), population (4.44 million) data from 2007, and birth rates in the UAE were determined from the Central Intelligence Agency World Factbook (2008). The infant population and baseline mortality rate is based on data from the UAE Health Statistics Yearbook for 2006, which is published annually by the UAE Ministry of Health (2007). Low and high estimates were derived from 95% confidence intervals around the β values (Pope et al. 2002; Bell et al. 2004; Woodruff et al. 2006). We summed the attributable deaths calculated from the formula, assuming no correlation in exposure and attributable risks for $PM_{2.5}$ and ozone, based on the low correlation rates found between PM_{10} and ozone (Bell et al. 2004), although it is not entirely clear that the effects of PM_{10} and ozone found in epidemiologic studies are perfectly separable and additive. To calculate the lowest mortality estimate, we assumed 100% correlation between the mortality risk of $PM_{2.5}$ and mortality risk of ozone so that there was no additional risk from ozone separate from that of $PM_{2.5}$ alone. For the highest mortality estimate, we assumed zero correlation between the mortality risk of $PM_{2.5}$ and mortality risk of ozone so that each pollutant acted independently to confer risk for mortality. This was the sum of the highest estimate for $PM_{2.5}$ and highest estimate for ozone. We recognize that the summation method is a simplification given that the bounds for each risk estimate were calculated for 95% confidence intervals around each specific pollutant. We express the low and high estimates to a single significant figure given our assumptions such as correlation between $PM_{2.5}$ and ozone, the ratio of $PM_{2.5}$ to PM_{10}, the estimation of background levels, the sparse coverage of air quality monitors within Abu Dhabi, lack of monitoring data across the entire UAE, use of annual instead of daily ozone levels, and use of population-based data from the U.S. that may not necessarily have the same distribution of variables that influence mortality risk as in the UAE.

Chance in a Million of Death per Year for the Average Resident. This is the average annual risk of death for a randomly chosen resident of the UAE as a result of exposure to a given hazard for 1 year. This figure is calculated by taking the number of deaths over the total population of the UAE from 2007, per one million individuals. We note that although the dose-response coefficients from epidemiologic literature apply to infants and adults over age 30, this is applied to the entire population, therefore potentially underestimating the chance of death for the average resident of any age.

Chance in a Million of Death per Year for the Resident at Highest Risk. For outdoor air pollution, the residents at greatest risk are smokers, approximately 25% of the male population. Smokers are more at risk than nonsmokers due to the synergistic effect between smoking and other air pollutants. The chance in a million for the high-risk group, then, is calculated by assuming all the deaths occur in the high-risk group: number of deaths divided by the high-risk population per million individuals. The Central Intelligence Agency World Factbook (2008) estimates a 2.74 male-to-female ratio among adults, which we applied to the 2007 mortality estimates above.

Greatest Number of Deaths in a Single Episode. Some of the hazards kill only one person at a time, whereas other hazards can kill a number of people at once. The low estimate of the number of deaths from a single event of acute outdoor air pollution poisoning is assumed to be one. The high estimate would be a Bhopal-like disaster (Broughton 2005), in which case approximately 6,000 residents might be affected. We assumed that a catastrophic, large-scale disaster occurring in the UAE would be more similar to a London-fog-scale disaster than a Bhopal-scale disaster, as industrial sites are located away from cities, and thus failure would affect fewer residential areas. An estimated 4,000 people died in the London fog, or "Great Smog of 1952" (Trivedi 2002). Historically, the Bhopal industrial disaster and London fog were the two largest-impact events resulting from outdoor air pollutant exposures. However, there were important differences in the geographical and chemical contexts surrounding these two events, compared with those that might occur in the desert environment of the UAE, in which there are no equivalently large chemical plants similar to those in Bhopal or coal-burning pollution such as was involved in the cold fog in London. Thus, similar events with equivalent magnitudes of mortality in the UAE are highly unlikely to occur.

Time between Exposure and Health Effect. Some hazards, such as exposure to ozone, have fairly immediate impacts, whereas hazards such as lead exposure have health effects that do not manifest until years or decades into the future. Exposure to PM has been associated with both short- and long-term mortality.

Quality of Scientific Understanding. There are two sources of uncertainty in estimating risks for the UAE population. One involves how well scientists know the relationship between exposure to a hazard and its resulting health impacts. The other involves how well we can predict the exposure of UAE residents to a particular hazard. This statistic characterizes the former. For instance, scientists still do not know whether exposure to electromagnetic fields from power systems causes cancer, but scientists understand very well the physical and biological processes leading to injury from auto accidents. Three categories are used to rate scientific understanding: high, moderate, and low. In the case of outdoor air pollution, exposures to some hazards (e.g., particulate matter or lead) are very well understood, while others (e.g., volatile organic chemicals) are much less characterized. As a whole, outdoor air pollutants are moderately characterized since it is often not clear which pollutants in the ambient mix are causing the poor health outcomes. For instance, in combustion products, it is clear that PM_{10} poses a serious mortality risk, but the evidence is less clear for nitrogen oxides. These are often co-pollutants, and the research does not always make distinctions at the specific pollutant level.

Combined Uncertainty in Deaths, Illness, and Injury. This statistic reflects uncertainty in both the scientific understanding of the risk and about the extent of exposure or susceptibility of UAE residents to the particular hazard. The table entry gives the amount of uncertainty in deaths, illness, and injury, expressed qualitatively with respect to other risks in UAE.

Ability of Resident to Control Exposure. Some hazards that UAE residents encounter can be avoided partly or entirely by measures they can take on their own. For instance, they can choose to stay indoors on heavy pollution days or wear filtering masks when they exercise. Three categories are used to rate this controllability: high, moderate, and low.

Indoor Air Pollution

Summary

Indoor air pollutants are found in a number of forms, including environmental tobacco smoke, combustion by-products, volatile organic chemicals, particulate matter, radon, asbestos, heavy metals such as lead and mercury, and mold and other biological pollutants. Exposure generally occurs through inhalation and may result in a wide range of health conditions, ranging from acute and chronic respiratory conditions (e.g., sinusitis, asthma) to cancers of the respiratory tract (e.g., lung cancer).

Risk attribute	Low estimate	Best estimate	High estimate
Fatalities			
Number of deaths per year	60	200	300
Chance in a million of death per year for the average resident	20	50	70
Chance in a million of death per year for the resident at highest risk	30	100	140
Greatest number of deaths in a single event	1	7	10
Illness or injury			
More serious long-term cases per year	Not reported	0	Not reported
Less serious long-term cases per year	Not reported	3,000	Not reported
More serious short-term cases per year	Not reported	300,000	Not reported
Less serious short-term cases per year	Not reported	200	Not reported
Other factors			
Time between exposure and health effects	Immediate (nausea, asthma) to 30 years (lung cancer, mesothelioma)		
Quality of scientific understanding	Moderate		
Combined uncertainty in death, illness, and injury	High		
Ability of resident to control exposure to hazard	High		

The amount of pollutant exposure differs drastically based on several factors:

- Patterns of indoor use of consumer products such as cigarettes, aerosol sprays, pesticides, particleboard, and treated textiles
- Amount of time a person spends indoors and proximity to sources
- Extent of ventilation within the occupied space
- Individual variation in vulnerability to indoor pollutants

The health risks associated with indoor air pollutants can be reduced by limiting the use of products containing harmful compounds; controlling dust generation; regular inspection and maintenance of equipment such as water heaters, dehumidifiers, and heating, ventilation and air-conditioning systems; and most importantly, ensuring adequate ventilation. Smokers, the very young, the very old, pregnant women, and individuals with pre-existing respiratory problems are the most vulnerable to the health effects of indoor air pollutants. Although there is currently little specific information on indoor air exposures in the United Arab Emirates, it is the subject of a major, two-year epidemiologic study in 2008–2010.

What Is Known About the Risk from Indoor Air Pollution?

Environmental tobacco smoke, also called secondhand smoke, is a major source of indoor air pollution and consists of a complex mixture of more than 4,000 chemicals, of which 50 are known or suspected carcinogens. Secondhand smoke contributes to lung cancer, pneumonia, bronchitis, ear infections, asthma, obstruction of peripheral arteries, low birth weight, sudden infant death syndrome, changes to the body's immune system, and aggravation of existing respiratory and cardiovascular disease.

Combustion by-products consist of a variety of chemicals, such as carbon monoxide, nitrogen dioxide, sulfur dioxide, and particulate matter, which arise from combustion sources such as stoves, ovens, water heaters, furnaces, and fireplaces. The health effects of these chemicals include respiratory tract irritation, pneumonia, worsening asthma symptoms, increased heart rate, asphyxiation, and decreased lung function.

Volatile organic compounds such as formaldehyde, benzene, and perchloroethylene (widely used in dry cleaning) are emitted from many household products such as paints, solvents, building materials, aerosol sprays, adhesives, furnishings, and pesticides. Exposure to these pollutants can result in eye and upper respiratory irritation, rash, headache, vomiting, asthma, and damage to the liver, kidneys, and central nervous system.

Radon is an odorless, colorless, tasteless, and naturally occurring radioactive gas that originates from the radioactive decay of radium. Radon becomes harmful when it is trapped in buildings (particularly basements) lacking adequate ventilation. Exposure to radon is the leading cause of lung cancer among nonsmokers and the second leading cause of lung cancer after smoking in the United States.

Asbestos is a naturally occurring mineral that has been used in numerous applications, including thermal system insulation, acoustic insulation, and tiles and shingles in many buildings. Exposure, which occurs when asbestos-containing material degrades or is damaged, is associated with several lung diseases, including asbestosis (primarily from occupational exposures), lung cancer, and mesothelioma.

Airborne lead indoors comes primarily from chipped or flaking paint in homes with leaded paint and from intrusions of leaded-gas emissions from outdoor air. Lead is a potent neurotoxin, exposure to which results in cognitive and developmental deficits, particularly in children.

Airborne mercury exposure occurs primarily through phenylmercuric acetate, present in latex paint. Mercury can cause serious and permanent nerve and kidney damage, rapid heartbeat, irritability, withdrawal, memory loss, peeling of skin on the hands and feet, leg pain, difficulty with fine motor control, sleeplessness, and headaches.

Mold and other biological pollutants such as mildew, dust mites, and animal dander can cause infections, allergic reactions, asthma, and nonspecific respiratory symptoms.

What Is the Exposure to Indoor Air Pollutants in the UAE?

Exposure depends on the type of pollutant, the amount of time that individuals are indoors, and the degree of risk-reduction measures already in place to limit harmful exposures. There is currently little specific information on indoor air exposures in the UAE. Based on data from other industrialized countries, however, individuals residing in more urban, industrialized areas of the UAE are more likely to spend the majority (90%) of their time indoors (U.S. Environmental Protection Agency 1994; Kaynakli and Kilic 2005), use more consumer products that emit pollutants, and may have higher exposures than residents in other industrialized countries.

Environmental Tobacco Smoke. The health and lifestyle survey conducted in the UAE during 2000 (Badrinath et al. 2002) found that at least one person smoked inside the house in over a third of households. Indoor smoking was more frequent in urban areas than rural areas, potentially because of the increased time spent indoors in urban areas.

Asbestos. In most structures, asbestos is unlikely to pose significant health risk. However, this source of indoor air pollution risk may increase in the future as asbestos begins to degrade in buildings constructed before the 2006 UAE federal ban on asbestos production and use (Kelly 2007). In addition, asbestos materials are still being used in migrant workers' housing and illegal home additions.

Airborne Lead and Mercury. Whether and how much lead- or mercury-based paint has been used historically in the UAE is not reported. Studies on indoor lead exposure in the UAE focus on occupational exposure in lead workers. Thus, there is little empirical evidence on which to base standards or develop regulations for nonoccupational indoor or ambient exposure to lead in the UAE.

Mold and Other Biological Contaminants. Mold is a risk whenever humidity levels are regularly above 40–50%. Coastal areas of the UAE experience average levels of ambient humidity between 50 and 60% year-round, peaking at 90% during the summer.

Radon. Publicly reported monitoring data are not available for exposure to radon. Based on the health effects and experiences in other countries, the health effects of exposure to radon in the UAE warrant further study.

What Has the UAE Already Done About the Risk from Indoor Air Pollution?

Smoking in public places was banned effective June 2008 in Sharjah (except in private homes). Other emirates have imposed similar restrictions. This action will to lower exposure to secondhand smoke in public places.

Asbestos importation, production, and use in the UAE were banned in 2006. Prior to this regulation, the UAE developed several federal and local (Dubai) regulations and laws concerning the production, management, and handling of asbestos by occupational asbestos workers and the management and discarding of asbestos during abatement.

Sick building syndrome and the health effects of indoor air pollutants have been addressed indirectly through a new initiative to implement green building guidelines by the Environment Agency–Abu Dhabi (EAD). Sick building syndrome refers to symptoms among a group of people in a building temporarily associated with being in that building. Symptoms include eye irritation, stuffy nose, inability to concentrate, headache, nausea, and feeling tired.

In addition to the new EAD green building initiative, the government of Dubai issued a requirement in October 2007 that all new buildings be constructed with green technologies that comply with globally accepted standards of certification, including the LEED rating system (Leadership in Energy and Environmental Design) used by the U.S. Green Building Council. Old buildings will have to use clean technologies and comply with the same standards.

Notes on Indoor Air Pollution Risk Calculation

Number of Deaths per Year. This is the average number of deaths expected per year among the population of the UAE as the result of lifetime exposure to indoor air pollution. The low and high mortality estimates are extrapolated from U.S. data by scaling the estimated deaths proportionally to population size. High estimates for secondhand-smoke-related deaths are from the National Cancer Institute's (1999) estimates. Baseline mortality rate (2.16 deaths/1,000) and population (4.44 million) data from 2007 were determined from the Central Intelligence Agency World Factbook (2008). The high and low estimates of risk show the range in absolute terms. The "best" estimate is based on figures reported for deaths due to secondhand smoke by Mokdad et al. in 2004.

Chance in a Million of Death per Year for the Average Resident. This is the average annual risk of death for a randomly chosen resident of the UAE as a result of exposure to a given hazard for 1 year. This figure was calculated by taking the number of deaths from indoor air pollution divided by the total population of the UAE from 2000 and dividing by 1 million.

Chance in a Million of Death per Year for the Resident at Highest Risk. For indoor air pollution, the residents at greatest risk are women, who tend to be in the home more and are generally in charge of the cooking, in addition to smokers, who represent approximately 25% of the male population. Smokers are more at risk than nonsmokers due to the synergistic effect between smoking and some pollutants. The chance in a million for the high-risk group is calculated by assuming all the deaths occur in the high risk group: number of deaths divided by high-risk population over 1 million.

Greatest Number of Deaths in a Single Episode. Some of the hazards kill only one person at a time, whereas other hazards can kill a number of people all at once. The low estimate of death from a single event of acute air pollution poisoning (such as from a natural gas leak or carbon monoxide poisoning) would be one for a single person, seven family members living in the same household for the median urban UAE, or 11 for a large family living in one household (Badrinath et al. 2002). We assumed that most gas leaks or other events are not likely to extend beyond one apartment or house, even if the others are attached, because the spaces have separate ventilation systems and are separated by walls and hallways. That is, we assumed that acute residential indoor air pollution exposures are fairly contained and localized to a household.

Illness or Injury. All of the hazards of indoor air pollution also present nonfatal risks, which vary in both duration and severity. The table describes four categories of cases of nonfatal illness or injury per year expected among the average residents of the UAE. The less serious, long-term morbidity stems mainly from the onset or exacerbation of asthma in children due to secondhand-smoke exposure, of which there are an estimated 200,000 cases in the United States (National Cancer Institute 1999). Less serious, short-term illnesses are dominated by allergies to molds, dust, and other biological pollutants. The UAE figure is based on the Asthma and Allergy Foundation of America's estimate[1] that the allergies of approximately 20 million of the 50 million American allergy sufferers are attributable to indoor molds, dust, and biological pollutants. We assumed that UAE residents experience allergies at similar levels and applied that attributable fraction to cases of allergies among UAE residents. Health-care facility visits due to asthmatic episodes or other acute respiratory distress are counted as more serious, short-term morbidity, and in the United States there are an estimated 15,000 hospitalizations annually. Because of a lack of reporting of morbidity numbers for indoor air pollutants other than secondhand smoke, each of these three figures contains a high level of uncertainty. To calculate our figures, we scaled the number of illnesses in the United States to the population of the UAE. Both the United States and the UAE have a similar percentage of their population under age 15 (around 20%, according to the Population Reference Bureau 2007).[2] Because these estimates have high uncertainty due to a lack of reported data, we only reported "best" estimates, which serve as an order of magnitude approximation of the number of illnesses in each category.

Time between Exposure and Health Effect. Some hazards, such as exposure to allergens—mold, dust, etc.—have immediate impacts, whereas hazards such as asbestos have effects that do not manifest for years or decades.

Quality of Scientific Understanding. There are two sources of uncertainty in estimating risks for the UAE population. One involves how well scientists understand the relationship between exposure to a hazard and its resulting health impacts. The other involves how well we can predict the exposure of UAE residents to a particular hazard. This statistic characterizes the former. Three categories are used

[1] http://aafa.org/index.cfm

[2] http://www.prb.org/DataFinder.aspx

to rate scientific understanding: high, moderate, and low. In the case of indoor air pollution, exposures to some hazards (e.g., secondhand smoke, particulate matter, or asbestos) are very well understood, while others (e.g., volatile organic chemicals and combustion products) are much less characterized. It is often unclear which pollutants in the mix of all indoor air pollutants are causing poor health effects since they are frequently correlated, and existing research does not always make distinctions at the specific pollutant level.

Combined Uncertainty in Deaths, Illness, and Injury. This statistic reflects both uncertain scientific understanding about the risk and uncertainty about the extent of exposure or susceptibility of UAE residents to the particular hazard. The table entry cites the amount of uncertainty in deaths, illness, and injury, expressed qualitatively with respect to other risks in UAE.

Ability of Resident to Control Exposure. Some hazards that UAE residents encounter can be avoided partly or entirely by measures they can take on their own. For instance, residents can increase ventilation in their home, install high-efficiency particulate air filters, or use cleaner-burning fuels for cooking and heating. Three categories are used to rate this controllability: high, moderate, and low.

Occupational Exposures in Agriculture

Summary

Agricultural workers can be exposed to a number of hazardous contaminants and relatively hazardous working conditions. The most serious hazard is exposure to pesticides. The effects from exposure can be acute or chronic, or both. Depending on the particular chemical and the level and duration of exposure, pesticides have been associated with respiratory, dermal, gastrointestinal, and reproductive problems; various types of cancer; and effects on the central nervous system.

Farming is a relatively new industry in the United Arab Emirates, and the workforce is relatively uneducated and may not be adequately trained in the use of pesticides. In general, the risk can be reduced by lowering the level of exposure to toxic substances through better practices, decreased use, and/or using less hazardous alternate substances.

Risk characteristic	Low estimate	Best estimate	High estimate
Fatalities			
Number of deaths per year	0	65	100
Chance in a million of death per year for the average worker	0	340	525
Chance in a million of death per year for the worker at highest risk	0	>340	>525
Greatest number of deaths in a single event		1	

(continued)

(continued)

Risk characteristic	Low estimate	Best estimate	High estimate
Illness or injury			
More serious long-term cases per year	0	15	~20
Less serious long-term cases per year	0	20,000	~120,000
More serious short-term cases per year	0	20,000	<120,000
Less serious short-term cases per year	0	20,000	~120,000
Other factors			
Time between exposure and health effects	Immediate to 10–30 years		
Quality of scientific understanding	Moderate		
Combined uncertainty in death, illness, and injury	Moderate		
Ability of worker to control exposure to hazard	High		

What Is Known About Occupational Risk for Agricultural Workers?

Exposure to agricultural contaminants is dependent on the amount and type of contaminants present/utilized, and the level of protection used, and the hygiene practiced.

Agricultural workers may be exposed to a number of toxic substances. Sources of hazardous substances in the agricultural environment include fertilizers, pesticides, engine exhausts, solvents, dusts, microbes, and endotoxins. Evidence suggests that the effects of these exposures can be seen in elevated rates of cancer, including leukemia, non-Hodgkin's lymphoma, multiple myeloma, soft-tissue sarcoma, and cancers of the skin, lip, stomach, brain, and prostate among farmers worldwide (Blair and Zahm 1995). However, due to the numerous exposures and general complexities of the diseases, a clear cause-and-effect relationship can be difficult to establish. Immune system deficiencies and other acute and chronic health problems may also result from these exposures.

The primary toxic substances of concern for agricultural workers are pesticides. Exposure can occur via direct inhalation, ingestion, or skin contact, or through contact with or ingestion/inhalation of contaminated soil, water, and/or food in the farming environment. Skin is considered to be a significant route of absorption (Zhang et al. 1991). The primary health concerns from exposure to these pesticides are effects on actions of the central nervous system that control heart rate, breathing rate, and intestinal functioning. Health effect symptoms associated with these chemicals, even at low levels include headaches, dizziness, weakness, sweating, stomach cramps, and vomiting. Pesticide exposure also is recognized as an important contributing risk factor to cancer development, including cancers of the prostate, pancreas, and liver (Jaga and Dharmani 2005). Agricultural workers throughout the world are a high-risk group for developing cancer from pesticide exposure. Farm worker exposure to pesticides has also been linked to Parkinson's disease (Gorell et al. 1998).

Dust and chemicals other than pesticides (e.g., fertilizers) also cause health effects. High morbidity and mortality rates from respiratory diseases are observed for agricultural workers as well (Linaker and Smedley 2002). More serious (but more rare) illnesses include hypersensitivity pneumonitis and respiratory infections. Skin effects, ocular problems, and reproductive risks may also result from pesticide exposure. On the other hand, some nonpesticide farm exposures may have a protective effect against allergies, asthma, and respiratory sensitization.

Exposure to pesticides can often be reduced by education/information, wearing protective clothing/equipment, and improvements in hygiene practices, but it may be difficult or impossible to completely eliminate exposure. Alternatively, usage of fewer or different chemicals on crops can reduce risk.

What Is the Exposure to Contaminants for Agricultural Workers in the UAE?

The UAE agricultural sector employed 193,000 people, or 6.8% of the workforce, in 2006 (UAE Ministry of Economy 2006). The primary crops include dates, green fodder, vegetables, citrus fruits, and mangos. In addition, the UAE agricultural sector raises livestock in the form of goats, sheep, camels, cows, horses, and poultry. Agricultural production has increased from 15,000 ha in 1971 to approximately 260,000 ha in 2007 and now accounts for more than 7% of land in the UAE (UAE Interact 2007).

Farming is a relatively new industry in the UAE. The workforce is relatively uneducated and has often not been trained in the use of pesticides (Gomes et al. 1999). The specific chemicals used, the frequency of use, and the general practices (as well as the types of crops, soil, and pests) appear to be fairly homogeneous within a geographic region of the UAE and differ widely between different geographic regions (Gomes et al. 1997). Based on publicly reported statistics from 1994 to 1995, 4,095 and 3,558 tons of pesticides were used in the UAE, respectively (Beshwari et al. 1999a, b). Given the increase in agricultural production, updated statistics would be useful. However, there is no publicly available information on the complete list of pesticides used in the UAE, including in which regions, on which crops, and in what amounts they are used.

In terms of conditions and practices, as of 1999 only a minority of UAE farm workers used protective equipment and had been trained properly to minimize their exposure to pesticides (Gomes et al. 1999). For example, the mostly expatriate agricultural workforce is not likely to be able to read the warning labels on pesticide containers. They may be completely unaware of the risks and not inclined to seek protective equipment (or medical treatment for exposure) even if they are, due to their lower status. Storage of pesticides near or in living quarters, lack of protective equipment by a majority of workers, lack of knowledge of the risk of exposure, and lack of training in pesticide use and application has been documented in the UAE (Gomes et al. 1999).

What Has the UAE Already Done About Occupational Risk for Agricultural Workers?

Despite generally high levels of government involvement in agriculture, recent federal and regional regulations related to pesticides, and indications that pesticide awareness and safe practices among farm workers are improving, there remain a number of steps that can help improve agricultural worker safety in the UAE.

Farmers in the UAE face considerable challenges related to climate, and, accordingly, the government provides much assistance, including granting land and supplying pesticides. As such, the government is in an excellent position for both knowledge and control of risk to agricultural workers. Research on biological control methods as alternatives to pesticides, such as introduction of predator species or use of insect pheromones to inhibit insect populations, is encouraged and supported by the government. Interest in organic farming is increasing in the UAE. The government has a number of experimental organic farms and recently certified the first privately owned organic farm (UAE Interact 2007).

The federal government has passed a number of regulations relevant to the use of pesticides in the UAE. For example, at least 93 separate pesticides have been outlawed or banned (Al Asram 2006). In addition, the manufacture and formulation of pesticides is prohibited in the UAE. In addition, in 2004, the National Consultative Council urged monitoring the import of pesticides and setting measures to prevent hazardous chemicals from reaching the local markets. Regionally, Abu Dhabi Municipality registered 597 products in 2004, and pesticides require registration and import permits. There are also regulations for return of excess or expired pesticides and efforts to minimize the amount of pesticide waste generated.

As noted, evidence indicates there is little awareness of pesticide risks among UAE agricultural workers and that use of protective measures and good hygiene are not common (Gomes et al. 1999). However, this literature is dated, and this situation may have improved in recent years. For example, according to the Environment Agency–Abu Dhabi, a recent survey conducted to gauge environmental awareness and behavior among the general public in Abu Dhabi indicated that most farmers were aware of the precautions and problems with pesticide use and that the level of self-protective behaviors correlated with the level of awareness.

Notes on Occupational Hazards from Agriculture Risk Calculations

Number of Deaths per Year. This is the average number of deaths expected per year among the agricultural population in the UAE as the result of lifetime exposure to pesticides. In the absence of specific data for the UAE, we used information for the United States related to lung cancer risk among agricultural workers exposed to chlorpyrifos (Lee et al. 2004). For details on how we determined these numbers among agricultural workers, see the "Illness or Injury" section below. In short, the numbers reported here are the lung cancer cases from this exposure that are expected to result in death (the remaining annual lung cancer cases are tabulated as "more serious long-term cases per year").

Chance in a Million of Death per Year for the Average Agricultural Worker Related to Pesticide Exposure. This is the average annual risk of death for a randomly chosen agricultural worker in the UAE as a result of exposure to a given hazard for one year. This is based on the number of deaths (65 and 102 for best and high estimates, respectively) and the total population of 190,000.

Chance in a Million of Death per Year for the Agricultural Worker at Highest Risk. In the absence of publicly available information regarding heterogeneities in the farm work population, we estimated this to be greater than or equal to the risk for the average worker.

Greatest Number of Deaths in a Single Episode. This is the greatest number of deaths resulting from a single cancer case (one).

Illness or Injury. Exposure to pesticides through agricultural employment presents a nonfatal risk of inhibition of the action of acetylcholinesterase in nerve cells. This risk is a less serious, short-term detriment to neurotransmission (manifesting itself as sweating, pinpoint pupils, leg weakness, and other effects). Other chronic problems such as respiratory symptoms, skin disorders, etc., would be less serious, long-term impacts. For the high estimate for both of these less serious risks, we assumed that all agricultural workers not wearing protective clothing would be subject to these risks. Based on percentages of unprotected workers in Gomes et al. (1999), if all workers not wearing gloves, coveralls, or scarves were at risk of these less serious health problems, the number of workers at risk would have been ~125,000, ~124,000 and ~118,000 (~60–65% of 193,000 total), respectively, in 2007 (i.e., ~120,000). This is the worst-case scenario. The low estimate is that no UAE agricultural workers are at risk despite exposure. The best guess is based on the percentages of agricultural workers actually manifesting symptoms of exposure compared with a comparable control group. Across all symptoms in the study, an average of 13% more farm workers experienced symptoms of exposure relative to the control group in Bener et al. (1999) and 10% more experienced symptoms in Beshwari et al. (1999a, b). This implies ~26,000 and ~19,000 farm workers, respectively; we accordingly listed ~20,000 as the best guess number for less serious long- and short-term cases based on these estimates. We then assumed that the number of more serious short-term cases would be less than the numbers estimated for the less serious cases.

For more serious long-term cases, we used these base numbers of 20,000 and 120,000 assumed to be exposed to pesticide (of a total of 193,000 farm workers in the UAE) and assumed an incidence of lung cancer that is 2.18 times the incidence in lung (and bronchus) cancers among men in the United States for this fraction of the agricultural worker population. This is based on exposure to chlorpyrifos in the United States among "highly exposed" farm workers and their increases in incidence of lung cancer relative to the general population (Lee et al. 2004). Approximately 0.04% of men in the United States were diagnosed with lung (and bronchus) cancer in 2008 (American Cancer Society 2008), which implies that ~0.08% of highly exposed male farm workers could be similarly diagnosed in the UAE. So for the best estimate, the number of cancer cases would be ~0.08% of the 20,000 most "highly exposed" workers plus ~0.04% of the balance of workers, 173,000 (i.e., (20,000×0.00081)+(173,000×0.00037)=~80). For the high estimate, the number of cancer cases would be ~0.08% of the 120,000 most "highly exposed" workers plus ~0.04% of the balance of workers, 73,000 (i.e., (120,000×0.00081)+(73,000×0.00037)=~125). This is the total number of cancer cases per year. We further assumed 82% of these 80 and 125 farm workers diagnosed each year would die based on U.S. percentages of lung cancer deaths relative to lung cancer diagnoses in 2008, giving best and high estimates of number of deaths per year of ~65 and ~102, respectively. The balance of the cases will be the best and high estimates of the "more serious long-term cases," or 15 and 23, respectively.

Time between Exposure and Health Effect. This is highly dependent on the exposure and the particular effect. Pesticide impacts on the nervous system (i.e., inhibition of the action of acetylcholinesterase in nerve cells) would have an immediate impact. Longer-term illness such as cancer would manifest on a multiyear timeframe (i.e., 10–30 years).

Quality of Scientific Understanding. There are several sources of uncertainty in estimating risks for the UAE population. One involves how well scientists know the relationship between exposure to a hazard and its resulting health impacts. The other involves how well we can predict exposure of UAE residents to a particular hazard. This statistic characterizes the former. In this case, the causality between pesticide intake and acetylcholinesterase inhibition is well-established, but the exact correlation between dose and response is less so. Similarly, the relationship between cancer and pesticide exposure is known, but the exact dose-response ratio is unclear due to a number of factors, including the frequent presence of multiple pesticide exposures. Three categories are used to rate scientific understanding: high, moderate, and low.

Combined Uncertainty in Deaths, Illness, and Injury. This statistic reflects both uncertain scientific understanding about the risk and uncertainty about the extent of exposure or susceptibility of UAE residents to the particular hazard. Sources of uncertainty specific to data in the UAE include: (1) lack of knowledge of the types and distribution/amounts of pesticides used across agricultural sites, (2) lack of knowledge of the current level of worker education and protection measures, and (3) unknown quantitative levels of exposure in the UAE. The table entry gives the amount of uncertainty in deaths, illness, and injury, expressed qualitatively with respect to other risks in UAE. The combined uncertainty is a weighted average of uncertainties in risks of death and injury.

Ability of Worker to Control Exposure. Some hazards that UAE workers encounter can be avoided partly or entirely by measures they can take on their own. For instance, they can wear personal protective equipment and practice good hygiene. However, clearly the risk cannot be completely eliminated due to the nature of the job and proximity to high concentrations of the hazardous substance relative to the general population. Three categories are used to rate this controllability: high, moderate, and low.

Occupational Exposures in Industry

Summary

Industrial workers in the United Arab Emirates may face numerous occupational hazards that increase the risk of death, injury or illness. These exposures differ according to industry, production methods, specific task(s) within each industry, and use of personal protective equipment. Health outcomes differ depending on individual characteristics and behaviors such as age, gender, and smoking status, in addition to the variation in individual vulnerability to exposures. Occupational-noise-induced hearing loss due to long-term exposure to high noise levels is the most prevalent irreversible industrial disease (Smith 1998). Workers may also be exposed to volatile organic compounds, crystalline silica, aluminum dust, cement dust, metalworking fluids, and heat stress.

Risk characteristic	Low estimate	Best estimate	High estimate
Fatalities			
Number of deaths per year	5	10	20
Chance in a million of death per year for the average industrial worker	1	2	5
Chance in a million of death per year for the industrial worker at highest risk	75	150	300
Greatest number of deaths in a single event		Not reported	
Illness or injury			
More serious long-term cases per year	Not reported	0	Not reported
Less serious long-term cases per year	Not reported	21,000	Not reported
More serious short-term cases per year	Not reported	28,000	Not reported
Less serious short-term cases per year	Not reported	30,000	Not reported
Other factors			
Time between exposure and health effects	Immediate (dermatitis) to 30 years (silicosis)		
Quality of scientific understanding	High		
Combined uncertainty in death, illness, and injury	High		
Ability of worker to control exposure to hazard	High		

What Is Known About the Occupational Risk for Industrial Workers?

In the UAE, industries that are associated with the greatest occupational hazards are oil and gas production, metal manufacturing, and cement making. Other risks are present due to exposure to asbestos, hot environments, and noise in the workplace.

In oil and gas production and distribution, the greatest health risks tend to result from exposure to volatile organic compounds (e.g., benzene), hydrocarbons, and inorganic chemicals (e.g., hydrogen sulfide) that are produced during petroleum treatment. Much of the exposure to these compounds occurs through inhalation and skin contact. The main health effects that are correlated with oil refinery and oil distribution jobs are skin cancer, mesothelioma, and leukemia, although the evidence for leukemia is not as strong. In addition, the assorted health effects associated with exposure to airborne chemicals can include cancer, respiratory irritation, damage to the nervous system, and hearing loss.

Aluminum smelter and steel workers are exposed to polycyclic aromatic hydrocarbons such as benzene-soluble material and benzo(a)pyrene, both of which are strongly associated with carcinogenic processes at the cellular level and increased risk for bladder and lung cancer. Aluminum workers also are exposed to aluminum dust, which can result in eye and respiratory tract irritation. Chronic exposure affects shortness of breath, weakness, and cough. Stainless steel workers are also exposed to hexavalent chromium and metalworking fluids, which can lead to increased cancer risks as well as respiratory and skin diseases.

Cement industry workers are exposed to cement dust, which has been linked with skin problems, lung functioning, and respiratory tract disorders such as chronic obstructive pulmonary disease, although not all studies have reported these associations. Portland cement, the most common type of cement used worldwide, is caustic and abrasive and is used as a strong adhesive in concrete, mortar, plaster, grout, stucco, and terrazzo. It contains a trace amount of hexavalent chromium, which is toxic to the skin and lungs. Cement workers are also exposed to crystalline silica, which has been known to cause silicosis and chronic obstructive pulmonary disease.

Other industrial exposures include occupational asbestos exposure, which occurs primarily in mining and construction; very hot environments, which have serious health implications, particularly heatstroke; and exposure to noise pollution, which has been most commonly associated with hearing impairment as well as hypertension and high blood pressure.

What Is the Exposure to Industrial Occupational Hazards in the UAE?

Oil and Gas Industry. While it is a major contributor to gross domestic product of the UAE and the other Gulf Cooperation Council countries, the oil and gas industry only employs about 1% of the workforce for those countries. This does not include people who work in transportation and distribution of oil and gas products, who may experience more serious exposures from petroleum-associated risks described above. Because the UAE has a strict no-flaring policy, in which oil and gas refineries are prohibited from burning off excess natural gas that arises from the refining process, there is a reduced presence of combustion-related carbon dioxide at the ambient and occupational-exposure levels.

Metals Manufacturing. Aluminum production is the UAE's main industry other than oil, and the Dubai Aluminum Company plant, owned by the Dubai government, provides 12% of Dubai's gross domestic product and 50% of nonoil-related revenues (UAE Ministry of Public Works 2006). Lung functioning was examined in a small group of Dubai iron foundry workers who were exposed to dusts, fumes, and gases (Gomes et al. 2001). Certain jobs (furnace and fabrication) with the highest concentrations of exposure had higher rates of respiratory symptoms. Smoking did not modify the health outcomes associated with exposure to dusts, fumes, and gases within the iron foundry workers. In addition, researchers found that noise levels in the iron foundry exceeded 90 decibels and the thermal stress index was high (Gomes et al. 2002). Workers at the iron foundry had higher rates of visual defects, hearing disability, and muscle cramps than workers at a bottling company. Although linkages between individual exposure and health were not examined, there were dose-response relationships between working in a location with higher noise exposures and hearing loss, and working in a location with higher heat exposure and muscle cramps.

Cement Industry. By 1998, there were a total number of nine cement factories throughout the UAE. Eight produced Portland cement, and one produced white cement. These factories currently employ 2,999 workers (UAE Ministry of Public Works 2006).

There have been at least two occupational studies of cement workers in the UAE. In one study, workers mainly reported chronic cough, bronchitis, burning and itching eyes, headache, and fatigue (Abou-Taleb et al. 1995). Another UAE occupational study found higher rates of cough, phlegm, wheezing, shortness of breath, sinusitis, bronchitis, asthma, poorer lung function, and obstructive respiratory disease in cement workers than in unexposed retail sales workers (Al Neaimi et al. 2001). Smoking increased the risk of decreased lung function. These workers reported that noise and dust were the primary exposures. Factory workers, factory supervisors, and machine operators who were directly involved in the production process were exposed to higher concentrations of dust than machine maintenance workers, although these subgroups did not differ in the extent to which dust was associated with respiratory health, suggesting the dust was pervasive across the factory. There were no dust controls or noise abatement systems in place, and workers did not wear any protective equipment, although some used a head cloth to cover their nose and mouth to protect them from dust exposure.

What Has the UAE Already Done About the Occupational Risk for Industrial Workers?

Governmental regulatory agencies in the UAE have established several federal maximum exposure guidelines for all industrial operations, including those for lead, silica, and asbestos. The law also provides guidelines for screening, routine monitoring and reporting of the health of all workers, setting exposure limits for certain occupational exposures, and providing training, education, and protective equipment in the workplace to prevent exposures. No public reports were found that assess the impact or status of implementation of this recent law. However, such studies may prove to be an important part of future risk management efforts.

While multinationals and the oil companies in the UAE have established guidelines for using personal protective equipment, small private sector businesses frequently operate under hazardous conditions. However, the Ministry of Environment and Water recently released guidelines that limit the amount of noise pollution and set standards for health and safety practices for the 90 quarries and associated crushing plants located in Fujairah and Ras Al Khaimah (2008).

Workers who are semi-skilled or unskilled tend to accept hazardous working conditions as part of the job, since they are being paid better than in their home countries. The UAE federal government faces challenges in establishing regulations and safety guidelines for numerous small companies, as well as internationally-owned or managed private companies.

Notes on Occupational Exposures in Industry Risk Calculations

Number of Deaths per Year. The total number of deaths was calculated using the attributable fractions of deaths related to asthma and chronic obstructive pulmonary disease (COPD) from workplace airborne pollutants (e.g., silica and asbestos in mining, construction, manufacturing) in the World Health Organization's Eastern Mediterranean B (EMR-B) region in 2000 (Driscoll et al. 2005). Driscoll et al. estimated that 12% of total asthma deaths (18% of male deaths) and 11% of COPD deaths (17% of male deaths) were attributable to workplace exposures for the total population in the EMR-B region (of which the UAE represents 1.56%). Using these attributable fractions for both males and females, we estimated the number of workplace deaths by applying these fractions to the WHO-reported asthma and COPD deaths for the UAE (121 in 2000). This conservatively yielded a low estimate of 15 workplace-attributed asthma and COPD deaths or about 22 deaths if we considered that the majority of those working in dangerous jobs in the UAE are men. We then added work-related air pollution deaths to cancer deaths reported by Driscoll et al. (2004). An estimated 1,000 cancer deaths in the EMR-B region are workplace attributable, yielding another 16 deaths when scaled to the UAE fraction of the total EMR-B population. Finally, if approximately 14% of the workforce was in industry in 2007 (UAE Ministry of Economy 2008a) and we assumed the disease burden was spread evenly over the entire workforce, then there should have been approximately six deaths due to industrial exposures in 2007. This assumes that the exposures were the same for UAE as they were across the EMR-B region, and the same from when the estimates were produced (2000) to 2007. It also assumes that workers have the same distribution of confounders (e.g., smoking, age, and pre-existing health conditions) and same distribution of occupations in the EMR-B as they do in the UAE. The estimate for males and females is low because it does not account for other risks encountered in the workplace such as deaths from accidents or cancer. It is also an underestimate since the majority of deaths and illnesses are more likely to be concentrated in industry and construction, rather than evenly spread throughout the workforce. Occupational-exposure-related mortality often has a long latency period after exposure and the high turnover rate of exposed workers would mean that the exposed population is greater than the current worker population. Thus, we have likely underestimated the mortality risk to the extent that there is high turnover in the industrial sector in the UAE. The best estimate available is provided using attributable fractions from Driscoll for males since men make up the vast majority of industrial workers in the UAE. No studies with data on mortality risks from occupational exposures other than airborne pollutants in the UAE exist. Lung cancer and leukemia risk resulting from occupational carcinogenic exposures is not expected to exceed the mortality estimates due to occupational airborne exposures based on the relative magnitude of lung cancer and leukemia deaths to asthma and COPD deaths (Driscoll et al. 2004). Thus we provide a high estimate that is twice the mortality burden from airborne pollutants in the workplace.

Chance in a Million of Death per Year for the Average Worker. The number of deaths was applied to calculate the chance of death per million workers for a randomly chosen UAE resident in 2007 (4.44 million total, according to the Central Intelligence Agency World Factbook 2008). Industrial workers made up 430,440 workers, or 14% of the total population of workers (3,096,000), in 2007, according to the UAE Ministry of Economy's Annual Social and Economic Report (2008).

Chance in a Million of Death per Year for the Worker at Highest Risk. The chance in a million for death in a worker at highest risk assumes that this death rate is applied for older male workers above age 40. Since there are no studies that report risk for all UAE workers with varying characteristics, we used the rate of 15.5% of workers who were 40 and older and smokers from a sample of 304 randomly chosen cement industry workers in the UAE (Abou-Taleb et al. 1995). There were 433,440 industrial workers in the UAE in 2007 (UAE Ministry of Economy 2008a). Thus, the denominator for individuals at highest risk is $0.155 \times 433{,}440 = 67{,}200$. The chance of death in a million industrial workers is therefore the number of deaths (see section above) divided by 67,200.

Greatest Number of Deaths in a Single Episode. Some hazards kill only one person at a time, whereas other hazards can kill a number of people all at once. A catastrophic event such as a chemical poisoning or accident (chemical explosion) within a factory would lead to the greatest number of deaths in an occupational setting. Thus, we assumed the largest estimate is for a chemical poisoning or accident in the largest oil and gas refinery and the lowest estimate for a small iron foundry. The best estimate would be for an event at an oil and gas refinery because this is the most common type of industry in the UAE.

Illness or Injury. Exposure to industrial pollutants also presents nonfatal risks. These morbidity risks vary in both duration and severity. The table describes four categories of cases of nonfatal illness or injury per year expected among average industrial workers in the UAE. Occupational-related asthma and COPD are classified as more serious long-term cases. Overall, 15% of asthma cases are thought to be due to occupational dust exposure (Driscoll et al. 2004), and the rate increases to 29%

for men. However, publicly reported figures for asthma and COPD prevalence in the UAE population are not available. Thus, the reported estimates here stem from excess illnesses reported by Bener et al. (2001) for industrial workers in the UAE.

There are no comparable risk estimates for other more serious long-term cases such as cancer or less serious short-term cases such as hypertension from noise exposure in the literature that can be combined with the Bener et al. extrapolated data in a valid way. Using the rates found in Bener et al. (2001) we calculated morbidity risk by averaging the excess illnesses reported for each category and applying them to the total industrial population, which is 14% of 3,096,000 total workers in 2007 (UAE Ministry of Economy 2008a). Bener's population of industrial workers included construction workers (estimated at 60% of the study sample industrial workforce), so we have removed them from the population to calculate morbidity among industrial workers not involved in construction work. Bener's study had fairly small sample sizes and only included workers in Abu Dhabi emirate. Thus, these estimates contain a high level of uncertainty if we assume the same distribution of morbidity risk across workers across all emirates and if we apply these morbidity numbers to the population of industrial workers in the UAE in 2007 (UAE Ministry of Economy 2008a). Further, it is important to note that the level of uncertainty varies to the extent that many of these occupational-related illnesses are co-occurring.

Time between Exposure and Health Effect. Some hazards, such as exposure to chromium in cement dust, have immediate impacts such as dermatitis, whereas hazards such as asbestos or silica exposure have effects that do not manifest for decades.

Quality of Scientific Understanding. There are two sources of uncertainty in estimating risks for the UAE working population. One involves how well scientists understand the relationship between exposure to a hazard and its resulting health impacts. The other involves how well we can predict the exposure of UAE residents to a particular hazard. This statistic characterizes the former. Three categories are used to rate scientific understanding: high, moderate, and low. In the case of industrial exposures, our quality of understanding is high since most health effects are well characterized due to the observation of occupational versus population-based exposure levels, dose-response effects, and the mitigation of effects after removal of workers from the industrial workplace.

Combined Uncertainty in Deaths, Illness, and Injury. This statistic reflects both uncertain scientific understanding about the risk and uncertainty about the extent of exposure or susceptibility of UAE workers to the particular hazard. The table entry cites the amount of uncertainty in deaths, illness, and injury, qualitatively with respect to other exposures in industry.

Ability of Worker to Control Exposure. Three categories are used to rate this controllability: high, moderate, and low. Many hazards that UAE industrial workers encounter can be avoided partly by using personal protective equipment and following safety guidelines for reducing risk of exposure, accidents, and resulting health risks. UAE Federal Law 8 specifies regulations for providing personal protective equipment. Thus, we classify a worker's potential ability to control his or her exposure to be moderate, if provided the necessary personal protective equipment and education for reducing exposures. It is important to note, however, that given the high exposure levels found in industrial factories, it is impossible to avoid exposure entirely even with the use of personal protective equipment because of errors in protection use, take-home exposures, and other routes of exposure. Moreover, most workers are uneducated and not made aware of the health risks associated with industrial exposures. Workers may not be in a position to demand safe working conditions and protective equipment from their employers if their employers do not comply with federal regulations.

Occupational Exposures in Construction

Summary

Construction workers face numerous occupational risks from breathing dust and debris, skin contact with dangerous chemicals, and exposure to dangerous levels of heat and noise. The 500,000 construction workers in the United Arab Emirates face serious risks of developing chronic obstructive pulmonary disease (COPD) and asthma from breathing dust on the job, as well as stomach cancer, lung cancer, and mesothelioma from asbestos exposure. Construction workers experience significantly higher pneumoconiosis mortality due to silica and asbestos. There are several less serious dangers from noise pollution, heat stress, skin contact with chromium in cement, and allergies due to biological pollutants such as pollen and dust.

Risk characteristic	Low estimate	Best estimate	High estimate
Fatalities			
Number of deaths per year	10	15	30
Chance in a million of death per year for the average construction worker	2	3	6
Chance in a million of death per year for the construction worker at highest risk	100	150	300
Greatest number of deaths in a single event		Not reported	
Illness or injury			
More serious long-term cases per year	Not reported	0	Not reported
Less serious long-term cases per year	Not reported	33,000	Not reported
More serious short-term cases per year	Not reported	43,000	Not reported
Less serious short-term cases per year	Not reported	47,000	Not reported
Other factors			
Time between exposure and health effects	Immediate to 30 years		
Quality of scientific understanding	High		
Combined uncertainty in death, illness, and injury	High		
Ability of construction worker to control exposure to hazard	High		

Construction workers face the most risk when they are uninformed about the dangers of exposure at their worksite, when their employers do not provide or enforce measures to reduce dust and chemical exposures, or when they are unable to self-pace their work or use other self-protection measures. The recently revised UAE labor law outlines several federal regulations across all types of occupations to monitor occupational exposure-related health, ensure safe work environments, protect against exposures, and provide education in the workplace.

What Is Known About the Occupational Risk for Construction Workers?

Construction workers are exposed to a number of pollutants in the course of their work that can cause asthma, COPD, pneumoconiosis, heat stress, and other diseases. Many of these are due to inhalation of dusts and pollutants, while others are due to exposure by skin contact.

By the nature of their work outdoors, construction workers are exposed to outdoor air pollutants such as particulate matter and gaseous combustion products, which can lead to the risk of health conditions such as asthma, lung cancer, and COPD.

Construction workers can be exposed to biological irritants such as pollens, insects, or fungi; natural dusts such as asbestos, crystalline silica, or coal; and chemical agents such as chlorofluorocarbons, alcohols, metals, salts, and welding fumes. Exposures differ among workers depending on whether they are engaging in new construction, renovation, or demolition. For renovations or demolitions, construction workers may be exposed to more fungi, mold, and asbestos.

Exposure to asbestos can cause numerous health problems, including asbestosis, fibrotic lesions on lining of the lungs, lung cancer, and mesothelioma. However, one study of Finnish construction workers showed that asbestos exposure had a significant influence on mesothelioma but not on lung cancer. In addition, construction workers exposed to asbestos are also at higher risk of stomach cancer.

Construction workers can be exposed to respirable crystalline silica when working with rock, concrete, or masonry. Crystalline silica is classified as a suspected carcinogen by the International Agency for Research on Cancer and also increases the risk for lung infections such as tuberculosis, as well as COPD and rheumatoid arthritis.

Construction workers also experience an increased risk for a chromium allergy from skin contact with water-soluble chromium present in cement. Although the allergy is not a serious illness, it is uncomfortable. In addition to hexavalent chromium in cement, construction workers are also exposed to plasters, epoxy resins, hardeners, and solvents that can cause skin problems and other health issues.

There is a risk of heat stress morbidity for construction workers working outdoors in the heat of the desert, especially during the summer and afternoon. The most serious health problem associated with working in a hot environment is heatstroke, in which an individual becomes mentally confused, delirious, and perhaps experiences convulsions. Some research suggests workers who are allowed to self-pace their work are better able to regulate their core body temperature (Brake and Bates 2002).

Noise pollution is a common exposure in construction, where workers often use loud machinery. Ironworkers, masons, and carpenters are most affected by high noise levels. Ironworkers, carpenters, and electricians experience the most variability in noise. Exposure to noise pollution has been most commonly associated with hearing impairment.

In construction, there is also the risk of catastrophic exposure events that kill multiple workers, mainly chemical poisonings that occur in confined spaces. In a survey of such accidents in the United States, 62% of the observed fatalities could have been prevented by enforcing a standard for adequate ventilation and risk communication for small spaces.

In addition to substances mentioned above, construction workers are also exposed to wood dusts, acids, organic solvents, isocyanates, metals, and fumes such as those from welding, each with very different toxicological properties and diverse health risks such as airway inflammation, asthma, dermatitis, and cancer.

What Is the Exposure to Health Risks for Construction Workers in the UAE?

The number of construction workers in the UAE has more than doubled, from 287,000 in 2000 to 650,160 in 2007, or 21% of the total labor force. The UAE construction workforce is expected to continue increasing as the UAE carries out its development plans. A number of large-scale endeavors, such as the eco-cities of Masdar in Abu Dhabi and Xeritown in Dubai, and numerous off-shore islands, are slated to begin construction by 2010, which will likely require an increased population of construction laborers, as well as associated skilled labor.

No empirical studies of the health effects due to exposures in construction work in the UAE have been published. One study on the respiratory health of UAE cement factory workers indicated higher levels of cough, phlegm, breathing difficulty, sinusitis, and bronchitis among exposed workers (Al Neaimi et al. 2001). Construction workers mixing cement may face the same risks as workers in the cement factory but probably to a lesser scale. Health effects due to cement dust exposure in construction workers are probably even smaller than those from smoking, which was a greater contributor to respiratory symptoms than cement dust exposure in the factory.

What Has the UAE Already Done About the Risks for Construction Workers?

Federal Law 8 was passed in 1995 to govern safety in all occupations in a general manner. No other laws specifically address the protection of UAE construction workers in such areas as regulations for concrete and masonry work, steel erection, tunnels, caissons, cofferdams, the use of explosives, and power transmission and distribution. The law was updated in 2007 to provide guidelines for screening, routine monitoring, and reporting of the health of all workers; setting exposure limits for certain occupational exposures; and providing training, education, and protective equipment in the workplace to prevent exposures. Companies that do not provide their workers with personal protective equipment violate the law, and the law states that equipment should be supplied along with instruction to protect workers from risks.

In addition to Federal Law 8, the UAE Ministry of Labor enacted a resolution in 2006 to reduce the risk of heat stress in construction workers. This resolution states that construction workers are not allowed to work between 12:30 and 3 p.m. during the months of July and August. Firms violating the rule could be fined and denied new work permits for three additional months.

The main problems with construction site safety in the UAE include the lack of orientation for new employees, education about hazardous exposures, and access to medical information. The lack of employer-provided training about normal safety procedures translates to higher exposures, more accidents, and higher risk of health conditions due to chemical and biological dangers in the workplace.

Aside from general occupational exposure guidelines provided in Federal Law 8, no other publicly available federal regulation exists to ensure the protection and safety of construction workers. The Abu Dhabi Executive Council is considering new laws to ensure construction site safety, harsher penalties for companies that do not follow those laws, and increased numbers of inspectors responsible for monitoring building site safety standards. As of 2008, proceedings are under way to design a federal legal framework for the construction industry.

Notes on Occupational Exposures in Construction Risk Calculation

Number of Deaths per Year. The total number of deaths in construction was calculated using the same method as the total number of deaths in industry, described in detail in the previous risk summary sheet. Since approximately 21% of the workforce was in construction in 2007 (UAE Ministry of Economy 2008a) and we assumed the disease burden was spread evenly over the entire workforce, then there should have been approximately eight deaths due to construction-related exposures in 2007.

Chance in a Million of Death per Year for the Average Construction Worker. The number of deaths is used to calculate the chance of death for a randomly chosen UAE construction worker (total 650,160 in 2007) per one million workers.

Chance in a Million of Death per Year for the Construction Worker at Highest Risk. The chance in a million for death in a worker at highest risk assumes that this death rate is applied for older male workers above age 40. Since there are no studies that report risk for all UAE workers with varying characteristics, we used the rate 15.5%, or workers who were 40 and older and smokers, from a sample of 304 randomly chosen cement industry workers in the UAE (Abou-Taleb et al. 1995). There were 650,160 construction workers in 2007 (UAE Ministry of Economy 2008a). Thus, the denominator for individuals at highest risk is $0.155 \times 650{,}160 = 100{,}800$. The chance of death in a million is therefore the number of deaths (see section above) divided by 100,800, per million residents.

Greatest Number of Deaths in a Single Episode. Some hazards kill only one person at a time, whereas other hazards can kill a number of people all at once. A catastrophic event such as a large construction site accident would lead to the greatest number of deaths in a construction setting. Thus, we assume the largest estimate for a construction accident would be an entire construction team and the smallest a single person. However, deaths due to construction-related pollutants are likely to be small since the most likely causes of death among construction workers (e.g., in the United States) are falls, electrocutions, and motor vehicle accidents (Jackson and Loomis 2002).

Illness or Injury. The table describes four categories of cases of nonfatal illness or injury per year expected among the average worker in the UAE. Occupational-related asthma and COPD are classified as more serious long-term cases. Overall, 15% of asthma cases are thought to be due to occupational dust exposure (Driscoll et al. 2004), and the rate increases to 29% for men. However, publicly reported figures for asthma and COPD prevalence in the UAE population are not available. Thus, the reported estimates here stem from excess illnesses reported by Bener et al. (2001) for construction workers in the UAE. Since Bener's study had fairly small sample sizes, these estimates are highly uncertain. To calculate morbidity attributable to construction-related exposures, we took an average of the excess illnesses reported for each category and applied them to the total industrial population (UAE Ministry of Economy 2008a). Then we estimated that construction-related morbidity made up 60% of the illnesses, as construction workers were approximately 60% of the "industrial" workforce studied by Bener et al.

Time between Exposure and Health Effect. Some hazards, such as exposure to chromium in cement dust, have immediate impacts such as dermatitis, whereas hazards such as asbestos or silica exposure have effects that do not manifest for decades.

Quality of Scientific Understanding. There are two sources of uncertainty in estimating risks for the UAE working population. One involves how well scientists understand the relationship between exposure to a hazard and its resulting health impacts. The other involves how well we can predict exposure of UAE residents to a particular hazard. This statistic characterizes the former. Three categories are used to rate scientific understanding: high, moderate, and low. In the case of construction-related exposures, the quality of our understanding is high since most health effects are well characterized due to the observation of occupational versus population-based exposure levels, dose-response effects, and the mitigation of effects after removal of workers from the construction workplace.

Combined Uncertainty in Deaths, Illness, and Injury. This statistic reflects uncertain scientific understanding about the extent of exposure or susceptibility of UAE workers to the particular hazard. The table entry cites the amount of uncertainty in deaths, illness, and injury, expressed qualitatively with respect to other exposures in industry. The combined uncertainty is a weighted average of uncertainties in risks of death and injury.

Ability of Worker to Control Exposure. Three categories are used to rate this controllability: high, moderate, and low. Many hazards that UAE industrial workers encounter can be avoided partly by using personal protective equipment and following safety guidelines for reducing risk of exposure, accidents, and resulting health risks. Federal Law 8 specifies regulations for providing personal protective equipment. Thus, we classify a worker's potential ability to control his or her exposure to be moderate, if provided the necessary personal protective equipment, and education for reducing exposures. It is important to note that it is impossible to avoid exposure entirely even with the use of personal protective equipment because of errors in protection use, take-home exposures, and other routes of exposure that are not protected. Moreover, most workers are uneducated and not made aware of the health risks associated with construction-related exposures. Workers may not be in a position to demand safe working conditions and protective equipment from their employers if their employers do not comply with federal regulations.

Drinking Water Contamination

Summary

Access to clean drinking water is essential for health. However, drinking water quality is a serious problem worldwide. Problems with drinking water are seen more often in developing countries than in industrialized countries like the United Arab Emirates. Drinking water hazards include microbial contamination that can lead to diarrheal diseases, compounds that may cause acute toxicity, compounds that may cause cancer, and radiological contamination. The UAE has established drinking water quality guidelines that are comparable to international guidelines, and a World Health Organization (WHO) estimate of risks associated with water, sanitation, and hygiene indicates that these risks in the UAE are comparable to those in other industrialized nations.

Risk characteristic	Low estimate	Best estimate	High estimate
Fatalities			
Number of deaths per year	0	Not zero but low	147
Chance in a million of death per year for the average citizen	0	Not zero but low	33
Chance in a million of death per year for the citizen at highest risk	0	Not zero but low	33
Greatest number of deaths in a single event		1	
Illness or injury			
More serious long-term cases per year	0	0	Not reported
Less serious long-term cases per year	0	0	Not reported
More serious short-term cases per year	0	Not zero but low	Not reported
Less serious short-term cases per year	Not reported	Not zero but low	Not reported

(continued)

(continued)

Risk characteristic	Low estimate	Best estimate	High estimate
Other factors			
Time between exposure and health effects	Immediate to 10–30 years		
Quality of scientific understanding	High		
Combined uncertainty in death, illness, and injury	High		
Ability of resident to control exposure to hazard	Moderate		

What Is Known About the Risk from Contaminated Drinking Water?

Drinking water can be contaminated in several different ways. These include microbial, chemical, and radiological contamination. The amount of these hazards in drinking water depends significantly on the source of the drinking water and how it is processed (including disinfection), stored, and distributed, such as via piping (tap water), bottles, or directly from surface waters or wells.

Poor water quality and sanitation take a heavy toll on public health, particularly in developing nations and on the health of children. This is due mostly to microbial contamination in drinking water, which is the focus of many water quality guidelines and standards. Lack of safe drinking water contributes to a variety of intestinal infections that can cause malnutrition and anemia in children. Chronic diarrheal disease can also exacerbate malnutrition. Early childhood malnutrition, anemia, and associated diarrheal disease can permanently affect brain development and cognitive ability.

While microbial contamination is the largest public health threat of water for drinking and sanitation, chemical contamination can be a major health concern in some cases. WHO lists guideline values for nearly 200 chemicals, ranging from naturally occurring arsenic and fluoride to synthetic chemicals found only in industrial settings. However, it neither practical nor necessary to test water for all of the chemicals that could cause health problems. Most potential water contaminants occur rarely, and many result from human contamination in limited areas, only affecting a few water sources. Water can be chemically contaminated through natural causes (e.g., arsenic and other elements) or through human activity (e.g., nitrate, heavy metals, pesticides resulting from agriculture or industry). This contamination can have significant effects.

Drinking water also can be contaminated by radioactivity. The contribution of drinking water to overall radioactive exposure is very small and is principally due to the presence of naturally occurring elements in the uranium and thorium decay series. Groundwater typically contains more radioactivity, such as from radon, than surface water does, so a country that receives most of its drinking water from desalination sources is likely at low risk for radiological contamination.

What Is Known About the Risk from Contaminated Drinking Water in the UAE?

The Gulf Coast countries, including the UAE, have the lowest supplies of fresh water per capita in the world. In the past, most drinking water in the UAE came from groundwater and a few surface water sources, both natural and anthropogenic. Recently, however, the groundwater extraction rate has become unsustainable, and desalinated water has become the main source of drinking water, either through piping (tap water) or as bottled water. Desalinated water meets 95% of the domestic water demand in Abu Dhabi emirate; however, it is unclear if this figure is representative of the rest of the UAE. The UAE is currently the world's largest consumer of bottled water per capita.

Bottled water has been reported to present additional risk due to contaminants in the bottles being introduced to the water during storage. A 2007 study in the UAE showed 100 parts per billion (ppb) of bromate (a potentially carcinogenic by-product of a particular desalination process used in the UAE) at one desalination and bottling plant, whereas the WHO recommends consumption of no more than 10 ppb at any time. The Environmental Agency–Abu Dhabi has recognized the risk posed by bromates in all drinking water sources and has taken steps to reduce bromate levels in water production plants to below the WHO-recommended level.

Microbial contamination is another potential risk posed by bottled water, as with other water sources. One study of UAE commercial bottled-water samples showed that 75% of 20-l bottles were contaminated by 10 different species of bacteria. However, this study did not calculate the health risk of this contamination, nor did it determine the source of the contamination.

There is some concern that desalination (i.e., demineralization) results in drinking water that is lacking in essential nutrients. The potential risk of consuming desalinated drinking water appears to be twofold: (1) this water may lack essential dietary ions, and (2) desalinated water could cause potentially harmful compounds to leach from distribution and storage systems. The literature on the health effects associated with long-term consumption of demineralized water is inconclusive because of factors that confound the mechanisms of harm. In the first case, essential ions may be obtained from other dietary sources. In both cases, reintroduction of essential ions to desalinated water could reduce the risk.

It is unclear whether blending or reintroduction of ions is performed routinely in the UAE. Bottled water labels and some information from two desalination plants (Ruwais and Fujairah) provide some indication of such practices from the UAE. Thus, data on the actual level of essential ions in tap and bottled water and the overall nutritional health of the country are needed to better assess the specific risk faced by the UAE.

Groundwater drinking sources may present risks not posed by desalinated or bottled water. Of 228 water samples collected in Abu Dhabi in 2005, 80% had concentrations of nitrate that exceeded the emirate's guideline limit. For most people, consuming small amounts of nitrate is not harmful, but even short-term exposure to high levels of nitrate can cause health problems for infants.

An additional potential risk to drinking water is the possibility of contamination from oil spills at desalination plants.

WHO has estimated the combined risks of water, sanitation and hygiene for the UAE and other countries. Based on data from 2002, WHO estimated that 200 deaths are caused in the UAE each year by this combined risk. This accounts for approximately 2% of all deaths in the UAE. However, no deaths in 2002 in the UAE were attributed to diarrheal disease, parasitic infections, or malnutrition; water-related deaths were attributed mostly to drowning.

What Has the UAE Already Done About the Risk from Contaminated Drinking Water?

Without information about specific exposures to contaminants in drinking water in the UAE, it is difficult to provide an estimate of risk from drinking water to either the general population or to specific, potentially sensitive populations. However, one might approach the question of risk by comparing the UAE's water quality guideline criteria against the established guidelines of other nations. The guidelines from other countries include both mandatory and recommended water quality criteria. The UAE guidelines generally compare well to those of other industrialized countries: they are at or below levels that other countries have declared to be of low risk. There are some potential drinking water contaminants for which most other countries have established guidelines but the UAE does not, including bromate, acrylamide, epichlorohydrin, and vinyl chloride, which pose cancer risks as well as some risk of acute toxicity. However, the UAE has stated that concentrations of any water constituents not included in their quality guidelines should be governed by the internationally accepted WHO guidelines.[3] Therefore, if UAE water quality is monitored and its guidelines are enforced, it is likely that the overall risk from drinking water in the UAE will be low.

Notes on Drinking Water Risk Calculation

The figures in the table are based on 2002 WHO data for reported mortality and morbidity due to the combined exposures of water, sanitation, and hygiene (WSH) (Prüss-Üstün et al. 2008).

Number of Deaths per Year. WHO reported a total of 200 WSH-related deaths in the UAE in 2002, and it reported the UAE population that year as approximately three million. Of those deaths, 100 were due to drowning. The Central Intelligence Agency World Factbook (2008) reports a UAE population of 4.4 million.

[3]Utilizing the water quality standards of other countries or international organizations is a standard practice. A number of countries make available their national drinking-water standards, which can serve as points of reference when developing national drinking-water standards.

Extrapolating 2002 WSH-related deaths, not including drowning, to the 2007 population results in an upper-bound estimate of 147 WSH-related deaths. This estimate is considered high for drinking-water risks because the WSH grouping includes more than just drinking water. In fact, WHO reported no deaths due to diarrheal, parasitic, or other similar diseases that might often be linked to drinking water, so the low estimate for deaths per year is zero. However, the WHO data did not include deaths from cancer due to WSH problems. Because water quality data and information about the disinfection and distribution of water in UAE is unreported, it is possible that some risk exists due to long-term consumption of carcinogens such as arsenic or chlorination by-products in drinking water. The best estimate of "not zero but low" reflects this uncertainty.

Chance in a Million of Death per Year for Residents. This estimate is similar for low- and high-risk populations because these populations are not easily distinguished for drinking water without further characterization of the water sources, contaminants, and distribution. These figures are based on the number of deaths per year described above and calculated for the 2007 population.

Illness or Injury. Most drinking-water-related illnesses are gastrointestinal illnesses of short duration. For 2004, WHO reported that the UAE has among the world's lowest rates of illness due to inadequate WSH (WHO 2009). Therefore, in the absence of reported information about water quality in the UAE, it is estimated that the risk of illness due to drinking water is probably not zero, but low.

Time between Exposure and Health Effect. Most health effects related to drinking water are acute and occur within days, but some contaminants may be carcinogenic and not exert effects for more than 10 years.

Quality of Scientific Understanding. Much is known about water quality, and this knowledge has supported the development of water quality standards in many nations. However, some uncertainty exists about specific cause-and-effect relationships, including, for example, the potential health effects of chlorination by-products and of chemicals present in water bottles.

Ability of Resident to Control Exposure. Drinking water is a basic requirement, and some residents may have little control over its sources or quality. However, some risk avoidance is possible when residents are able to choose water sources or provide additional treatment such as home filtration or boiling.

Coastal Recreational Water Pollution

Summary

As in many other countries with ample coastline and favorable climate, bathing (e.g., swimming) in the sea is a popular form of recreation in the UAE. However, using the UAE's coastal waters for recreation may pose some health risks as a result of exposure to contamination by pathogenic bacteria and viruses as well as chemical pollution from storm water runoff and industrial wastewater. The association between bathing in contaminated recreational water and illness is difficult to establish, but some studies have indicated such a link may exist. Indeed, many nations have established guidelines for recreational water quality as a safeguard against the possibility of illness. The most likely illnesses to result from coastal bathing include short-term gastrointestinal, skin, or ear, nose, and throat infections.

Risk characteristic	Low estimate	Best estimate	High estimate
Fatalities			
Number of deaths per year	0	0	147
Chance in a million of death per year for the average resident	0	0	33
Chance in a million of death per year for the resident at highest risk	0	0	33
Greatest number of deaths in a single event		1	
Illness or injury			
More serious long-term cases per year	0	0	0
Less serious long-term cases per year	0	0	0
More serious short-term cases per year	0	Not zero but low	Not reported
Less serious short-term cases per year	0	Not zero but low	Not reported
Other factors			
Time between exposure and health effects	Immediate		
Quality of scientific understanding	Low to moderate		
Combined uncertainty in death, illness, and injury	High		
Ability of resident to control exposure to hazard	High		

What Is Known About the Risk from Coastal Recreational Water Pollution?

The most commonly reported health effects are diseases caused by bathing in recreational waters that are contaminated with bacterial and viral pathogens. Pathogens often come from discharge of sewage into coastal water, storm water runoff from agricultural lands, and contact with other people who may be sick or carriers of pathogens. The risk of disease depends on which pathogens are present, which in turn depends on the source of pollution.

The most frequently reported adverse health outcome associated with exposure to contaminated recreational water involves intestinal illnesses and diarrhea (such as gastroenteritis). Coastal water pollution has also been associated with acute febrile respiratory illness (AFRI), which presents symptoms similar to influenza. AFRI is a more severe health outcome than the more frequently seen (and self-limiting) gastrointestinal symptoms. Little evidence supports an association between the quality of recreational water and other, nonintestinal-related health effects, such as those associated with the skin, ears, or eyes (Prieto et al. 2001).

Some studies have suggested a link between illness and exposure to recreational waters, although the links discovered have been of varying strength. For example, one study included interviews with 1,858 bathers seven days after they were at a beach in Spain and asked about respiratory, gastrointestinal, eye, and ear symptoms, as well as fever. Incidence rates of gastrointestinal, skin, and respiratory tract symptoms were higher in bathers than those who visited the beach but did not bathe, but the differences were not significant.

Another study conducted in the United Kingdom over four summers compared 548 bathers ("bathing" being defined as total immersion of the head) to 668 beachgoers who did not bathe (Fleisher et al. 2006). This study found that rates of gastroenteritis were significantly higher in the bathing group. Further, concentrations of fecal streptococci found in chest-deep water correlated in a dose-response fashion with illness. It was not suggested that these bacteria caused the illnesses, but they seemed to be a good indicator.

However, the association between bathing in coastal waters and illness is not definitively confirmed. While some studies, including those mentioned above, have shown an association between beach bathing and illness, it has also been found that nonbathing-related factors such as the ingestion of foods related to transmission of gastroenteritis may complicate estimations of swimming risk. The possibility that illness may be associated with bathing even in clean waters has also been proposed. Moreover, another study points out that a possible differential health status of bathing and nonbathing groups could be responsible for their choice of beach activities; this difference may result in an underestimation of bathing-associated risk if individuals chose not to swim due to an existing illness. Therefore, it is difficult to associate levels of illness directly with the amount of time a person spends bathing in coastal waters.

What Is Known About the Risk from Coastal Recreational Water Pollution in the UAE?

Monitoring of coastal water quality in the UAE indicates periodic episodes of contamination. One study found a seasonal contamination in UAE coastal waters (and the few creeks that feed them), where bacterial numbers peaked from April to May, followed by a dramatically sharp decrease in the summer months and a minimum in August (Banat et al. 1998). This was followed by a second peak in October and a subsequent drop during winter. More importantly, total and fecal coliforms fluctuated in numbers at different sites depending on several factors, including the presence of nearby drains and wastewater outlets or recreational areas. This study concluded that these coastal areas had a small degree of microbial pollution.

Coastal waters in the UAE may also be contaminated by wastewater streams from industrial processes that discharge to the sea. These include, for example, discharge from petroleum processing industries and from power generation (and concurrent desalination) plants. A review of one petroleum refinery at Al Ruwais found that its discharge included high biochemical oxygen demand and chemical oxygen demand levels as well as polyaromatic hydrocarbons and phenolic compounds in the major wastewater streams. Polychlorinated biphenyls were also detected in some waste streams. At this refinery (likely as with others), dilution of the wastewater with process cooling water serves as the main treatment approach before the effluent wastewater is disposed into the sea (samples taken from the sea where bathing may occur were not conducted as part of this study). This study suggested that primary as well as secondary treatment units are thought to be essential and strongly recommended to ensure pollutant levels are below UAE standards for marine discharge. However, the study did not include an assessment of the risk posed by this wastewater stream.

An additional potential risk to bathers in Arabian Gulf coastal waters is the potential for contamination from oil spills. Given the large amount of oil transport that takes place in the Gulf, accidental oil spills are unfortunate realities, there more than in any other body of marine water in the world. While limited data have been reported on the distribution of such spills and resulting contamination, it is also recognized that much more data collection is required to support any risk estimation. It is unclear, for example, if oil spills have affected coastal areas during times of sea bathing; large spills would likely deter sea bathing for aesthetic reasons, but smaller amounts of pollutants that are not obvious via sight or smell could pose some risk.

What Has the UAE Already Done About the Risk from Coastal Recreational Water Pollution?

Most measures to control beach pollution focus on preventing sewage from contaminating recreational coastal waters. However, the limited evidence available from cost-benefit studies of point-source pollution control suggests that direct health benefits alone rarely justify the proposed investments and may be ineffective, particularly in cases where pollution results from other sources. While some limited studies have been performed, more complete monitoring data is necessary to describe the extent of coastal pollution in the UAE. Furthermore, reports are lacking on the steps taken to control such pollution in the UAE.

Notes on Coastal Recreational Water Risk Calculations

Number of Deaths per Year. WHO reported a total of 200 deaths related to water, sanitation, and hygiene (WSH) in the UAE in 2002 (WHO 2006). Of those, 100 were due to drowning. Extrapolating 2002 WSH-related deaths, not including drowning, to the 2007 population would result in a high estimate of 147 WSH-related deaths. This estimate is considered high for coastal bathing illness-related risks, because the WSH grouping includes more than coastal bathing. In fact, WHO reported no deaths in the UAE in 2002 due to diarrheal, parasitic, or other similar diseases that might often be linked to coastal bathing, and the expected potential illnesses are generally self-limiting infections. Therefore, the low and best estimate for deaths per year is zero. However, the WHO data also reported approximately 100 drownings in the UAE in 2002, and this number is the basis for the high estimate (extrapolated to the 2007 population of 4.4 million as reported by the Central Intelligence Agency World Factbook). Because data are unreported on coastal water quality and about which residents bathe and how often, the additional risk of illness cannot be accurately estimated.

Chance in a Million of Death per Year for Residents. These figures are based on the number of deaths per year due to drowning and extrapolated to the 2007 population.

Illness or Injury. Most reported coastal-bathing-related illnesses are gastrointestinal illnesses of short duration. For 2002, WHO reported disability-adjusted life years (DALYs) related to WSH causes for the UAE (WHO 2006). These causes included diarrheal diseases and other infectious diseases. It is likely that few of these illnesses were caused by coastal bathing, but no studies documenting coastal water exposure or associated illnesses in the UAE have been reported. DALYs due to parasitic diseases were not reported from WSH-related causes except for trachoma, which is unlikely to be transmitted by coastal bathing. Therefore, in the absence of reported information about water quality in the UAE, it is estimated that the risk of illness due to coastal bathing is not zero but is probably low.

Time between Exposure and Health Effect. Most coastal bathing-related illnesses are gastrointestinal illnesses of rapid onset and short duration.

Quality of Scientific Understanding. Some epidemiologic studies have correlated illnesses with coastal water quality. While the pathogenicity of microbial contaminants sometimes found in coastal waters is well documented, the direct cause-effect relationship between bathing and illness is less well established.

Ability of Resident to Control Exposure. Residents may choose not to bathe in coastal waters.

Exposure to Residential Soil

Summary

Threats to health from residential soil may exist in the UAE due to industrial sources and poorly regulated waste disposal. Pollutant exposures include heavy metals and trace elements, inorganic compounds, aromatics, hydrocarbons, and pesticides. Pollutants in residential soil can be brought into homes via dirty shoes, agricultural produce, or the wind. Exposures are fairly simple to remedy by removing shoes at the door, washing or peeling fruits and vegetables, and keeping doors and windows shut if it is particularly dusty outside. The exact sources and magnitude of possible soil contamination has not been reported.

Risk characteristic	Low estimate	Best estimate	High estimate
Fatalities			
Number of deaths per year		Not reported	
Chance in a million of death per year for the average resident		Not reported	
Chance in a million of death per year for the resident at highest risk		Not reported	
Greatest number of deaths in a single event		Not reported	
Illness or injury			
More serious long-term cases per year		Not reported	
Less serious long-term cases per year		Not reported	
More serious short-term cases per year		Not reported	
Less serious short-term cases per year		Not reported	
Other factors			
Time between exposure and health effects	10–30 years		
Quality of scientific understanding	Moderate		
Combined uncertainty in death, illness, and injury	High		
Ability of resident to control exposure to hazard	High		

Exposures in children are often much higher than in adults due to children's tendency to spend more time on the ground in playgrounds or parks. Additionally, even if they receive less absolute pollution than adults, it is often greater on a scale relative to weight.

What Is Known About the Risk from Residential Soil?

There are many different possible contaminants in residential soil: heavy metals and trace elements, inorganic compounds, aromatics, hydrocarbons, pesticides, and others. Of those, some of the most dangerous to human health and development (and most common) are heavy metals (which particularly affect young people), hydrocarbons, and pesticides.

Heavy metals can include lead, cadmium, mercury, and arsenic. Lead can come from cars burning leaded gasoline or from industrial sources and presents risk through inhalation or ingestion via the food chain. Cadmium can often come from incorrect disposal of nickel-cadmium batteries, industrial sources, or application of fertilizers and sewage sludge to farmland; however, the main pathway for exposure among non-smokers is through ingestion of food that takes up the cadmium in the soil. Mercury and arsenic are also usually from industrial sources—particularly nonferrous smelting and energy production from fossil fuels—or could just be prevalent in soil naturally.

Hydrocarbons (particularly polycyclic aromatic hydrocarbons) are known to be animal carcinogens and mutagens, although specific characterization of their effects on humans from soil exposure has not been quantified. Hydrocarbons in the soil and water most often come from industrial sources and other combustion processes, such as heating or cooking.

Pesticides, depending on their concentration, can cause a number of health outcomes such as diarrhea, nausea, vomiting, rash, ocular irritation, anxiety, dizziness, headache, muscular pain, memory loss, fatigue, shortness of breath, insomnia, and contact dermatitis. Pesticides can enter the home by blowing in on contaminated dirt or by being transferred from produce.

Research on residential soil exposure suggests that children may be more affected by chemical exposure because their systems are still developing (with more porous bones and membranes). Additionally, they spend more time in the dirt than their adult counterparts. Further, by bodyweight, their exposure is relatively larger. This is certainly true for lead and hydrocarbons. However, in a study of arsenic exposure in children living near a pesticide factory, no significant correlation was shown between soil arsenic levels and levels of arsenic measures in the children's urine.

Mitigation measures for exposure to contaminated soils include wiping shoes on a mat outside the house and leaving shoes near the door, vacuuming and mopping, minimizing carpeting, keeping windows and doors closed on windy days, and washing agricultural products.

What Is the Exposure to Residential Soil in the UAE?

Multiple threats to health from residential soil may exist in the UAE, but the exact nature of the contamination and degree of human exposure is not known because reported data do not provide the information needed to assess the associated health risks.

According to a July 2008 news release by the Environment Agency–Abu Dhabi (EAD), there are six extremely large, unregulated landfills outside of Abu Dhabi that the government plans to begin cleaning up (Kwong 2008). The landfills may contain medical, chemical, household, industrial, construction, and agricultural waste, and even, at one site, discarded military weapons. The largest site is Al Dhafra, which is 16 km^2 and receives 20,000 tons of waste daily. Another 8 km^2 landfill in Al Gharbia receives 1,800 tons of municipal and 5,000 tons of construction waste daily. Additionally, there are numerous small-scale landfills near smaller settlements that also lack appropriate waste treatment facilities. Although the exact nature of the chemical exposures due to these fills is unclear, they are known to contain oil sludge that can lead to hydrocarbon and heavy metal contamination of the soil.

Additionally, there has been widespread and fairly unregulated use of pesticides, particularly organophosphates and carbamates, which have been shown to be a significant health risk to farmers working with them and significant sources of occupational agricultural cancer. However, it is not clear to what extent these pollutants are transferred to nonagricultural areas.

There have not been any large-scale oil spills in the UAE, but there have been a number of minor ones, mainly in ports. Although the UAE deals harshly with oil spills and promptly cleans them up, the efficiency of cleaning operations is not clear. Beaches are especially dangerous in this regard because people generally lie in the sand and are more likely to ingest contaminants or absorb them through the skin.

What Has the UAE Already Done About the Risk of Residential Soil Exposure?

There has been some research into bioremediation of crude oil-contaminated soils in the UAE, and EAD plans to clean up landfills, but beyond that, there has been no significant movement to address the problem of residential soil exposure.

Eating Contaminated Seafood

Summary

Eating seafood has documented health benefits, but it also exposes people to risk if the fish contains hazardous contaminants. These contaminants include pathogens, which can be eliminated with proper handling and preparation, and toxic metals and organic compounds, which cannot be removed in preparation. Depending on the chemical and the level of exposure, toxic substances can have effects on cognition, the immune system, and neurological functions and may cause cancer. The effects can be long-term and are generally the result of consuming contaminated fish over a long time period. Furthermore, pregnant women may be at risk because of possible health effects of mercury on fetuses. People can reduce these risks by limiting the amount of fish they eat.

Risk characteristic	Low estimate	Best estimate	High estimate
Fatalities			
Number of deaths per year	0	4	10
Chance in a million of death per year for the average resident	0	2	3
Chance in a million of death per year for the resident at highest risk	0	>2	>4
Greatest number of deaths in a single event		1	
Illness or injury			
More serious long-term cases per year	0	7	15
Less serious long-term cases per year	0	27,000	67,000
More serious short-term cases per year	0	0	0
Less serious short-term cases per year	0	0	0
Other factors			
Time between exposure and health effects	Immediate to 10–30 years		
Quality of scientific understanding	Moderate		
Combined uncertainty in death, illness, and injury	Moderate		
Ability of resident to control exposure to hazard	High		

Note: Fatalities and serious long-term cases are due to the carcinogenicity of dioxins in seafood. Less serious long-term cases are due to the effects of mercury consumption on fetuses/unborn children

What Is Known About the Risk of Eating Contaminated Seafood?

The two types of contaminants that may be found in seafood are human pathogens and toxic substances. Exposure to contaminants in seafood is dependent on two factors: the amount of fish consumed; and the level of contamination in the fish consumed.

Typical human pathogens found in fish are *Clostridium botulinum* type E, which causes slurred speech and muscle weakness as symptoms of muscle paralysis, and *Vibrio parahaemolyticus*, which causes acute gastrointestinal effects such as diarrhea, nausea, vomiting, abdominal cramps, and sometimes fever. These health risks can be minimized or eliminated through proper handling (i.e., using proper refrigeration and good hygiene practices) and making sure the fish is thoroughly cooked before eating.

Toxic substances such as mercury (as methylmercury) and other metals (e.g., cadmium, nickel, and lead), PCBs, and dioxins present potential concern for consumption of seafood because they tend to bioaccumulate in fish and other aquatic animals, can have long-term health effects, and cannot be removed by preparation methods such as cleaning and cooking. The levels of these contaminants in seafood depend on their concentrations in the aquatic environment and on the lifespan, diet, and level of fatty tissue of the fish themselves; mercury contamination is highest in fish that are highest in the food chain, and dioxins are fat soluble and therefore increase with fatty tissue content.

Many metals are naturally occurring, but chronic overexposure can lead to adverse health effects. For example, toxic levels of cadmium and nickel can cause dizziness, headache, vomiting, vertigo, and intestinal irritation. Excessive amounts of lead can cause anemia, renal tubular nephrosis, diminished intellectual capacity and developmental delays in children, headache, drowsiness, and gastrointestinal upset. Exposure to mercury can cause pulmonary, brain, kidney, liver, and gastrointestinal damage. Methylmercury at high concentrations can induce sensory abnormalities, paresthesias, and ataxia in adults, and can delay cognitive and neuromuscular development in children. PCBs and dioxins have been found to impair the immune system and neurological functions. Depending on the substance, bioaccumulation can occur in fish tissue (especially in larger or older fish), as well as in the tissue of humans who consume them. Even so, the health benefits from fish consumption often outweigh the risk of adverse health effects.

What Is the Exposure to Contaminants in Seafood in the UAE?

Because seafood holds an important place in the diet of many residents in the UAE, consumption levels are relatively high compared with the world average. According to one source, the apparent per capita fish consumption in the UAE was 24 kg per year in 2005, including both citizens and noncitizens. Other sources indicate higher consumption levels for UAE citizens, namely that the average UAE citizen consumes 33 kg of fish per year; and up to 90% of UAE citizens consume fish at least once a week. UAE citizens typically eat rabbit fish, grouper, mullet, sea bream, and shrimp.

Studies of contamination in seafood off the UAE coast indicate that the UAE has relatively low metal contamination in its domestically produced fish. Regional studies also indicate a comparatively low level of an even wider range of contaminants. However, studies are limited and dated, and changes in domestic industries and patterns in urbanization affect pollution that reaches fish. No data are publicly available for contaminant levels in shrimp, commonly consumed in the UAE. However, shrimp is generally relatively low in mercury and is not an "oily" fish so it also will have relatively low levels of dioxins. Approximately 50% of seafood in the UAE is imported, mainly from India, Thailand, Oman, Pakistan, Tanzania, China (both mainland and Taiwan), Uganda, Malaysia, Iran, and Yemen. There is no publicly available information on the levels of contaminants in imported fish, nor on any systematic governmental monitoring of either domestic or imported seafood contamination.

There is no publicly available information on specific industrial sources of mercury and/or dioxins in the UAE, or monitoring data from industrial waste and emissions in the UAE. The Environment Agency–Abu Dhabi has recently reported that mercury levels were nondetectable in Abu Dhabi coastal waters and that PCB levels were lower than 0.4 ppm at all sites except one (at 1 ppm). However, it is difficult to predict the levels of contaminant in fish tissue based on concentrations in the water, making direct studies of fish tissues necessary to assess exposure. Therefore, in the absence of additional current and specific seafood contamination data, it is advisable for pregnant women to limit intake of older/predatory fish, and for pregnant women and the general population to limit consumption of oily fish to two and four servings per week, respectively.

What Has the UAE Already Done About the Risk of Contamination in Seafood?

A number of actions and activities have been initiated in the UAE and the region to protect water from pollution, especially oil spillage. In particular, the UAE has:

- Established the Federal Environmental Agency in 1993 to set federal plans and policies that prevent pollution, specifically addressing the marine environment
- Established relevant laws to regulate dumping in and around the marine environment
- Participated in regional organizations and ratified a number of conventions with the goal of protecting the marine environment (e.g., the Regional Organization for the Protection of the Marine Environment; the London Convention)
- Formed public and private organizations to protect the environment from pollution
- Conducted experimental studies to find out the effects of oil on certain fish
- Given instructions to fishermen on careful handling of fuel and avoiding spills

Currently, most of the UAE regulations regarding oil spills are responsive, not preventive, and no specific regulations are targeted at seafood safety protection.

An overview of environmental regulations and food control infrastructure indicates a need to:

- Strengthen relevant regulations on water pollution and seafood safety control
- Establish a comprehensive framework for integrated planning and management of the coastal zone at the federal level
- Establish a system to classify and assess the potential risks associated with seafood consumption and encourage the development of national monitoring or surveillance schemes for contaminants that cover seafood
- Enhance the emergency response to accidents such as oil spills or power plant and/or industrial discharges

Notes on Eating Contaminated Seafood Risk Calculations

Number of Deaths per Year. This is the average number of deaths expected per year among the population of the UAE as the result of lifetime exposure to contamination in seafood. Assuming proper handling and preparation of seafood, this number is zero from pathogens. In fact, consumption of seafood is known to decrease a number of health risks, including coronary heart disease, which may actually enhance quality and length of life. The risk of death reported herein is related to the carcinogenicity of dioxin exposure. The details of this calculation are described below in the "Illness or Injury" section. In brief, the numbers reported here are the cancer cases resulting from this exposure that are expected to end in death in a given year (the remaining annual cancer cases are tabulated as "more serious long-term cases per year").

Chance in a Million of Death per Year for the Average Resident. This is the average annual risk of death for a randomly chosen resident of the UAE as a result of exposure to a given hazard for one year. Again, this number is related to dioxin exposure, and average consumption levels are assumed for this calculation.

Chance in a Million of Death per Year for the Resident at Highest Risk. Higher-risk residents would be those with higher levels of consumption of oily fish or higher-risk groups such as children. Lacking specific knowledge of the distribution of fish consumption across the population and lacking a scientific basis for cancer risk assessment across different demographic groups, we have simply estimated these numbers to be equal to or greater than for the average resident.

Greatest Number of Deaths in a Single Episode. This is the greatest number of deaths resulting from a single cancer case, or one.

Illness or Injury. Exposure to dioxins causes a risk of cancer, which was used as the estimate for more serious long-term cases of illness. Because the use of PCBs has largely been banned and contaminant levels have been decreasing over the past few decades, we focused on the risks associated with dioxin exposure.

The best estimate assumes moderate levels of oily fish consumption, and the high estimate assumes higher levels of oily fish consumption. However, because these values are uncertain due to the absence of specific information on dioxin/PCB contamination of seafood in the UAE, data are based largely on U.S. and U.K. information, as detailed in the following paragraphs.

The implied average contamination level of fatty fish (<6 ng dioxins/kg fish), as well as the proportion of total fish consumption which is fatty fish (<25%) assumed herein was based on U.K. information for the best estimate values for the best estimate values (Scientific Advisory Committee on Nutrition 2004). An assumption of 50% of total seafood consumption being fatty fish yielded the high estimate values. A body weight of 65 kg (143.3 lbs) was assumed in both cases. Exposure was then determined as follows:

$$Exposure = \left(Oily\ fish\ (g)\,/\,day\right) \times \left(Contaminant\ in\ fish\left(\mu g\,/\,g\right)\right) \div \left(Body\ weight\left(kg\right)\right)$$

This exposure was then multiplied by the carcinogenicity, which for dioxins was established by the U.S. Environmental Protection Agency (EPA) as 1.56×10^5 per mg/kg-day (1999), and divided by a typical lifespan of 70 years to yield the percentage of the population in a given year expected to be diagnosed with cancer. The values used are summarized in the table below.

Dioxin in seafood exposure for the UAE: "More Serious Long-Term Cases per Year" and "Fatalities" calculations

Estimate	Assumptions	Exposure, μg/kg-day	Percentage of population with cancer annually
Best	Oily fish consumption: 15 g/day (i.e., 23% of total consumption) Contamination: 5.8 ng/kg	1.5×10^{-6}	3.3×10^{-4}
High	Oily fish consumption: 21 g/day (i.e., 50% of total consumption) Contamination: 5.8 ng/kg	3.2×10^{-6}	7.1×10^{-4}

These percentages of the total population were used to determine the number of annual cancer cases: 11 and 25 people per year for the lower and higher levels of oily fish consumption, respectively. Based on U.S. statistics (American Cancer Society 2008), it was assumed that ~40% of the annual number of cases would result in death in a given year; this allowed calculation of the expected number of deaths per year (best estimate = 4, high estimate = 10), as well as the chance in a million of death for the average resident. The remaining cancer cases that did not result in death in a given year were tabulated as the "more serious, long-term illnesses" (i.e., 60% of the 11 and 25 annual cancer cases, or 7 and 15 cases for the best and high estimates, respectively).

Exposure to mercury through consumption of contaminated fish presents a nonfatal risk to the offspring of mothers who consume this contaminated fish during pregnancy. This risk is a more serious long-term detriment to cognitive development. The high estimate is that all newborns of both citizens and noncitizens in the UAE (i.e., all residents of the UAE) are at risk for cognitive impacts related to methylmercury exposure.

The best estimate is that only Emirati newborns are at risk due to the relatively higher levels of fish consumption by citizens relative to noncitizens. The impact on cognitive ability is immediate and lifelong. Details of the calculation follow.

The reported mercury contaminant levels were used to determine whether the reference dose is being exceeded, that is, whether or not consumption of fish is risky for the average UAE citizen. The exposure calculation performed was as follows:

$$Exposure = \left(Fish(g)/day\right) \times \left(Contaminant\ in\ fish\left(\mu g/g\right)\right) \div \left(Body\ weight\left(kg\right)\right)$$

The estimates reported in the table below are based on the assumption that all of the average 24 or 33 kg of annual fish consumed has the minimum and maximum level of contamination, respectively, indicated in the above table. They also assume a bodyweight of 65 kg. The ranges presented below represent (1) the lowest observed level of contamination and the lower estimated consumption level and (2) the highest observed level of contamination and the higher estimated consumption level. The exposure level was then compared with the reference dose, and the results of the comparison are shown in the third column.

The reference dose established for pregnant women is 0.1 μg/day (U.S. National Research Council 2000; U.S. EPA 2004). Note that while average levels of contamination are low compared to U.S. numbers, the fact that fish consumption levels are ~3.5–5 times higher than they are in the United States puts all citizen and resident births (under the "best estimate" and "high estimate" cases, respectively) at risk for at least some cognitive impact.

Determination of "less serious long-term cases per year" was based on the scenarios in the table below. The "low estimate," "best estimate," and "high estimate" were determined for both average UAE residents and citizens separately, with assumptions detailed below.

Mercury in seafood exposure for the UAE: "Less-Serious Long-Term Cases per Year" calculation

Exposure scenario	Assumptions	Exposure, μg/kg/day	Exceeds reference dose?
Low, resident	Fish consumption: 66 g/day Contamination: 0.036 μg/g	0.036	No
Low, citizen	Fish consumption: 90 g/day Contamination: 0.036 μg/g	0.050	No
Median, resident	Fish consumption: 66 g/day Contamination: 0.072 μg/g	0.072	No
Median, citizen	Fish consumption: 90 g/day Contamination: 0.072 μg/g	0.099	Approximately equal
High, resident	Fish consumption: 66 g/day Contamination: 0.11 μg/g	0.11	Yes
High, citizen	Fish consumption: 90 g/day Contamination: 0.11 μg/g	0.15	Yes

Assuming that 1 μg/kg/day of exposure to mercury results in 10 μg of mercury per gram of maternal hair, and that the IQ loss associated with mercury exposure is –0.2 points per μg mercury per gram maternal hair, the risk for each of these exposure levels can be quantified (Cohen et al. 2005). For example, "low exposure" in children born of Emirati women would be expected to result in a loss of approximately 0.1 IQ points, whereas "high exposure" would result in a loss of approximately 0.3 points. Note that Cohen states that the range of IQ point impacts is 0–1.5 IQ points per μg mercury per gram of maternal hair and that this range is large relative to the uncertainty in other numbers in our analysis. So, the magnitude of the cognitive impacts of these exposure levels is quite uncertain.

Based on the UAE census of 2005, the average annual number of births for UAE residents (i.e., both citizens and noncitizens) is approximately 67,100 births per year, and the average annual number of births for citizens is approximately 26,800 births per year (UAE Ministry of Economy 2008a).

Time between Exposure and Health Effect. The fatal and more serious long-term effects from dioxin, namely cancer, may take many years to appear. As noted, cognitive impacts from mercury are immediate and persistent.

Quality of Scientific Understanding. There are several sources of uncertainty in estimating risks for the UAE population. One involves how well scientists know the relationship between exposure to a hazard and its resulting health impacts. The other involves how well we can predict exposure of UAE residents to a particular hazard. This statistic characterizes the former. For instance, scientists still do not know the exact impacts on cognition from prenatal exposure to methylmercury (e.g., the range of possible loss of IQ related to exposure is very large), but scientists understand very well the physical and biological processes leading to injury from auto accidents. In the case of dioxin exposure, the current available evidence suggests that dioxins may cause cancer in humans. Three categories are used to rate scientific understanding: high, moderate, and low.

Combined Uncertainty in Deaths, Illness, and Injury. This statistic reflects both uncertain scientific understanding about the risk and uncertainty about the extent of exposure or susceptibility of UAE residents to the particular hazard. Sources of uncertainty for mercury exposure in the UAE in particular include: (1) the distribution of consumption levels across the population, (2) the distribution of fish types, (3) the contamination levels in all types of fish consumed, and (4) the contamination of imported fish. For dioxin exposure in the UAE, the uncertainty is greater due to a lack of publicly available information on the dioxin contamination levels, in addition to these other factors. The table entry gives the amount of uncertainty in deaths, illness, and injury, expressed qualitatively with respect to other risks in UAE.

Ability of Resident to Control Exposure. Some hazards that UAE residents encounter can be avoided partly or entirely by measures they can take on their own. For instance, decreasing the total amount of fish consumption, or more advisably, limiting certain types of fish in the diet (especially for high-risk groups such as women of childbearing years) are well within an individual's control. Three categories are used to rate this controllability: high, moderate, and low.

Eating Contaminated Fruits and Vegetables

Summary

Eating fruits and vegetables has documented health benefits, but it also exposes people to risk if the fruits and vegetables contain hazardous contaminants. The two primary potential contaminants are human pathogens, which can be eliminated or greatly reduced with proper handling and preparation, and pesticides. Assuming fruits and vegetables are not exposed to pathogens in production, there is no risk from this contaminant. The washing and/or peeling of uncooked produce reduces this risk, as does cooking. Assuming the UAE is following standard practices for pesticide use similar to those, for example, in the United States and the United Kingdom, there is minimal acute risk and virtually no chronic risk from pesticide exposure via consumption of fruits and vegetables in the UAE.

Risk characteristic	Low estimate	Best estimate	High estimate
Fatalities			
Number of deaths per year	0	0	Low but nonzero
Chance in a million of death per year for the average resident	0	0	Low but nonzero
Chance in a million of death per year for the resident at highest risk	0	0	Low but nonzero
Greatest number of deaths in a single event		1	
Illness or injury			
More serious long-term cases per year	0	0	0
Less serious long-term cases per year	0	0	Low but nonzero
More serious short-term cases per year	0	0	0
Less serious short-term cases per year	0	0	<35,000–89,000
Other factors			
Time between exposure and health effects	10–30 years or immediate		
Quality of scientific understanding	Moderate		
Combined uncertainty in death, illness, and injury	Low		
Ability of resident to control exposure to hazard	High		

What Is Known About the Risk of Contaminants in Fruits and Vegetables?

Cooking fruits and vegetables eliminates virtually any risk of exposure to pathogens, but when fruits and vegetables are eaten raw, the risk of exposure to pathogens depends on handling in production and in preparation. The most common pathogens of concern are *Salmonella* and *E. coli*; these are generally only a problem if they are introduced during production (e.g.), in use of nondisinfected reclaimed water. Pathogens of particular concern for pregnant women due to possible harm to the unborn child, such as *Toxoplasma gondii* (which can cause miscarriage or eye and brain damage to child), *Listeria monocytogenes* (which can cause stillbirth, miscarriage, or physical retardation in child), and aflatoxins (which can cause growth faltering in child after birth), can also be virtually eliminated with thorough washing, peeling, and/or cooking.

The main toxic substances of concern in fresh produce are pesticides that typically come from the organophosphate and carbamate chemical groups. The primary health concerns from exposure to these pesticides are effects on actions of the central nervous system that control heart rate, breathing rate, and intestinal functioning. Organophosphates produce chronic effects (i.e., it takes 4–6 months to return to normal functioning after exposure), and carbamates cause acute effects (i.e., it takes 48–72 h to return to normal). Health effect symptoms, even at low levels, associated with these chemicals are headaches, dizziness, weakness, sweating, stomach cramps and vomiting. Studies on the carcinogenicity related to these pesticides are limited, and in most cases these pesticides are not considered carcinogens. Exposure to pesticides can often be reduced by washing and peeling food prior to consumption, but contaminants may remain even after such preparation.

What Is the Exposure to Contaminants in Fruits and Vegetables in the UAE?

Exposure to contaminants depends on both the amount of fruits and vegetables consumed and the level of contamination in the fruits and vegetables consumed.

The average UAE citizen consumes 124 kg of fruits and 113 kg vegetables per year (~340 and 310 g per day, respectively) (Dehghan et al. 2005). This is comparable to consumption levels in the United States and Europe and exceeds the minimal recommendation of more than 400 g/day of fruit and vegetables by the (World Health Organization 2008). The percentage of UAE citizens who report consuming cooked vegetables and fresh fruit 6 or 7 days a week is 50 and 45%, respectively; only 8 and 10% report rarely or never consuming vegetables and fruit, respectively (Dehghan et al. 2005).

The main fruit in the traditional UAE diet is dates, which continue to be a relatively important part of the diet. However, the modern diet of UAE citizens also includes a broader range of fruits and vegetables.

Fruit and vegetable production in the UAE includes dates, green fodder, vegetables, and fruit (mainly citrus and mangoes). The country produces 100% of the dates it consumes and 50–60% of fruits and vegetables. Tomatoes are also a major crop, and all salad crops are produced domestically for much of the year. Import bans and government incentives and subsidies encourage domestic production.

There is some publicly available data regarding pesticide residues on fruits and vegetables in the UAE (for Abu Dhabi) and in the region. Of 185 samples collected for one study in Abu Dhabi between 1998 and 2001, eight samples exceeded the recommended maximum residue limits (MRLs) of target pesticides; another study analyzed 26 samples for numerous pesticides and found that three samples (of the same vegetable, corn) exceeded MRLs for one of the pesticides, pirimicarb.

The types of pesticides used, and the quantitative residue levels of those pesticides that remain on fruit and vegetables, do not appear to be publicly available for other emirates. Also, while information on banned pesticides is publicly available, there is no publicly available list of all pesticides in use.

Numerous pesticides have been banned from entry into the UAE. But because imports account for a significant fraction of fruit and vegetables in the UAE, exposure to some of these pesticides may nevertheless occur. However, based on one recent study, assuming even a very homogenous diet (for example, a diet that consisted entirely of potatoes exclusively imported from the United States), less than 1% of the population would be exposed to levels of a given pesticide that would exceed the acute reference dose (in this example, the acute reference dose for aldicarb). Given that diets in general are much more diverse and given that only 50–60% of vegetables are imported, no citizens are likely to be exposed to risky levels of a given pesticide, either acute or chronic. However, note that the effects of combinations of pesticides are generally unknown, so these are not considered here even though they may well impact health.

There is no publicly available information regarding ongoing, systematic governmental monitoring of pesticide residues on either domestically produced or imported fruits and vegetables.

What Has the UAE Already Done About the Risk of Contaminants in Fruits and Vegetables?

The Federal Environmental Agency was established in 1993 to set federal policies, specifically addressing human health and agricultural crops. The UAE government has passed a number of regulations on the use of pesticides in the UAE. For example, at least 93 pesticides have been banned as a result. Additionally, the manufacture and formulation of pesticides is prohibited in the UAE, and only pesticides registered by the Ministry of Environment and Water (formerly the Ministry of Agriculture and Fisheries) can be imported and used. There is no publicly available information on established "best practices" for minimization of pesticide residues on fruits and vegetables.

Local municipalities also conduct monitoring activities. For example, in 2002 the Pesticide Residue Analysis Section at the Food and Environment Control Centre of Abu Dhabi Municipality found that 68% of tested samples of locally grown vegetables had no unacceptable levels of pesticide residues, but 25% were found to have quantities above the MRL. Results were better for samples of locally produced fruits: 45% were found to have no unacceptable levels of pesticide residues and none were above the legal acceptable level (UAE Interact 2007).

Individual emirates have food control agencies to establish rules and regulations for handling and standards for contaminant levels. In addition to those agencies, there are a number of additional steps that can be taken in the UAE to reduce pathogen contamination of fruits and vegetables. In particular, standards should be established for farming practices, including the use of reclaimed water. Washing and proper hygiene is important in order to minimize the risk of pathogens from fruits and vegetables in the home; this includes separation of food items to avoid cross-contamination, thorough cooking, and proper refrigeration. These efforts are particularly important for pregnant women. In terms of processed fresh fruits and vegetables, standards for handling and production facilities are appropriate, and worker training is essential.

Notes on Eating Contaminated Fruits and Vegetables Risk Calculations

Number of Deaths per Year. This is the average number of deaths expected per year among the population of the UAE as the result of lifetime exposure to contamination in fruits and vegetables. Assuming proper handling and preparation, this number is zero due to pathogen exposure. In fact, consumption of fruits and vegetables is known to decrease a number of health risks and may actually enhance quality and length of life. The high estimate of "low but nonzero" is based on possibly carcinogenic effects of exposure to pesticide residue; the number cannot be exactly calculated due to lack of information on carcinogenicity and limited sample size.

Chance in a Million of Death per Year for the Average Resident. This is the average annual risk of death for a randomly chosen resident of the UAE as a result of exposure to a given hazard for 1 year. Again, this number is zero for pathogens. The high estimate of "nonzero" is based on the possibly carcinogenic effects of exposure to pesticide residue; the number cannot be calculated due to lack of information on carcinogenicity and limited sample size.

Chance in a Million of Death per Year for the Resident at Highest Risk. The number is estimated as zero for pathogens. The high estimate of "nonzero" is based on the possibly carcinogenic effects of exposure to pesticide residue; the number cannot be calculated due to lack of information on carcinogenicity and limited sample size.

Greatest Number of Deaths in a Single Episode. The number is estimated as zero.

Illness or Injury. Exposure to pesticides through consumption of contaminated fruits and vegetables generally presents a nonfatal risk of inhibition of the action of acetylcholinesterase in nerve cells. This risk is a less serious, short-term detriment to neurotransmission, manifesting itself as sweating, pinpoint pupils, leg weakness, and other effects. While pesticides can also have long-term effects and have been linked to cancer and other life-threatening problem, there is no evidence that eating fruits and vegetables alone exposes an individual to sufficient quantities of pesticides to experience a risk of these impacts. As such, the "less serious short-term cases" table entry refers to cases of nonfatal illness or injury per year expected among residents of the UAE related to acute pesticide exposure. The high estimate assumes that 92% of residents (i.e., all except the 8% who say they "rarely" consume vegetables) are consuming large amounts of a single crop (namely, that all of their vegetable intake is of a single type on which a single type of pesticide is used, and this being one of the two most contaminated vegetables that we surveyed: either a parathion-methyl-contaminated vegetable from the UAE or an aldicarb-contaminated potato from the United States) and are therefore exposed to acute risk with the same probability of a given sample of that crop exceeding acceptable limits for acute exposure (one sample in 100). So the numbers in the range presented here are 2.6 and 1%, for the two contaminants respectively, of the 92% of the total population. Further, we assume no reduction in pesticide residue due to washing, peeling, and/or cooking. As this is clearly the absolute worst-case scenario and not highly plausible, the best estimate is that no UAE citizens will be at risk for acute exposure due to both diversity of diet and proper handling of produce. Even in the worst case, acute impacts from pesticides at these relatively low levels are reversible.

One set of samples (parathion-methyl-contaminated vegetables from the UAE) may indicate that chronic reference doses are being exceeded based on the median value of all the samples. However, since only a range of all sample residues is given in the reference to which we had access, without the average or even the total number of samples included, we are unable to state this with confidence. As such, we have stated only that the high estimate for the less serious long-term cases is "nonzero." Access to the details of the results of the study would presumably allow this number to be estimated to a specific nonzero (or even zero) value.

Time between Exposure and Health Effect. Long-term illnesses such as cancer would, in theory, manifest on a multiyear timeframe (i.e., 10–30 years). Short-term nervous system impacts from pesticide exposure would be immediate.

Quality of Scientific Understanding. There are several sources of uncertainty in estimating risks for the UAE population. One involves how well scientists know the relationship between exposure to a hazard and its resulting health impacts. The other involves how well we can predict exposure of UAE residents to a particular hazard. This statistic characterizes the former. In this case, the causality between pesticide intake and acetylcholinesterase inhibition is well-established, but the exact correlation between dose and response is less so. Three categories are used to rate scientific understanding: high, moderate, and low.

Combined Uncertainty in Deaths, Illness, and Injury. This statistic reflects both uncertain scientific understanding about the risk and uncertainty about the extent of exposure or susceptibility of UAE residents to the particular hazard. Sources of uncertainty specific to data in the UAE include: (1) lack of knowledge of the distribution of consumption levels across the population, (2) lack of knowledge of the distribution of produce types consumed, (3) unknown levels contamination in all types of produce consumed and across all regions of the UAE, and (4) unknown levels of contamination and proportion of imported produce. The table entry gives the amount of uncertainty in deaths, illness, and injury, expressed qualitatively with respect to other risks in UAE.

Ability of Resident to Control Exposure. Some hazards that UAE residents encounter can be avoided partly or entirely by measures they can take on their own. For instance, they can wash and peel fruits and vegetables that might have pesticide residues. However, the risk cannot be completely eliminated due to the potential of the pesticide to contaminate other parts of the fruit or vegetable, and there is nutritional value in eating vegetable peels. Three categories are used to rate this controllability: high, moderate, and low.

Electromagnetic Fields

Summary

Electromagnetic fields (EMF) are produced by an electric charge or current, such as those found in power cords to household appliances or as a result of high-voltage power lines. The strength of both electric and magnetic fields drops off steeply with distance from their source.

The science linking exposure to EMF to health effects remains unconvincing. The only biological impact for which there is any correlation is childhood leukemia, but no cause and effect relationship has actually been shown. Therefore, these findings remain controversial, earning extremely low frequency (ELF) magnetic fields the International Agency for Research on Cancer's "possibly carcinogenic in humans" classification.

Currently there is scant information available on the typical EMF levels for different locations in and around the United Arab Emirates, workplaces, schools or residences. According to the Dubai Electricity and Water Authority, the utility does not allow buildings constructed within 50 m of high-voltage power lines, which are thought to be particularly strong producers of magnetic fields.

Risk characteristic	Low estimate	Best estimate	High estimate
Fatalities			
Number of deaths per year	0	0.16	3.9
Chance in a million of death per year for the average resident	0	0.06	1.4
Chance in a million of death per year for the resident at highest risk	0	0.21	5.0
Greatest number of deaths in a single event		1	
Illness or injury			
More serious long-term cases per year	0.5	2	14
Less serious long-term cases per year	0	0	0
More serious short-term cases per year	0	0	0
Less serious short-term cases per year	0	0	0
Other factors			
Time between exposure and health effects	10–30 years		
Quality of scientific understanding	Low		
Combined uncertainty in death, illness, and injury	Low		
Ability of resident to control exposure to hazard	Moderate		

What Is Known About the Risk of Exposure to Electromagnetic Fields?

Electromagnetic fields occur both naturally and from anthropogenic sources. Electric fields occur whenever an electric charge is present, even if a device is not on. Magnetic fields, on the other hand, are only created when an electric current is flowing, such as when an appliance is switched on. A higher current drawn through the device will produce a stronger magnetic field. Electric fields may be easily shielded by objects such as buildings, whereas magnetic fields are not as easily blocked.

The frequency or its corresponding wavelength is the characteristic that defines how much energy the electromagnetic field will contain. Fields that do not have enough energy to break molecular bonds are called "nonionizing" radiation. The fields produced by modern technology, including radio and television antennas, microwave ovens, and electricity all fall into this spectrum. Since the radiation from these fields declines rapidly with distance from the source, typical human exposures to EMF are generally in the ELF range, as classified by the World Health Organization.

Although controlled experiments have shown that exposure to very strong, high frequency levels of EMF can cause biological effects such as nerve and muscle stimulation, those levels have not typically been found in the community at large. Thus, most attention from the public health sector has focused on ELF-EMF, the form of EMF to which humans are most typically exposed. For ELF-EMF, the connection between exposure and biological effect is more tenuous. The World Health Organization (WHO) recently reviewed all of the scientific research on ELF-EMF and found there to be little statistically significant correlation with disease (World Health Organization WHO 2007b).

Many health effects have been studied, including childhood and adult cancers; mental health outcomes such as headaches, anxiety, suicidal intentions, and depression; reproductive problems, including spontaneous abortions, low birth weight, and congenital malformations; and other immunologic and neurological outcomes. Individuals reporting "electromagnetic hypersensitivity" and depression have inconsistent reactions under controlled exposure environments, and there is no accepted biologic mechanism that supports a hypersensitivity reaction. None of the outcomes mentioned above support an association between ELF-EMF exposure and disease.

Childhood leukemia is one health outcome for which the evidence of a causative link from EMF remains highly controversial. Human studies have consistently shown a pattern of a twofold increase in childhood leukemia associated with average residential EMF exposure. On a worldwide basis, this translates into 100–2,400 additional cases per year that could be attributable to ELF-EMF exposure. A 2,000 study by the Electric Power Research Institute shows a potential connection between contact current and childhood leukemia (Brain et al. 2003). This is related to the way homes in the United States are grounded (usually to pipes) and only affects children while they are bathing. Although it is not a constantly present danger, it is—in contrast with ambient EMF—a significant physical threat to children.

International research on health effects from EMF exposure is now focused more prominently on the potential effects from mobile phone use. Although research in this area is at an early stage, thus far the overall evidence does not point to any excess risk from that source, and sophisticated computer models of the head show that the energy absorbed from a mobile phone is not in excess of current guidelines.

What Is the Electromagnetic Field Exposure in the UAE?

Many sources in the everyday environment contain ELF-EMF. In the outside environment, common sources include high-voltage power lines and various utility substations and antenna base stations. In the indoor environment, sources include any household appliance, including computers, mobile phones, electric blankets, air conditioners, microwave ovens, electric shavers, etc. It is also unclear how UAE residences are grounded or how children bathe, so it is not certain whether contact current is a significant concern.

Currently there are no known surveys or studies that have been conducted in the UAE to document typical residential, school, workplace, or other exposures. The Dubai Water and Electricity Authority and other utilities in the UAE have taken some measurements just within proximity to their equipment, and the highest level measured immediately outside of the 50-m radius of the high-voltage power lines was below levels associated with reported health effects.

There are, however, new high-voltage power lines being planned to connect all the emirates into a single power grid. This could mean a growing population that could potentially be affected by EMF radiation in the future.

What Has the UAE Already Done About the Risk from Electromagnetic Fields?

Although short-term ELF-EMF exposure guidelines exist in the United States and other nations to protect workers and the public from acute exposure, no long-term exposure guidelines have been set internationally because of the weak evidence on the links of ELF-EMF to health effects. Different countries have radically different codification of EMF protection. For instance, Turkey regulates frequencies between 10 kHz and 60 GHz, while Italy regulates 50 Hz to 300 GHz. However, according to the WHO, Turkey and Israel are the only two Middle Eastern countries that have EMF regulations. In 2002, the UAE passed a federal law that regulates ionizing radiation, but currently there are no laws regulating nonionizing radiation levels such as EMF.

As of 2008, the UAE had taken limited action to assess EMF exposure for its population. This risk assessment found no evidence that the government either had developed risk communication materials or had undertaken efforts to prepare baseline measurements of EMF to begin to understand the typical levels to which the population is exposed.

Notes on Electromagnetic Fields Risk Calculations

Number of Deaths per Year. This is the average number of deaths expected per year among the population of the UAE as the result of lifetime exposure to extremely low frequency electromagnetic field radiation (ELF-EMF). The high and low estimates of risk show the range in absolute terms. The high and best estimates come from WHO estimates that ELF-EMF could be responsible for 0.2 (best) to 4.9% (high) of leukemia deaths worldwide. The number of annual deaths from leukemia in the UAE is approximately 81 (WHO 2007a). The low estimate is 0 because it has not been proven definitely that EMFs influence mortality at all.

Chance in a Million of Deaths per Year for the Average Resident. This is the average annual risk of death for a randomly chosen UAE resident as a result of exposure to a given hazard for 1 year. The risk to the average resident is the number of deaths in the UAE due to EMFs divided by the entire population of the UAE, according to WHO's GBD figures.

Chance in a Million of Death per Year for the Resident at Highest Risk. For some hazards, certain residents are known to be more exposed or more susceptible than others. For EMF, children are thought to be most at risk, so the calculation here is the number of deaths attributed to EMF divided by the population of children under the age of 15 from UAE population data, approximately 26% of the total population.

Greatest Number of Deaths in a Single Episode. There are not really any large-scale acute incidents of EMF exposure, especially since the proposed mechanism for health effects is low-level exposure over extended periods. Thus, we estimate that the most likely number to die in an "episode" is zero, but since the mechanism for mortality is not well understood, the greatest number of people could alternately be a single person.

Illness or Injury. The main illness risk from EMF is leukemia. We applied the risk of contracting nonfatal leukemia to the estimated 2005 population of the UAE.

Time between Exposure and Health Effects. Science is unsure of the mechanism for health effects and thus also unclear about the time between exposure and effect. However, the estimates are mainly around a decade for children and possibly longer for adults.

Quality of Scientific Understanding. There are two sources of uncertainty in estimating risks for the UAE population. One involves how well scientists know the relationship between exposure to a hazard and its resulting health impacts. The other involves how well we can predict exposure of UAE residents to a particular hazard. This statistic characterizes the former. For instance, scientists still do not know whether exposure to EMF from power systems causes cancer, but scientists understand very well the physical and biological processes leading to injury from auto accidents. Three categories are used to rate scientific understanding: high, moderate, and low.

Combined Uncertainty in Deaths, Illness, and Injury. This statistic reflects uncertain scientific understanding about the risk the extent of exposure or susceptibility of UAE residents to the particular hazard. The table entry gives the amount of uncertainty in deaths, illness, and injury, expressed both qualitatively with respect to other risks in the UAE.

Ability of Resident to Control Exposure. Some hazards that UAE residents encounter can be avoided partly or entirely by measures they can take on their own. For instance, in the case of ELF-EMF exposure, residents can avoid living near power lines, or they can take household measurements and spend less time in rooms with high EMF or arrange the power switches in their rooms to minimize EMF. Three categories are used to rate this controllability: high, moderate, and low.

Ambient Noise

Summary

Noise is unwanted sound that results from many activities. In urban settings, common sources of noise include airplanes, trains, trucks and automobiles, construction and demolition activity, and industrial facilities. The most obvious effect from high noise levels is hearing impairment and its associated interference with general communication. Other impacts from continued noise exposure include cardiovascular effects and sleep disturbances. In a 2006 ambient noise survey (Muskett and Bohler 2006), the noise level at monitoring sites in Abu Dhabi appears to be in excess of national and international standards, raising the concern that residents may be exposed to ambient noise pollution. But no systematic studies have been done to assess the noise levels in other emirates.

Risk characteristic	Low estimate	Best estimate	High estimate
Fatalities			
Number of deaths per year	0	Nonzero but low	Not reported
Chance in a million of death per year for the average resident	0	Nonzero but low	Not reported
Chance in a million of death per year for the resident at highest risk	0	Nonzero but low	Not reported
Greatest number of deaths in a single event		1	
Illness or injury			
More serious long-term cases per year		Not reported	
Less serious long-term cases per year		Not reported	
More serious short-term cases per year		Not reported	
Less serious short-term cases per year		Not reported	
Other factors			
Time between exposure and health effects	Weeks to months		
Quality of scientific understanding	Low		
Combined uncertainty in death, illness, and injury	High		
Ability of resident to control exposure to hazard	Moderate		

What Is Known About the Health Risks of Noise Pollution?

The effects of noise on health are complex and remain poorly understood. It is believed that the effects of noise exposure depend on both the level of sound as well as the length of time to which an individual is exposed. The most obvious impact of noise exposure relates to hearing loss, however there are a wide array of other identified health effects that raise concerns. The best way to describe the health effects of noise exposure is to separate what is currently described in the occupational setting as opposed to the community or "environmental" setting.

Occupational noise exposure measured over an 8-h period shows that at 75 dBA (*A* represents a weighting filter that has been widely adopted for environmental noise measurement) and lower, even prolonged exposures over many months or years will not result in noise-induced hearing impairment. However, environmental noise standards are much lower than occupational requirements because they also factor in the issue of annoyance and other quality-of-life issues. The U.S. Environmental Protection Agency (EPA) and the World Health Organization (WHO) have recommended standards between 30 and 70 dBA depending on the area of concern (e.g., industrial, commercial, or residential).

It has been much more difficult to measure and link the health effects of noise in the environmental than in the occupational setting. Measurable occupational impacts associated with noise, such as hearing loss and hypertension (though less definitively) are not documented at levels of noise typically measured over an average day in a community environment. Many effects in an environmental setting are more closely linked with annoyance and quality-of-life issues, rather than with primary health effects such as hearing damage. The major issues relate to interference with speech and the impact on sleep, which may have numerous secondary effects, including stress, loss of productivity, increased fatigue, mood changes, decreased cognitive performance, and an increase in accidents.

The literature has shown mixed results of a causal relationship between noise and serious health problems such as heart disease and hypertension. Some literature suggests that chronic noise exposure, even at low levels, has the potential to cause chronic stress hormone increases in humans that accelerate the aging of the myocardium and vascular walls. There is scant information available regarding which type of noise might have an effect (e.g., noise that is continuous versus noise that pulses on and off) and the length of exposure required to produce such effect. In addition, in many cases, individuals adopt coping strategies that may reduce the impact of noise over time and thereby reduce the potential associated health effects. Certain subgroups—in particular, fetuses and children, the elderly and those that already have pre-existing conditions such as high blood pressure—may be more vulnerable to noise in community settings.

What Ambient Noise Exposure Levels Are Documented in the UAE?

In 2006, the UAE conducted an ambient noise survey of various sources around Abu Dhabi, which helped provide baseline data for levels in the city. The noise level in Abu Dhabi appears to be in excess of national and international standards, and residents may be at risk from one or more of the health effects described above. Much of the available information about environmental noise exposure levels in the UAE is based upon the Abu Dhabi noise strategy document (Muskett and Bohler 2006). According to this report, road traffic is the major source of noise in urban Abu Dhabi, while industrial noise is more localized, affecting areas with heavy industry (e.g., power stations and oil refineries). Aircraft noise appears to be confined to the regional area of the airport, and construction zones vary by what is being built and how long it takes to construct.

A review of the 2006 Abu Dhabi noise survey shows that large areas of Abu Dhabi city experience noise levels primarily due to road traffic in excess of 65 dBA, the noise level above which research shows a significant percentage of the population will be disturbed. The measurements in this survey were only done for 15-min intervals, and it is not certain what the time weighted averages would be. Noise levels measured away from traffic were generally documented at much lower levels. Given these baseline results, it is likely that in many parts of Abu Dhabi city and other regions within the emirate, residents are being exposed to noise substantially in excess of established national and international standards and are therefore experiencing many of the primary and secondary health effects described for environmental noise exposure. However, insufficient data are available to document the health effects, or to document noise levels in other emirates.

What Has the UAE Already Done to Reduce the Risk of Noise Exposure?

There are minimally developed noise regulations in Abu Dhabi and elsewhere in the UAE. Inspection and enforcement procedures are also limited. However, the Environment Agency–Abu Dhabi has drafted recommended standards, drawn from international guidelines. These guidelines will be used to evaluate and guide new development proposals and direct future noise management issues in the UAE.

The table below presents a brief comparison of noise-level guidelines across three types of areas in different countries. (The UAE values are recommended in a draft regulation.) In general, the UAE values are lower than those in other countries and may reflect the UAE's preference for stricter criteria for allowable noise levels.

Comparison of International Guidelines for Noise in Residential, Industrial, and Commercial Areas

	Residential areas[a]			
Country	Day	Night	Industrial areas[a]	Commercial areas[a]
UAE	40–60	30–50	50–70	45–65
U.S. (EPA)[b]	45–55		70	60
Japan	55	45	50–60	50–60
India (Delhi)	50	40	70–75	55–65
Malaysia	55–60	45–50	70	65
Global (WHO)	55	50	70	70

[a]Measured in decibels (dB); the range of maximum acceptable levels are reported, due to variation in guidelines between federal versus local governments *within* a country and/or the definition of 'area type' *by* a country

[b]In 1981, the noise regulation authority of the United States shifted from the federal to the state level. However, the Noise Control Act of 1972 and the Quiet Communities Act of 1978 remain in effect. U.S. numbers reflect the federal guidelines

Currently, a strategy is in place to establish a framework for the management of noise in Abu Dhabi emirate through 2015. The first phase of this strategy entails the collection of sufficient noise monitoring and mapping data as a baseline for implementing further noise control management and action plans in the future. Once that data is collected, the noise management approaches will consider:

- Introduction of traffic management schemes to restrict heavy vehicles traveling through noise-sensitive areas at certain times of the day
- Use of acoustic barriers to protect noise-sensitive buildings
- Establishment of quiet zones around sensitive buildings such as hospitals and schools
- Restricting noise-generating industry to specific zones, away from residential or other noise sensitive development

Notes on Ambient Noise Risk Calculations

Number of Deaths per Year. Although recent studies conducted by European researchers have shown a relationship between noise exposure and premature deaths in Europe, the evidence is not strong. Overall the literature has shown mixed results of a causal relationship between noise and serious health problems such as heart disease and hypertension.

More Serious Long-Term Cases per Year. This reflects the heart attack cases attributable to road traffic in UAE annually only. We used the following formulas (Babisch 2006) to calculate the illnesses cases:

$$AR = \frac{RR-1}{RR} \times 100$$

$$PAR\% = \frac{P_e}{100} \times \frac{RR-1}{{P_e}/{100} \times (RR-1)+1} \times 100$$

$$PAR = PAR\% \times N_d$$

Where:

AR = Attributable fraction

RR = Relative risk (odds ratios are estimates of the relative risk)

$PAR\%$ = Population attributable risk percentage

P_e = Percentage of the population exposed

PAR = Absolute numbers of affected subject

N_w = Number of subjects with disease occurrence

The relative risks of exposure to different noise levels were estimated by Babisch (2006).

The WHO estimates that 20% of the population worldwide is annoyed by road noise, and we used this percentage as the low estimate for the UAE population exposed to traffic noise. Considering that the population of the UAE is overwhelmingly urban, with more than 90% of people living in cities, at the high end we assumed that 85% of UAE population is exposed. This is roughly the population of Abu Dhabi, Dubai, and Sharjah in total. The UAE population in 2008 was 4,621,399 (UAE Ministry of Economy 2008b).

In the United States, there are 900,000 annual heart attack cases, which represent about 0.3% of the whole population (Centers for Disease Control and Prevention 2009). We assumed that the occurrence in the UAE is similar, based on the understanding that currently the trend of such diseases in UAE as diabetes, hypertension, and heart attack is very like that in developed countries, as a consequence of higher living standards.

Using the above formulas, relevant data, and assumptions, we got the following numbers for noise level between 70 and 75 dBA:

$$AR = \frac{RR-1}{RR} \times 100 = \frac{1.19-1}{1.19} \times 100 = 15.97$$

$$PAR\%(Low) = \frac{P_e}{100} \times \frac{RR-1}{\frac{P_e}{100} \times (RR-1)+1} \times 100 = \frac{20}{100} \times \frac{1.19-1}{\frac{20}{100} \times (1.19-1)+1} \times 100 = 3.66$$

$$PAR\%(High) = \frac{P_e}{100} \times \frac{RR-1}{\frac{P_e}{100} \times (RR-1)+1} \times 100 = \frac{85}{100} \times \frac{1.19-1}{\frac{85}{100} \times (1.19-1)+1} \times 100 = 13.90$$

$$PAR(Low) = PAR\% \times N_d = 3.66\% \times (4{,}621{,}399 \times 0.3\%) = 507$$

$$PAR(High) = PAR\% \times N_d = 13.90\% \times (4{,}621{,}399 \times 0.3\%) = 1{,}927$$

Risk estimation (Risk of myocardial infarction due to road traffic noise) low/high

Average noise level during the day (dBA)	Percentage of population exposed	Relative risk	Attributable fraction	Population attributable risk percentage PAR%	Number of subjects per year
<60	80/15	1.00	0.00	0.00	0
>70–75	20/85	1.19	15.97	3.66/13.90	500/1900

Time between Exposure and Health Effect. Some health effects, such as hypertension, can be fairly immediate, whereas health effects such as hearing loss or heart diseases may not manifest for years or decades into the future.

Quality of Scientific Understanding. The effects of noise on health are complex and remain poorly understood.

Combined Uncertainty in Deaths, Illness, and Injury. This statistic reflects both uncertain scientific understanding about the risk and uncertainty about the extent of exposure or susceptibility of UAE residents to the particular hazard.

Ability of Resident to Control Exposure. The main sources of noise in the UAE are road traffic, industry, aircraft, and construction. People living close to these sources generally cannot control their exposure.

Global Climate Change

Summary

The Earth is warming, as evidenced by increasing air and ocean temperatures, widespread melting of ice and snow, and rising global average sea levels. The increase in the global average temperature since the mid-twentieth century is very likely caused by the increased concentration of greenhouse gases (GHG), such as carbon dioxide, methane, and nitrogen oxides, in the atmosphere. There is a scientific consensus that the primary reason for this increase is human activities—the combustion of fossil fuels, including coal, oil, and natural gas. Evidence is mounting that climate change has been associated with heat-related deaths and disease; deaths and illnesses from extreme climate-related events such as flooding and droughts; and malnutrition, diarrheal diseases, and infectious diseases. The risks of climate change can be managed by a combination of international and local government policies, such as reductions in carbon emissions, and by community and individually adaptive strategies.

Risk characteristic	Low estimate	Best estimate	High estimate
Fatalities			
Number of deaths per year	0	20	50
Chance in a million of death per year for the average resident	0	6	13
Chance in a million of death per year for the resident at highest risk	0	6	13
Greatest number of deaths in a single event		1	
Illness or injury			
More serious long-term cases per year	0	0	0
Less serious long-term cases per year	0	0	0
More serious short-term cases per year	0	0	0
Less serious short-term cases per year	0	54,000	110,000
Other factors			
Time between exposure and health effects	Immediate to 30 years		
Quality of scientific understanding	Low		
Combined uncertainty in death, illness, and injury	Moderate		
Ability of resident to control exposure to hazard	Low		

What Is Known About the Potential Health Risks of Climate Change?

Over the next 50–100 years, global warming is projected to be five times greater than what has been experienced over the last 25 years. Importantly, it is predicted that the effects of global warming will not be uniform and will affect various populations and regions of the world differently. Because hot regions of the world suffer more from temperature increases, the regions most seriously affected by climate change will be Africa and the poorer areas of the Eastern Mediterranean and Southeast Asia.

Weather can have a major influence on human health in a number of ways, including the direct effects of extreme events such as heat waves, floods, and storms; the indirect effects of infectious diseases; and the limited availability of clean freshwater, which can cause many diseases, and food, which can lead to malnutrition and starvation.

The table below presents the potential pathways through which climate change may affect human health.

Potential pathways through which climate change may adversely impact human health

Direct impacts	Intermediate factors	Potential human health effects
Increasing temperature and heat waves		Heat-related mortality and disease, cardiovascular and respiratory diseases
Affecting precipitation	Food production (yield)	Malnutrition
Increasing extreme climate-related events	Ozone concentration increasing, air pollution	Air pollution related mortality and illness such as respiratory diseases
Hurricanes, wildfire, droughts, floods, etc.	Microbial contamination and transmission in food and drinking water	Diarrheal diseases, infectious diseases, water- and foodborne diseases, vector- and rodent-borne diseases
	Sea level rising	Mortality and injury from flooding, malaria during droughts, health-related problems with displaced populations

However, uncertainty remains with direct attribution of these health consequences to climate change. If climate change progresses, its effects on weather can result in cascading health consequences. The extent to which climate change affects human health in one region/nation is determined by the baseline conditions of the exposure, a population's vulnerability to changes in climate, as well as the capacity of populations to respond to change.

What Are the Current and Predicted Impacts of Global Climate Change on Health in the UAE?

The United Arab Emirates' arid and semi-arid regions are considered moderately to highly vulnerable to the effects of climate change, which are expected to affect temperatures, extreme climate-related events, precipitation, water resources, and sea levels. In turn, these effects can then have a serious impact on public health through the various pathways shown above. No studies on deaths or diseases directly attributable to climate change in the UAE have been done, but the UAE's baseline climate-sensitive diseases (malaria, diarrhea, and malnutrition) are at comparatively low levels that are similar to those in developed countries, and the health effects of climate change through this path might be limited. However, climate change may pose moderate to high risks to the UAE population's health condition, as potential deaths or illnesses associated with climate change could include heat-related illness, respiratory symptoms, vector-borne infectious diseases, and deaths during flooding.

An increase in temperature and decrease in precipitation may influence the UAE's freshwater supply. Currently the entire UAE population (both urban and rural) has sustainable access to improved drinking water sources, and drinking water quality is good. However, the UAE's water supply has been a critical issue due to its arid climate, fast population growth and economic development, and expanding agricultural activities. Higher average temperatures and greater evaporation will worsen the condition of the UAE's water supply, which is largely produced by the desalination of sea water, a process that consumes a large amount of energy and is responsible for large amounts of carbon dioxide.

Rising sea level is a significant risk for the UAE's coastal zones, which comprise approximately 1,318 km of coastline, with 85% of the population living within 100 km of the coast. This large coastal population is likely to be particularly vulnerable to floods, as are many buildings and industrial facilities (including oil and gas) that operate along the coast.

What Has Already Been Done About Climate Change in the UAE?

Because the climate system is affected by global emissions of GHGs, the UAE does not have the ability to control the climate change it will experience. By taking action to reduce its own emissions, the UAE, as an energy-exporting nation, can set an important example on climate-change mitigation efforts for its neighbors as well as other nations around the world.

There are generally two approaches for countries to prevent or lessen the negative impacts of global climate change on its people: address the drivers of climate change, or mitigate the effects. The UAE has focused on the first approach by taking a number of steps:

Shifting away from high-energy processes. The major source of GHG emissions is the burning of fossil fuels, and the UAE is one of the major emitters of GHG per capita, ranking fourth in the world. The UAE is actively trying to shift away from high-energy processes and is also transitioning to natural gas, which produces less carbon dioxide per unit energy than oil.

Participating in international treaties and coordinated efforts to set targets to reduce emissions. The UAE hosted the first World Summit on Energy for the Future in January 2008. It also ratified the Kyoto Protocol in January 2005, which aims at reducing global emissions of GHG to 5% below 1990 levels during the period 2008–2012.

Developing clean energy technologies and alternative energy source. For example, a "green city"—Masdar City in Abu Dhabi—will be built in 8 years and will be mostly powered by solar energy. Another example began in January 2008 when Abu Dhabi announced a $15 billion initiative to develop clean energy technologies using hydrogen.

Promoting research and developing preparedness plans for future defined risks. The UAE has published its initial national communication to the United Nations, conducted its own GHG inventory, defined methods to improve data quality, and identified vulnerabilities and adaptation options.

Setting improved efficiency standards. The UAE outlined "greener" building codes and appliance standards in its report to the United Nations.

Despite taking the above actions, the UAE still needs a national and regional strategy for climate change to deal with the various health consequences, especially those caused by increasing temperatures, shrinking fresh water supply, and sea level rise.

Notes on Global Climate Change Risk Calculations

We calculated the fatalities and cases of illness due to climate change based on the relative risks estimated by McMichael et al. (2004). The following health effects (mortality/morbidity) are estimated because (1) they are sensitive to climate change; (2) it is important to predict these outcomes in the future; and (3) data are available:

- Mortality: deaths from cardiovascular disease and from floods (inland and coastal floods)
- Morbidity: malnutrition, malaria, and diarrheal diseases

Number of Deaths per Year. This is the average number of excess deaths (from cardiovascular disease and floods) among the UAE population as a result of exposure to global climate change. To calculate this number, we assume that (1) deaths attributable to climate change are distributed proportionally to the populations across all EMR-B countries; (2) UAE accounts for 1.87% population in all EMR-B countries; and (3) the annual deaths from cardiovascular disease and floods in the region are constant. The low estimates are zeroes, reflecting the most optimistic scenario under which the UAE completely adapts to the change.

We used the following formula to calculate best and high estimates of excess deaths from cardiovascular disease and floods attributable to climate change:

$$\textit{Excess deaths due to climate change} = \textit{Numbers of current death} \times (\textit{Relative risk} - 1)$$

Relative risks under three climate scenarios for EMR-B region are estimated by McMichael et al. (2004). These data are presented in the table below.

Relative risks for EMR-B region, 2005

Climate	Cardiovascular disease			Inland floods/ landslides			Coastal floods		
change scenario	Low	Mid	High	Low	Mid	High	Low	Mid	High
s550[a]	1.000	1.001	1.001	1.00	1.98	2.49	1.11	1.22	1.45
s750[b]	1.000	1.001	1.002	1.00	2.37	3.09	1.12	1.24	1.48
Unmitigated[c]	1.000	1.001	1.003	1.00	2.41	3.15	1.16	1.31	1.63

[a]The future exposure scenario that assumes GHG emission reduction achieving stabilization at 550 ppm CO_2-equivalent by 2170

[b]The future exposure scenario that assumes GHG emission reduction achieving stabilization at 750 ppm CO_2-equivalent by 2210

[c]Unmitigated emission trends

For our calculations, we assumed unmitigated emission trends. We used the "mid estimates of relative risks" to calculate our best estimates and high estimates of relative risks to calculate our high estimates.

The number of cardiovascular disease deaths in 2002 in the UAE was 369 per 100,000 population, and we have assumed this death rate to remain constant (WHO 2006). The 2005 UAE population was 4,104,695. Therefore the number of cardiovascular disease (CVD) deaths in 2005 in UAE is 15,144.

The annual incidence of deaths per 10,000,000 population in 2000 in all EMR-B countries caused by floods, in the absence of climate change, was 13.8 for inland floods and landslides and 0 for coastal floods. The population of the EMR-B region in 2000 was 139,070,000 (United Nations Population Division 2000).

Examples of calculations:

Excess CVD deaths due to climate change (high) = 15,144 × (1.003–1) = 46
Excess CVD deaths due to climate change (best) = 15,144 × (1.00 –1) = 16
Excess deaths from inland floods/landslides due to climate change (high) = (13.8 × 13.9) × (3.15–1) × 1.87% = 8
Excess deaths from inland floods/landslides due to climate change (best) = (13.8 × 13.9) × (2.41–1) × 1.87% = 6
Excess deaths from coastal floods due to climate change (high and best) = 0

Chance in a Million of Death per Year for the Average Resident. This is the average annual risk of death for a randomly chosen UAE resident as a result of exposure to global climate change. We calculated this number by dividing "number of deaths per year" by the total population (in millions) of the UAE in 2005.

Chance in a Million of Death per Year for the Resident at Highest Risk. This is the average annual risk of death for a UAE resident at highest risk as a result of exposure to global climate change. We assumed it is the same with chance of death for the average resident.

Greatest Number of Deaths in a Single Episode. This is the greatest number of deaths resulting from a single cardiovascular disease case, or one.

Illness or Injury. These are cases of nonfatal illness or injury per year expected among the UAE population resulting from the exposure to climate change. Three types of illnesses are estimated using the same method described above: malnutrition (more serious long-term), malaria (less serious long-term), and diarrhoeal diseases (less serious short-term). Data are given as follows.

Relative Risks for EMR-B Region, 2005

	Malnutrition			Malaria			Diarrhea		
Scenario	Low	Mid	High	Low	Mid	High	Low	Mid	High
s550	1.00	1.01	1.03	1,00	1.00	1.00	0.99	1.02	1.04
s750	1.00	1.03	1.06	1.00	1.00	1.00	0.99	1.02	1.04
Unmitigated	1.00	1.00	1.00	1.00	1.00	1.00	0.99	1.03	1.06

The numbers of baseline cases of malnutrition, malaria, and diarrheal diseases in 2002 estimated by WHO in all EMR-B countries were 585,000, 363,000, and 96,324,000, respectively.

Time between Exposure and Health Effect. Some health effects, such as diarrheal diseases caused by microbial contamination, have fairly immediate impacts, whereas health effects such as malnutrition may not manifest for years or decades into the future.

Quality of Scientific Understanding. Climate change may affect human health via many mechanisms. The extent of health impacts depends on the magnitude of climate change that occurs, the links that are identified between climate and public health, and the ability of people to adapt to climate change and avoid its effects. But scientific understanding is improving, with the publication of the Intergovernmental Panel on Climate Change's Fourth Assessment Report in 2007 (IPCC 2007).

Combined Uncertainty in Deaths, Illness, and Injury. This statistic reflects both uncertain scientific understanding about the risk and uncertainty about the extent of exposure or susceptibility of UAE residents to the particular hazard.

Ability of Resident to Control Exposure. Climate change is a global issue and the UAE as an individual country does not have the ability to control the climate change that it will experience. Although to some extent human beings have the ability to adapt to temperature change, the hazards cannot be removed or avoided. Extreme events such as flooding and wildfire cannot be controlled by individuals.

Stratospheric Ozone Depletion

Summary

The depletion of the ozone layer, which protects the planet from potentially damaging amounts of ultraviolet (UV) radiation reaching the surface, is the source of numerous adverse health effects caused by increased exposure to ultraviolet radiation. Substances including chlorofluorocarbons (CFCs), halons, methyl bromide, and hydrochlorofluorocarbons (HCFCs), which are used for air-conditioning and refrigeration systems, have been shown to deplete the ozone layer if released into the air.

Risk attribute	Low estimate	Best estimate	High estimate
Fatalities			
Number of deaths per year	16	20	24
Chance in a million of death per year for the average resident	6	8	10
Chance in a million of death per year for the resident at highest risk	7	9	11
Greatest number of deaths in a single event		1	

(continued)

(continued)

Risk attribute	Low estimate	Best estimate	High estimate
Illness or injury			
More serious long-term cases per year	2,100	2,600	3,200
Less serious long-term cases per year	4,600	5,700	6,900
More serious short-term cases per year	800	1,000	1,200
Less serious short-term cases per year	4,700	5,800	7,000
Other factors			
Time between exposure and health effects	Immediate to 30 years		
Quality of scientific understanding	High		
Combined uncertainty in death, illness, and injury	Low		
Ability of resident to control exposure to hazard	High		

As the ozone layer is depleted, more ultraviolet radiation from the sun passes through the atmosphere. The health effects of increased exposure to ultraviolet radiation include cancers of the skin and eyes, cataracts, corneal damage, reduced resistance to infectious diseases, and the diminished effectiveness of vaccines. In response to this risk, 190 countries, including the UAE, signed the Montreal Protocol to formalize international agreement on plans to phase out the production and consumption of ozone-depleting compounds. However, the recovery of the ozone layer remains uncertain because it is dependent on the continued international agreement and commitment to limit emissions of these harmful substances into the atmosphere.

What Is Known About the Risk of Increased UV Exposure as the Result of Ozone Depletion?

Increased exposure to UV rays due to ozone depletion can lead to a variety of harmful health effects, from sunburn to malignant cancers. Acute exposure of the skin to UV rays causes sunburn. The amount of exposure depends on the amount of protective pigment in the skin, genetic factors and the level of UV-B radiation, the most harmful part of the UV spectrum, contained in the sun exposure. Longer-term exposure may also cause wrinkling, thinning, and loss of elasticity in the skin.

UV radiation may also cause nonmalignant or malignant skin cancer by altering important genes that control cell division and cell death. Exposure to radiation can cause both malignant and nonmalignant cancers, with nonmalignant carcinomas occurring at highest frequencies in Caucasians living in sunny environments. These are generally fairly easily treated and are rarely fatal. The malignant form of skin cancer, although much more dangerous, occurs at a much lower frequency than the other, nonmalignant forms of skin cancer. Some evidence suggests that melanomas are correlated with acute sunburns and high exposures during childhood. It should be noted, however, that the rate of some nonmalignant cancers has been increasing in most countries over time, and in a pattern correlated with regions affected by ozone depletion.

UV radiation also has varying effects on the human eye, depending on the type of exposure (acute and intense versus chronic and lower-level intensity). The effects from ozone-depleting chemicals broadly falls into the latter category and the resulting concern is for the development of cataracts, which may be irreversible and ultimately lead to blindness if not corrected. Because cataracts are so highly associated with advanced age and certain diseases such as diabetes, there have been few studies that have been able to distinctly correlate any additional impacts from UV exposure.

Immune effects have also been documented by UV exposure in numerous ways. "Antigen-presenting cells" of the immune system, which are present in human skin, are responsible for bringing invading microorganisms into the lymph nodes, where the immune response begins. Damage to these cells by UV radiation can limit or alter this response. Also, UV-B radiation has been shown to stimulate cell mediators and activate chemical responses that increase immune suppression instead of activation. While it is not clear how this ultimately affects the response to diseases in humans, there is some evidence to show it increases both the susceptibility and severity of infections and reduces the effectiveness of vaccines. This remains an area of important research.

What Are UV Exposure Levels in the UAE?

Because the ozone layer is mainly depleted around the Earth's poles, ultraviolet radiation exposure due to ozone depletion varies by the latitude of the area of study. The UAE lies within the low latitudes or "tropics," which experience a lower risk gradient due to ozone depletion for harmful UV exposure than those closer to the polar regions. However, although the increase in UV exposure is smaller in the UAE than in northern latitudes, the baseline level of exposure is higher because of the angle of the sun and relatively shorter layer of atmosphere they must pass through for most of the year. Additionally, the clear, hot weather in the UAE increases the likelihood of more high UV radiation days in the region, as compared to other parts of the world where air pollution and clouds significantly reduce the radiation that reaches the ground.

The nature of ozone depletion means that the quantity of different wavelengths of radiation reaching Earth increase by different amounts. Because of this, levels of UV-B radiation have not changed significantly for the tropics (23°N to 23°S). Even though levels of UV-B radiation have surpassed pre-1980 levels in other regions, they still have not reached the overall levels experienced by the UAE.

Despite these risks, there is also a risk of too little UV exposure, especially for women in the UAE. A large portion of the population spends much time indoors in factories and businesses, and too little sun exposure can result in a vitamin D deficiency, which can cause skeletal diseases such as rickets, osteomalacia, and osteoporosis, as well as possibly even decreased immunity to cancer and autoimmune disorders such as multiple sclerosis or Type 1 diabetes.

In the UAE, twice as many men as women are affected by diseases and deaths related to excess UV radiation. This could be partially because, according to a UAE study, many Muslim women spend the bulk of their time indoors or under conservative dress. These women, if they work, are also more likely to work indoors, rather than in some of the more manual-labor-intensive jobs.

Risks in the UAE are particularly high for migrant workers, who tend to make up the bulk of those working outdoors. However migrants also tend to have darker skin, which mitigates their risk to some extent. Because the UAE population is primarily comprised of darker skinned populations (with less than 10% identified as ethnicities with fair skin), the risk for skin cancers in this population is relatively lower than for fairer-skinned populations.

What Has the UAE Already Done to Reduce the Risks of UV Exposure?

It is hoped that if emissions are controlled on a global scale, the ozone layer will return to normal around the year 2050. The UAE is complying with its role as an international participant in the Montreal agreement to reduce emissions. Currently, there is no production of ozone-depleting substances in the UAE. Import and consumption of ozone-depleting substances remained stable between 1990 and 2001 in the UAE, with 448 tons of CFCs consumed in 1990 and 423 tons in 2001. Since then, there was a phaseout of 50% of these substances by 2005. As of 2007, CFCs and halons should have reached 85% below 1989 levels in Abu Dhabi and the UAE. The phaseout of these substances should be complete by January 2010 when the UAE will ban all imports of ozone-depleting substances.

Until the ozone layer has been restored, behavior modification is very effective at reducing human exposure to excess UV rays. However, in the UAE most of the population already covers up their entire body except for hands and faces. Much of the population already employs protective measures against UV exposure such as wearing sunscreen, covering skin with clothes, wearing sunglasses, and reducing exposure during the hottest part of the day. All are effective means of protection from the sun's harmful rays. Although it is unknown whether the UAE has implemented any specific educational campaigns to reduce the public's exposure to the sun, these approaches have been shown to be successful and if necessary, may be further tailored to fit the specific cultural norms of the UAE.

References

Abou-Taleb, A.N.M., A.O. Musaiger, and R.B. Abdelmoneim. 1995. Health status of cement workers in the United Arab Emirates. *The Journal of the Royal Society for the Promotion of Health* 115: 378.

Al Asram, O. 2006. *Wastes and pollution sources of Abu Dhabi emirate*. AGEDI Sector Paper. Environment Agency–Abu Dhabi. http://www.agedi.ae/sectorpapers/default.aspx

Al Neaimi, Y.I., J. Gomes, and O.L. Lloyd. 2001. Respiratory illnesses and ventilatory function among workers at a cement factory in a rapidly developing country. *Occupational Medicine* 51(6): 367–373.

Ali, F. 2008. Quarries in UAE must follow new environmental law says minister. *Gulfnews.com* 17:20, June 18.

American Cancer Society. 2008. Cancer Statistics 2008.

Babisch, W. 2006. *Transportation noise and cardiovascular risk: Review and synthesis of epidemiological studies dose-effect curve and risk estimation*. Berlin: Federal Environmental Agency.

Badrinath, P., Q.A. Al Shboul, T. Zoubeidi, A.S. Gargoum, R. Ghubash, and O.E. El Rufaie. 2002. *Measuring the health of the nation: United Arab Emirates health and lifestyle survey 2000*. Al Ain: UAE University.

Banat, I.M., E.S. Hassan, M.S. El Shahawi, and A.H. Abu Hilal. 1998. Post-Gulf-War assessment of nutrients, heavy metals, hydrocarbons, and bacterial pollution levels in the United Arab Emirates coastal waters. *Environment International* 24(1–2): 109–116.

Bell, M.L., A. McDermott, S.L. Zeger, J.M. Samet, and F. Dominici. 2004. Ozone and short-term mortality in 95 U.S. urban communities, 1987–2000. *Journal of the American Medical Association* 292(19): 2372–2378.

Bener, A., G.G. Lestringant, M.M. Beshwari, and M.A. Pasha. 1999. Respiratory symptoms, skin disorders and serum IgE levels in farm workers. *Allergy and Immunology* 31(2): 52–56.

Bener, A., A.M. Almehdi, R. Alwash, and F.R.M. Al Neamy. 2001. A pilot survey of blood lead levels in various types of workers in the United Arab Emirates. *Environment International* 27(4): 311–314.

Beshwari, M.M., A. Bener, A.M. Almehdi, A. Ameen, A. Ibrahim, M.A. Pasha, and H.Z. Ouda. 1999a. Aminoacid profiles in farm workers. *Environment International* 25(4): 411–416.

Beshwari, M.M., A. Bener, A. Ameen, A.M. Almehdi, H.Z. Ouda, and M.A. Pasha. 1999b. Pesticide-related health problems and diseases among farmers in the United Arab Emirates. *International Journal of Environmental Health Research* 9: 213–221.

Blair, A., and S.H. Zahm. 1995. Agricultural exposures and cancer. *Environmental Health Perspectives* 103(Suppl 8): 205–208.

Brain, J.D., R. Kavet, D.L. McCormick, C. Poole, L.B. Silverman, T.J. Smith, P.A. Valberg, R.A. Van Etten, and J.C. Weaver. 2003. Childhood leukemia: Electric and magnetic fields as possible risk factors. *Environmental Health Perspectives* 111(7): 962–970.

Brake, D.J., and G.P. Bates. 2002. Deep body core temperatures in industrial workers under thermal stress. *Journal of Occupational and Environmental Medicine* 44: 125–135.

Broughton, E. 2005. The Bhopal disaster and its aftermath: A review. *Environmental Health* 4: 6.

Campbell-Lendrum, D.H., and R. Woodruff. 2007. Climate change: Quantifying the health impact at national and local levels. *World Health Organization Environmental Burden of Disease Series*, no. 14. http://whqlibdoc.who.int/publications/2007/9789241595674_eng.pdf

Centers for Disease Control and Prevention. 2009. Smoke-free air. http://www.cdc.gov/Features/Smoke-FreeAir/

Central Intelligence Agency. 2008. The world factbook: United Arab Emirates. https://www.cia.gov/library/publications/the-world-factbook/geos/ae.html

Cohen, J.T., D.C. Bellinger, W.E. Connor, P.M. Kris-Etherton, R.S. Lawrence, D.A. Savitz, B.A. Shaywitz, S.M. Teutsch, and G.M. Gray. 2005. A quantitative risk-benefit analysis of changes in population fish consumption. *American Journal of Preventive Medicine* 29(4): 325–334.

Dehghan, M., N. Al Hamad, A. Yusufali, F. Nusrath, S. Yusuf, and A.T. Merchant. 2005. Development of a semi-quantitative food frequency questionnaire for use in United Arab Emirates and Kuwait based on local foods. *Nutrition Journal* 4(18). http://www.nutritionj.com/content/4/1/18

Driscoll, T., K. Steenland, D.I. Nelson, and J. Leigh. 2004. *Occupational airborne particulates: Assessing the environmental burden of disease at national and local levels*. Environmental burden of disease series, No. 7. Geneva: World Health Organization.

Driscoll, T., D.I. Nelson, K. Steenland, J. Leigh, M. Concha-Barrientos, M. Fingerhut, and A. Prüss-Üstün. 2005. The global burden of non-malignant respiratory disease due to occupational airborne exposures. *American Journal of Industrial Medicine* 48(6): 432–445.

Fleisher, J.M., D. Kay, R.L. Salmon, F. Jones, M.D. Wyer, and A.F. Godfree. 2006. Marine waters contaminated with domestic sewage: Nonenteric illnesses associated with bather exposure in the United Kingdom. *American Journal of Public Health* 86(9): 1228–1234.

Gomes, J., O. Lloyd, D.M. Revitt, and J.N. Norman. 1997. Erythrocyte cholinesterase activity levels in desert farm workers. *Occupational Medicine* 47(2): 90–94.

Gomes, J., O. Lloyd, and D.M. Revitt. 1999. The influence of personal protection, environmental hygiene and exposure to pesticides on the health of immigrant farm workers in a desert country. *International Archives of Occupational and Environmental Health* 72(1): 40–45.

Gomes, J., O. Lloyd, N.J. Norman, and P. Pahwa. 2001. Dust exposure and impairment of lung function at a small iron foundry in a rapidly developing country. *Occupational and Environmental Medicine* 58(10): 656–662.

Gomes, J., O. Lloyd, and N. Norman. 2002. The health of the workers in a rapidly developing country: Effects of occupational exposure to noise and heat. *Occupational Medicine* 52(3): 121–128.

Gorell, J.M., C.C. Johnson, B.A. Rybicki, E.L. Peterson, and R.J. Richardson. 1998. The risk of Parkinson's disease with exposure to pesticides, farming, well water, and rural living. *Neurology* 50: 1346–1350.

Intergovernmental Panel on Climate Change (IPCC). 2007. Summary for policymakers. In *Climate change 2007: The physical science basis. Contribution of working group I to the fourth assessment report of the Intergovernmental Panel on Climate Change*, ed. M.L. Parry, O.F. Canziani, J.P. Palutikof, P.J. van der Linden, and C.E. Hanson, 7–22. Cambridge: Cambridge University Press.

Jackson, S.A., and D. Loomis. 2002. Fatal occupational injuries in the North Carolina construction industry 1978–1994. *Applied Occupational and Environmental Hygiene* 17(1): 27–33.

Jaga, K., and C. Dharmani. 2005. The epidemiology of pesticide exposure and cancer: A review. *Reviews on Environmental Health* 20(1): 15–38.

Kaynakli, O., and M. Kilic. 2005. Investigation of indoor thermal comfort under transient conditions. *Building and Environment* 40(2): 165–174.

Kelly, R. 2007. Handling and managing asbestos in the United Arab Emirates. *Zawya.com*. http://www.zawya.com/story.cfm/sidZAWYA20070407054030

Kwong, M. 2008. Protecting the earth from ourselves. *The National*, July 31.

Lee, W.J., A. Blair, J.A. Hoppin, J.H. Lubin, J.A. Rusiecki, D.P. Sandler, M. Dosemeci, and M.C.R. Alamanja. 2004. Cancer incidence among pesticide applicators exposed to chlorpyrifos in the agricultural health study. *Journal of the National Cancer Institute* 96(23): 1781–1789.

Linaker, C., and J. Smedley. 2002. Respiratory illness in agricultural workers. *Occupational Medicine* 52(8): 451–459.

McMichael, A.J., D. Campbell-Lendrum, S. Kovats, S. Edwards, P. Wilkinson, T. Wilson, R. Nicholls, et al. 2004. Global climate change. In *Comparative quantification of health risks: Global and regional burden of disease attributable to selected major risk factors*, vol. 1, ed. M. Ezzati, A.D. Lopez, A. Rodgers, and C.J.L. Murray, 1543–1649. Geneva: World Health Organization.

Mokdad, A.H., J.S. Marks, D.F. Stroup, and J.L. Gerberding. 2004. Actual causes of death in the United States, 2000. *Journal of the American Medical Association* 291(10): 1238–1245.

Muskett, C., and T. Bohler. 2006. *Abu Dhabi noise strategy: Abu Dhabi air monitoring network supervision*. Kjeller: Norwegian Institute for Air Research.

National Cancer Institute. 1999. *Monograph 10: Health effects of exposure to environmental tobacco smoke*. Bethesda: National Cancer Institute, U.S. National Institutes of Health.

Ostro, B. 2004. *Outdoor air pollution: Assessing the environmental burden of disease at national and local levels*. Environmental burden of disease series, No. 5. Geneva: World Health Organization.

Pope, C.A. III, R.T. Burnett, M.J. Thun, E.E. Calle, D. Krewski, K. Ito, and G.D. Thurston. 2002. Lung cancer, cardiopulmonary mortality, and long-term exposure to fine particulate air pollution. *Journal of the American Medical Association* 287(9): 1132–1141.

Prieto, M.D., B. Lopez, J.A. Juanes, J.A. Revilla, J. Llorca, and M. Delgado-Rodríguez. 2001. Recreation in coastal waters: Health risks associated with bathing in sea water. *Journal of Epidemiology and Community Health* 55: 442–447.

Prüss-Üstün, A., R. Bos, F. Gore, and J. Bartram. 2008. *Safer water, better health: Costs, benefits and sustainability of interventions to protect and promote health*. Geneva: World Health Organization.

Scientific Advisory Committee on Nutrition. 2004. *Advice on fish consumption: Benefits & risks: U.K. Food Standards Agency and the Department of Health. Committee on Toxicity*. London: The Stationery Office.

Smith, A.W. 1998. The World Health Organization and the prevention of deafness and hearing impairment caused by noise. *Noise & Health* 1: 6–12.

Trivedi, C. 2002. The great smog of London. *BBC News Online*, December 5. http://news.bbc.co.uk/2/hi/uk_news/england/ 2545759.stm

UAE Interact. 2007. *UAE Yearbook 2007*. http://www.uaeinteract.com/uaeint_misc/pdf_2007/index.asp

UAE Ministry of Economy. 2006. Annual Economic and Social Report.

UAE Ministry of Economy. 2008a. *Annual social and economic report, United Arab Emirates, 2007*. UAE Ministry of Economy Central Statistics Department.

UAE Ministry of Economy. 2008b. United Arab Emirates: Population and vital statistics 2008.

UAE Ministry of Energy. 2006. *Initial national communication to the United Nations framework convention on climate change*. http://unfccc.int/resource/docs/natc/arenc1.pdf

UAE Ministry of Health. 2007. Annual report 2006. Ministry of Health, Preventive Medicine Sector, United Arab Emirates.

UAE Ministry of Public Works. 2006. UAE Federal E-Government Portal. http://www.mopw.gov.ae/gov/en/biz/industry/about.jsp

U.S. Environmental Protection Agency. 1994. *Indoor air pollution: An introduction for health professionals*. EPA-402-R-94-007. http://www.epa.gov/iaq/pubs/hpguide.html

U.S. Environmental Protection Agency. 1999. *Polychlorinated dibenzo-p-dioxins and related compounds update: Impact on fish advisories. EPA Fact Sheet, Office of Water*. EPA-823-F-99-015. http://www.epa.gov/waterscience/fish/files/dioxin.pdf

U.S. Environmental Protection Agency. 2004. *What you need to know about mercury in fish and shellfish. Fish Advisories*. EPA-823-F-04-009. http://www.epa.gov/waterscience/fish/advice/index.html

U.S. National Research Council. 2000. *Toxicological effects of methylmercury*, Committee on the toxicological effects of methylmercury, board on environmental studies and toxicology. Washington, DC: National Academy Press.

United Nations Population Division. 2000. *Total population by sex and sex ratio, by country*. http://www.un.org/spanish/esa/population/wpp2000at.pdf

Whitford, J. 2008. Environment agency–Abu Dhabi Ambient air quality monitoring network 2007 annual report

Woodruff, T.J., J.D. Parker, and K.C. Schoendorf. 2006. Fine particulate matter ($PM_{2.5}$) air pollution and selected causes of postneonatal infant mortality in California. *Environmental Health Perspectives* 114(5): 786–790.

World Health Organization (WHO). 2006. *The global burden of disease and risk factors*. Geneva: World Health Organization.

World Health Organization (WHO). 2007a. *Country profiles of environmental burden of disease: United Arab Emirates*. Geneva: World Health Organization.

World Health Organization (WHO). 2007b. *Environmental health criteria 238: Extremely low frequency fields*. Geneva: World Health Organization.
World Health Organization (WHO). 2008. *Promoting fruit and vegetable consumption around the world*. http://www.who.int/dietphysicalactivity/fruit/en/ index.html
World Health Organization (WHO). 2009. *Country profiles of environmental burden of disease: United Arab Emirates*. Geneva: World Health Organization. http://www.who.int/quantifying_ehimpacts/countryprofiles/en/.
Zhang, Z.W., J.X. Sun, S.Y. Chen, Y.Q. Wu, and F.S. He. 1991. Levels of exposure and biological monitoring of pyrethroids in spraymen. *British Journal of Industrial Medicine* 48(2): 82–86.

Appendix B: How to Use the *UAE Environmental Burden of Disease Model*

This appendix provides a brief overview of the software package *Analytica* and instructions on how to open the *UAE Environmental Burden of Disease Model*, view input and output variables, view model results, and edit variables.

Overview of the Analytica *Software*

Analytica is a powerful forecasting software package and a visual tool for creating, analyzing, and communicating decision models. You will need to have *Analytica* (version 4.1 or higher) installed on your computer to open, view, and edit the *UAE Environmental Burden of Disease Model.* It is available from Lumina Decision Systems at http://www.lumina.com. A free download is available for viewing and running models (*Analytica Player*); however, note that the normal *Analytica* software is needed should you wish to directly edit variables, nodes, or internode relationships. The *UAE Environmental Burden of Disease Model* itself is available by request from the Department of Environmental Sciences and Engineering at the University of North Carolina–Chapel Hill. For additional details, please contact the department directly.

Viewing and Editing the Model in Analytica

Opening the Model

The model can be opened in two ways: (1) by double-clicking the icon for the model file, or (2) by selecting the *File* pulldown menu at the top left of the *Analytica* application window, then selecting the *Open Model* tab. A directory browser dialog appears, from which you can locate the model file. Figure B.1 shows the top-level diagram window that appears when the model is opened.

The first seven nodes at the left of the diagram depict the seven modules that calculate the burden of disease attributable to seven environmental health risks in the UAE for this project. The last node depicts the module that contains definitions and descriptions of all the global variables in the model.

The burden of disease calculation results can be viewed directly from the top-level diagram window shown in Fig. B.1. To view the disease burden estimates from a risk module, click that module's corresponding *Calc* button. A *Result* window will appear, showing mean values of the burden of disease estimates in table format. It may take a few seconds or longer—depending on your computer's CPU model and speed—for the window to open as the model computes the results. The *Result* window can also display the statistics of an estimate, such as its minimum and

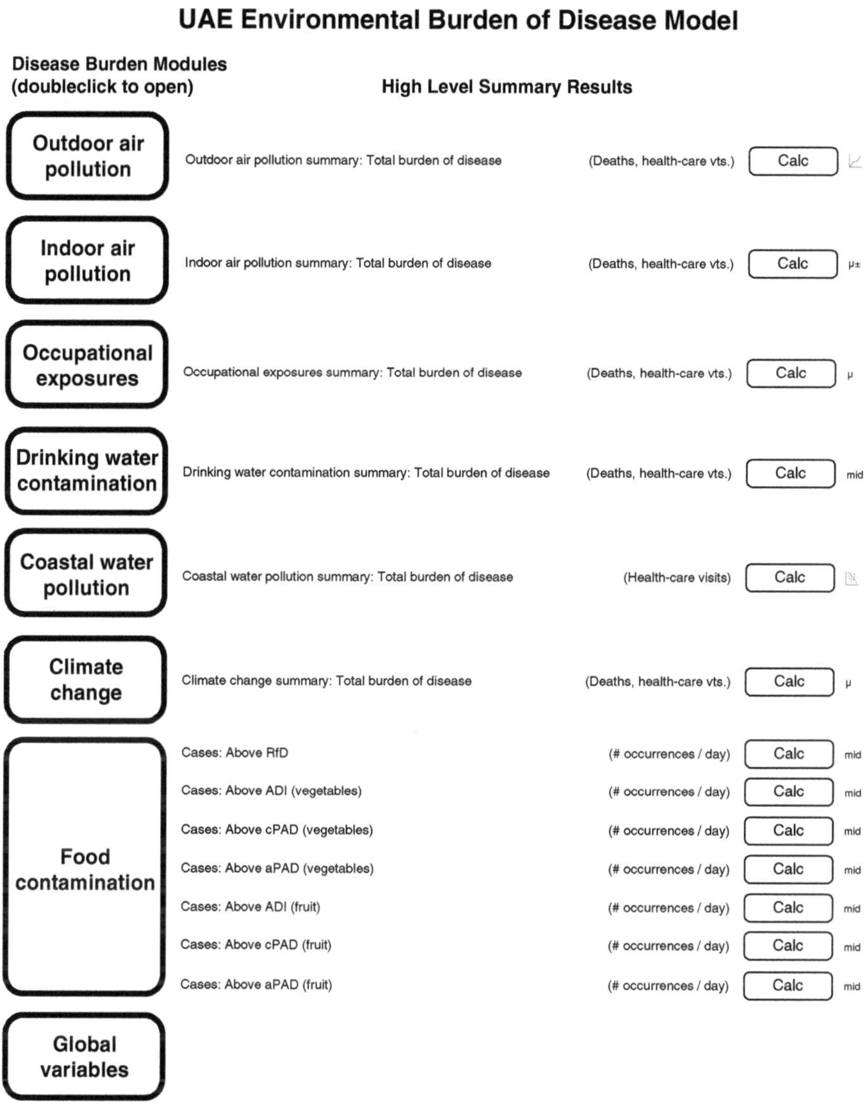

Fig. B.1 Top-level diagram window of the *UAE Environmental Burden of Disease Model*

maximum values, and results can alternatively be viewed in graphical format. (See the section below on viewing result tables and graphs for more details.)

Opening Modules

To see details of an environmental risk module, double-click the module node in the top-level diagram window. This opens the next level of detail of the model.

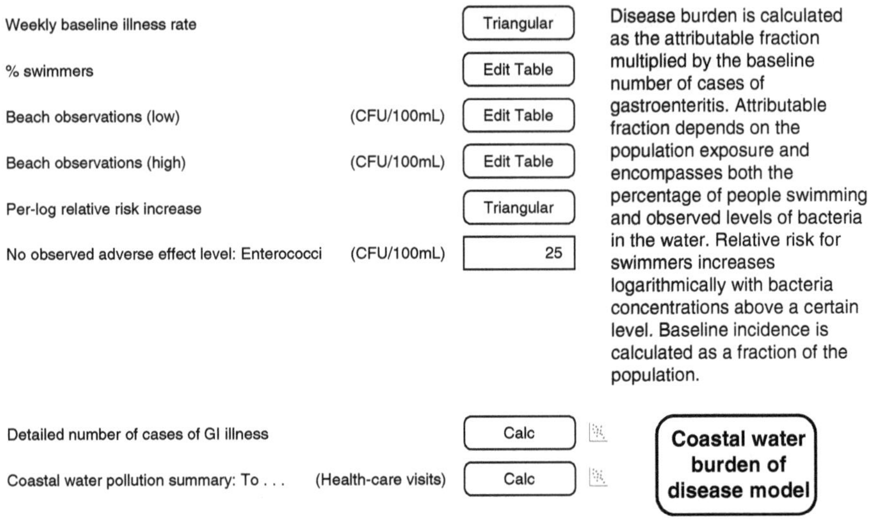

Fig. B.2 Top-level diagram window of the ***Coastal Water Contamination*** module

For example, double-clicking the ***Coastal Water Contamination*** module opens the diagram window seen in Fig. B.2.

This diagram window shows the input nodes (the first six rows on the left side) and the output nodes (the last two rows on the left side) of the ***Coastal Water Contamination*** module. It also contains a brief description of the method used to estimate the disease burden in this module (upper right) and a module node that contains the details of the model (lower right).

Most of the seven modules have submodules that depict the calculation of the disease burden attributed to various contaminants in a risk area. For example, the ***Indoor Air Pollution*** module has eight submodules, such as ***Benzene***, ***Formaldehyde***, etc. Double-clicking a submodule node opens a diagram window that depicts input and output nodes similar to the one shown in Fig. B.2.

Viewing a Variable

To view an input or output variable, double-click its node, which will open an *Object* window that shows the attributes of an object, commonly including its class (e.g., chance, objective, etc.), identifier (a unique name of up to 20 characters assigned to a variable), title, units, description, definition, value, inputs (if the variable has input variables), outputs (if the variable is an input variables of other variables), and others such as sources and references. Users can manage the display of attributes of a variable in its object window, for example, by creating user-defined attributes. This can be done by using the attributes dialog. To open the attributes dialog, select *Attributes* from the *Object* menu. You can display optional attributes, create new attributes, and rename an attribute.

For example, double-clicking the first input variable, ***Weekly baseline illness rate***, opens its *Object* window, which shows the variable's class (*chance*, which means the variable is uncertain and its definition contains a probability distribution), identifier (Weekly_baseline_illn), title (Weekly baseline illness rate), description, definition (triangular (8 m, 0.014, 0.024)), value (0.01506), outputs (Baseline_cases2), references, and units (percentage).

To view the value of a variable from the diagram window shown in Fig. B.2, click the colored button in a node. For example, in Fig. B.2 above, clicking the *Triangular* button next to ***Weekly baseline illness rate*** opens an *Object Finder* showing that the variable has a triangular distribution with a minimum value of 8 m (i.e., 8×10^{-3}; *Analytica* uses the suffix "m" for 10^{-3}), a mode value of 0.014, and a maximum value of 0.024. When the input variable consists of a table, the button is shown as *Edit Table* (rows 2, 3, and 4 in Fig. B.2). Click the button to open the table, and view the values in the table. More details of viewing and editing a multidimensional table are discussed below in the section on viewing and editing multidimensional tables. In the case that a variable has only a single value, such as the variable ***No observed adverse effect level***: enterococci in Fig. B.2, the value is shown directly in the button.

Viewing Result Tables and Graphs

To open a *Result* window for an output node, such as the node ***Detailed Number of Cases of GI Illness*** in Fig. B.2, simply click its *Calc* button. It may take a few seconds or longer for the window to open as the model computes the results. When you open the *Result* window, the default view is set to display the mean values of the burden of disease estimates as a table. You may use the controls in the upper-left corner of the *Result* window to change the view of the results. Specifically, click the 1.2 button to display the result as a table or the button to display the result as a graph.

Most of the variables defined in the *UAE Environmental Burden of Disease Model* have an uncertain or probabilistic value. As a result, the model computes the mean values and ranges of its output variables. The *Result* window offers seven uncertainty views, including the *mid* value (for variables that have a certain or deterministic value) and six ways to display a *prob* value (including *Mean Value*, *Statistics*, *Probability Bands*, *Probability Density*, *Cumulative Probability*, and *Sample*, for variables that have an uncertain or probabilistic value). Click the mid button in the upper-left corner of the *Result* window to open the uncertainty view menu and select an option.

Editing an Input Variable

The current version of the model estimates the disease burden attributable to the seven primary risk areas in the UAE based on the best available data we could find when the model was developed. In cases when UAE data were not available, the model relies on data from other countries. Therefore, the model should be updated as new information becomes available. A key advantage of this computer-based model is that the values of input variables can be easily refined to incorporate new information, and the model will generate new predictions accordingly.

To begin editing a variable, select the *Edit Tool* in the navigation toolbar at the top of an *Analytica* window by clicking it (when the file is opened, *Analytica* automatically selects the *Browse Tool*, which does not allow editing). The edit tool will be highlighted to show that it is selected, and the cursor will switch from to an arrow.

Use the steps to view a variable described above to open its *Object* window. Double-click the *Definition* attribute and then type in the new values of the input variable. If the input is in table format, double-click the *Edit Table* button to open the *Edit Table* window. Now you can edit the input values in cells. To replace the value in a cell, select the cell, click three times to get a cursor in the cell, and then type. Press *Enter* to accept the value and to select the next cell, or click in another cell. When you are done editing, click to accept all the changes you have made. If you close a table, it also accepts the changes unless you click to cancel changes you have made to the table since you opened it or last clicked. Click to open the *Result* window to see the new estimates based on updated input values.

You may also copy and paste the data in a table directly from or to a spreadsheet. For further assistance with editing a table in *Analytica*, please refer to the "Editing a table" section in the *Analytica User Guide* (Lumina Decision Systems 2008).

Multidimensional Tables or Graphs

As mentioned at the beginning, one of the key advantages of *Analytica* is that it allows users to create and manage multidimensional tables. Many of the input and result tables in the *UAE Environmental Burden of Disease Model* have more than two dimensions. For example, the input variable ***% Swimmers*** in the *Diagram* window in Fig. B.2 has three dimensions identified by three indexes (in *Analytica*, an index is the identifier of a dimension of a multidimensional table): Population Category, Emirate, and Month. The output variable ***Detailed Number of Cases of GI Illness*** also has the same three indexes. In an *Edit Table* window or a *Result* window, the index selection area is the top part of the window, in which there are two buttons. The first button shows which index goes down the rows, and the second shows which index goes across the columns. If the table has too many dimensions to display directly, the index selection area also shows the indexes that are currently not displayed in the table (these are called a slicer index in *Analytica*). For example, in the *Edit Table* window of the ***% Swimmers*** variable, if the two buttons are *Emirate* and *Month*, and the textbox next to the slicer index ***Population Category*** says *Tourist*, it means that the current table displays the values corresponding to each emirate and each month for the population category *Tourist*. Press for a popup menu from which you can select the other value of the slicer index: *Resident*. Click or to switch the slicer value between *Tourist* and *Resident*.

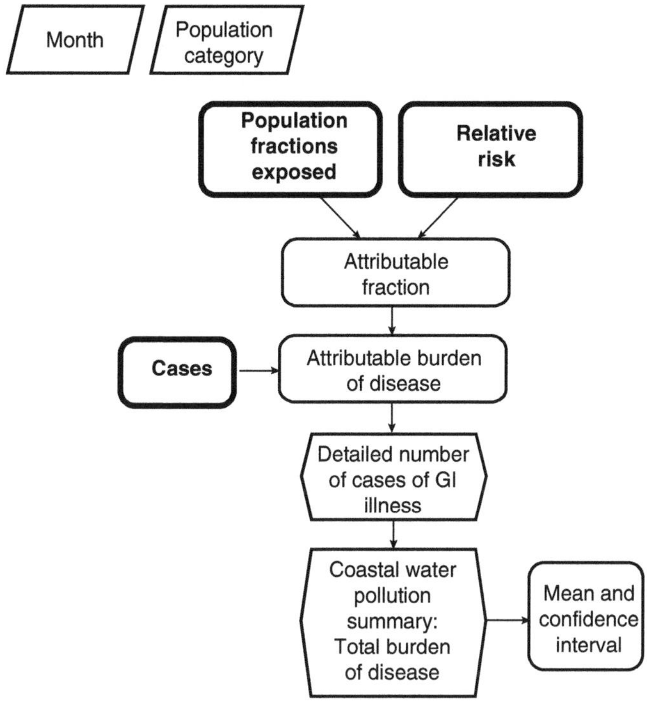

Fig. B.3 Influence diagram of the ***Coastal Water Contamination*** module

Viewing the Influence Diagrams of the Model

Each top-level diagram in the burden of disease model, such as the one shown in Fig. B.2, contains a module node that depicts the details of the model (e.g., the ***Recreational Water Burden of Disease*** module in Fig. B.2). Double-clicking the node opens an influence diagram showing details of the model as depicted in Fig. B.3.

An influence diagram in *Analytica* is a qualitative representation of a model that shows the variables and their dependencies. The rounded nodes with thick outline (i.e., the nodes ***Cases, Population Fractions Exposed***, and ***Relative Risk***) are modules that include separate influence diagrams. Double-clicking a module node will open its influence diagram. The parallelogram-shaped nodes (i.e., the nodes ***Month*** and ***Population Category***) depict the index variables that are used to define dimensions of tables in the model. The two hexagon-shaped nodes depict objective variables. Lastly, the remaining three rounded nodes depict general variables.

The arrows in an influence diagram that link two nodes are influence arrows. The *Analytica User Guide* states that "an influence arrow from variable A to variable B means that the value of A influences B because A is in the definition of B" (Lumina Decision Systems 2008). In this case (the influence diagram in Fig. B.3), the two arrows from ***Population Fractions Exposed*** and ***Relative Risk*** to ***Attributable***

Fraction mean that the fraction of each population exposed to coastal pollution through swimming and the relative risk of gastroenteritis for those exposed to microbial coastal pollution both affect the fraction of disease attributable to coastal water contamination. Therefore, when those values change, it changes the attributable fraction estimates.

The influence diagram shown in Fig. B.3 illustrates the essential qualitative structure of the model used to compute the disease burden attributable to an environmental health risk. The influence diagrams and details underlying the structure, including numbers and mathematical formulas, are described in the corresponding chapters of this report.

Overview of the *Global Variables* Module

The ***Global Variables*** module depicts the variables that are used by more than one module in the model. Some index variables, such as the ***Emirate*** index (a list of the seven emirates in the UAE), are used in most of the seven modules since we estimated the disease burden in each emirate wherever possible. This module also depicts the population data (the node ***Population***) and the baseline health data (the module node ***Baseline Health Endpoints***, including mortality and hospital visits due to various causes) used in different modules in the model. In order to reduce the complexity of the model, we did not link the input values of the population and baseline health endpoints in this module to the modules that use them as inputs. Instead, we list the data in the ***Global Variables*** module as references for users. Therefore, changing the values in the ***Global Variables*** module will not change the values used by the model to compute the results. Users need to go to an individual module to change input values.

Further Assistance

The purpose of these brief instructions is to get users started using the computer-based *UAE Environmental Burden of Disease Model*. Individual chapters in this report describe the details of the methods, assumptions, and data used in each module in the model. The *Analytica User Guide* also contains more detailed information about the software itself. It can be accessed from within the software by pressing the function key F1 at any time.

Reference

Lumina Decision Systems. 2008. *Analytica user guide*.

Appendix C: Model Input Parameters

Table C.1 Exposure variables, values, probability distribution functions (PDFs), and sources of data for each risk factor considered in this study

Outdoor air pollution									
		Mean annual average concentration by emirate[a]							
Exposure variable	Unit	Abu Dhabi	Dubai	Sharjah	Ajman	Umm Al Quwain	Ras Al Khaimah	Fujairah	Sources of data
Measurement-based approach									
PM_{10}	μg/m³	107.2	105.6	104.1	105.6	109.0	105.4	104.3	Detailed PM (PM_{10} and $PM_{2.5}$) data in 2007 were obtained from 10 monitoring stations, and detailed ozone data in 2007 were obtained from seven monitoring stations. All stations are located in Abu Dhabi emirate
$PM_{2.5}$	μg/m³	37.5	36.9	36.4	37.0	38.1	36.9	36.5	
Ozone: Daily avg.	ppb	29.1	28.5	28.5	28.5	28.5	28.5	28.5	
Ozone: Daily 1-h max	ppb	50.9	50.2	50.2	50.2	50.2	50.2	50.2	
Community Multiscale Air Quality Modeling System (CMAQ)-based approach									
PM_{10}	μg/m³	33.5	31.8	33.9	38.1	42.6	39.1	36.6	Concentration estimates from CMAQ modeling by the UNC air quality team
$PM_{2.5}$	μg/m³	14.0	14.0	14.1	14.2	14.3	13.9	13.5	
Ozone: Daily 1-h max	ppb	60.7	68.9	67.8	68.8	70.2	65.3	67.1	

[a]1,409 grid cells (resolution: 55 km²) were created across the UAE to represent the spatial variations of pollutant concentrations (the numbers of grid cells in each emirate: Abu Dhabi–1,164; Dubai–79; Sharjah–55; Ras al Khaimah–51; Umm al Quwain–18; Ajman–75; and Fujairah–37). Pollutant concentrations in each grid cell are lognormally distributed and characterized by a mean and a standard deviation. The mean concentration and standard deviation in each grid cell are estimated using two different approaches: measurement-based and CMAQ model-based

Indoor air pollution			
Exposure variable	Unit	PDF/exposure category	Sources of data
PM_{10}	μg/m³	Lognormal (92.8, 144.9)	18 studies[a] worldwide, including countries such as the United States, Belgium, Austria, India, Korea, and China
$PM_{2.5}$	μg/m³	Lognormal (30.6, 34.36)	26 studies[a] worldwide, including countries such as the United States, Belgium, Singapore, Austria, Italy, Greece, Switzerland, Finland, Czech Republic, France, and Mexico
Benzene	μg/m³	Lognormal (9.5, 9.46)	8 studies[a] worldwide, including countries such as the United States, the United Kingdom, China, Korea, and some European countries
Formaldehyde	μg/m³	Lognormal (47.4, 36.2)	12 studies[a] worldwide, including countries such as the United States, Canada, Japan, Riyadh, Saudi Arabia, China, Turkey, and some European countries
Radon	Bq/m³	Abu Dhabi city: lognormal (13.8, 6.6); Sharjah: triangular (8, 50.3, 164)	Measured radon concentration data from the emirates of Abu Dhabi and Sharjah were obtained for this analysis. The data from Abu Dhabi consisted of 202 measurements in 111 residential dwellings. The data from Sharjah were provided only as a mean, a minimum value, and a maximum value
Environmental tobacco smoke (ETS)		Exposed to secondhand smoke in a residential environment, or unexposed; uniform (0.15, 0.75)	
Mold		Exposed to mold in a residential environ ment, or unexposed; uniform (0.05, 0.5)	
Incense use		Exposed to incense use in a residential environment, or unexposed; citizens: uniform (0, 0.9); noncitizens: uniform (0, 0.5)	

[a]See Appendix D for details on mean concentration estimates

Occupational exposures		
Exposure variable	Exposure category	Sources of data[a]
Carcinogens[b] and leukemogens[c]	Background, low, and high. High exposure refers to exposures above the relevant U.S. Permissible Exposure Limit (PEL), and low exposure below it. The WHO estimates that in Eastern Mediterranean Region B countries, 50% of the workforce is exposed to carcinogens at high level and 50% at low level	Driscoll et al. (2004b). Estimated proportion of UAE workforce exposed to each carcinogen or leukemogen within each economic subsector was derived from the Carcinogen Exposure (CAREX) database (FIOH 2006)
Particulate matter	*Asthma:* Exposure and relative risk categorized by occupational group including administration, technical, sales, agriculture, mining, transportation, manufacturing, and services	Driscoll et al. (2004a)
	Chronic obstructive pulmonary disease: Background (includes combined proportions of workers in trade, finance, and services), low (agriculture, electricity, and transportation), medium/high (mining, manufacturing, and construction)	Korn et al. (1987)
	Asbestosis and silicosis: 100% attributable fraction	Driscoll et al. (2004a)
Noise	Moderately high (85–90 dB(A)) and high (>90 dB(A))	Concha-Barrientos et al. (2004)

[a]The proportion of the UAE workforce in each occupational group or economic subsector by gender was derived from the UAE Ministry of Economy
[b]Asbestos, arsenic, beryllium, cadmium, chromium, diesel fumes, nickel, and silica
[c]Benzene and ethylene oxide

Drinking water contamination			
Exposure variable	Unit	PDF/exposure category	Sources of data
Trihalomethanes (THMs)	μg/L	Normal (30.08, 13.57)	Statistical summary of drinking water samples taken in Kuwait (Al Mudhaf et al. 2009)
Microbial contamination		WHO defined water quality scenarios: Scenario II – Population having access to improved water supply and sanitation services in countries where more than 98% of the population is served by those services Scenario Vb – Population having access to improved water supply but not served with improved sanitation in countries which are not extensively covered by those services (less than 98% coverage), and where water supply is not likely to be routinely controlled	Fewtrell et al. (2007)

Coastal water contamination										
			Range of 12-month observed beach concentrations							
Exposure variable	Unit	Exposure category	Abu Dhabi	Dubai	Sharjah	Ajman	Umm Al Quwain	Ras Al Khaimah	Fujairah	Sources of data
Enterococci	CFU/100 mL	Low	0–5	0–10	0–5	0–5	0–5	0–5	0–5	Observed 2006 concentrations, Al Raha Beach, EAD
		High	0–250	0–500	0–250	0–250	0–250	0–250	0–250	Observed 2006 concentrations, Abu Dhabi public beach, EAD

Produce and seafood contamination			
Exposure variable	Unit	PDF/range	Source of data
Methylmercury in seafood	mg/kg	Triangular (0.033, 0.068, 0.098)	Kosanovic et al. (2007), data from 3 sampling points on UAE coast

(continued)

(continued)

Produce and seafood contamination				
Exposure variable	Unit	PDF/range		Source of data
Pesticide residue on vegetables and fruit	mg/kg	Vegetables	Fruit	The range refers to maximum residue levels (MRL) on various vegetables or fruit. Strict adherence to MRLs is assumed
Acephate		0.02–50	0.02	
Bifenthrin		0–2	0–3	
Bromopropylate		0–1	0–2	
Chlorfenvinphos		0	0	MRL data from Codex Alimentarius Commission for all pesticides except those listed below:
Chlorpyrifos		0–20	0–3	
Dichlorvos		0	0	• MRL data from EC (European Commission) and ESFA (European Food Safety Authority) for acephate, bifenthrin, bromopropylate, dimethoate, quintozene, cyhalothrin, methamidophos, procymidone, vinclozlin, methadathion, chlorpyrifos
Dimethoate		0–2	0–5	
Ethion		0	0	
Fenitrothion		0	0–0.5	• MRL data from USDA/FAS (MRLDatabase.com) for malathion
Cyhalothrin		0–1	0–0.2	
Malathion		0–8	0–8	
Methamidophos		0–2	0.01–0.5	
Methidathion		0.02–0.1	0–5	
Phenthoate		0	0	
Pirimicarb		0–20	0–3	
Procymidone		0–50	0–5	
Quintozene		0–0.1	0.02	
Vinclozolin		0–30	0.05–10	

Table C.2 Relative risk and sources of data for each risk factor considered in this study

Exposure variable	Unit	Health endpoint	Relative risk	Sources of data
Outdoor air pollution				
PM_{10}	μg/m³	All-cause mortality due to short-term exposure	Percentage increase in incidence per 1 μg/m³: Normal (0.08, 0.001)	Mortality: Ostro (2004)
		Respiratory mortality in children <5 due to short-term exposure	Normal (0.166, 0.007)	Morbidity: Ostro and Chestnut (1998)
		Respiratory health-care facility visits due to short-term exposure	Normal (0.084, 0.0019)	
		Cardiovascular disease health-care facility visits due to short-term exposure	Normal (0.03, 0.0031)	
$PM_{2.5}$	μg/m³	All-cause mortality due to long-term exposure in adults > 30	Percentage increase in incidence per 1 μg/m³: Normal (0.6, 0.2)	Pope et al. (2002)
		Cardiopulmonary mortality due to long-term exposure in adults >30	Normal (0.9, 0.3)	
		Lung cancer mortality due to long-term exposure in adults > 30	Normal (1.4, 0.5)	
Ground-level ozone	ppb	Total non-accidental mortality due to short-term exposure	Percentage increase in incidence per 1 μg/m³: Normal (0.052, 0.0128)	Bell et al. (2004)
		Cardiovascular and respiratory mortality due to short-term exposure	Normal (0.064, 0.017)	Bell et al. (2004)
		Respiratory mortality in adults >30 due to long-term exposure	Normal (0.4, 0.1)	Jerrett et al. (2009)

(continued)

Table C.2 (continued)

Exposure variable	Unit	Health endpoint	Relative risk	Sources of data
		Respiratory health-care facility visits due to short-term exposure	Normal (0.34, 0.06)	Levy et al. (2001)
Indoor air pollution				
PM_{10}	μg/m³	Asthma in children < 6	Lognormal (mean = 1.06, standard deviation = 0.0255) per 10 μg/m³	McCormack et al. (2009)
$PM_{2.5}$	μg/m³	Asthma in children < 6	Lognormal (mean = 1.03, standard deviation = 0.0204) per 10 μg/m³	McCormack et al. (2009)
Benzene	μg/m³	Asthma in children < 6	Lognormal (mean = 1.085, standard deviation = 0.0141) per 10 μg/m³	Rumchev et al. (2004)
Formaldehyde	μg/m³	Asthma in children < 6	Lognormal (mean = 1.003, standard deviation = 0.0005) per 10 μg/m³	Rumchev et al. (2002)
Radon	Bq/m³	Lung cancer mortality	Lognormal (1.0008, 1.000255) per Bq/m³	Darby et al. (2006)
		Lung cancer		Darby et al. (2006)
Environmental tobacco smoke (ETS)		Cardiovascular disease mortality	Male: Normal (1.25, 0.112) Female: Normal (1.35, 0.087)	Hill et al. (2007)
		Lung cancer mortality	Male: Normal (1.1, 0.255) Female: Normal (1.2, 0.204)	Cardenas et al. (1997)
		Lung cancer	Normal (1.25, 0.051)	Boffetta (2002)
		Cardiovascular disease	Normal (1.25, 0.041)	He and Whelton (1999)
		Lower respiratory tract infection in children <6	Normal (1.57, 0.148)	Li et al. (1999)
		Asthma in children <18	Normal (1.48, 0.816)	Vork et al. (2007)
		Leukemia	Normal (2.28, 0.576)	Kasim et al. (2005)
Bio-aerosols (mold)		Asthma in adultsx	Normal (1.54, 0.2704)	Jaakkola et al. (2002)
		Asthma in children 6–12	Normal (1.35, 0.0765)	Antova et al. (2008)
Incense use		Respiratory tract cancer mortality and morbidity	Normal (1.8, 0.306)	Friborg et al. (2008)

Occupational exposures			
Occupational carcinogens and leukemogens	Lung cancer mortality and morbidity	*Background*: 1 *Low exposure*: Normal (1.21, 0.0148) *High exposure*: Normal (1.77, 0.0316)	Driscoll et al. (2004b); (2005); Steenland et al. (1996, 2003); Nurminen and Karjalainen (2001)
	Leukemia mortality and morbidity	*Background*: 1 *Low exposure*: Normal (1.9, 0.15) *High exposure*: Normal (4, 0.2)	Steenland et al. (2003); IARC (1997); Lynge et al. (1997); Driscoll et al. (2004b)
	Malignant mesothelioma	A relative risk for malignant mesothelioma in exposed versus non-exposed population is not available since mesothelioma does not occur in populations that have not been exposed to asbestos. It has been estimated in the literature that 90% of mesothelioma in males and 25% in females is related to occupational exposure to asbestos.	Nurminen and Karjalainen (2001); Steenland et al. (2003)
Particulate matter	Asthma mortality and morbidity	Occupation group *Administration* Male: 1 Female:1 *Technical* Male: Normal (1.05, 0.0357) Female: Normal (1.06, 0.0204) *Sales* Male: Normal (1.1, 0.0663) Female: Normal (1.13, 0.0255)	Karjalainen et al. (2001, 2002); Kogevinas et al. (1999); Driscoll et al. (2004a)

(continued)

Table C.2 (continued)

Exposure variable	Unit	Health endpoint	Relative risk	Sources of data
			Agriculture Male: Normal (1.41, 0.3112) Female: Normal (1.41, 0.3112) *Mining* Male: Normal (1.95, 0.2296) Female: Normal (1, 1.5408) *Transportation* Male: Normal (1.31, 0.0459) Female: Normal (1.22, 0.0459) *Manufacturing* Male: Normal (1.56, 0.0459) Female: Normal (1.33, 0.0306) *Services* Male: Normal (1.53, 0.1226) Female: Normal (1.41, 0.0255)	
		Chronic obstructive pulmonary disease (COPD) mortality and morbidity	*Background*: 1 *Low exposure* Male: 1.2 Female: 1.1 *High exposure* Male: 1.6 Female: 1.4	Korn et al. (1987); Driscoll et al. (2004a)
		Asbestosis and silicosis	100% attributable fraction (asbestosis and silicosis are almost solely caused by occupational exposure to asbestos and silica, respectively)	Driscoll et al. (2004a)

Noise	dB(A)	Hearing loss	Age groups *15–29 years* Moderately high (85–90 dB(A)): 1.96 High (>90 dB(A)): 7.96 *30–44 years* Moderately high: 2.24 High: 5.62 *45–59 years* Moderately high: 1.91 High: 3.83 *60–69 years* Moderately high: 1.66 High: 2.82 *70–79 years* Moderately high: 1.66 High: 2.82	Concha-Barrientos et al. (2004)
Climate change				
Heat exposure		Cardiovascular disease mortality and morbidity	Climate change scenario (Projection year) *UnmitB* 2007: Triangular (1, 1.001, 1.003) 2020: Triangular (1, 1.003, 1.005) 2030: Triangular (1, 1.003, 1.007) *S750* 2007: Triangular (1, 1.001, 1.002) 2020: Triangular (1, 1.002, 1.004) 2030: Triangular (1, 1.002, 1.004) *S550* 2007: Triangular (1, 1.001, 1.001) 2020: Triangular (1, 1.001, 1.003) 2030: Triangular (1, 1.002, 1.004)	McMichael et al. (2004)

(continued)

Table C.2 (continued)

Exposure variable	Unit	Health endpoint	Relative risk	Sources of data
Drinking water contamination				
Trihalomethanes (THMs)	μg/L	Bladder cancer (mortality and morbidity)	*Relative risk method: Drink chlorinated water versus not*	Morris et al. (1992)
			Male: Normal (1.24, 0.1378)	
			Female: Normal (1.17, 0.07143)	
			Relative risk method: Increasing RR with increasing exposure	Villanueva et al. (2003)
			Male:	
			(Exposure levels)	
			0: 1	
			0–1: 1	
			1–5: Normal (1.1, 0.0918)	
			5–25: Normal (1.26, 0.1071)	
			25–50: Normal (1.25, 0.1071)	
			>50: Normal (1.44, 0.1224)	
			Female: 1	
		Rectal cancer	Male: Normal (1.24, 0.1939)	
			Female: Normal (1.1, 0.102)	
		Colon cancer	Male: Normal (1.09, 0.1429)	
			Female: Normal (1.19, 0.1327)	
Microbial contamination		Gastroenteritis mortality and morbidity	Scenario II: Uniform (1, 4)	Fewtrell et al. (2007)
			Scenario Vb: Uniform (7.2, 10.2)	
Coastal water contamination				
Enterococci	CFU/100 mL	Gastroenteritis	Increase in relative risk for every log increase in enterococci concentration: Triangular (1.00, 1.34, 1.75)	Wade et al. (2003)

Produce and seafood contamination					
Methylmercury in seafood	mg/kg/day	Reference Dose (RfD): 0.0001			U.S. EPA (2002)
Pesticide residue on vegetables and fruit		Acceptable daily intake (ADI)	Chronic population adjusted dose (cPAD)	Acute population adjusted dose (aPAD)	ADI from WHO; cPAD, aPAD from U.S. EPA (1997, 2009)
Acephate		0.005	0.0012	0.005	
Bifenthrin		0.015	0.015	0.015	
Bromopropylate		0.030	0.2	0.2	
Chlorfenvinphos		0.002	0.0007	0.0007	
Chlorpyrifos		0.010	0.0003	0.005	
Dichlorvos		0.004	0.0005	0.008	
Dimethoate		0.010	0.0022	0.013	
Ethion		0.002	0.0005	0.001	
Fenitrothion		0.005	0.0013	0.130	
Cyhalothrin		0.020	0.001	0.005	
Malathion		0.020	0.07	0.140	
Methamidophos		0.004	0.0001	0.0001	
Methidathion		0.001	0.0015	0.0015	
Phenthoate		0.003	0.003	0.003	
Pirimicarb		0.020	0.02	0.020	
Procymidone		0.100	0.035	0.035	
Quintozene		0.007	0.003	0.003	
Vinclozolin		0.070	0.0012	0.006	

References

Al Mudhaf, A.F., F.A. Alsharifi, and A.-I. Abu Shady. 2009. A survey of organic contaminants in household and bottled drinking waters in Kuwait. *Science of the Total Environment* 407: 1658–1668.

Antova, T., S. Pattenden, B. Brunekreef, J. Heinrich, P. Rudnai, F. Forastiere, H. Luttmann-Gibson, et al. 2008. Exposure to indoor mould and children's respiratory health in the PATY study. *Journal of Epidemiology and Community Health* 62(8): 708–714.

Bell, M.L., A. McDermott, S.L. Zeger, J.M. Samet, and F. Dominici. 2004. Ozone and short-term mortality in 95 U.S. urban communities, 1987–2000. *Journal of the American Medical Association* 292(19): 2372–2378.

Boffetta, P. 2002. Involuntary smoking and lung cancer. *Scandinavian Journal of Work, Environment and Health* 28(Suppl 2): 30–40.

Cardenas, V.M., M.J. Thun, H. Austin, C.A. Lally, W.S. Clark, R.S. Greenberg, and C.W. Heath Jr. 1997. Environmental tobacco smoke and lung cancer mortality in the American Cancer Society's cancer prevention study II. *Cancer Causes & Control* 8(1): 57–64.

Concha-Barrientos, M., D. Campbell-Lendrum, and K. Steenland. 2004. *Occupational noise: Assessing the burden of disease from work-related hearing impairment at national and local levels*, Environmental Burden of Disease Series, no. 9. Geneva: World Health Organization.

Darby, S., D. Hill, H. Deo, A. Auvinen, J.M. Barros-Dios, H. Baysson, F. Bochicchio, et al. 2006. Residential radon and lung cancer: Detailed results of a collaborative analysis of individual data on 7,148 persons with lung cancer and 14,208 persons without lung cancer from 13 epidemiologic studies in Europe. *Scandinavian Journal of Work, Environment and Health* 32(Supplement 1): 1–83.

Driscoll, T., K. Steenland, D. Imel Nelson, and J. Leigh. 2004a. *Occupational airborne particulates: Assessing the environmental burden of disease at national and local levels*. Environmental burden of disease series, No. 7. Geneva: World Health Organization.

Driscoll, T., K. Steenland, A. Prüss-Üstün, D. Imel Nelson, and J. Leigh. 2004b. *Occupational carcinogens: Assessing the environmental burden of disease at national and local levels.* Environmental burden of disease series, No. 6. Geneva: World Health Organization.

Driscoll, T., D.I. Nelson, K. Steenland, J. Leigh, M. Concha-Barrientos, M. Fingerhut, and A. Prüss-Üstün. 2005. The global burden of disease due to occupational carcinogens. *American Journal of Industrial Medicine* 48(6): 419–431.

Fewtrell, L., A. Prüss-Üstün, R. Bos, F. Gore, and J. Bartram. 2007. *Water, sanitation, and hygiene: Quantifying the health impact at national and local levels in countries with incomplete water supply and sanitation coverage*, Environmental Burden of Disease Series, no. 15. Geneva: World Health Organization.

FIOH. 2006. *CAREX.* International information system on occupational exposure to carcinogens. http://www.ttl.fi/Internet/English/Organization/Collaboration/Carex/

Friborg, J.T., J.M. Yuan, R. Wang, W.P. Koh, H.P. Lee, and M.C. Yu. 2008. Incense use and respiratory tract carcinomas: A prospective cohort study. *Cancer* 113(7): 1676–1684.

He, J., and P.K. Whelton. 1999. Passive cigarette smoking increases risk of coronary heart disease. *European Heart Journal* 20(24): 1764–1765.

Hill, S.E., T. Blakely, I. Kawachi, and A. Woodward. 2007. Mortality among lifelong nonsmokers exposed to secondhand smoke at home: Cohort data and sensitivity analyses. *American Journal of Epidemiology* 165(5): 530–540.

Jaakkola, M.S., H. Nordman, R. Piipari, J. Uitti, J. Laitinen, A. Karjalainen, P. Hahtola, and J.J. Jaakkola. 2002. Indoor dampness and molds and development of adult-onset asthma: A population-based incident case-control study. *Environmental Health Perspectives* 110(5): 543–547.

Jerrett, M., R.T. Burnett, C.A. Pope III, K. Ito, G. Thurston, D. Krewski, Y. Shi, E. Calle, and M. Thun. 2009. Long-term ozone exposure and mortality. *The New England Journal of Medicine* 360(11): 1085–1095.

Karjalainen, A., K. Kurppa, R. Martikainen, T. Klaukka, and J. Karjalainen. 2001. Work is related to a substantial portion of adult-onset asthma incidence in the Finnish population. *American Journal of Respiratory and Critical Care Medicine* 164(4): 565–568.

Karjalainen, A., K. Kurppa, R. Martikainen, J. Karjalainen, and T. Klaukka. 2002. Exploration of asthma risk by occupation—extended analysis of an incidence study of the Finnish population. *Scandinavian Journal of Work, Environment and Health* 28(1): 49–57.

Kasim, K., P. Levallois, B. Abdous, P. Auger, and K.C. Johnson. 2005. Environmental tobacco smoke and risk of adult leukemia. *Epidemiology* 16(5): 672–680.

Kogevinas, M., J.M. Anto, J. Sunyer, A. Tobias, H. Kromhout, and P. Burney. 1999. Occupational asthma in Europe and other industrialised areas: a population-based study. European Community Respiratory Health Survey Study Group. *Lancet* 353(9166): 1750–1754.

Korn, R.J., D.W. Dockery, F.E. Speizer, J.H. Ware, and B.G. Ferris Jr. 1987. Occupational exposures and chronic respiratory symptoms: A population-based study. *American Review of Respiratory Disease* 136(2): 298–304.

Kosanovic, M., M.Y. Hasan, D. Subramanian, A.A.F. Al Ahbabi, O.A.A. Al Kathiri, and E.M.A.A. Aleassa. 2007. Influence of urbanization of the western coast of the United Arab Emirates on trace metal content in muscle and liver of wild Red-spot Emperor (*Lethrinus lentjan*). *Food and Chemical Toxicology* 45(11): 2261–2266.

Levy, J.I., T.J. Carrothers, J.T. Tuomisto, J.K. Hammitt, and J.S. Evans. 2001. Assessing the public health benefits of reduced ozone concentrations. *Environmental Health Perspectives* 109(21): 9–20.

Li, J.S., J.K. Peat, W. Xuan, and G. Berry. 1999. Meta-analysis on the association between environmental tobacco smoke (ETS) exposure and the prevalence of lower respiratory tract infection in early childhood. *Pediatric Pulmonology* 27(1): 5–13.

McCormack, M.C., P.N. Breysse, E.C. Matsui, N.N. Hansel, D. Williams, J. Curtin-Brosnan, P. Eggleston, and G.B. Diette. 2009. In-home particle concentrations and childhood asthma morbidity. *Environmental Health Perspectives* 117(2): 294–298.

McMichael, A.J., D. Campbell-Lendrum, S. Kovats, S. Edwards, P. Wilkinson, T. Wilson, R. Nicholls, et al. 2004. Global climate change. In *Comparative quantification of health risks: Global and regional burden of disease attributable to selected major risk factors*, vol. 1, ed. M. Ezzati, A.D. Lopez, A. Rodgers, and C.J.L. Murray, 1543–1649. Geneva: World Health Organization.

Morris, R.D., A. Audet, I.F. Angelillo, T.C. Chalmers, and F. Mosteller. 1992. Chlorination, chlorination by-products, and cancer: A meta-analysis. *American Journal of Public Health* 82(7): 955–963.

Nurminen, M., and A. Karjalainen. 2001. Epidemiologic estimate of the proportion of fatalities related to occupational factors in Finland. *Scandinavian Journal of Work, Environment and Health* 27(3): 161–213.

Ostro, B. 2004. *Outdoor air pollution: Assessing the environmental burden of disease at national and local levels*, Environmental burden of disease series, no. 5. Geneva: World Health Organization.

Ostro, B.D., and L. Chestnut. 1998. Assessing the health benefits of reducing particulate matter air pollution in the United States. *Environmental Research (Section A)* 76: 94–106.

Pope, C.A., III, R.T. Burnett, M.J. Thun, E.E. Calle, D. Krewski, K. Ito, and G.D. Thurston. 2002. Lung cancer, cardiopulmonary mortality, and long-term exposure to fine particulate air pollution. *Journal of the American Medical Association* 287(9): 1132–1141.

Rumchev, K.B., J.T. Spickett, M.K. Bulsara, M.R. Phillips, and S.M. Stick. 2002. Domestic exposure to formaldehyde significantly increases the risk of asthma in young children. *European Respiratory Journal* 20(2): 403–408.

Rumchev, K.B., J.T. Spickett, M.K. Bulsara, M.R. Phillips, and S.M. Stick. 2004. Association of domestic exposure to volatile organic compounds with asthma in young children. *Thorax* 59(9): 746–751.

Steenland, K., D. Loomis, C. Shy, and N. Simonsen. 1996. Review of occupational lung carcinogens. *American Journal of Industrial Medicine* 29(5): 474–490.

Steenland, K., C. Burnett, N. Lalich, E. Ward, and J. Hurrell. 2003. Dying for work: The magnitude of U.S. mortality from selected causes of death associated with occupation. *American Journal of Industrial Medicine* 43(5): 461–482.

U.S. Environmental Protection Agency (EPA). 1997. *Reference dose tracking report*. Office of Pesticide Programs. http://npic.orst.edu/tracking.htm

U.S. Environmental Protection Agency (EPA). 2002. *Integrated risk information system: Methylmercury* (CASRN 22967-92-6). http://www.epa.gov/iris/subst/0073.htm

U.S. Environmental Protection Agency (EPA). 2009. *Pesticide reregistration status*. http://www.epa.gov/opp00001/reregistration/status.htm

Villanueva, C.M., F. Fernández, N. Malats, J.O. Grimalt, and M. Kogevinas. 2003. Meta-analysis of studies on individual consumption of chlorinated drinking water and bladder cancer. *Journal of Epidemiology and Community Health* 57: 166–173.

Vork, K.L., R.L. Broadwin, and R.J. Blaisdell. 2007. Developing asthma in childhood from exposure to secondhand tobacco smoke: Insights from a meta-regression. *Environmental Health Perspectives* 115(10): 1394–1400.

Wade, T.J., N. Pai, J.N.S. Eisenberg, and J.M. Colford. 2003. Do U.S. Environmental Protection Agency water quality guidelines for recreational waters prevent gastrointestinal illness? A systematic review and meta-analysis. *Environmental Health Perspectives* 111(8): 1102–1109.

Appendix D: Literature Review for Indoor Air Concentrations

At the time Chap. 5 was written, observational data on indoor air quality only were available for radon. Hence, for the other indoor air pollutants evaluated in the chapter, means and standard deviations of indoor air contaminant observations were collected from an exhaustive literature review (Tables D.1, D.2, D.3, D.4, and D.5, below), then used in the *UAE Environmental Burden of Disease Model* to simulate potential indoor concentrations in the UAE (see Appendix C). This appendix lists the studies included in the review.

Table D.1 PM_{10} input concentration data

References	Location	Notes	Mean (μg/m³)	Standard deviation (μg/m³)
Leaderer et al. (1999)	Virginia, connecticut	Air-conditioned homes, summer	28.9	18.7
		Non-air-conditioned homes, summer	33.3	14.2
		Kerosene heater, winter	44.36	30.37
		No kerosene heater, winter	25.71	21.12
Stranger et al. (2009)	Belgium		28.9	27.2
Breysse et al. (2005)	Baltimore	72-h samples from bedroom	56.5	40.7
		Smoking households	71.2	46.7
		Nonsmoking households	37.7	18.8
Simons et al. (2007)	Baltimore	Suburbs	23	17
		Inner city	57	41
Komarnicki (2005)	Vienna	Indoor, day	61	32
		Indoor, night	45	23
		Outdoor, day	34	11
		Outdoor, night	37	23
Williams et al. (2003)	North Carolina		27.7	19.6
Liu et al. (2003)	Seattle		14.1	6.6
			12.6	7.8
			19.4	11.1
			16.2	11.3
Suh (2003)	Los Angeles	Winter	30.6	21.2
		Summer to fall	29	14.7
Keeler et al. (2002)	Detroit		52.2	30.6
Williams et al. (2000)	Baltimore	Summer	13.5	5.9
Long et al. (2000)	Boston		19.4	12.7
Abt et al. (2000)	Boston		19.6	16.1
Rojas-Bracho et al. (2000)	Boston	Winter	37.3	23.2
		Summer	28.3	25.4

(continued)

Table D.1 (continued)

References	Location	Notes	Mean (μg/m³)	Standard deviation (μg/m³)
Khillare et al. (2004)	India	Summer	178.8	14.91
		Summer	171.19	12.32
Jo and Lee (2006)	Korea	Winter, lower floor	35	33
		Summer, lower floor	36	17
		Winter, higher floor	36	99
		Summer, higher floor	33	19
Houyin et al. (2005)	Beijing	Smoker's home	122	
			97	
Cheng et al. (2007)	Guiyang City, China	Smoking	130	35.6
		Nonsmoking	106	24.4
Lung et al. (2003)	Taiwan	Incense burning, unventilated room	723	
			601	
			385	
		Incense burning, ventilated room	178	
			136	
			119	

Table D.2 $PM_{2.5}$ input concentration data

References	Location	Notes	Mean (μg/m³)	Standard deviation (μg/m³)
Leaderer et al. (1999)	Virginia and Connecticut	Air-conditioned homes, summer	18.7	13.2
		Non-air-conditioned homes, summer	21.1	7.5
		Kerosene heater, winter	29.97	23.58
		No kerosene heater, winter	17.43	23.63
Stranger et al. (2009)	Belgium	Houses 1–15	36	13
		Houses 16–19	41	31
Baxter et al. (2007)	Urban Boston	Indoors	20.3	12.5
Breysse et al. (2005)	Baltimore	μg/m³	45.1	37.5
Simons et al. (2007)	Baltimore	Suburbs	12	8.6
		Inner city	45	37
Meng et al. (2005)	California		16.2	9.4
	New Jersey		20.1	15.5
	Texas		17.1	12.7
See et al. (2007)	Singapore	Cooking (μg/m³)	38.8	14.9
		Indoors	18.2	5.2
		Incense	142.6	16.3
		Cigarettes	227.2	37.3
Sarnat et al. (2002)	Boston	Spring–summer	12.5	7.1
		Fall–winter	7.2	2.5

(continued)

Table D.2 (continued)

References	Location	Notes	Mean (μg/m³)	Standard deviation (μg/m³)
Komarnicki (2005)	Vienna	Day	44	21
		Night	44	22
Wallace et al. (2006)	North Carolina	Indoors	19.4	16
Allen et al. (2007)	Seattle		8.25	2.31
Suh and Koutrakis (2004)	Los Angeles		17.6	11.4
Williams et al. (2003)	North Carolina		19.3	8.4
Liu et al. (2003)	Seattle	COPD patients	8.5	5.1
		Healthy	7.4	4.8
		Asthmatic	9.2	6
		Coronary	9.5	6.8
Suh (2003)	Los Angeles	Winter	16.9	11.7
	COPD patients	Summer–fall	18.1	11.1
Keeler et al. (2002)	Detroit		34.4	21.7
Long et al. (2000)	Boston		11.9	9.8
Abt et al. (2000)	Boston		13.9	15.2
Lachenmyer and Hidy (2000)	Birmingham, Alabama	Summer	16.1	9.6
		Winter	11.2	5.4
Rojas-Bracho et al. (2000)	Boston	Winter	17.2	13
		Summer	17.7	14.9
Wallace et al. (2003)	Seven U.S. cities		27.7	35.9
Simoni et al. (2004)	Italy	Urban winter	67	38
		Urban summer	47	20
		Rural winter	76	35
		Rural summer	50	21
Gotschi et al. (2002)	Athens, Greece		35.6	29.4
	Switzerland		21	16.7
	Finland		9.5	6.1
	Czech Republic		34.4	28.7
Zmirou et al. (2002)	France	Grenoble	28.7	26.3
		Paris	25.3	18.7
		Nice	20	10.2
Brown et al. (2008)	Boston	Winter	10.1	4.6
		Summer	12	7.3
Cortez-Lugo et al. (2008)	Mexico City	Winter	35	20
		Spring	31	15
		Summer	26	13
		Fall	26	13

Table D.3 Benzene input concentration data

References	Location	Notes	Mean (μg/m³)	Standard deviation (μg/m³)
Kim et al. (2001)	Birmingham, U.K.		13.9	13.8
Wallace (1996)	Virginia (winter 1987)	Day, living room	9.9	8.4
		Day, kitchen	11	16
		Night, kitchen	15	13.2
	Virginia (summer 1987)	Day, living room	6.5	5.7
		Day, kitchen	5.5	4.9
		Night, kitchen	6.5	7.3
	Virginia (summer 1990)	Day	13	21.9
		Night	18	28.5
	Virginia (winter 1991)	Day	26	23.7
		Night	24	23.9
	Virginia (spring 1990)	24-h	4.7	7.6
Sax et al. (2004)	New York	Winter	5.3	4.7
		Summer	1.7	1.2
	Los Angeles	Winter	4.9	2.8
		Fall	15	6.2
Van Winkle and Scheff (2001)	Chicago		4.1	4.8
Guo et al. (2003)	Hong Kong		4.4	2.5
Son et al. (2003)	Seoul, Korea		43.7	36.9
	Asan, Korea		20.3	12.6
Batterman et al. (2007)	Michigan		2	1.9
Bruinen de Bruin et al. (2008)	European Union cities		3.2	2.1
	Helsinki		1.7	
	Leipzig		1.3	
	Brussels		2.7	
	Arnhem		1.9	
	Budapest		5	
	Dublin		2	
	Nijmegen		1.5	
	Athens		6.5	
	Nicosia		6.3	

Table D.4 Formaldehyde input concentration data

References	Location	Notes	Mean (μg/m³)	Standard deviation (μg/m³)
Quackenboss et al. (1989)	United States		32	14
			47	27
Sax et al. (2004)	New York	Winter	12	4.7
		Summer	21	11
	Los Angeles	Winter	21	11
		Fall	16	6.2

(continued)

Table D.4 (continued)

References	Location	Notes	Mean (μg/m³)	Standard deviation (μg/m³)
Sherman and Hodgson (2004)	United States	New homes	40	15
Gilbert et al. (2005)	Canada		39	22.4
Park and Ikeda (2006)	Japan	New homes:	120.1	100.5
		1st year	134	93
		New homes: 2nd year	112	105
		New homes: 3rd year	86	58
		Older homes: 1st year	88	115
		Older homes: 2nd year	89	107
		Older homes: 3rd year	90	98
Al Rehaili (2002)	Saudi Arabia		29.5	2.5
			20.9	2.5
			20.9	4.9
			25.8	3.7
			34.4	9.8
			27	7.4
			28.2	11.1
			24.6	3.7
			4.9	17.2
			3.7	7.4
Gilbert et al. (2006)	Quebec City		32.7	15.3
Bruinen de Bruin et al. (2008)	Helsinki		28.6	9.3
	Leipzig		18.6	13.4
	Brussels		19.5	3
	Arnhem		30.7	17.8
	Budapest		24.4	8.2
	Dublin		14.4	4.9
	Nijmegen		30.1	24.2
	Athens		24.1	12.9
Guo et al. (2009)	Hong Kong		112.3	90.3
Li et al. (2008)	China		60	30
			60	20
			60	40
			40	40
			110	130
Mentese and Gullu (2006)	Turkey		109.3	134.9

Table D.5 Radon concentration data

Emirate	μ (Bq/m³)	Σ (Bq/m³)	Minimum (Bq/m³)	Maximum(Bq/m³)
Abu Dhabi	13.91	7.00	3.5	41.3
Sharjah	50.3	N/A	8	164

References

Abt, E., H.H. Suh, P. Catalano, and P. Koutrakis. 2000. Characterization of indoor particle sources: A study conducted in the metropolitan Boston area. *Environmental Health Perspectives* 108(1): 3579–3587.

Al Rehaili, A.M. 2002. Outdoor-indoor air quality in Riyadh: SO_2, NH_3, and HCHO. *Environmental Monitoring and Assessment* 79(3): 287–300.

Allen, R., L. Wallace, T. Larson, L. Sheppard, and L.J. Liu. 2007. Evaluation of the recursive model approach for estimating particulate matter infiltration efficiencies using continuous light scattering data. *Journal of Exposure Science & Environmental Epidemiology* 17(5): 468–477.

Batterman, S., C. Jia, and G. Hatzivasilis. 2007. Migration of volatile organic compounds from attached garages to residences: A major exposure source. *Environmental Research* 104(2): 224–240.

Baxter, L.K., J.E. Clougherty, F. Laden, and J.I. Levy. 2007. Predictors of concentrations of nitrogen dioxide, fine particulate matter, and particle constituents inside of lower socioeconomic status urban homes. *Journal of Exposure Science & Environmental Epidemiology* 17(5): 433–444.

Breysse, P.N., T.J. Buckley, D. Williams, C.M. Beck, S.J. Jo, B. Merriman, S. Kanchanaraksa, et al. 2005. Indoor exposures to air pollutants and allergens in the homes of asthmatic children in inner-city Baltimore. *Environmental Research* 98(2): 167–176.

Brown, K.W., J.A. Sarnat, H.H. Suh, B.A. Coull, J.D. Spengler, and P. Koutrakis. 2008. Ambient site, home outdoor and home indoor particulate concentrations as proxies of personal exposure. *Journal of Environmental Monitoring* 10(9): 1041–1051.

Bruinen de Bruin, Y., K. Koistinen, S. Kephalopoulos, O. Geiss, S. Tirendi, and D. Kotzias. 2008. Characterisation of urban inhalation exposures to benzene, formaldehyde and acetaldehyde in the European Union: Comparison of measured and modelled exposure data. *Environmental Science and Pollution Research International* 15(5): 417–430.

Cheng, Y.L., Y. Min, Y.H. Bai, et al. 2007. Study on the distribution and sources of indoor and outdoor air pollutants in domiciles in Guiyang City. *Environmental Monitoring in China* 5(23): 55–58.

Cortez-Lugo, M., H. Moreno-Macias, F. Holguin-Molina, J.C. Chow, J.G. Watson, V. Gutiérrez-Avedoy, F. Mandujano, M. Hernández-Avila, and I. Romieu. 2008. Relationship between indoor, outdoor, and personal fine particle concentrations for individuals with COPD and predictors of indoor-outdoor ratio in Mexico City. *Journal of Exposure Science & Environmental Epidemiology* 18(1): 109–115.

Gilbert, N.L., M. Guay, J.D. Miller, S. Judek, C.C. Chan, and R.E. Dales. 2005. Levels and determinants of formaldehyde, acetaldehyde, and acrolein in residential indoor air in Prince Edward Island, Canada. *Environmental Research* 99(1): 11–17.

Gilbert, N.L., D. Gauvin, M. Guay, M.E. Héroux, G. Dupuis, M. Legris, C.C. Chan, R.N. Dietz, and B. Lévesque. 2006. Housing characteristics and indoor concentrations of nitrogen dioxide and formaldehyde in Quebec City, Canada. *Environmental Research* 102(1): 1–8.

Gotschi, T., L. Oglesby, P. Mathys, C. Monn, N. Manalis, K. Koistinen, M. Jantunen, O. Hanninen, L. Polanska, and N. Kunzli. 2002. Comparison of black smoke and $PM_{2.5}$ levels in indoor and outdoor environments of four European cities. *Environmental Science and Technology* 36(6): 1191–1197.

Guo, H., S.C. Lee, W.M. Li, and J.J. Cao. 2003. Source characterization of BTEX in indoor microenvironments in Hong Kong. *Atmospheric Environment* 37: 73–82.

Guo, H., N.H. Kwok, H.R. Cheng, S.C. Lee, W.T. Hung, and Y.S. Li. 2009. Formaldehyde and volatile organic compounds in Hong Kong homes: Concentrations and impact factors. *Indoor Air* 19(3): 206–217.

Houyin, Z., S. Longyi, and Y. Qiang. 2005. Microscopic morphology and size distribution of residential indoor PM_{10} in Beijing City. *Indoor and Built Environment* 14(6): 513–520.

Jo, W.-K., and J.-Y. Lee. 2006. Indoor and outdoor levels of respirable particulates (PM_{10}) and carbon monoxide (CO) in high-rise apartment buildings. *Atmospheric Environment* 40(32): 6067–6076.

Keeler, G.J., T. Dvonch, F.Y. Yip, E.A. Parker, B.A. Isreal, F.J. Marsik, M. Morishita, et al. 2002. Assessment of personal and community-level exposures to particulate matter among children with asthma in Detroit, Michigan, as part of Community Action Against Asthma (CAAA). *Environmental Health Perspectives* 110(Suppl. 2): 173–181.

Khillare, P.S., R. Pandey, and S. Balachandran. 2004. Characterisation of indoor PM_{10} in residential areas of Delhi. *Indoor and Built Environment* 13: 139–147.

Kim, Y.M., S. Harrad, and R.M. Harrison. 2001. Concentrations and sources of VOCs in urban domestic and public microenvironments. *Environmental Science and Technology* 35(6): 997–1004.

Komarnicki, G.J. 2005. Lead and cadmium in indoor air and the urban environment. *Environmental Pollution* 136(1): 47–61.

Lachenmyer, C., and G.M. Hidy. 2000. Urban measurements of outdoor-indoor $PM_{2.5}$ concentrations and personal exposure in the deep South, Part I: Pilot study of mass concentrations for nonsmoking subjects. *Aerosol Science and Technology* 32(1): 34–51.

Leaderer, B.P., L. Naeher, T. Jankun, K. Balenger, T.R. Holford, C. Toth, J. Sullivan, J.M. Wolfson, and P. Koutrakis. 1999. Indoor, outdoor, and regional summer and winter concentrations of PM_{10}, $PM_{2.5}$, SO_4(2)-, H+, NH_4+, NO_3-, NH_3, and nitrous acid in homes with and without kerosene space heaters. *Environmental Health Perspectives* 107(3): 223–231.

Li, T.T., Z.R. Liu, and Y.H. Bai. 2008. Human cancer risk from the inhalation of formaldehyde in different indoor environments in Guiyang City, China. *Bulletin of Environmental Contamination and Toxicology* 81(2): 200–204.

Liu, L.-J.S., M. Box, D. Kalman, J. Kaufman, J. Koenig, T. Larson, T. Lumley, L. Sheppard, and L. Wallace. 2003. Exposure assessment of particulate matter for susceptible populations in Seattle. *Environmental Health Perspectives* 222(7): 908–918.

Long, C.M., H.H. Suh, and P. Koutrakis. 2000. Characterization of indoor particle sources using continuous mass and size monitors. *Journal of the Air and Waste Management Association* 50:1236–1250.

Lung, S.C., M.C. Kao, and S.C. Hu. 2003. Contribution of incense burning to indoor PM_{10} and particle-bound polycyclic aromatic hydrocarbons under two ventilation conditions. *Indoor Air* 13(2): 194–199.

Meng, Q.Y., B.J. Turpin, A. Polidori, J.H. Lee, C. Weisel, M. Morandi, S. Colome, T. Stock, A. Winer, and J. Zhang. 2005. $PM_{2.5}$ of ambient origin: Estimates and exposure errors relevant to PM epidemiology. *Environmental Science and Technology* 39(14): 5105–5112.

Mentese, S., and G. Gullu. 2006. Variations and sources of formaldehyde levels in residential indoor air in Ankara, Turkey. *Indoor and Built Environment* 15(3): 273.

Park, J.S., and K. Ikeda. 2006. Variations of formaldehyde and VOC levels during 3 years in new and older homes. *Indoor Air* 16(2): 129–135.

Quackenboss, J.J., M.D. Lebowitz, J.P. Michaud, and D. Bronnimann. 1989. Formaldehyde exposure and acute health effects study. *Environment International* 15: 169–176.

Rojas-Bracho, L., H.H. Suh, and P. Koutrakis. 2000. Relationships among personal, indoor, and outdoor fine and coarse particle concentrations for individuals with COPD. *Journal of Exposure Analysis and Environmental Epidemiology* 10: 294–306.

Sarnat, J.A., C.M. Long, P. Koutrakis, B.A. Coull, J. Schwartz, and H.H. Suh. 2002. Using sulfur as a tracer of outdoor fine particulate matter. *Environmental Science and Technology* 36(24): 5305–5314.

Sax, S.N., D.H. Bennett, S.N. Chillrud, P.L. Kinney, and J.D. Spengler. 2004. Differences in source emission rates of volatile organic compounds in inner-city residences of New York City and Los Angeles. *Journal of Exposure Analysis and Environmental Epidemiology* 14(Supplement 1): S95–S109.

See, S.W., Y.H. Wang, and R. Balasubramanian. 2007. Contrasting reactive oxygen species and transition metal concentrations in combustion aerosols. *Environmental Research* 103(3): 317–324.

Sherman, M.H., and A.T. Hodgson. 2004. Formaldehyde as a basis for residential ventilation rates. *Indoor Air* 14(1): 2–8.

Simoni, M., A. Scognamiglio, L. Carrozzi, S. Baldacci, A. Angino, F. Pistelli, F. Di Pede, and G. Viegi. 2004. Indoor exposures and acute respiratory effects in two general population samples from a rural and an urban area in Italy. *Journal of Exposure Analysis and Environmental Epidemiology* 14(Suppl 1): S144–S152.

Simons, E.J., T. Curtin-Brosnan, P.Breysse Buckley, and P.A. Eggleston. 2007. Indoor environmental differences between inner city and suburban homes of children with asthma. *Journal of Urban Health* 84(4): 577–590.

Son, B., P. Breysse, and W. Yang. 2003. Volatile organic compounds concentrations in residential indoor and outdoor and its personal exposure in Korea. *Environment International* 29(1): 79–85.

Stranger, M., S.S. Potgieter-Vermaak, and R. Van Grieken. 2009. Particulate matter and gaseous pollutants in residences in Antwerp, Belgium. *Science of the Total Environment* 407(3): 1182–1192.

Suh, H. 2003. *Characterization of the composition of personal, indoor, and outdoor particulate matter exposures*. Final report to California Air Resources Board for contract no. 98-330.

Suh, H., and P. Koutrakis. 2004. *Detailed characterization of indoor and personal particulate matter concentrations*. Final report to CARB for contract no. 00-302.

Van Winkle, M.R., and P.A. Scheff. 2001. Volatile organic compounds, polycyclic aromatic hydrocarbons, and elements in the air of ten urban homes. *Indoor Air* 11(1): 49–64.

Wallace, L. 1996. Environmental exposure to benzene: An update. *Environmental Health Perspectives* 104(6): 1129–1136.

Wallace, L.A., H. Mitchell, G.T. O'Connor, L. Neas, M. Lippmann, M. Kattan, J. Koenig, et al. 2003. Particle concentrations in inner-city homes of children with asthma: The effect of smoking, cooking, and outdoor pollution. *Environmental Health Perspectives* 111(9): 1265–1272.

Wallace, L., R. Williams, A. Rea, and C. Croghan. 2006. Continuous weeklong measurements of personal exposures and indoor concentrations of fine particles for 37 health-impaired North Carolina residents for up to four seasons. *Atmospheric Environment* 40(3): 399–414.

Williams, R., J. Suggs, G. Evans, J. Creason, R. Kwok, C. Rodes, P. Lawless, and L. Sheldon. 2000. The 1998 Baltimore particulate matter epidemiology-exposure study, Part 1: Comparison of ambient, residential outdoor, indoor and apartment particulate matter monitoring. *Journal of Exposure Analysis and Environmental Epidemiology* 46(10): 518–532.

Williams, R., J. Suggs, A. Rea, K. Leovic, A. Vette, C. Croghan, L. Sheldon, et al. 2003. The Research Triangle Park particulate matter panel study: Modeling ambient source contribution to personal and residential PM mass contributions. *Atmospheric Environment* 37: 5365–5378.

Zmirou, D., S. Gauvin, I. Pin, I. Momas, J. Just, F. Sahraoui, Y. Le Moullec, et al. 2002. Five epidemiological studies on transport and asthma: Objectives, design and descriptive results. *Journal of Exposure Analysis and Environmental Epidemiology* 12(3): 186–196.

Appendix E: Authors and Contributors

This book represents the synthesis of research carried out by a large, interdisciplinary team from several institutions and multiple nations between June 2008 and June 2011. The lead authors are responsible for weaving together the pieces prepared by the team. Nonetheless, this book would not have been possible without major contributions from each team member. The list below shows contributors to each chapter. Following this list are biographies of all of the authors and contributors

Lead Authors

Jacqueline MacDonald Gibson, Angela S. Brammer, Christopher A. Davidson, Tiina Folley, Frederic J. P. Launay, Jens T. W. Thomsen

Additional Contributors (By Chapter)

Chapter 2: Prioritizing Environmental Risks to Health
Henry H. Willis, Aimee Curtright, Gary Cecchine, Zeinab S. Farah, Sandra A. Geschwind, Jianhui Hu, Ying Li, Melinda Moore, Sarah Olmstead, Hanine Salem, Regina A. Shih, J. Jason West

Chapter 3: Assessing the Environmental Burden of Disease: Method Overview
Elizabeth S. Harder, Mejs Hasan

Chapter 4: Burden of Disease from Outdoor Air Pollution
Ying Li, Gavino Puggioni, Prahlad Jat, Mejs Hasan, Marc Serre, Kenneth G. Sexton, J. Jason West, Saravanan Arunachalam, Uma Shankar, William Vizuete, Mohammed Zuber Farooqui

Chapter 5: Burden of Disease from Indoor Air Pollution
Chris B. Trent

Chapter 6: Burden of Disease from Occupational Exposures
Leena A. Nylander-French

Chapter 7: Burden of Disease from Climate Change
Richard N.L. Andrews, Leslie Chinery, Elizabeth S. Harder, J. Jason West

Chapter 8: Burden of Disease from Drinking Water Contamination
Gregory W. Characklis, Joseph N. LoBuglio

Chapter 9: Burden of Disease from Coastal Water Pollution
Gregory W. Characklis, Leigh-Anne H. Krometis, Joseph N. LoBuglio

Chapter 10: Burden of Disease from Soil and Groundwater Contamination
Chidsanuphong Chart-asa, Stephanie Soucheray-Grell

Chapter 11: Burden of Disease from Produce and Seafood Contamination
Leigh-Anne H. Krometis, Leslie Chinery

Author Biographies

Richard N.L. Andrews is professor and chair of the Department of Public Policy at the University of North Carolina–Chapel Hill (UNC), with joint appointments in Environmental Sciences and Engineering and City and Regional Planning, the Curriculum in Environment and Ecology, and the Carolina Institute for the Environment. He earned a bachelor's degree at Yale University in 1966, and both a master's degree in regional planning and a doctorate in environmental policy and planning at UNC. He has written on and taught environmental policy in the United States and elsewhere for more than 38 years, particularly the history of U.S. environmental policy; the National Environmental Policy Act and environmental impact assessment; the use of risk and cost-benefit analysis in government decision-making; the use of environmental management systems by businesses and government; and innovations in energy, environment, and climate change policies. He also has studied environmental policy in Cameroon, the Czech Republic, Nepal, Thailand, and the United Arab Emirates. Andrews serves as a member of the National Research Council's Committee on Human Dimensions of Global Change and its Panel on Addressing the Challenges of Climate Change through the Behavioral and Social Sciences, and the North Carolina Legislative Commission on Global Climate Change. He is a Fellow of the National Academy of Public Administration and of the American Association for the Advancement of Science, and a member of Sigma Xi, the Delta Omega Public Health Honor Society, and the Association for Public Policy Analysis and Management.

Saravanan Arunachalam is a research associate professor at the UNC Institute for the Environment who performs research related to air quality modeling and analyses to better understand the formation of air pollution at local and regional scales. He earned his doctorate and master's degree in chemical engineering from Rutgers University. During the past 10 years, he has received research funding as principal investigator or co-investigator from several federal, state or private organizations, including among others, the U.S. Environmental Protection Agency (EPA), Federal Aviation Administration (FAA), National Aeronautics and Space Administration (NASA), and National Science Foundation. Arunachalam is an expert air quality modeler with extensive experience in developing modeling applications for regulatory support using both existing and evolving regional air quality models. At UNC, he directs the FAA-funded Partnership for Air Transportation Noise and Emissions Reduction Center of Excellence, developing methodologies for performing a non-scale assessment of the impact of aviation emissions on air quality to support health-based risk assessment. His research for the FAA in the recent past has made critical contributions to understanding the impacts of aviation on air quality. Arunachalam is the principal investigator on an Emissions, Meteorological, and Air Quality Modeling Assistance contract with the U.S. EPA in support of developing their regulatory programs. He was the recipient of the 2008 FAA Centers of Excellence Faculty of the Year award.

Angela S. Brammer is a copy editor and graphic designer. She earned her master's degree in entomology and nematology and her bachelor's degree in journalism from the University of Florida. Prior to venturing into self-employment, she was a copy editor for the *Seattle Times*, the *Winston-Salem Journal*, and the *Independent Florida Alligator*. She also has taught journalism at the secondary and undergraduate levels in the United States and English at the secondary level in Japan. Under her advisement, Brammer's students won the Columbia Scholastic Press Association Gold Medal for yearbook and for literary magazine. Brammer also served as editor and graphic designer of the 2010 UAE *National Strategy and Action Plan for Environmental Health*, part of the UAE National Environmental Health Project.

Gary Cecchine is a natural scientist with RAND. He received his doctorate in biology and public policy at the Georgia Institute of Technology and a bachelor's degree in marine science and biology at the University of Tampa. Cecchine's research interests include biology, toxicology, environmental and energy policy, biological and chemical terrorism, sustainability, and management systems. He has a background in national security. Before joining RAND he was a management systems consultant and auditor for Det Norske Veritas and an officer in the U.S. Army. Cecchine's recent projects have involved the role of the military in emergency response, infectious disease and national security, countering bioterrorism threats, occupational health and wellness in the military, and establishing research funding organizations internationally

Gregory W. Characklis is an associate professor in the Department of Environmental Sciences and Engineering at UNC. His primary research interests involve integrated planning of water supply and treatment strategies through the consideration of both engineering and economic criteria. Specific areas of interest include the use of water transfers in mitigating drought risk, the impacts of water quality on resource value and allocation, and developing minimum cost strategies for water-related infrastructure. He also directs several laboratory and field studies that explore the role particles play in pathogen and indicator organism transport, research with particular relevance in the development of water quality models used to evaluate the location and severity of public health risks posed by microbial contamination. Prior to joining UNC, Characklis served as Director of Resource Development and Management at Azurix Corporation, where his responsibilities included assessing the technical and financial merits of water supply development projects throughout the United States, including most of the western states. Before entering the private sector, he spent 2 years in Washington, DC, as a fellow with the National Academy of Engineering, where he co-authored a study on industrial environmental performance metrics for U.S. manufacturers and conducted work related to market-based reform of environmental policy. Characklis earned his doctorate and master's degree in environmental science and engineering at Rice University and his bachelor's degree in materials science and engineering at Johns Hopkins University.

Chidsanuphong Chart-asa is a research assistant and doctoral student in the Department of Environmental Sciences and Engineering at UNC. His dissertation focuses on exposure mapping and environmental risk. Chart-asa earned his master's degree in information management on environment and resources from Mahidol University, Thailand. In 2008 he was granted a full scholarship by the Royal Thai Government to pursue doctoral study in a field that emphasizes environmental impact assessment or related fields. His past research interests have included geographical information systems and natural resource management.

Leslie Chinery earned her master's degree in environmental sciences and engineering at UNC, with a specialization in climate change and transportation policy. She has a bachelor's degree in environmental studies with a concentration in economics from the University of Tennessee. Chinery was a top graduate of the University of Tennessee in 2007 and received the Nissan-World Wildlife Fund Environmental Leadership Award in 2006. Chinery compiled a greenhouse gas emissions inventory of the University of Tennessee–Knoxville campus as an undergraduate honors thesis in 2007. For her master's thesis, Chinery modeled the potential emissions reductions and cost-effectiveness of various changes in the UAE's passenger vehicle fleet.

Aimee Curtright is an associate physical scientist in RAND's Pittsburgh office. She works primarily in the areas of energy and environmental policy and technology assessment. Presently she is working on a number of projects examining the environmental implications and technical challenges to the use of biomass for energy production. She has a doctorate in chemistry from the University of California–Berkeley and a bachelor's degree in chemistry from the University of Miami.

Christopher A. Davidson is the project manager for the UAE National Environmental Health Project at UNC. He puts his engineering and teaching background to work for optimizing the flow, storage, and presentation of information for the project. Davidson holds a master's degree in agricultural and biological engineering and a bachelor's degree in computer engineering from the University of Florida. Related work includes analyzing air quality and agricultural data in the UAE, establishing and maintaining electronic document repositories, and minimizing barriers to understanding scientific data for faculty and graduate students involved in this interdepartmental project.

Zeinab S. Farah earned her doctorate at The London School of Hygiene and Tropical Medicine, University of London, and her bachelor's degree in microbiology, majoring in medical microbiology at the University of East Anglia, United Kingdom. She worked in the medical services in Abu Dhabi for more than 20 years and assumed various responsibilities in the laboratory of one of the major hospitals in Abu Dhabi. In 2008 she was appointed as consultant and in-country coordinator for the UAE National Environmental Health Project at UNC.

Mohammed Zuber Farooqui received his doctorate in environmental engineering from Texas A&M University–Kingsville. His research interests include air quality modeling, emission inventory, meteorology and environmental information systems. Farooqui's dissertation focused on developing an air quality modeling framework for South Texas, and he was subsequently awarded first place for best overall doctoral research efforts in the TAMUS Annual Pathway Student Research Symposium in 2007. His research addresses ambient air quality issues within the UAE, specifically on enhancing estimates of dust emissions due to dust storms within the Arabian peninsula.

Tiina Folley is a research associate in the Department of Environmental Sciences and Engineering at UNC. She earned a master's degree in occupational hygiene from the University of Kuopio, Finland, and a master's degree in public health from UNC as a Fulbright Scholar. Before joining the occupational exposures team at UNC, Folley worked as a researcher at the Department of Food and Environmental Hygiene at the University of Helsinki. Her research interests include exposure assessment in environmental and occupational settings, and her current research involves quantitative modeling of occupational exposures to chemical hazards.

Sandra A. Geschwind is an environmental epidemiologist with more than 25 years of experience in the environmental health field. Her research areas at RAND have spanned studying the impact of endocrine disrupting chemicals on wildlife for the White House Office of Science and Technology Policy to evaluating the effects of pesticide exposures on Gulf War veterans. She is currently involved in helping set environmental health risk priorities for countries within the Middle East. Additionally, she serves as a reviewer for the American Journal of Epidemiology. Prior to her work at RAND, Geschwind led the Cancer Cluster Program at the Connecticut Department of Health. She also spent 5 years as a hazardous waste field investigator for the U.S. Environmental Protection Agency, Region IX. Geschwind earned her bachelor's degree at the University of California–Berkeley and both her master's degree in environmental health sciences and her doctorate environmental epidemiology from Yale University. She has been recognized and received awards for her work by the Environmental Protection Agency and the Agency for Toxic Substances and Disease Registry under the Centers for Disease Control.

Elizabeth S. Harder is a research assistant at UNC. Her research areas include renewable energy, climate change, economics, and energy policy. Previous graduate work included mapping water pollutant concentrations with geographic information systems software, conducting a life cycle analysis of solar photovoltaic energy, and analyzing the barriers of entry for renewable energy in the United States. Harder earned her master's degree in environmental science at UNC. Her thesis research focused on the economic feasibility, energy production benefits, green house gas mitigation, and air quality health benefits for constructing large-scale photovoltaic power plants in Abu Dhabi.

Mejs Hasan earned her master's degree in environmental sciences and engineering and her bachelor's degree in economics at UNC. She taught secondary math for 2 years in Philadelphia. Hasan is interested in geographic information systems, spatial analysis, contaminant flow in water and related health effects, and remote sensing.

Jianhui Hu is a research assistant and a doctoral fellow at the Pardee RAND Graduate School. Hu's research interests are in areas of public health, quality of care, environmental health, disease prevention, and drug abuse and HIV/AIDS. She is especially interested in global health care policies. Hu has been providing research assistance to a variety of health-care projects at RAND, focusing on qualitative research and program evaluation. Hu earned her bachelor's degree in business administration at Renmin University of China and her master's degree in public policy at Pepperdine University.

Prahlad Jat is a graduate research assistant and doctoral candidate in the Department of Environmental Sciences and Engineering at UNC. Jat's research focus is on space and time mapping and analysis of air pollutants, integration of economic and environmental models and the estimation of environmental burden of disease. Currently he works in the Bayesian Maximum Entropy lab on software updates and development. Jat holds a master's degree in biological engineering from the University of Arkansas and a master's degree in energy studies from the Indian Institute of Technology.

Leigh-Anne H. Krometis received her doctorate in environmental sciences and engineering from UNC and her master's degree in biological systems engineering from Virginia Polytechnic Institute and State University. Her dissertation research on the fate and transport of pathogenic microorganisms in urban stormflow won the 2009 Carolina Impact Award in recognition of its potential to protect the health of state residents. Following completion of her postdoctoral work on the *UAE Environmental Burden of Disease Model* in the summer of 2009, Krometis accepted a position as a research assistant professor in the Biological Systems Engineering Department at Virginia Polytechnic Institute. Her current work focuses on improving scientific understanding of the waterborne transport of indicator organisms and pathogens, particularly in support of nonpoint source regulatory programs.

Frederic J.P. Launay is senior advisor to the Secretary General, Environment Agency–Abu Dhabi (EAD). Dr. Launay began his career as a wildlife biologist in Saudi Arabia, where he established protected areas and conducted reintroduction programs for species such as the houbara bustard, Arabian oryx, and various species of indigenous gazelles. Subsequently, he moved to the National Avian Research Centre in Abu Dhabi to lead the organization in its work on the study and conservation of the houbara. In addition to his role at EAD, he also is director general of the Mohammed bin Zayed Species Conservation Fund, United Arab Emirates, and is chair of the International Union for Conservation of Nature/Species Survival Commission (IUCN/SSC) Reintroduction Specialist Group, United Arab Emirates. Launay earned his doctorate in wildlife management and ecology from the Université de Rennes. He earned his master's degree in ecology from the Université Paul Sabatier.

Ying Li is a postdoctoral research associate at the Department of Environmental Sciences and Engineering, UNC. She obtained her doctorate in public policy from UNC and her master's and bachelor's degrees in environmental sciences from Peking University. Her main research interests include environmental risk assessment and health benefits of pollution control policies. Since June 2008, she has been working to assess and prioritize various environmental health risks in the UAE. Her dissertation, which focused on the health benefits of traffic-related particulate matter control policies in Bangkok, was honored with the 2008 Association for Public Policy Analysis and Management Best Dissertation in Asia Award. Li has also won several prestigious academic awards in the past few years, such as the UNC Society of Fellows William Neal Reynolds Fellowship from 2002 to 2007, a student merit award for the Dose Response Specialty Group of Society for Risk Analysis in 2007, and the Graduate Education Advancement Board North Carolina Impact Award in 2008.

Joseph N. LoBuglio is a research specialist working with the Water Institute at UNC. LoBuglio has more than 10 years of experience in environmental management, water supply planning, and drinking water quality and additional expertise in uncertainty analysis and modeling. He has worked for the Massachusetts Water Resources Authority, the agency responsible for drinking and wastewater services in the Boston metropolitan area, helping develop processes for obtaining, managing, and reporting on river and marine water quality data associated with the Boston Harbor Cleanup Project. He has contributed to projects for the North Carolina state legislature promoting North Carolina's economic development through strategic water resource management. His recent work helped frame and analyze policy options for water allocation considering social, ecological, economic, and institutional systems. He has received a bachelor's degree in mechanical and aerospace engineering from Princeton University and a master's degree in aeronautics and astronautics from Stanford University. He is currently pursuing a doctorate in environmental sciences and engineering from UNC. His dissertation focuses on managing uncertainty in water resource and water quality models and means of using uncertain information to facilitate policy decisions. He has been a licensed professional engineer for more than 15 years.

Jacqueline MacDonald Gibson is an assistant professor in the Department of Environmental Sciences and Engineering at UNC. MacDonald Gibson conducts interdisciplinary research on the quantification of risks due to environmental contamination and on the quantitative comparison policy options for controlling environmental risks. She earned dual doctorate degrees from the Department of Engineering and Public Policy and the Department of Civil and Environmental Engineering at Carnegie Mellon University. She was previously a senior engineer at RAND. She also previously was associate director of the Water Science and Technology Board, a unit of the National Research Council of the National Academy of Sciences. In these previous positions, she studied a range of issues at the interface between environmental science and public policy. Research topics included assessment of options for improving potable water service to small U.S. communities, evaluation of regulatory requirements for the remediation of contaminated groundwater, assessment of research priorities for new environmental remediation technologies, evaluation of research on alternative methods for detecting and cleaning up landmines, and evaluation of risk assessment methods for sites contaminated with unexploded military ordnance. She has given briefings on these and other topics to a variety of federal officials, members of Congress and their staffs, and institutional advisory boards. She earned a master's degree from the Department of Civil and Environmental Engineering at the University of Illinois–Urbana-Champaign and a bachelor's degree (magna cum laude) in mathematics from Bryn Mawr College.

Melinda Moore is a public-health physician and epidemiologist who joined RAND as a senior natural scientist in 2005 following a 25-year career in government. She is RAND's associate director for global public health. Her research at RAND has focused on infectious disease surveillance, public-health and pandemic influenza preparedness, global health, military health, and environmental health. She led the RAND collaboration with UNC to develop the *National Strategy and Action Plan for Environmental Health* for the UAE. She is co-leading an ongoing project aimed at developing a tool for local civilian and military disaster preparedness planning and contributing to the development and implementation of the U.S. National Health Security Strategy. Moore earned her medical degree and master's degree in public health from Harvard University and is board certified in pediatrics and preventive medicine. Prior to joining RAND, she served at the U.S. Department of Health and Human Services for 25 years. She has worked in over 50 countries (including Saudi Arabia, Kuwait, Jordan, and the UAE) and lived in the Democratic Republic of the Congo for nearly 5 years. She is a retired medical officer of the U.S. Public Health Service.

Leena A. Nylander-French is a professor of occupational and environmental health at UNC. She obtained her doctorate in occupational and industrial hygiene from the Royal Institute of Technology in Sweden and joined the UNC faculty in 1997. Nylander-French has served as the director of the industrial hygiene program in the National Institute of Occupational Safety and Health Educational Resource Center since 2002. In 2006, she became the director of the Exposure and Biomarkers Research Core under the Center for Environmental Health and Susceptibility in the Gillings School of Public Health at UNC. She is certified by the American Board of Industrial Hygiene and has expertise in exposure assessment, biomarkers, and toxicokinetics. Nylander-French's research and teaching program is focused on understanding the consequences of human exposure to toxic substances. She is particularly interested in the relationship between dermal and inhalation exposure to hazardous environmental or occupational agents and the effect of individual genetic differences on the function of enzymes that detoxify these agents and that affect the development of disease. Nylander-French engages in research projects that are full-scale occupational and environmental studies incorporating both methods development in the laboratory as well as health surveys where individual exposures are monitored using a battery of exposure measurement techniques (inhalation, dermal, and biological monitoring) multiple times over an extended period. Her research group has pioneered approaches to quantitatively measure skin and inhalation exposures to toxicants. Additionally, her group has developed sophisticated exposure modeling tools using mathematical and statistical principles in an effort to standardize and improve exposure and risk assessment.

Sarah Olmstead is an assistant policy analyst at RAND and doctoral fellow at the Pardee RAND Graduate School. She works on a wide variety of topics, including energy and environment issues and conflict prevention. Her dissertation looks at the ways in which water management can affect economic development and local conflict. Prior to joining RAND, Olmstead was employed at the Science and Human Rights Program of the American Association for the Advancement of Science, where she worked on topics such as using geospatial technologies to identify large-scale human rights violations and assessing the effectiveness of transitional justice mechanisms. She also worked with the Parliamentary Monitoring Group and the Black Sash in South Africa and on issues of health and education with the Movimiento de Mujeres Dominico-Haitiana in the Dominican Republic. Olmstead holds a master's degree in physics from the University of Minnesota.

Gavino Puggioni is a postdoctoral researcher at Emory University in the Center for Disease Ecology and the Department of Biostatistics and Bioinformatics. His research interests include spatiotemporal models, stochastic differential equations, Bayesian statistics, environmental applications and econometrics. Puggioni received his doctorate in statistical science from Duke University. Before working for Emory, he was a research associate working with space-time analysis of air and water pollutants at UNC.

Hanine Salem is an associate director for Middle East Development at the RAND Education Unit. Much of her work is concerned with results-oriented public management methods such as strategic planning, organizational performance management and measurement, program evaluation, and quality improvement methods. Prior to joining RAND, Salem was a United National Development Programme Advisor to two Ministers of Administrative Reform in Lebanon. She introduced concepts of organizational performance management and measurement in the Lebanese public sector as well as developed a national performance-based reporting system. Her projects focused on the introduction of modern public management methods that promote transparency and accountability such as organizational performance measurement and strategic performance management. Salem is a doctoral candidate at the University of Strathclyde, where her research focus is on performance measurement and strategic planning in the public sector. She earned her master's degree in business communication from the University of Oklahoma and her bachelor's degree in business administration from Kuwait University.

Uma Shankar has more than 20 years' experience developing aerosol models for multiscale applications in particulate matter air quality and its feedbacks to climate, and has led a number of projects developing and applying CMAQ modeling and its prototype, MAQSIP, under funding from a diverse group of sponsors. Shankar is currently a co-investigator leading the CMAQ modeling studies recently funded by the UAE to support their national environmental health strategy. She participates with NASA contractors on providing training on the use of satellite data in air quality model evaluations through the Community Modeling and Analysis System Center, for which she also serves as research coordinator. She has served on EPA and NASA proposal peer review panels, and reviewed submissions to the Journal of Geophysical Research and Atmospheric Environment. Shankar earned her doctorate in physics at UNC and her master's degree in nuclear engineering at North Carolina State University.

Marc Serre is an associate professor at in the Department of Environmental Sciences and Engineering at UNC. He has a doctorate in environmental space/time geostatistics from UNC. His scientific expertise includes spatiotemporal modeling and mapping of air and water pollutants, exposure analysis and human health risk assessment, temporal geographic information systems, and the numerical development of advanced mapping functions. He has been extensively involved in many research efforts in advanced spatiotemporal modeling of ambient air and water pollution, with wide applications to local, regional, national, and international settings. He currently directs the UNC-BMElab research group at the University of North Carolina, which is dedicated to the numerical development of the Bayesian maximum entropy method of modern geostatistics and its worldwide application to exposure, disease, and risk mapping. As part of this effort, he has been an integral part of the development of a well-known BMElab package of spatiotemporal geostatistics used by scientists and researchers in more than 30 countries.

Kenneth G. Sexton is a research associate professor at UNC, where he earned his master's degree in public health and his doctorate in atmospheric chemistry. His research interests include the atmospheric chemistry of urban systems of nitrogen oxides and hydrocarbons, with a focus on understanding the reactive chemistry producing ozone and other air toxics, using smog chambers. Recent efforts have focused on developing and demonstrating new technological systems to interface smog chambers and in vitro toxicological exposure systems for evaluating the effects of photochemistry on urban air mixtures and the resulting toxic potential for health effects. Recent projects have compared the inflammatory toxicity potential of exhaust from traditional diesel and biodiesel exhaust, the toxicity and fate of 1,3-butadiene and related air toxics and their degradation products, and development of field deployable biological methods for estimating risk from exposure to urban mixtures.

Regina A. Shih is an associate behavioral/social scientist at RAND. Trained in epidemiology, her interdisciplinary research focuses on the influence of particulate matter and the built environment on cardiovascular disease. She previously published research on the relationship between lead exposure and neurological functioning in adults and on the methodological issues of measuring cumulative lead dose. She is currently a co-investigator on a grant from the National Heart, Lung, and Blood Institute to examine the impact of neighborhood socioeconomic status, racial/ethnic composition, economic segregation, and particulate matter air pollution on the incidence of cardiovascular disease among women. Prior to her work at RAND, she was a postdoctoral fellow at the National Institute of Child Health and Human Development, where she helped design the scientific protocol for measuring nonlevel social, structural, and chemical environmental risk factors for poor neurobehavioral and developmental outcomes across the lifespan, for which she received a Director's Merit Award. She teaches measurement methods and structural equation modeling in the Department of Environmental and Occupational Health at the George Washington University School of Public Health and Health Services. Shih earned her doctorate in mental health from the Johns Hopkins Bloomberg School of Public Health and her bachelor's degree in neuroscience from Johns Hopkins University.

Stephanie Soucheray-Grell is an administrative support specialist at UNC. She has been the main assistant to the UAE National Environmental Health Project, as well as the USAID WaterSHED initiative. Before joining the Department of Environmental Sciences and Engineering, Soucheray-Grell worked as a freelance editor and English tutor in Germany, a production assistant for Minnesota Public Radio, and a newspaper reporter. She received her bachelor's degree in English and history from St. Olaf College.

Jens T.W. Thomsen is Section Head, Occupational and Environmental Health, for the Health Authority–Abu Dhabi (HAAD). Thomsen's primary role at HAAD, which he joined in 2007, is to lead and oversee the strategy development, planning, implementation and monitoring and evaluation of public health activities in the areas of occupational and environmental health and safety, injury prevention, and road safety in the community and emirate of Abu Dhabi. His and his team's responsibilities include developing the strategy and action plan for these areas; collecting and analyzing data and statistics; planning, conducting, monitoring and evaluating public health interventions; supporting targeted research; capacity building; developing policies and standards; organizing mass media and community outreach programs; and coordinating with stakeholders. Thomsen is a medical doctor, public health specialist, and German Medical Board-certified specialist in clinical microbiology and infectious disease epidemiology. He is also German and U.S. board-certified in health-care infection control. He holds a master's degree in public health from Emory University, with a major in epidemiology, and has 17 years of medical and public health experience in the United States, Germany, and the UAE.

Chris B. Trent received her master's degree in industrial hygiene from UNC and her bachelor's degree in biology from James Madison University. She is currently pursuing her doctorate in environmental health and risk assessment at UNC. Trent has more than a decade of practical work experience in the environmental and occupational health fields, including numerous years conducting field-level exposure assessments. She has held management positions for nonprofit organizations focusing on overarching programmatic issues in the occupational and environmental health fields. More recently, she worked for the U.S. EPA on issues related to the role of exposure science in environmental health assessments, with specific application to broad programs such as outdoor air quality standards. Trent's research interests include the characterization of exposures and related uncertainties for health-based risk assessments, and capitalizing on the role of science in policy decision-making.

William Vizuete is an assistant professor in the Department of Environmental Sciences and Engineering at UNC. He is an expert on the use of computer models in understanding the role of atmospheric chemistry in the formation of air pollutants. His current research projects include assessing air quality in the UAE and evaluating the regional scale air quality models used by regulators to guide air quality policy in Houston, Texas. He is also the recipient of a Gillings Innovation Lab award to fund a project that uses UNC outdoor smog chamber facilities, biosensors, and atmospheric modeling to understand the changing toxicity of chemically aged atmospheric pollutants. He has developed several advanced diagnostic tools that are currently being used to improve the model's ability to simulate the chemistry of the upper troposphere. He has also worked closely with the UNC Institute for the Environment in developing inputs for and evaluating regional scale atmospheric models. Vizuete received his doctorate and master's degrees in chemical engineering from the University of Texas–Austin and his bachelor's degree in chemical engineering from the University of Missouri–Rolla.

J. Jason West is an assistant professor in the Department of Environmental Sciences and Engineering at UNC, where he performs interdisciplinary research addressing air pollution and climate change. West recently worked as a research scientist at Princeton University and prior to that, at the U.S. EPA in Washington, D.C., under a fellowship from the American Association for the Advancement of Science. He has a doctorate from Carnegie Mellon University, earned jointly between civil and environmental engineering and engineering and public policy, a master's degree in environment and development from the University of Cambridge, a master's degree in civil and environmental engineering from Carnegie Mellon, and a bachelor's degree in civil and environmental engineering from Duke University. West is interested broadly in the relationships between air pollution and climate change and their relevance for environmental science and policy. Using computer models, he has investigated the effects of changes in emissions on global air quality, the international transport and health impacts of air pollutants (focusing on ozone and particulate matter), and the radiative forcing of climate. As a postdoctoral researcher at the Massachusetts Institute of Technology and a visiting scientist at the Mexican National Institute of Ecology, West conducted computer modeling of air pollution and analyzed the integrated mitigation of greenhouse gases and air pollutants in Mexico City. West has also been an expert consultant to the World Bank, studying source apportionment techniques for identifying the sources of urban particulate matter in developing nations. Recently, West's research has emphasized quantification of the climate, air quality, and human health benefits of reductions in global emissions of methane, black carbon, and carbon monoxide.

Henry H. Willis is an expert in environmental and health risk assessment, risk communication, and risk management. He has recently reviewed existing risk assessment tools used across government for managing former U.S. Army lands contaminated with chemicals and unexploded ordnance and developed a focus-group process for incorporating public participation into comparative risk management. He has also led the risk assessment task in the UAE National Environmental Health Project. Willis serves on the editorial board of *Risk Analysis* and is the secretary of the Society for Risk Analysis. He earned his doctorate in engineering and public policy at Carnegie Mellon University and holds a master's degree in environmental science from the University of Cincinnati as well as a bachelor's degree in chemistry and environmental studies from the University of Pennsylvania.

Index

J. MacDonald Gibson et al., *Environmental Burden of Disease Assessment*, Environmental Science and Technology Library 24, DOI 10.1007/978-94-007-5925-1,

MIX
Papier aus verantwortungsvollen Quellen
Paper from responsible sources
FSC® C105338

If you have any concerns about our products,
you can contact us on
ProductSafety@springernature.com

In case Publisher is established outside the EU,
the EU authorized representative is:
Springer Nature Customer Service Center GmbH
Europaplatz 3, 69115 Heidelberg, Germany

Printed by Libri Plureos GmbH
in Hamburg, Germany